55th Annual Loss Prevention Symposium (LPS 2021)

Topical Conference at the 2021 AIChE Spring Meeting and 17th Global Congress on Process Safety

Online
18-23 April 2021

ISBN: 978-1-7138-5728-0

Printed from e-media with permission by:

Curran Associates, Inc.
57 Morehouse Lane
Red Hook, NY 12571

Some format issues inherent in the e-media version may also appear in this print version.

AIChE
120 Wall Street, FL 23
New York, NY 10005-4020

Phone: (800) 242-4363
Fax: (203) 775-5177

www.aiche.org

Additional copies of this publication are available from:

Curran Associates, Inc.
57 Morehouse Lane
Red Hook, NY 12571 USA
Phone: 845-758-0400
Fax: 845-758-2633
Email: curran@proceedings.com
Web: www.proceedings.com

TABLE OF CONTENTS

621490 The latent triggers of disasters are alive - deep inside your company

Abstract:

Never in our history have our workplaces been so safe and our workers so well protected. Yet in recent years, something strange started to happen. While many companies have 'hit a wall' in safety performance, some have seen a slow but visible increase in serious accidents, and most alarmingly, a few 'best-in-class' companies suffered catastrophic events. It seems as if the closer we get to 'zero accidents', the more vulnerable we become for serious events to happen... Something is wrong. Making the systems safer, puts the organization at higher risk? This presentation will demonstrate how the few seconds before a disaster were years in the making – by well-intended, but dangerous ideas, beliefs and delusions that become destructive forces in the culture. It will outline the new roles of leaders and practitioners in a different world awaiting us.

619273 The Case for Management Systems in Preventing Combustible Dust Explosions

Abstract:

Warning! This is not a sexy topic. Housekeeping, record retention, and procedures may not excite industry professionals in the same way that the newest engineering controls might, but these are critical and often overlooked pieces of a facility's Combustible Dust Protection Program. How often does a dust control program exist with well-written procedures, yet the plant floor appears to have seen a powder storm? How many incidents and injuries could be avoided if contractors had the proper training?

This presentation seeks to inform the audience of the required Management Systems which must be in place for safe and compliant operation of a combustible dust handling facility, as described by NFPA 652. This includes Operating Procedures, Housekeeping, Hot Work Permitting, PPE, Inspection and Maintenance, Hazard Awareness Training, Emergency Planning, Incident Investigation, Management of Change, Document Retention, and Employee Participation.

Case studies will illustrate the importance and impact of Management Systems in both positive and negative ways:

- Case Study #1 – How attention to detail and an outstanding Preventative Maintenance reduces the need for Housekeeping and Electrical Classification
- Case Study #2 – The hidden cost of lost documentation and ad hoc process changes when the knowledge walks out the door

SPRING21
+17TH GCPS
A Joint AIChE and CCPS Meeting

Dust Explosions: A New Path To Estimate The Explosibility Characteristics Of An Organic Powder

Matteo Pietraccini
LRGP, CNRS-Université de Lorraine
1 rue Grandville 54001 Nancy (France)
Matteo.pietraccini@univ-lorraine.fr

Audrey Santandrea, Marie Verdonck, Anthony Dufour, Pierre-Alexandre Glaude, Olivier Dufaud
LRGP, CNRS-Université de Lorraine
1 rue Grandville 54001 Nancy (France)

Prepared for Presentation at
American Institute of Chemical Engineers
2020 Spring Meeting and 17th Global Congress on Process Safety
Virtual
April 18 - 22, 2021

Dust Explosions: A New Path To Estimate the Explosibility Characteristics of an Organic Powder

Matteo Pietraccini
LRGP, CNRS-Université de Lorraine
1 rue Grandville 54001 Nancy (France)
Matteo.pietraccini@univ-lorraine.fr

Audrey Santandrea, Marie Verdonck, Anthony Dufour, Pierre-Alexandre Glaude, Olivier Dufaud
LRGP, CNRS-Université de Lorraine
1 rue Grandville 54001 Nancy (France)

Keywords: organic dust, explosion, model, reaction mechanism

Abstract

Organic dust explosions can be analyzed by decomposing them into their fundamental steps: heat transfers, pyrolysis, oxidation and flame propagation. This work was focused on the first three phenomena. It aims at simplifying a complex system such as a dust explosion to obtain a model capable of estimating some important parameters for industrial safety (laminar flame speed, deflagration index). The model is based on ten lumped reactions for the pyrolysis and the oxidation steps. Two powders were chosen: wheat starch and micro-crystalline cellulose. A modified Godbert-Greenwald furnace was used to reproduce the fast pyrolysis and oxidation conditions. Gas products, tar and solid residues are collected and analyzed. The dependence over the reaction temperature is investigated. The main components of the gas products are H_2, CO, CO_2, CH_4, C_2H_2, C_2H_4 and benzene. As the temperature rises, for the cellulose sample the H_2-to-CO ratio does not seem to change significantly, going from 0.32 at 700°C to 0.34 at 900°C. This behavior suggests that no reaction between hydrogen and/or carbon monoxide activates between 700 and 900°C, but also that flash-pyrolysis reactions involving hydrogen do not strongly depend on the reaction temperature over this temperature range. On the other hand, for the starch sample this dependency seems stronger. The CO_2-to-CO ratio, on the other hand, decreases with the temperature, from 0.76 at 700°C to 0.13 at 900°C, showing a strong dependency on the temperature. CH_4-to-CO ratio slightly increases with the temperature, but not enough to assess a significant dependency on the temperature. The analysis of the tars shows that, whatever the temperature is, for the cellulose sample the levoglucosan is the main component of the condensable part. A global low conversion degree of the powder was identified. The semi-global model showed results matching with pyrolysis and combustion experimental data. It provides a thorough understanding of the explosive behavior of organic powders and will help to better model dust explosions.

1 Introduction

Dust explosion are multifaceted phenomena and the variables that intervene are multiple: e.g. dust concentration, ignition energy, turbulence level, particle size distribution, temperature, and they can hardly be studied in their entirety. Their study and modelling thus intrinsically needs a certain degree of simplification, such as neglecting steps of secondary importance for kinetic approaches, focusing on rate-limiting steps. Reproducing the complexity of an industrial environment on a laboratory scale is a challenge and one has to be aware of the simplifications and approximations made. However, the experimental study of dust explosion is still essential both to create and validate predictive models and to feed them.

This works proposes to investigate and to model the explosion of organic dusts through a semi-global lumped-reaction system, specifically chosen to describe the most important building blocks of the entire phenomenon. Numerous works have already been done to describe and to model organic powder pyrolysis [1], [2], [3], but unfortunately, they do not match the needs of the dust explosion field. In fact, these works are based on analyses and experimental setups whose operative parameters are very far from those characterizing a dust explosion test, *inter alia* heating rate, particle size distribution and temperature ranges. For this very reason, this work aims at representing the pyrolysis phase occurring during an organic dust explosion, i.e. for particle diameters ranging from 1 μm to 500 μm, temperatures ranging from 500°C to 1000°C and heating rates greater than 5000 $K.s^{-1}$.

The approach proposed is to divide the dust explosion phenomenon into fundamental steps: heat transfers, pyrolysis, oxidation and flame propagation. Each step is treated separately and, as a function of its importance in the explosion kinetics, will be neglected or integrated in the final model. The heat transfer study is based on the approach proposed in Piskorz et al [4] and Paulsen et al [5], and it aims to determine the temperature profile of a particle during its gasification. The pyrolysis step is investigated by the means of a modified Goldberg-Greenwald furnace; results are fitted and a semi-global model is elaborated for this reaction step. The same approach is used for the oxidation step, studied with the same experimental setup. In order to study the flame propagation step and to determine the explosion parameters, a further work is already being conceived [6]. However, it is beyond the purpose of this article, which is centered on heat exchanges, pyrolysis and oxidation steps.

2 Materials and experimental methods

2.1 Sample choice

Wheat starch (Sigma-Aldrich) and micro-crystalline cellulose (Avicel PH 101 - DuPont) were chosen as samples as they share the same monomer ($C_6H_{10}O_5$), but their structure is deeply different. In a starch chain, monomers are linked to each other by a α(1-4) and a α(1-6) glycosidic bond, while in cellulose, monomers are bonded together by the means of a β(1-4) glycosidic bond. This results in a branched structure for the starch and in a more linear one for cellulose, respectively leading to a more spherical and a more elongated and fibrous shape of the particles. This choice aims to elucidate a different thermal behavior, in terms of kinetics and reaction products, consequence of the different chemical structure, as well as the particle shape and the inter-particle interaction. Other organic powders and mixtures will be considered in the future, in order to approach the phenomenon with an increasing level of complexity.

2.2 *Sample characterization*

2.2.1 Particle size distribution

The particle size distribution (PSD) of the samples was determined by wet and dry dispersion. The wet PSD analysis is commonly used in an industrial context to easily and quickly identify a powder, for instance through the D_{10}, D_{50} and D_{90} values of the volume distribution. However, particles may interact with the solvent or with each other (by agglomeration and/or fragmentation phenomena), which is not representative of the operating conditions encountered during dust explosion or flash pyrolysis tests. Wet PSD analyses were performed in ethanol, preventing cellulose from significant swelling, using a Mastersizer 2000 laser diffraction sensor (Malvern).

Dry dust dispersion was performed with the same apparatus used for the tests (a modified Godbert-Greenwald furnace, detailed in Section 2.3). PSD analyses were done with a Helos laser diffraction sensor (Sympatec) at two different levels of the experimental setup (indicated with a yellow star in Figure [1]) in order to highlight potential agglomeration or fragmentation phenomena. Both analyses are performed thrice for reproducibility reasons.

2.2.2 Optic microscope imaging

Using a 5 Mp Dino-lite Pro HR digital microscope, samples were submitted to an optic imaging analysis necessary to collect information about the particle shape, their average shape factor (needed for PSD analysis) and the potential powder agglomeration.

2.3 *Experimental setup*

Pyrolysis and combustion experiments were carried out using a modified Godbert-Greenwald furnace, usually employed for the determination of the Minimum Ignition Temperature (MIT) of a dust cloud (according to the ISO/IEC 80079-20-2 standard). Its simple structure makes it suitable for other purposes, such as a classic plug flow reactor ([7]; [8]; [9]), which will be the case in this work. The standard Godbert-Greenwald furnace was equipped with two coaxial Inconel cylinders to ensure its air tightness and the correct circulation of the gases. As shown in Figure [1], the powder was dispersed into the vertical tubular furnace by a gas pulse (as in Dufaud et al. [10]). Argon was used for studying the pyrolysis step, whereas air was the gas vector for the combustion one. At the exit of the furnace, a solid residue trap allows the collection of the solid residue and its subsequent analysis. Tars condense in the U part of the cooling system, thanks to a Dewar vase filled with isopropyl alcohol at -30°C. Gas products were collected in a collapsible Tedlar bag for analysis. In order to determine the kinetic parameters of the selected reactions, three temperatures were chosen, to be consistent with the temperature range and heating rates defined for studying dust explosion: 700, 800 and 900°C. For each temperature level, 0.2 g of sample were dispersed in the heated chamber of the Godbert-Greenwald furnace. The corresponding dust concentration in the heated chamber is about the stoichiometric concentration. The test procedure is repeated fivefold (in all, 1 g of each sample was injected for a given temperature value) to obtain enough solid residue and tar to analyze. Gas products were analyzed only for the first three dispersions. Since the residence time of the dust cloud in the heated chamber lays between 150 and 200 ms (estimated using a high-frequency camera), the phenomenon that is occurring can be classified as a flash pyrolysis as its heating rate is greater than 1000°C.s^{-1} [11].

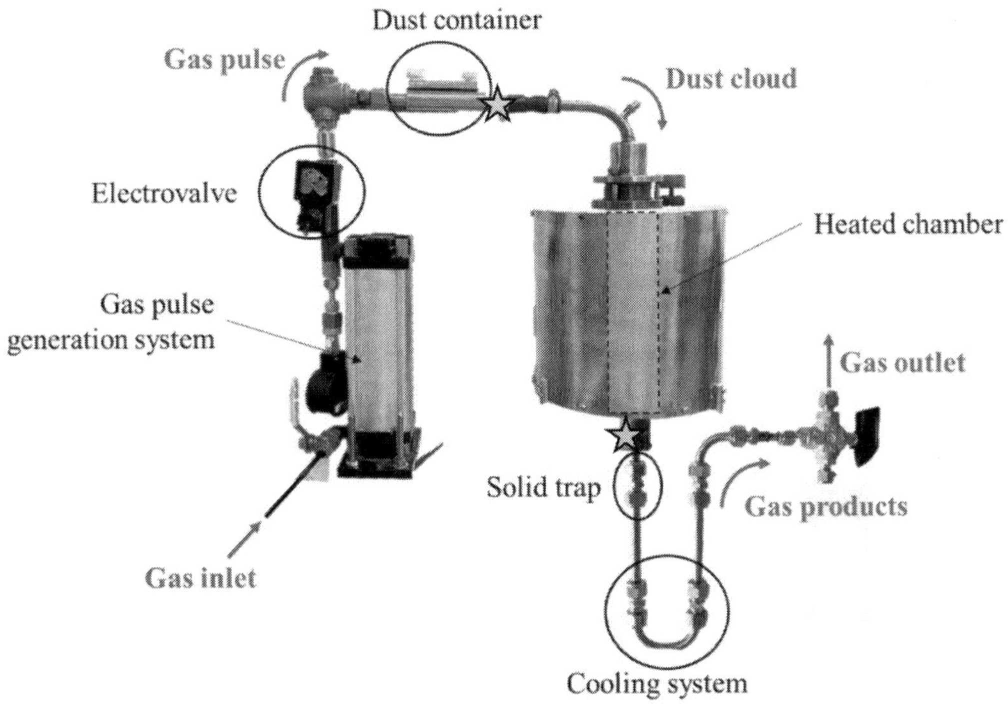

Figure [1] – Modified Godbert-Greenwald furnace. The yellow stars show where the in-situ PSD analysis was performed.

2.4 Pyrolysis products analysis

2.4.1 Solid residues

Solid residues were observed using a 5 Mp Dino-lite Pro HR digital microscope. The degree of conversion of the particles was qualitatively assessed by analyzing the color variation of the powder. Such analysis also allows the observation of potential agglomeration phenomenon occurring in the reactor.

2.4.2 Tar – Condensates

The U part of the cooling system is rinsed with methanol, in order to solubilize the condensed molecules. 1 µL of 1-tetradecene is then added as internal standard for the quantitative analysis of the components. The solution is then filtered with a 0.45 um – pore filter and analyzed by gas chromatography (Agilent 7890A System equipped of a 5975C Triple-Axis detector) to assure a qualitative and quantitative analysis of the tar and condensates. Results will be exploited to consider the most accurate tar-representing molecule in the model.

2.4.3 Gases

Gas products were analyzed by gas chromatography (SRA 3000 µGC equipped with a FID detector). Permanent gases (CO, CO_2, CH_4, H_2, O_2 and N_2) are identified, as well as some aromatics (benzene, toluene and xylene isomers) and some light hydrocarbons (C_2H_2, C_2H_4 and C_3H_6). This analysis will allow to identify the gas species as well as to determine the main reactions occurring during such a flash pyrolysis.

3 Results

3.1 Sample characterization

3.1.1 Particle size distribution

As it can be seen in Figure [2], the PSD of both powders dispersed in ethanol show a mono-modal distribution for cellulose, centered around 90 µm, and a bimodal distribution for wheat starch, with a larger and a smaller peak, centered respectively around 30 and 3 µm. More precisely, the D_{10}, D_{50} and D_{90} parameters are 11.8, 26.4, 49.5 µm and 22.2, 67.9 and 146.1 µm for starch and cellulose, respectively. The PSD of the cellulose cloud determined at the entrance of the furnace, i.e. just after the dust container, (Figure [3]) is consistent with that observed by wet measurement. Concerning starch, the main difference between PSD performed by wet or dry method is the absence of small particles (lower than 10 µm) when the dust is dispersed in air. This absence can be explained by strong particle-particle interactions leading to the 'coating' of small particles onto the surface of bigger structures.

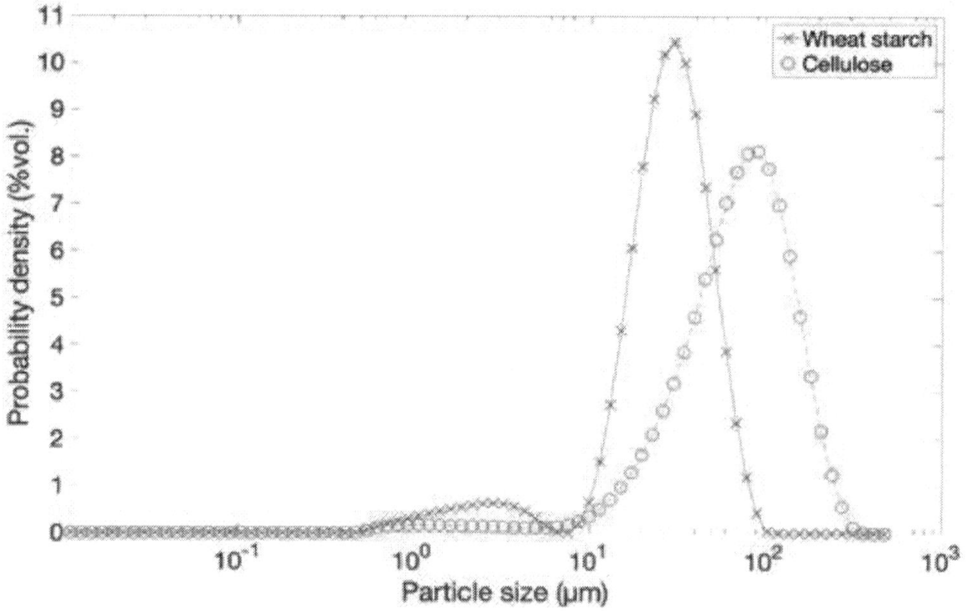

Figure [2] – In-liquid particle size distributions of the two samples.

Figure [4] enlightens similar interaction occurring during the cellulose dispersion within the furnace and leading to the formation of coarser structures identifiable as agglomerates. As previously said, the PSD is not significantly modified at the exit of the dust container; on the other hand, after the elbow and the heated chamber, particle-particle and particle-wall interactions are no longer negligible and powder agglomerates reach diameters even larger than 250 µm. A similar analysis was performed for starch powder and enlightened an agglomeration phenomenon of the same order of magnitude as the cellulose.

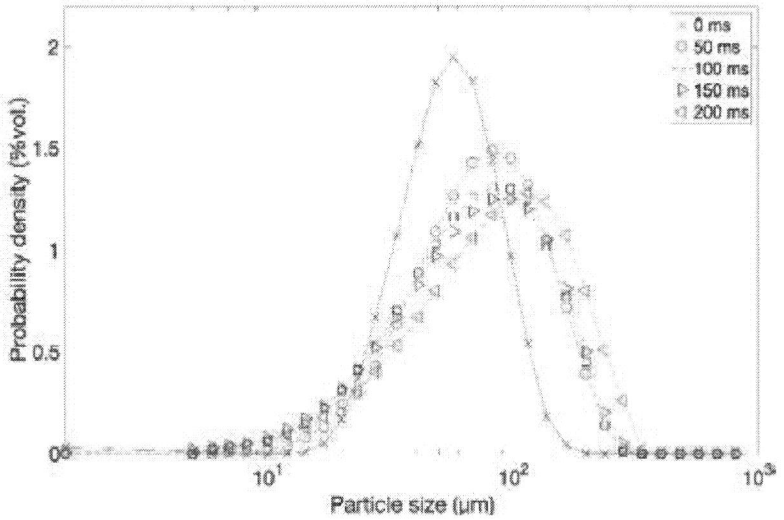

Figure [3] – Particle size distribution of the cellulose sample before and after the heated chamber of the Godbert-Greenwald furnace.

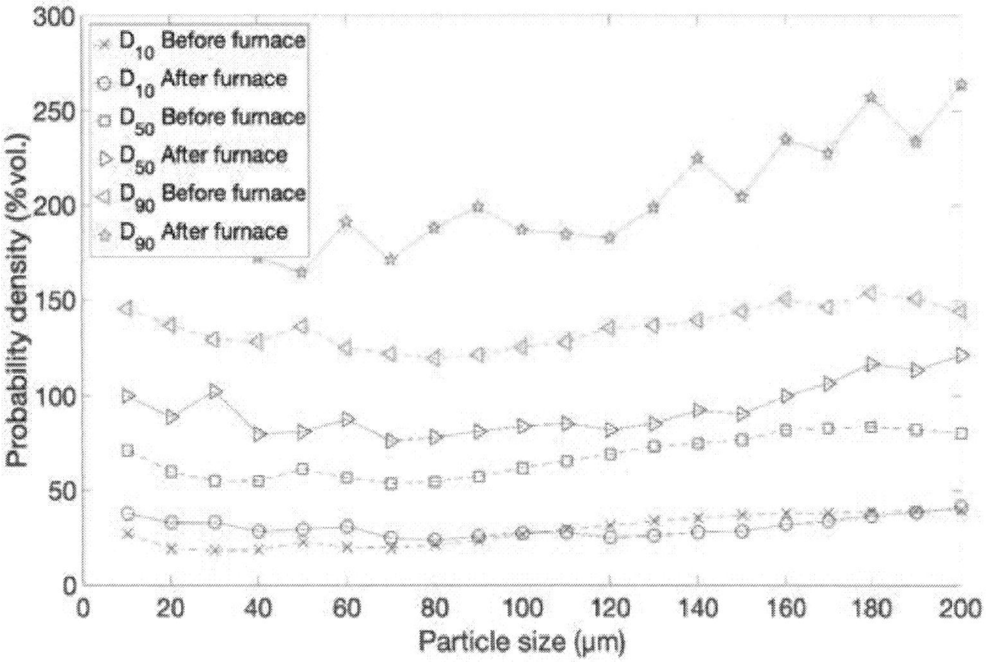

Figure [4] – D_{10}, D_{50} and D_{90} of the cellulose sample dispersed through the Godbert-Greenwald furnace, as a function of time.

3.1.2 Optic imaging

The optic imaging was useful to illustrate the typical characteristics of cellulose and starch particle shapes. In Figure [5], it can be noticed how different the two sample particles appear. Cellulose showed an important fibrous character, which leads to more elongated particles. On the contrary, wheat starch particles seemed to be associated to a significantly higher sphericity, which might confer a lower tendency to form agglomerates and fragments, with the exception of small particles adhesion/coating. Indeed, optic microscopy confirms the presence of starch particles with a diameter lower than 10 μm (Figure [5]). It must be emphasized that, if agglomeration and fragmentation phenomena strongly depend on the particles geometry, it is also related to other factors such as the shear stress applied during the dispersion [12]. It will be shown that the particle structure/shape can significantly impact its reactivity.

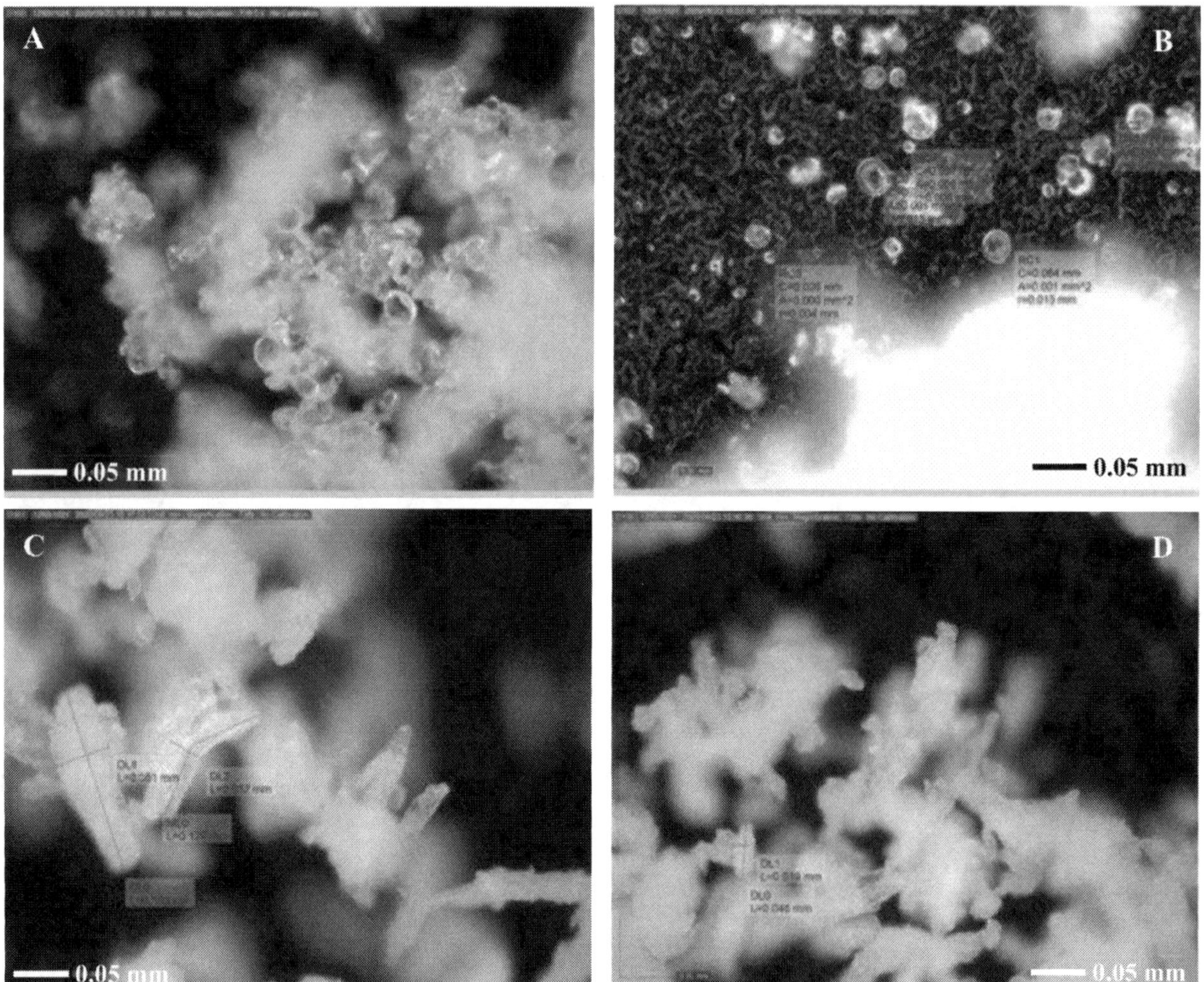

Figure [5] – Optical microscopy pictures of starch (A and B) and of cellulose (C and D).

3.2 *Pyrolysis step*

3.2.1 Solid residue

Results of the optical imaging analysis of the solid residue are shown in Figure [6] for cellulose; a similar analysis was performed for starch. Images clearly show in evidence a color difference of the three cellulose residues collected after pyrolysis under argon at 700, 800 and 900°C, compared to the original sample. While it is obvious that the solid conversion is not fully complete, a clear change in the color of the residues logically suggests that the conversion increases as a function of the furnace temperature. This modification is the results of two different contributions: i) the first one is associated with the unreacted particles (white powder), whose temperature never reaches the activation temperature for pyrolysis reactions, ii) the second one is associated with particles having undergone a partial pyrolysis reaction due to a small residence time. This qualitative analysis will support the evolution of the conversion degree considered in the model.

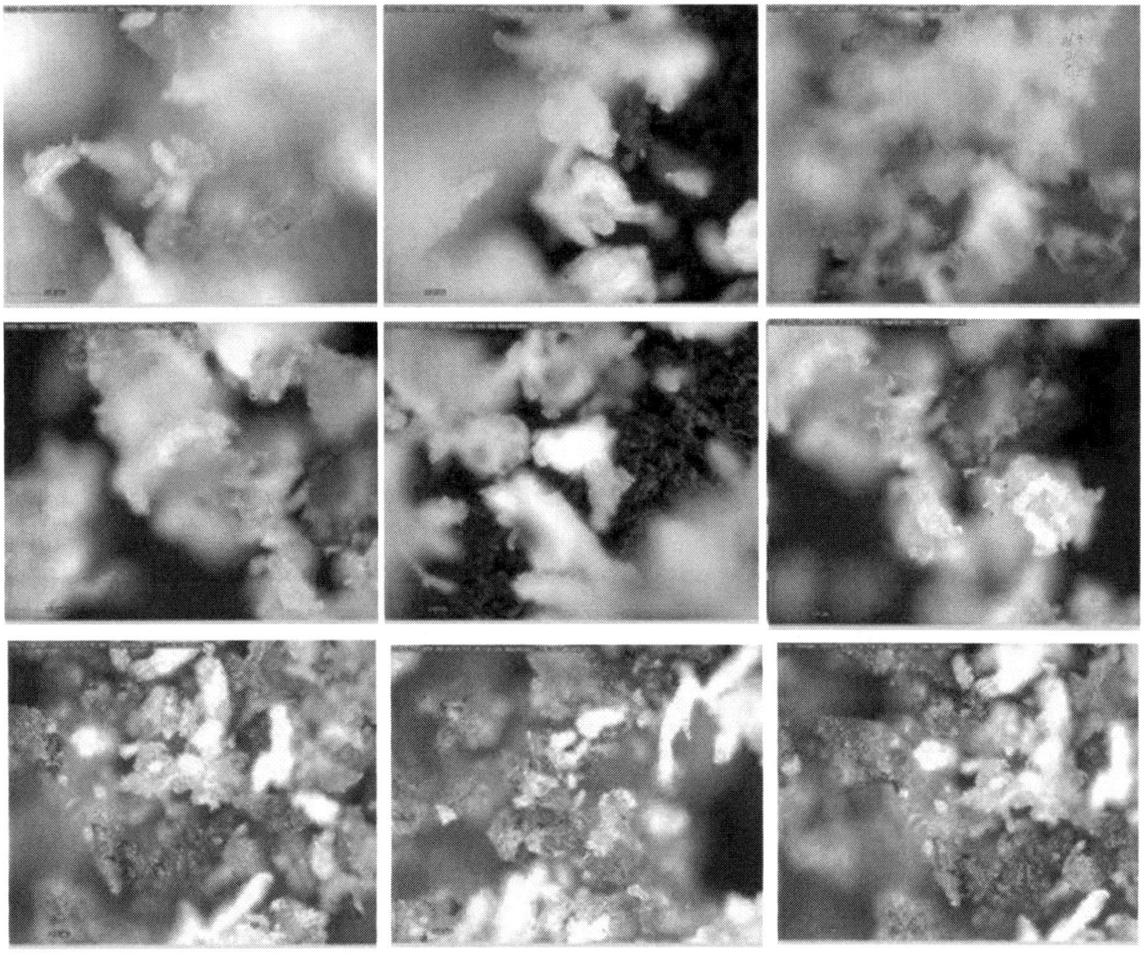

Figure [6] – Images of the solid residues collected at the three temperature (from the top, 700, 800 and 900°C) for cellulose powder.

3.2.2 Tar - Condensate part

Results of the gas chromatography – mass spectrometry analysis of the condensed part of the products are shown in Figure [7], for 700°C as an example. The first part of the graph not showing any major peaks, the figure was focused on the 25-35 min zone. The first peak corresponds to 1-tetradecene, the internal standard used for the quantitative analysis. The second peak is associated with levoglucosan, the most abundant compounds in flash pyrolysis products [13]. It should be added that a slight presence of hydroxyl- and dihydroxyl-acetone is recorded. A Relative Response Factor (RRF) of 4.47, experimentally determined for the association levoglucosan/1-tetradecene, was used for exploiting the data relative to levoglucosan. The levoglucosan yields, calculated considering the cellulose injected in the reactor, are shown in Table [1]. Since the levoglucosan mass yields are very low compared to the values in the literature [4], [13], the flash pyrolysis is consequently incomplete. Thus, over this temperature range, the effect of the temperature seems secondary compared to the residence time, the variable that seemed to really rule over the pyrolysis reactions. As for the starch sample, the GC/MS results showed different chemical species among the tars, such as 1-methoxyethyl benzoate, oxo-acetic acid methyl ester and a minor presence of levoglucosan (especially at 800 and 900°C). Starch fast pyrolysis reaction mechanisms seem to slightly differ from those of cellulose.

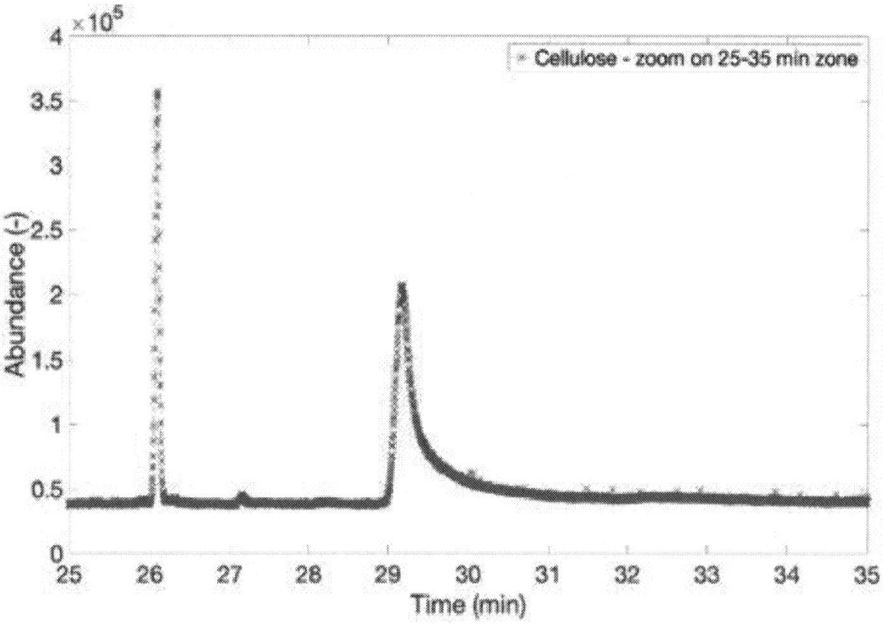

Figure [7] – GC/MS result for the test performed at 700°C. The first part of the graph not showing any major peaks, the figure was focused on the 25-35 min zone.

Table [1] – Levoglucosan mass yields, based on the cellulose sample injected in the reactor.

Reactor temperature, °C	Levoglucosan yield, %wt
700	0.55
800	0.91
900	0.73

3.2.3 Gas products

The composition of the pyrolysis permanent gases obtained for cellulose is shown in Figure [8]. They are expressed as CO-ratios, this gas compound being chosen as reference. In fact, by dividing all the molar concentrations of each compound by that of the carbon monoxide, it is possible to better emphasize potential reactions being activated in the temperature range 700-900°C. The main components of the gas products are H_2, CO, CO_2, CH_4, C_2H_2 (acetylene) and C_2H_4 (ethylene). Traces of light hydrocarbons, benzene and toluene were observed as well.

As the temperature rises, the H_2-to-CO ratio does not seem to change significantly, slightly increasing from 0.32 at 700°C to 0.34 at 900°C. This behavior suggests that no major reaction between hydrogen and/or carbon monoxide occurs between 700 and 900°C, but also that flash-pyrolysis reactions involving hydrogen do not strongly depend on the reactor temperature under such operating conditions. The CO_2-to-CO ratio, on the other hand, decreases strongly with the temperature, from 0.76 at 700°C to 0.13 at 900°C, showing a strong dependency on the reactor temperature. Moreover, CH_4-to-CO ratio slightly increases with the temperature, but not enough to assess a strong dependency on the reactor temperature. Finally, C_2H_4- and benzene-to-CO ratios slightly increase with temperature, whereas C_2H_2-to-CO ratio decreases; nevertheless, the low concentration of benzene and C_2H_2 allowed neglecting them in the model.

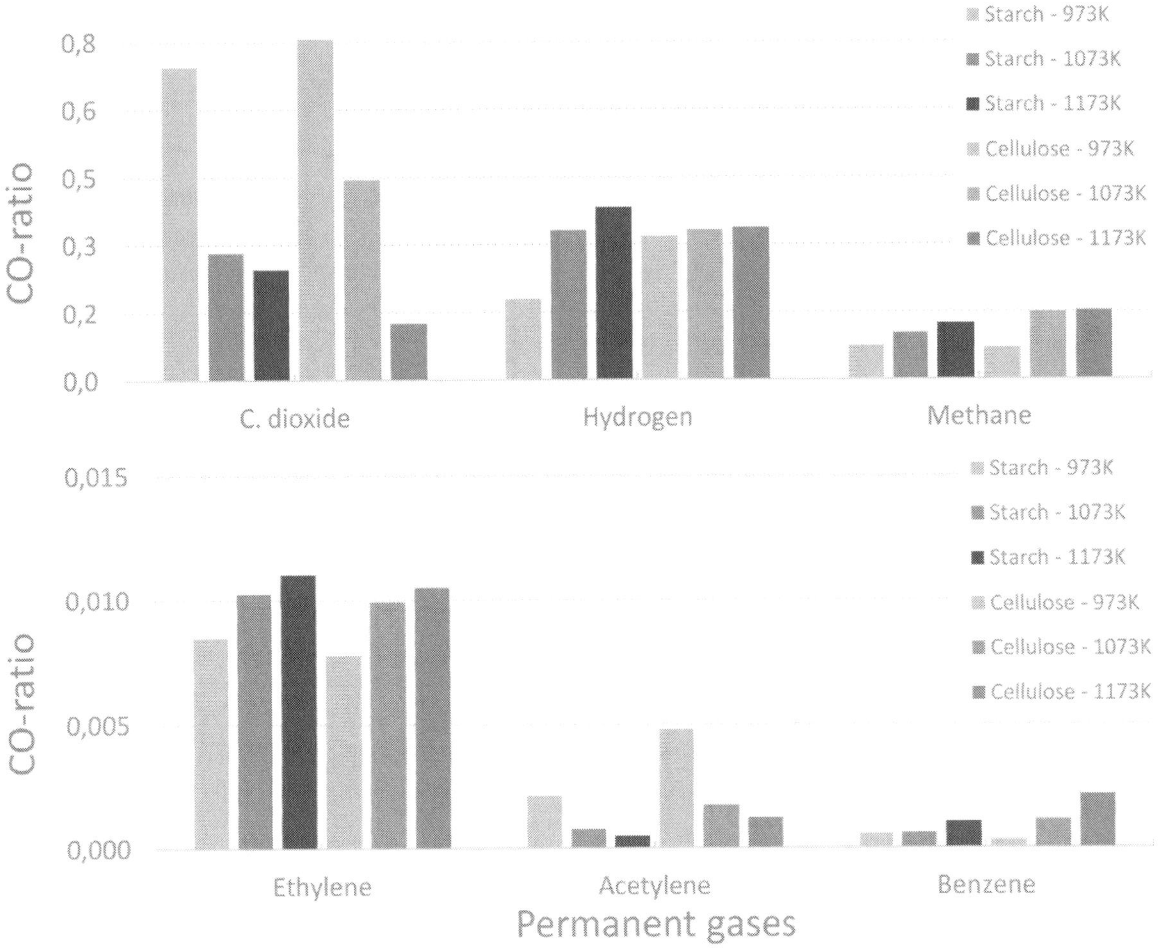

Figure [8] – Composition of the pyrolysis permanent gases obtained for cellulose, expressed as CO-ratios.

4 Modeling

4.1 Heat transfers

The Godbert-Greenwald furnace is almost exclusively employed for the Minimum Ignition Temperature determination and, as the standard defines strictly the procedure to follow, the thermal characteristics of a dust cloud passing through this set-up are rarely investigated. In order to understand the importance of the heat transport phenomenon, especially when compared to the reaction steps, the determination of selected dimensionless numbers can be useful. For instance, it is possible to define the following dimensionless numbers:

$$Bi = \text{Biot number} = \frac{h\,L}{\lambda} \qquad (1)$$

$$Py' = \text{First pyrolysis number} = \frac{\lambda}{\rho.C_p.L^2.k} \qquad (2)$$

$$Py'' = \text{Second pyrolysis number} = \frac{h}{\rho.C_p.L.k} \qquad (3)$$

where h and λ are the convective and conductive heat transfer coefficients, respectively; ρ is density of the gas phase and C_p, the specific heat at constant pressure. The last two parameters can be significantly different for starch and cellulose: L is a characteristic length of a particle and k, the global kinetic constant of the pyrolysis reactions in s^{-1}.

Piskorz et al [4] assessed 0.1 as critical value for Bi, under which the internal temperature profile is isothermal. As Bi attains 0.2, the particle starts to exhibit a rather important temperature difference between its core and surface. However, it still can be considered isothermal without introducing a high error, especially considering the shrinking effect during pyrolysis. As soon as Bi surpasses 0.2, the radial temperature gradient is too important to be neglected and the particle is not isothermal. As for the pyrolysis numbers Py' and Py'', they play a fundamental role comparing the reaction time scale to the internal conduction and external convection, respectively. When Bi is lower than unity, i.e. when the particle internal conduction is faster than the external convection, the value of Py'' allows to conclude whether the pyrolysis reaction or the external convection is the limiting step. On the contrary (Bi > 1), Py' has to be considered. In this work, both sample particles are associated to a Bi lower than 1, i.e. Py'' will be considered.

As shown in Figure [9], three regimes can be identified according to the cellulose particle size:

i) Regime A, for Bi < 0.1 and high values of Py''. The external convection is slower than the internal conduction (the particle is then isothermal) and the pyrolysis is slower than the external convection: the pyrolysis is the limiting step.
ii) Regime B, for 0.1 < Bi < 0.2 and low values of Py''. The particle can be considered isothermal and the pyrolysis rate is comparable with the external convection: either pyrolysis or the external convection are the limiting step.

iii) Regime C, for Bi > 0.2 and low values of Py''. The particle shows an important temperature gradient and the external conduction is slower than the pyrolysis: either the internal conduction or the external convection is the limiting step.

In this work, considering the PSD of the dust clouds, i.e. diameters lower than 300 μm, it can be concluded that there is no limitation from the internal conduction. Since the particle size lays mostly between 20 and 200 μm, the more important limiting step is likely the pyrolysis, which confirms the interest of modeling this reaction step.

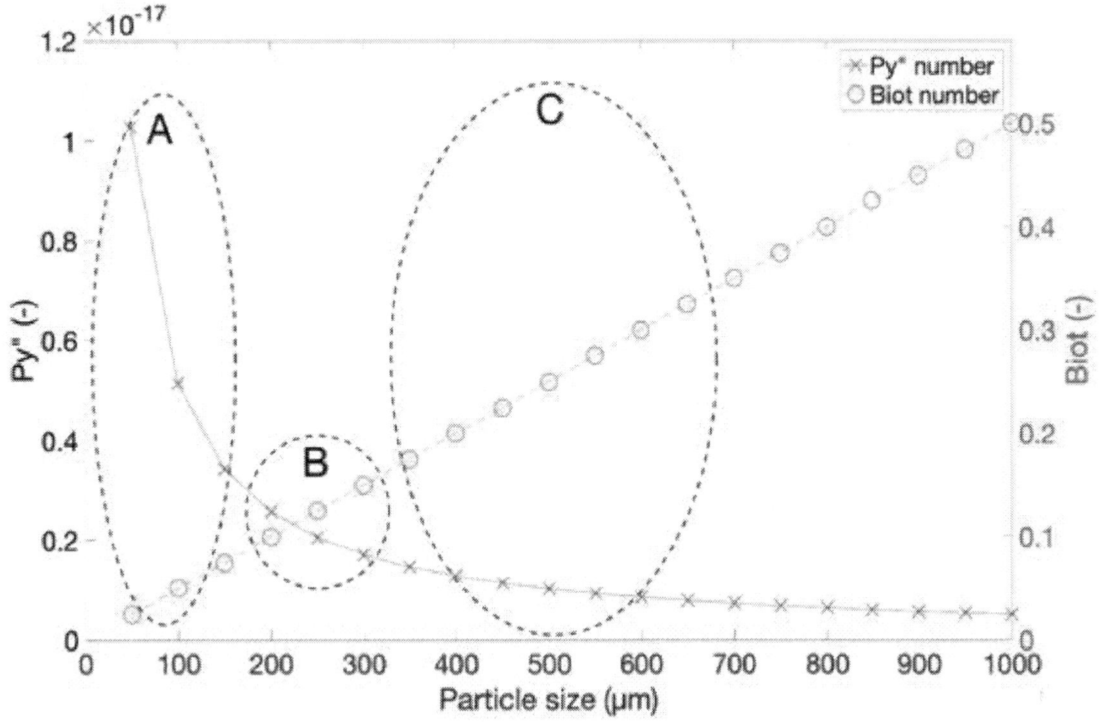

Figure [9] – Py'' and Bi as functions of the cellulose particle size. Three regimes (A, B and C) are identifiable according to these two dimensionless numbers.

4.2 Pyrolysis model

The main objective of this part is to elaborate a model able to determine the kinetic and reaction parameters (stoichiometric coefficients, activation energies E_a and pre-exponential factors A) of some specific chemical reactions, selected to represent the global flash pyrolysis phenomenon occurring in the Godbert-Greenwald furnace during the tests. The model provides for 3 main parts: the construction of the equations to solve, their resolution and a verification step. The codes thus elaborated were solved with Matlab.

4.2.1 Generator

The core of the model is "Generator", which is (at the moment) based on the following system of reactions:

Cellulose → Permanent gas + Water + Tar

Cellulose → Permanent gas + Water + Tar

Cellulose → Solid residue + Water

$$\text{Cellulose} \xrightarrow{k_1} n_1H_2 + n_2CO + n_3CO_2 + n_4CH_4 + n_5C_2H_4 + n_6H_2O + n_7C_6H_{10}O_5 \quad (4)$$

$$\text{Cellulose} \xrightarrow{k_2} m_1H_2 + m_2CO + m_3CO_2 + m_4CH_4 + m_5C_2H_4 + m_6H_2O + m_7C_6H_{10}O_5 \quad (5)$$

$$\text{Cellulose} \xrightarrow{k_3} 5H_2O + 6C \quad (6)$$

The three equations share the cellulose as same reactant, which converts to permanent gas, water, tar and char. Since the GC/MS analysis showed that the levoglucosan is the most abundant product, the entire class of compounds that form the condensable part of the pyrolysis products are represented by the molecule $C_6H_{10}O_5$. The first two equations are the same and they have each 7 coefficients (n_i and m_i, respectively). They have also two different kinetic constants (k_1 and k_2, respectively), in order to have the CO and CO_2 concentrations separately varying according to the temperature. In fact, with just one equation the result would be the CO and CO_2 concentration both increasing or decreasing with the temperature, which is far from what happens [14]. The last equation involves the solid residue, so the transformation of the sample in a different solid phase. The choice not to call it char is made on purpose, as the associated conversion degree is too low to lead to a veritable charring of the particle.

It is obvious that the model is very global, as it does not consider the large number of reactions actually occurring during a pyrolysis phenomenon (the direct solid-gas reactions which involves C and CO, the Water Gas Shift reaction, the methane oxidation, etc.). However, as the model does not aim to perfectly represent reality, but rather to estimate the main species produced and their concentration, this path is chosen to be investigated.

All kinetic constants follow the Arrhenius law, thus there is a pre-exponential factor A_i and an activation energy $E_{a,\,i}$ that are associated to each k_i. In the end: 2x7 coefficients (n and m) + 3x2 constants (A and Ea, at two temperature levels) = 20 unknowns to determine. In order to match these unknowns with an equal number of equations, atomic balances (carbon, hydrogen and oxygen) are elaborated. Furthermore, considering the furnace as a stirred tank reactor, it is possible to elaborated seven more equations for each temperature value, using the CO-ratios. For instance, considering the H_2-to-CO ratio:

$$\frac{[H_2]}{[CO]} = \frac{n_1k_1+m_1k_2}{n_2k_1+m_2k_2} \quad (7)$$

The CO-ratios are known by the means of the experimental analysis previously done. Unfortunately, C-to-CO and H_2O-to-CO ratios were not quantified. However, by applying the atomic balances (C, H and O) to the experimental results, they can be estimated.

4.2.2 Calculators

The codes that calculates the kinetic parameters, the stoichiometric coefficients and the products percentages are "Calculators". The first one calculates the composition of each considered species in the products (thus H_2, CO, CO_2, CH_4, C_2H_4, H_2O, $C_6H_{10}O_5$ and solid residue). The second one calculates the global composition of the products, in terms of permanent gases, H_2O, $C_6H_{10}O_5$ and solid residue. The third one determines the composition of the gas products (thus H_2, CO, CO_2, CH_4 and C_2H_4).

4.2.3 Solver

The tool that allows to solve the system is called "Solver" and it uses a non-linear least mean square fitting function to calculate the kinetic parameters. As the experimental data were collected from 700 to 900°C, the extreme temperature values chosen for the calculations are 600 and 1200°C. By doing so the error introduced with the fitting technique can be accepted. "Solver" exploits "Calculator" to respond to the inputs received by the operator. These can be three: "Given temperature", "Temperature range" and "Calculations". The first one allows to calculate all the products percentages for one given temperature value; the second one calculates the curves of the products as a function of the temperature; finally, the third one responds with the kinetic parameters and the stoichiometric coefficients calculated. Wheat starch tests lead to the results shown in Figure [10]. Table [2] resumes the kinetic parameters calculated by the code.

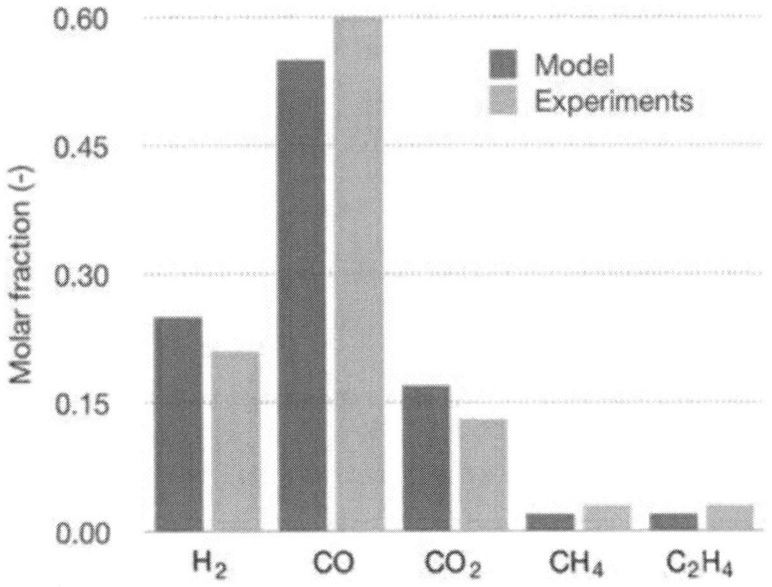

Figure [10] - Comparison between experimental and calculated gas molar fractions for starch at 800°C.

Table [2] - Parameters of the reaction rate constants k_i determined for starch pyrolysis

Reaction rate constant	Pre-exponential factor, s^{-1}	Activation energy, $kJ.mol^{-1}$
k1	1.2×10^{17}	318.6
k2	8.8×10^{17}	41.0
k3	2.8×10^{17}	72.6

4.3 Oxidation

The approach followed for the pyrolysis was adopted for the oxidation step too. The Generator was modified in order to take into account the oxygen and the associated combustion reactions. Each chemical species previously considered (apart from CO_2, water and C) is assumed to react with the oxygen, generating CO_2 and H_2O. The only assumption that was made is that NO_x generation is neglected. Other simplifications will be eventually introduced to fit better with experimental data. Considering thus five species potentially able to react with oxygen, to determine the associated ten new stoichiometric coefficients (n_1'- n_6' and m_1'- m_6') and two kinetic constants (k_1'and k_2'), the following equation system is elaborated:

$$cellulose \xrightarrow{k_1} [pyrolysis\ products]_1 \underset{+O_2}{\overset{k'_1}{\rightarrow}} n'_1H_2 + n'_2CO + n'_3CO_2 + n'_4CH_4 + n'_5C_2H_4 +$$

$$n'_6H_2O + n'_7C_6H_{10}O_5 \quad (8)$$

$$cellulose \xrightarrow{k_2} [pyrolysis\ products]_2 \underset{+O_2}{\overset{k'_2}{\rightarrow}} m'_1H_2 + m'_2CO + m'_3CO_2 + m'_4CH_4 + m'_5C_2H_4 +$$

$$m'_6H_2O + m'_7C_6H_{10}O_5 \quad (9)$$

$$C_6H_{10}O_5 \xrightarrow{k_3} 5H_2O + 6C \quad (10)$$

Preliminary results of pyrolysis and combustion step for starch powder are reported in Figure [11]. Table [3] shows the values of the pre-exponential factors and activation energies of the two new constants k_1'and. It appears that the dependence on the temperature of the gas species considered varies less significantly with temperature than during the pyrolysis step. As expected, carbon monoxide and dioxide seem to be the most abundant in the gas phase.

Table [3] - Values of the parameters of the reaction rate constants for the combustion model

Reaction rate constant	Pre-exponential factor, s^{-1}	Activation energy, $kJ.mol^{-1}$
k_1'	1.2×10^{17}	318.6
k_2'	8.8×10^{4}	41.0

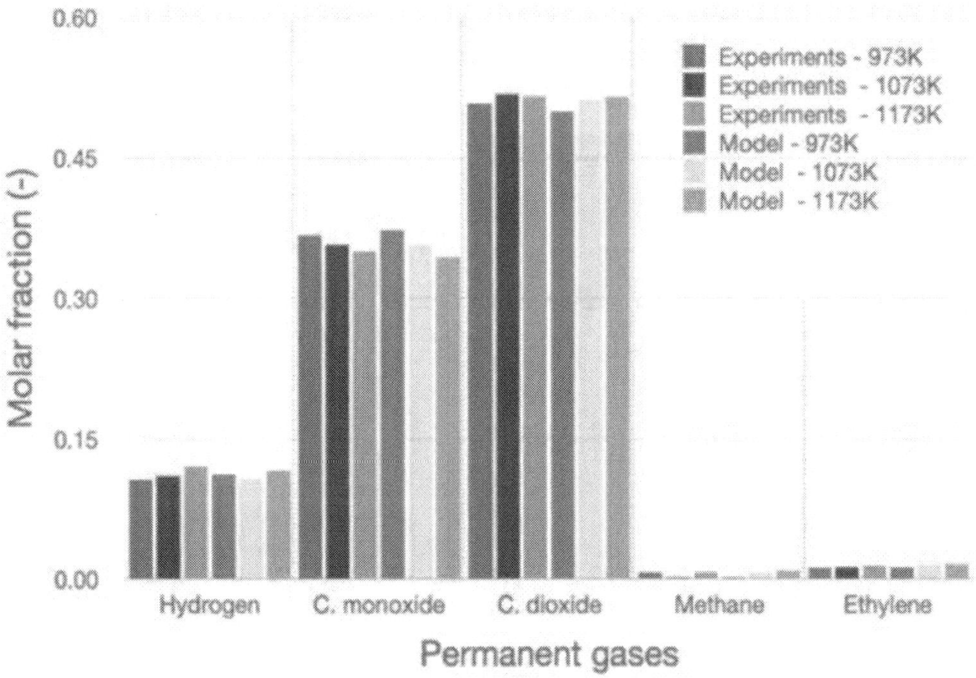

Figure [11] – Comparison between experimental data and modeling results for the combustion step of the starch sample.

5 Towards the determination of the laminar flame velocity

Determining the laminar burning velocity of a dust explosion flame is essential as this parameter is an intrinsic property of the oxidizer-fuel mixture. As a consequence, it is used as an input for CFD simulation aiming an assessing the effects of an explosion in an industrial context. Difficult to determine experimentally, as the dust dispersion implies a certain level of turbulence, the development of a numerical model can be a valuable alternative.

Torrado et al. [15] developed a 1-D numerical model of a flame propagation for dust-gas mixtures (carbon black-methane), considering the convection and radiative heat transfer, the diffusion of the species and the combustion reaction kinetics. A simulation domain was divided into three parts representing the preheat zone, the flame (or reaction zone) and the post-flame zone. The energy balance in each zone was performed assuming a constant pressure and neglecting the viscous forces. Thermal radiation was considered by solving the radiative transfer equation (RTE) and the emitted, absorbed and scattered energy is estimated by defining absorption and scattering coefficients as previously done by Haghiri and Bidabadi [16]. The black body emission has been neglected in the preheat zone. The mass balance of the numerical model was notably based on a semi-global kinetic mechanism representing the combustion of methane and on the addition of a single reaction representing the carbon oxidation in carbon monoxide. The finite volume method was used for the discretization of the balances.

If this approach proved its relevance and has been validated for hybrid mixtures, it needs to be modified, especially by taking into account more detailed pyrolysis-combustion reactions for organic dusts. Therefore, the pyrolysis-combustion mechanism elaborated in this study constitutes now the core of the modified burning velocity model. After implementing the new reaction mechanism, the model will be validated by performing flame propagation test in a semi-open tube. During such experiments, both spatial velocity and flame surface are recorded as a function of time. Following the approach developed by Cuervo et al. [17], the flame velocity can be expressed as a function of the stretching factor, known also as Karlovitz factor. It represents the evolution of the flame surface area over time, as shown in Figure [13] for cellulose powder. The intercept corresponds theoretically to the laminar flame velocity of this cellulose/air mixture. In this case, a value of 21.5 cm/s was obtained for cellulose during preliminary tests.

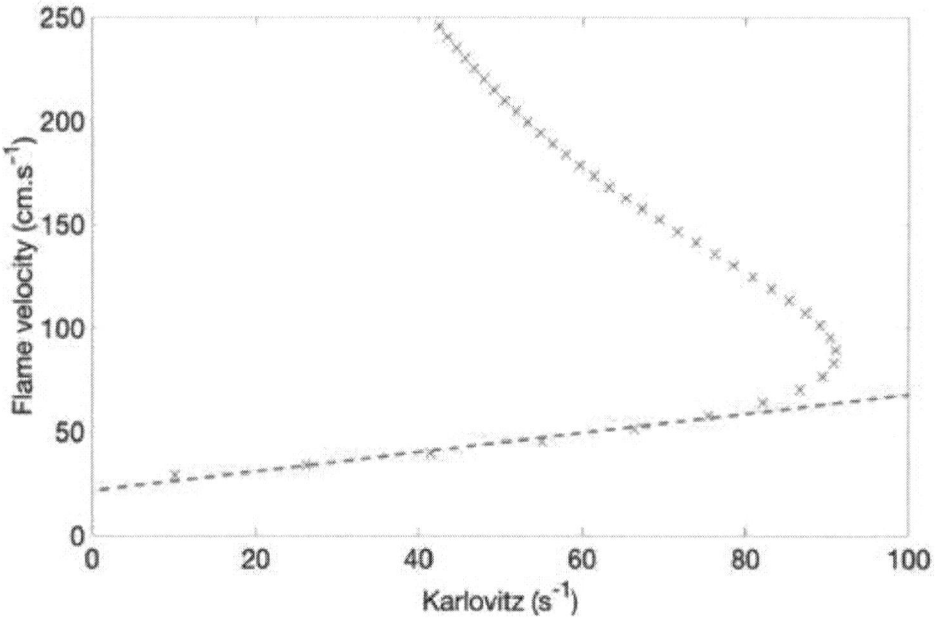

Figure [13] - Flame velocity as a function of the stretching factor for a cellulose/air mixture explosion (t_v =180 ms, E = 1 J, m = 1 g)

6 Conclusions

An organic dust explosion can be disassembled into its fundamental building blocks in order to be studied and to examine which phenomenon rules over the others. Considering cellulose and starch powder, pyrolysis seemed to be the controlling step. In fact, Bi and Py'' calculated for the two chosen sample showed that the internal conduction is unlikely the limiting one, and the external convection might rule over the global phenomenon only for coarser particle (> 200 μm). By using a modified Godbert-Greenwald furnace, the pyrolysis step was investigated. Collecting the solid residues and the condensate, in addition to the gas products, allowed to assess two things.

- The average conversion degree of particles in a Godbert-Greenwald furnace is far from 1: the short residence time of particles in the reactor, the low pyrolysis reaction rate and, eventually, the relatively low efficiency of the external convection are the most probable responsibles.

- Levoglucosan is the most abundant product: taking into account other tar-representative molecules in the model will not lead to a higher accuracy level or to a more important matching degree with the experimental data.

As for the modelling codes elaborated, results from pyrolysis and combustion simulations were in good agreement with the experimental results. The activation energies of the considered reactions seemed not to be influenced by taking into account the combustion step, in addition to the pyrolysis one. Moreover, the pre-exponential factors showed a slight difference between pyrolysis and combustion step.

The complete and optimized reaction mechanism will be introduced in a wider model for the determination of the laminar flame speed. This parameter thus obtained will eventually serve as a fundamental input for the estimation of the explosion severity and the associated explosion parameters (P_{max} and K_{St}).

These preliminary results enlightened the feasibility of this simplification approach, even though there is room for further improvement. Firstly, a complete comparison between wheat starch and micro-crystalline cellulose will be made, stressing out eventual difference in their thermal behaviors. Secondly, a deeper analysis of the particle conversion degree will be developed, in order to better consider this aspect in the elaboration of the model. In the end, the model will be integrated with one or two temperature levels, to enhance the accuracy level of the fitting procedure.

References

[1] A. Liu, J. Chen, X. Huang, J. Lin, X. Zhang, W. Xu. "Explosion parameters and combustion kinetics of biomass dust", Bioresource Technology, September 2019.

[2] G. Zhang, Y. Zhang, X. Huang, W. Gao, X. Zhang. "Effect of pyrolysis and oxidation characteristics on lauric acid and stearic acid dust explosion hazards", Journal of Loss Prevention in the Process Industries, December 2019.

[3] A. Dufour, B. Ouartassib, R. Bounaceura and A.Zoulalianb. "Modelling intra-particle phenomena of biomass pyrolysis", Chemical Engineering Research and Design, pp. 2136-2146, January 2011.

[4] J. Piskorz, P. Majerski, D. Radlein, A. Vladars-Usas and D. S. Scott. "Flash pyrolysis of cellulose for production of anhydro-oligomers", Journal of Analytical Pyrolysis, pp. 145-166, April 2000.

[5] A. D. Paulsen, M. S. Mettler and P. J. Dauenhauer. "The Role of Sample Dimension and Temperature in Cellulose Pyrolysis", Energy&Fuels, February 2013.

[6] Santandrea, A., Gavard, M., Pacault, S., Vignes, A., Perrin, L. and Dufaud, O. "Knock on nanocellulose': Approaching the laminar burning velocity of powder-air flames", Process Safety and Environmental Protection, pp. 247–259, 2020.

[7] X. Tan, M. Schmidt, P. Zhao, A. Wei, W. Huang, X. Qian and D. Wu. "Minimum ignition temperature of carbonaceous dust clouds in air with $CH_4/H_2/CO$ below the gas lower explosion limit", Fuel, December 2019.

[8] Y. Bu, Y. Yuan, S. Xue, P. Amyotte, C. Li, W. Yuan, Z. Ma, C. Yuan, G. Li. "Effect of admixed silica on dispersibility of combustible dust clouds in a Godbert-Greenwald furnace", Powder Technology, pp. 496-506, July 2020.

[9] Azam, S., Mishra D. P. "Effects of particle size, dust concentration and dust-dispersion-airpressure on rock dust inertant requirement for coal dust explosion suppression in underground coal mines", Process Safety and Environmental Protection, pp. 35-43, March 2019.

[10] O. Dufaud, M. Poupeau, I. Khalili, N. Cuervo, M. Christodoulou, R. Olcese, A. Dufour and L. Perrin. "Comparing pyrolysis gases and dusts explosivities: A clue to understanding hybrid mixtures explosions?", Industrial and Engineering Chemistry Research, pp. 7656-7662, 2012

[11] A. Pattiya. "Fast Pyrolysis", Direct Thermochemical Liquefaction for Energy Applications, pp. 3-28, 2018.

[12] Santandrea, A., Pacault, S., Bau, S., Oudart, Y., Vignes, A., Perrin, L. and Dufaud, O. "Safer and stronger together? Effects of the agglomeration on nanopowders explosion", Journal of Loss Prevention in the Process Industries, pp. 1-11, 2021.

[13] J. Piskorz, D. Radlein and D. S. Scott. "On the mechanism of the rapid pyrolysis of cellulose", Journal of Analytical and Applied Pyrolysis, pp. 121-137, 1986.

[14] A. Dufour, P. Girods, E. Masson, Y. Rogaume and A. Zoulalian. "Synthesis gas production by biomass pyrolysis: Effect of reactor temperature on product distribution",International Journal of Hydrogen Energy, pp. 1726-1734, 2009.

[15] D. Torrado, A. Pinilla, M. Amin, C. Murillo, F. Munoz, P.-A. Glaude and O. Dufaud. "Numerical study of the influence of particle reaction and radiative heat transfer on the flame velocity of gas/nanoparticles hybrid mixtures", Process Safety and Environmental Protection, pp. 211-226, 2018

[16] A. Haghiri, M. Bidabadi, "Modeling of laminar flame propagation through organic dust cloud with thermal radiation effect", Int. J. Therm. Sci. 49, pp. 1446–1456, 2010.

[17] N. Cuervo. "Influences of turbulence and combustion regimes on explosions of gas-dust hydrid mixtures", PhD Thesis – The University of Lorraine, France, 2015

SPRING21
+17TH GCPS
A Joint AIChE and CCPS Meeting

Impact of Simulation Methods on Column Relief Systems Analysis

Robert Foley
Smith & Burgess
190 S La Salle St. Ste. 550 Chicago, Illinois 60603
Robert.Foley@smithburgess.com

Smith & Burgess
Process Safety Consulting

Prepared for Presentation at
American Institute of Chemical Engineers
2020 Spring Meeting and 17th Global Congress on Process Safety
Virtual
April 18 - 22, 2021

Impact of Simulation Methods on Column Relief Systems Analysis

Robert Foley
Smith & Burgess
190 S La Salle St. Ste. 550 Chicago, Illinois 60603
Robert.Foley@smithburgess.com

Keywords: Simulations, Safety, Dynamic, Steady state

Abstract

Accurate required relief load calculation for column systems are critical in accurately sizing relief devices and relief load handling systems. Column systems, which can contain reboilers, condensers, pumparounds, refluxes, and other complexities, and have many overpressure scenarios that must be evaluated and for which many methods exist to estimate the required relief rate. The purpose of this article is to examine different relief estimation methods for a column system and discuss the results of each approach. These results are not meant to be prescriptive for future systems, but to provide additional information so that the most fitting approach can be used given the requirements and constraints of the project. Methods in this evaluation include unbalanced heat input, dynamic simulation, and several methods of steady state simulations of varying complexity from a single flash drum to a standard column simulation. The impact of rigorously modelling the column reboiler and overhead condensers vs using the normal operating duties was also evaluated. To simplify the analysis, I evaluate three common overpressure scenarios: (1) closed vapor outlets, (2) top tower reflux failure, (3) total power failure. Previous works in this area have compared the difference between the unbalanced heat input, steady state simulation, and dynamic simulation methods, but have not evaluated the impact of differing steady state simulation methodologies on the estimated relief rate. For the total power failure scenario relief rates were estimated by using both the normal reboiler duty and a rigorously modelled reboiler, as well as several different column liquids commonly used to evaluate this scenario. Simulations were performed using Aspen HYSYS V10 and existing process data for a Naphtha Stripper system. This article finds that for the evaluated system a series of flash drums closely matches the results of a standard column simulation. Decreasing the number of flash drums used resulted in a larger estimation. The results for the unbalanced heat input and dynamic simulation methods aligns with those of previous studies. The dynamic simulation method estimated lower required reliefs, while the unbalanced heat method is more conservative, resulting in higher required reliefs. Rigorously simulating the reboiler in a steady state model resulted in lower estimated relief for complex simulations, but when used with more simplified steady state methods was found to be more conservative than the unbalanced heat input method, making it the most conservative method analyzed. Rigorously simulating the overhead condensers in a steady state model resulted in the largest variance for the closed outlets scenario, as at relief pressure the column overhead was fully condensed resulting in no relief. For the total power failure scenario, the dynamic simulation method resulted in the lowest estimated relief and was found to match closely with the results of the steady state method using the normal reboiler duty applied to the column feed liquid.

1 Background

Accurate required relief load calculation for column systems are critical in accurately sizing the relief devices and relief load handling systems. Column systems, which can contain reboilers, condensers, pumparounds, refluxes, and other complexities, have many overpressure scenarios that must be evaluated and over time many methods have been created to estimate the required relief rates. API 521 provides generally accepted guidelines for determining required relief rates. In the past, the unbalanced heat method has been used widely; however, with the advent of process software more advanced methods such as steady state and dynamic simulations have become practicable. Due to the complexity of estimating a complex system at upset conditions and different levels of simplification and assumptions used, these methods can produce significantly different relief rate estimations and relief device requirements.

The methodologies evaluated in this study are the unbalanced heat input method, steady state simulation methods, and dynamic simulation method. The unbalanced heat input method uses an enthalpy balance created from the normal operation conditions. During upset conditions an enthalpy unbalance can occur, which is used to estimate the required relief. In steady state simulations, software (such as Aspen HYSYS, KBC Petro-Sim, SimSci Pro/II, and others) is used to create a steady state balance at normal operating conditions or use one provided by the client, which is then modified according to the relief scenario. There are multiple ways to simulate the multiple ways to estimate the distillation portion of the system, such as the column unit operation or simulating the column trays as flash drums. In dynamic simulations, software is used to create a transient analysis of the system at discrete time intervals, including equipment volumes, instrumentation, and response logic. This simulation is extensively verified, which allows for additional credits to be considered.

Steady state analyses can be performed more quickly than dynamic analyses but may provide more conservative results, potentially requiring additional protection. An overly conservative analysis can result in an oversized relief device and lead to negative impacts on relief handling system and may require unnecessary changes to the system or bottlenecks on future rate increases. Some of these impacts may be insignificant for certain systems, such as those not tied into the flare network, or for analysis of non rate-dependent scenarios. The impacts of conservative analysis should be identified for a system.

The purpose of this article is to examine different relief estimation methods for a specific column system and discuss the results of each approach. These results are not meant to be prescriptive for future systems, but to provide additional information so that the most fitting approach can be used given the requirements and constraints of the project. To simplify the scope of this study, three relief scenarios are examined in this article: closed vapor outlets, top tower reflux failure, and power failure.

The column system basis selected for this analysis is a Naphtha Splitter, which separates light and heavy naphtha. The column operates at 40 psig, and the relief devices located on the column overhead are set at 60 and 63 psig, respectively. The information used in this paper is based on actual process information.

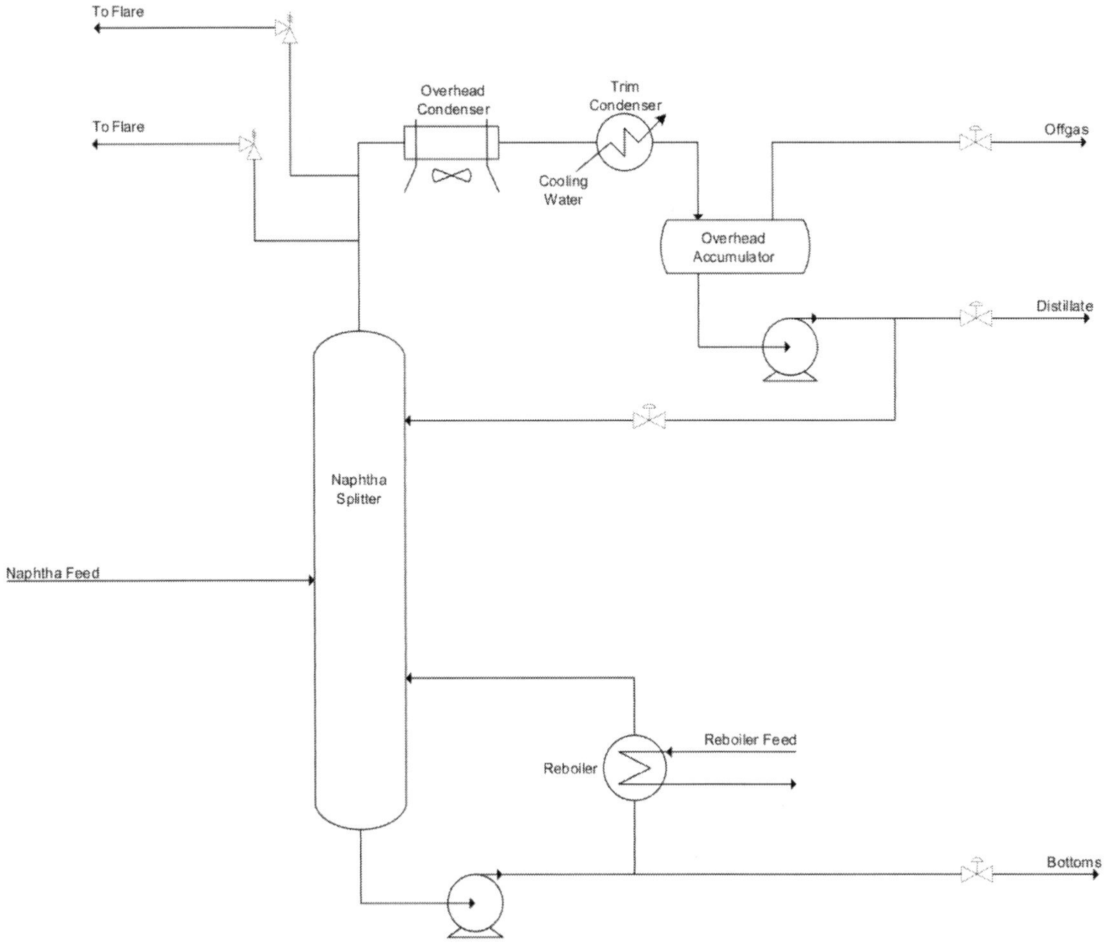

Figure 1. Naphtha Splitter System

2 Relief Rate Calculation Methodology

2.1 Method A: Unbalanced Heat Method

The unbalanced heat method is an approach to calculating relief loads based around an enthalpy balance of the system during upset conditions. During normal operation, the system's enthalpy is kept in balance by the energy entering the system in the form of feeds and reboiler heat input, and energy leaving the system in the form of outlets and cooling duty. During a relief scenario, this balance is upset, and excess energy is accumulated in the system. The unbalanced heat method calculates a relief rate based on the mass needed to remove this excess energy from the system. The unbalanced heat method requires minimal data on the instrumentation controls of the system and can be completed quickly without the need for simulation software, provided that the necessary stream properties are readily available.

Unbalanced heat is simplified compared to more rigorous methods such as simulation, and uses the following assumptions:

- Compositions are unaltered during relief.
- Flowrates will continue at normal rates limited by hydraulics at relief pressure.
- Any unbalanced heat is applied to the top tray liquid.
- There is enough liquid in the top tray to vaporize at relief conditions.

The following equation, based on CHEM-GEMS K-3D1, 18.3.9, is used to calculate the required relief rate.

$$W_{Relief} = \frac{\sum Q_{In} - \sum Q_{Out} + \sum W_F H_F - \sum W_P H_P - W_B H_B - (\sum W_F - \sum W_P - W_B)(H_R - L_R)}{L_R}$$

ΣQ_{In} = The sum of the heat inputs to the system from reboilers and heaters (btu/hr)

ΣQ_{Out} = The sum of the heat outputs from the system (coolers) (btu/hr)

W_F = Mass flow rate of feed

H_F = Mass enthalpy of feed

W_P = Mass flow rate of product

H_P = Mass enthalpy of product

W_B = Mass flow rate of accumulation in the system

H_B = Mass enthalpy of the system accumulation

H_R = Mass enthalpy of the top tray liquid at relief pressure

L_R = Mass heat of vaporization of the top tray liquid at relief pressure

2.1.1 Overpressure Scenarios

Top tower reflux scenario: In the event of a top tower reflux failure, the distillate outlet is blocked, and the vapor outlet and condenser duty are lost due to overfilling of the overhead accumulator and flooding of the condenser. The system feed and bottoms are assumed to continue at their normal rates, and the reboiler is expected to continue.

$$W_{Relief} = \frac{Q_{Reboil} - \cancel{Q_{cond}} + W_F H_F - (\cancel{W_{vap}} H_{vap} + \cancel{W_{dist}} H_{dist} + W_{bot} H_{bot}) - \cancel{W_B} H_B - (W_F - \cancel{W_{vap}} - \cancel{W_{dist}} - W_{bot} - \cancel{W_B})(H_R - L_R)}{L_R}$$

$$W_{Relief} = \frac{Q_{Reboil} + W_F H_F - (W_{bot} H_{bot}) - (W_F - W_{bot})(H_R - L_R)}{L_R}$$

Total power failure: In the event of a partial power failure, the system feed, outlets, and condensing are lost, while the reboiler heat circulation and heat input continue. The remaining equation calculates the required rate based on a boilup of the top tray liquid by the reboiler.

$$W_{Relief} = \frac{Q_{Reboil} - \cancel{Q_{cond}} + \cancel{W_F} H_F - (\cancel{W_{vap}} H_{vap} + \cancel{W_{dist}} H_{dist} + \cancel{W_{bot}} H_{bot}) - \cancel{W_B} H_B - (\cancel{W_F} - \cancel{W_{vap}} - \cancel{W_{dist}} - \cancel{W_{bot}} - \cancel{W_B})(H_R - L_R)}{L_R}$$

$$W_{Relief} = \frac{Q_{Reboil}}{L_R}$$

Closed outlets: In the event of a closed vapor outlet, the overhead vapor product flowrate is set to 0 lb/hr, and the overhead condenser duty is assumed to be lost due to vapor locking of the overhead. The remaining feeds and outlets are assumed to continue at their normal rates, and the reboiler heat input continues.

$$W_{Relief} = \frac{Q_{Reboil} - \cancel{Q_{cond}} + W_F H_F - (\cancel{W_{vap}} H_{vap} + W_{dist} H_{dist} + W_{bot} H_{bot}) - \cancel{W_B} H_B - (W_F - \cancel{W_{vap}} - W_{dist} - W_{bot} - \cancel{W_B})(H_R - L_R)}{L_R}$$

$$W_{Relief} = \frac{Q_{Reboil} + W_F H_F - (W_{dist} H_{dist} + W_{bot} H_{bot}) - (W_F - W_{dist} - W_{bot})(H_R - L_R)}{L_R}$$

2.2 Steady State Simulations

Several methods of steady state simulations were compared, varying in complexity, conservativeness, and rigorousness. The following assumptions are made when simulating the system at steady state:

- The system feeds continue at their normal rates unless hydraulically limited at relief pressure.
- The reflux continues at its normal conditions and composition.
- Equipment pressure drops are unchanged at relief pressure.

2.2.1 Method B: Standard Column

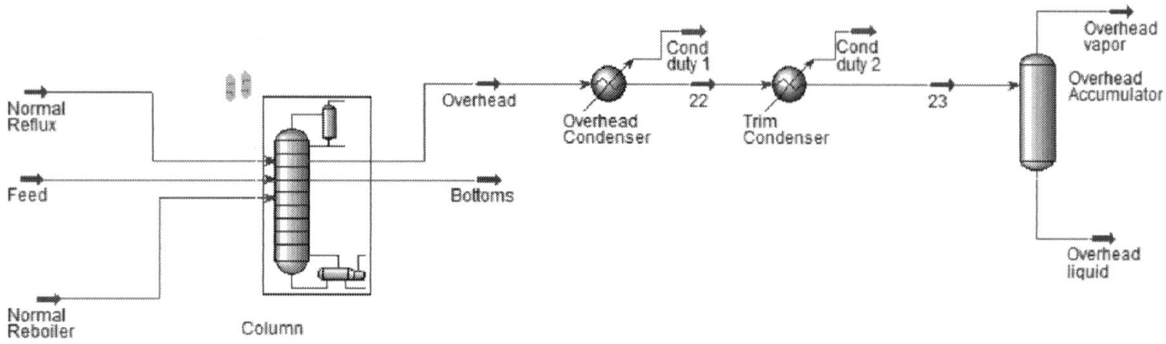

Figure 2. Column simulation

In the standard column unit operation, a series of equilibrium trays are used to simulate the distillation process, with several equations to solve the vapor and liquid flow through the system at steady state, including the Francis Weir and resistance equations. The column unit operation allows for flexibility in how the system is set up, and the embedded flow sheet allows for recycle streams without the need for extra unit operations.

2.2.2 Method C: Simple Flash Drums

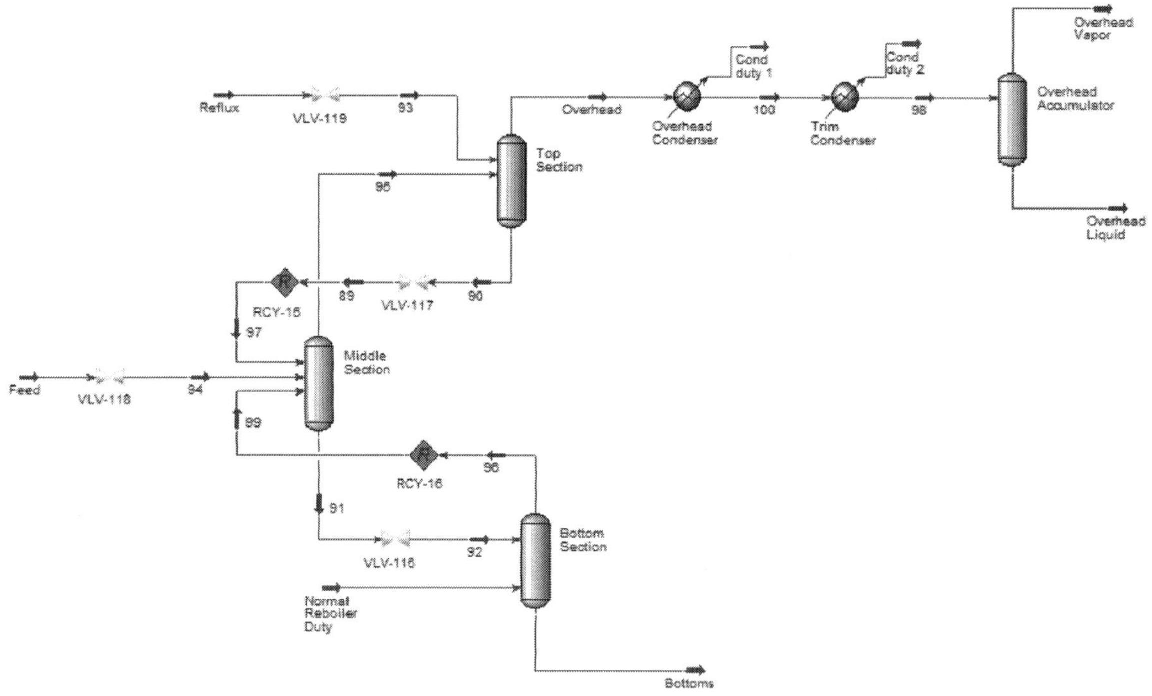

Figure 3. Simple flash drum simulation

The flash drum method of steady state simulation uses a series of separators taking the place of trays in the column unit operation. The separator unit operation uses a pressure-enthalpy flash performed at the lowest feed pressure with any additional heating or cooling duty to calculate the resulting vapor and liquid outlets.

The simplified flash drum method shown in figure 3 above uses three separators representing the top, middle, and bottom sections of the column. The normal column feed is fed to the middle section, and the normal reflux stream is isenthalpically flashed to relief pressure using a valve. The normal reboiler duty is applied to the bottom section, and valves are used to apply the normal pressure profile across the column system. For this method, the overhead condensers are modelled by limiting the cooling duty to the normal duty or the normal outlet temperature, whichever results in a higher outlet temperature.

This method of simulation can be useful in cases where the overpressure scenario is difficult to simulate at relief pressure using a column unit operation, such as cases resulting in a loss of liquid traffic through the column trays. Due to simplification the column into 3 stages, the separation achieved across the column is less than the steady state model and does not match when benchmarked against the column. This can be improved by increasing the number of stages in the model.

2.2.3 Method D: Complex Flash Drums

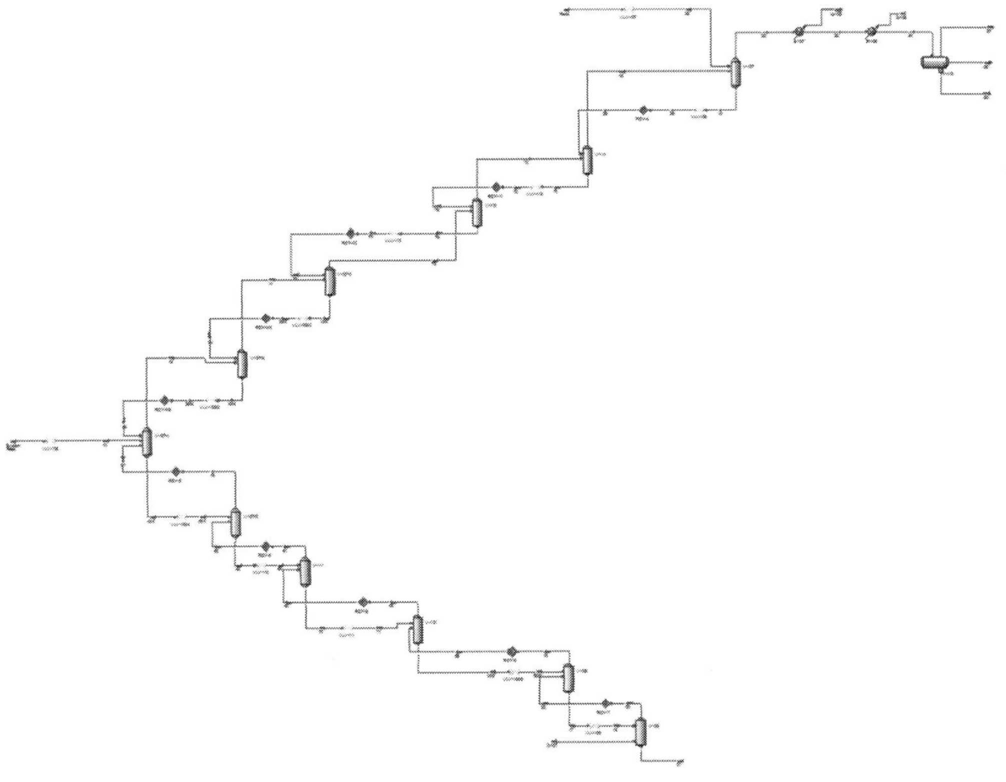

Figure 4. Complex flash drum simulation

Increasing the number of separators in the column improves the separation across the system, lowering the molecular weight of the overhead vapor stream. The system was benchmarked at operating conditions and it was found that a series of 11 separators most closely matched the steady state balance. Due to the added complexity and extra step of adjusting the number of stages to match the column at normal operation, this method is more time-intensive than other steady state simulation methods.

2.2.4 Method E: Single Flash Drum

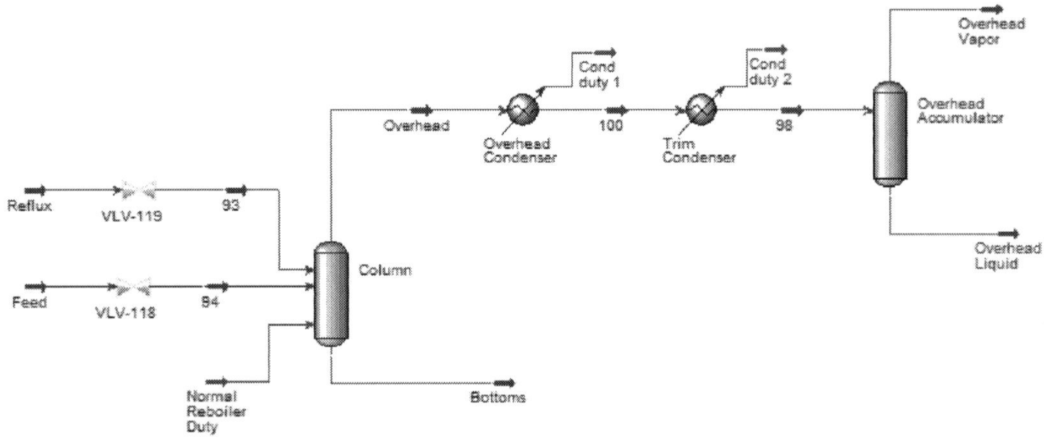

Figure 5. Single flash drum simulation

On the other end of the spectrum is the most simplified method to simulate a column system, which involves a single flash drum. The system feed and reflux are isenthalpically flashed to relief pressure and fed to a separator along with the normal reboiler duty.

2.2.5 Simulated Reboiler

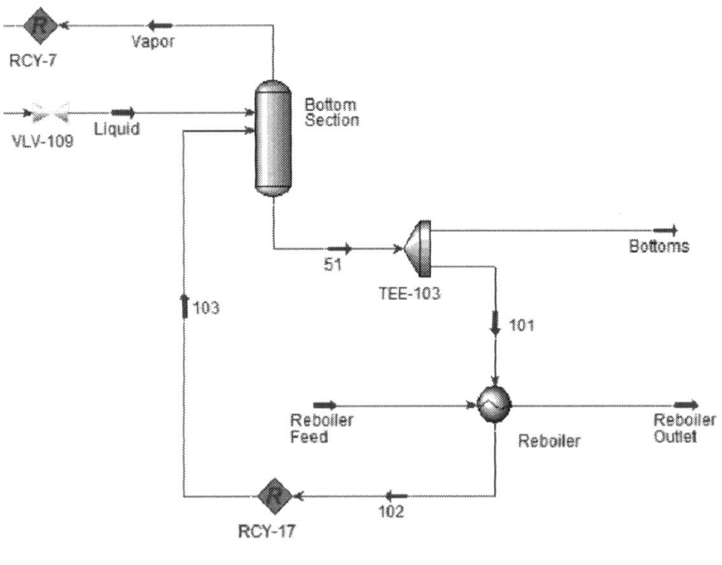

Figure 6. Simulated reboiler

Simulation of the column reboiler can be refined by modelling the exchanger rigorously. Modelling the reboiler in this way is simple but requires exchanger and utility data that may not be readily available. For this analysis, the reboiler clean UA was specified, and the reboiler hot side bypass was considered closed to maximize the reboiler duty. A splitter was used to send the normal volumetric rate of column liquid to the reboiler.

2.2.6 Simulated Condensers

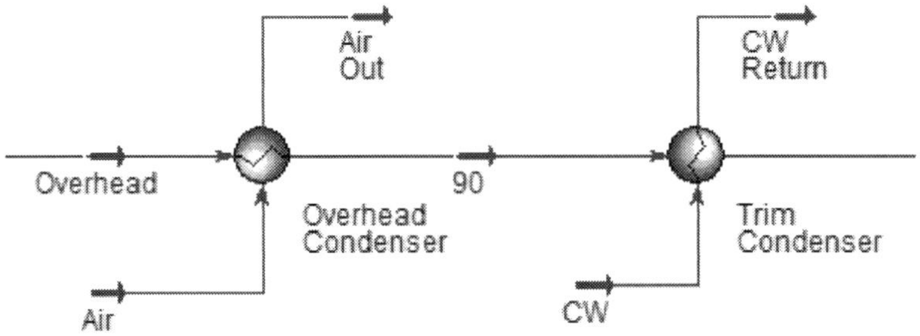

Figure 7. Simulated condensers

With additional information the overhead condensers can be rigorously modelled, allowing for credit to be taken to potentially lower the condenser outlet temperature below the normal value. When simulating the condensers in this way, the assumption that the normal condensing duty is not exceeded was still used, and it is important to be aware of potential temperature crosses and minimum approach temperatures in the exchangers. For the closed vapor outlet scenario, credit can be taken to reduce the required relief rate when the continuing forward flow of vapor, after the relief device takeoff, is fully condensed by the overhead coolers. If noncondensables are not present in the overhead cooler outlet stream, the resulting liquid can continue to flow through the overhead accumulator without an accumulation of vapor. If the overhead coolers cannot fully condense the remaining column overhead, the full column overhead must be relieved to prevent a buildup of vapor in the overhead accumulator. For this analysis, the overhead and trim condenser UAs were set to match the normal overhead duties during steady state operation using utility information obtained from the equipment datasheets.

2.2.7 Overpressure scenarios

Closed outlets:

The closed outlets scenario was simulated using a splitter on the column overhead to represent vapor leaving the system through the relief devices. The relief rate was increased until the vapor product flowrate was 0 lb/hr. For the condensers modelled using the normal discharge temperature, this results in relieving the full column overhead.

Top tower reflux failure scenario:

The top tower reflux failure scenario was simulated by minimizing the column reflux flow rate with the system at relief pressure. As overfilling of the overhead accumulator is expected to result in flooding of the condensers, the simulated condensers method was not evaluated for this scenario. The column overhead was taken as an estimate for the relief stream.

Total power failure:

As the power failure scenario causes a loss of feed, cooling and reflux to the system, the methods used above for simulating the system can no longer be used. The methodologies for evaluating boil up scenarios such as this are typically set by the user's guidelines and vary in the method of calculating the reboiler heat duty, as well as the liquid used in the vaporization. Some variations in boil up methodologies include using different streams to calculate the required rate, such as the column feed, 2nd tray liquid, top tray liquid, or the bottoms liquid. As part of this study, the column feed liquid, top tray liquid, 2nd tray liquid, reflux, and bottoms liquid were evaluated using both the reboiler normal duty and a simulated reboiler using the clean UA for a total of 10 different steady state methodologies. Each liquid was set to bubble point at relief pressure, and the boil up calculation was evaluated using the normal reboiler duty and a reboiler simulated with the clean UA value. For the normal reboiler duty cases, the flowrate of process liquid to the reboiler was based on the normal volumetric feed rate at normal operating conditions. For the simulated reboiler cases, the feed rate was based on the normal reboiler vaporization rate, and the normal reboiler duty was used when the simulated duty was calculated to exceed 100% of the normal duty.

2.3 Method F: Dynamic Simulation

Per API 521 7th edition §4.3.3, *"Conventional methods for calculating relief loads are generally conservative and can lead to overly sized relief and flare system designs. Dynamic simulation provides an alternative method to better define the relief load and improves the understanding of what happens during relief."*

Dynamic simulation is a transient analysis that evaluates pressure changes and equipment inventories to evaluate a system in discrete time intervals. This can allow for a more detailed analysis but requires significantly more information to set up, and a large amount of rigor to validate the simulation.

API requires that a dynamic simulation incorporate the physical features of the system such as liquid inventories of vessels and piping and closely match the steady-state model. Additionally, per API 521 7th edition §4.3.3, *"If dynamic simulation is used, sensitivity analyses shall be performed to assess factors such as the effect of PRDs with excess capacity, the action of automatic controls, controller tuning, heat integration with other columns, and operator intervention. These factors increase the number of run permutations that shall be performed."* These requirements added to the base complexity of dynamic simulation and make this method significantly more time and cost intensive than the steady state analyses. Because of these constraints, dynamic simulations are better suited for critical and/or complicated systems, such as fluid catalytic cracking towers.

To create a dynamic simulation, several additional pieces of data are required. The vessel and piping volumes, tray dimensions and spacing, instrumentation controls, exchanger dimensions and heat transfer information, and utility flowrates are all needed to properly model the system. Particular attention must be given to the sources of each feed. As the dynamic simulation calculates flowrates based on the pressures in the system, the feed rates will decrease as the system pressure increases. If the system is fed by a vessel with a level control valve, the valve will open to try to maintain this flow. If the system is pump fed, the pump will back up the curve and may not be able to supply the normal feed rate. Because of this, it should be evaluated how the system feeds will react to changes in the system in order to more accurately estimate the system at upset conditions.

In the dynamic simulation used for this study, only the Naphtha Splitter and its associated equipment were included to avoid complications. The reboiler clean UA was specified, and the reboiler hot side bypass was closed to maximize the reboiler duty. The overhead and trim condenser UAs were set to match the normal overhead duties during steady state operation.

2.3.1 Overpressure scenarios

Closed outlets:

The closed outlets scenario was simulated using the following actions, beginning after 2 minutes have elapsed:

- Off-gas pressure valve closed

Top tower reflux failure scenario:

The top tower reflux failure scenario was simulated using the following actions, beginning after 2 minutes have elapsed:

- Distillate outlet valve closed
- Reflux valve closed

And then on overfilling of the overhead accumulator:

- Off-gas pressure valve closed
- Overhead and Trim condensers duty set to 0 due to flooding of the condenser

Total power failure:

The power failure scenario was simulated using the following actions, beginning after 2 minutes have elapsed:

- Feed valve closed
- Trim condenser duty set to 0 btu/hr

- Overhead condenser set to 20% of steady state duty
- Distillate outlet valve closed
- Off-gas pressure valve closed
- Reflux pump shutdown
- Bottoms product valve closed

And then after an additional 2 minutes have elapsed:

- Overhead condenser duty set to 0 btu/hr due to flooding of the condenser.

3 Results

3.1 Closed Outlets

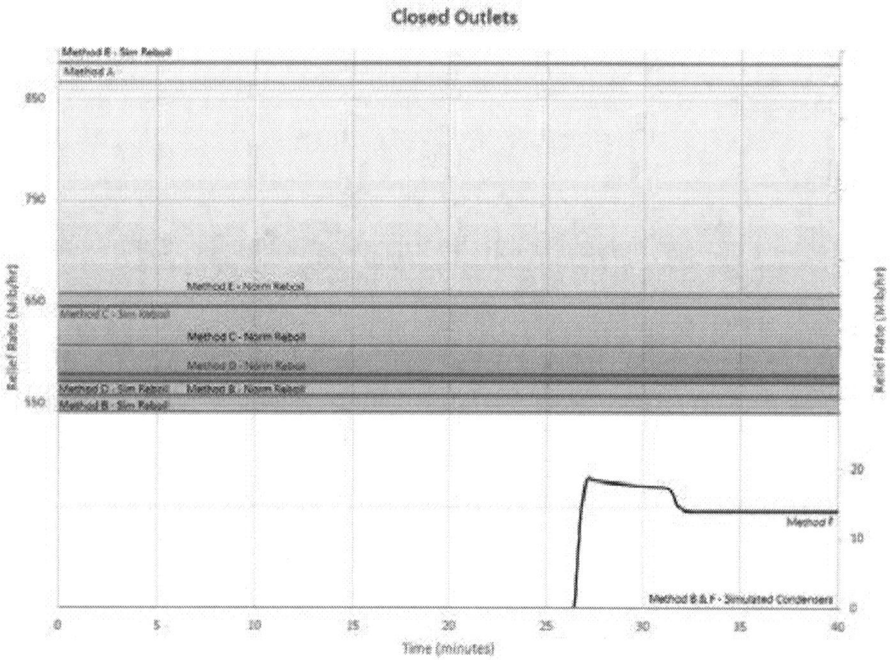

Figure 8. Estimated relief rates for closed outlets scenario

Figure 8 above shows the predicted relief rates for the closed outlet scenario for each of the 14 methods evaluated. The simulated reboiler methods are shown in red, the normal reboiler duty methods are shown in blue, and the simulated condenser methods are shown in purple. The dynamic simulation and more complex steady state simulations with rigorously simulated condensers predict significantly lower or no relief in the event of a blocked vapor relief. Taking credit for additional cooling by simulating the overhead condensers results in the overhead vapor being fully condensed, preventing accumulation of vapor in the receiver and eliminating any relief.

This credit allows for reducing the condenser outlet temperature below the normal steady state temperature (168 F), as shown in table 1 below. The condenser outlet of the steady state B and D methods were limited by the normal outlet temperature in the base simulation, while the more simplified C and E methods were limited by the normal condenser duty. For this reason, even when simulating the condensers, the assumption that the normal condnenser duty is not exceeded results in methods C and E being unchanged.

Table 1. Condenser outlet temperature

	Method			
	B	C	D	E
Normal Reboiler Duty	168.2 F	175.4 F	168.2 F	208.1 F
Simulated Reboiler	168.2 F	190.5 F	168.2 F	254.7 F
Simulated Condenser	145.4 F	151.5 F	146.6 F	160.3 F

The large variance between the simulated condenser and other calculation methods shows the impact that a single conservative assumption can make. For both the steady state simulations using the normal reboiler duty and those using a simulated reboiler, methods B and D match closely, with method C and single flash drum methods estimating a more conservative result. However, the difference between the various options between these methods is significantly larger for the simulated reboiler method with a difference of 63% between the upper and lower estimated rates, compared to a 15% difference for the normal reboiler duty method. Both the normal reboiler duty and simulated reboiler methods model the condensers by limiting the cooling duty to the normal duty or normal outlet temperature, whichever is more conservative. As a result of this assumption, the column overhead is not fully condensed at relief pressure, and the full column overhead must be relieved to prevent an accumulation of vapor in the receiver.

The dynamic simulation predicts a small relief rate due to leakage through the lower-set relief device but does not predict that overpressure will occur for longer than a minute, as shown below in figure 9. This aligns with the results of the steady state method B and D simulations which modelled the condensers, and also predicted no relief at 16% overpressure. Due to the gap between the set pressure and reseating pressure of the valve, there continues to be a small leakage rate through the relief valve. As the steady state simulations are evaluated at the allowable overpressure of the relief devices, this valve leakage is not captured in the simulated condenser simulations.

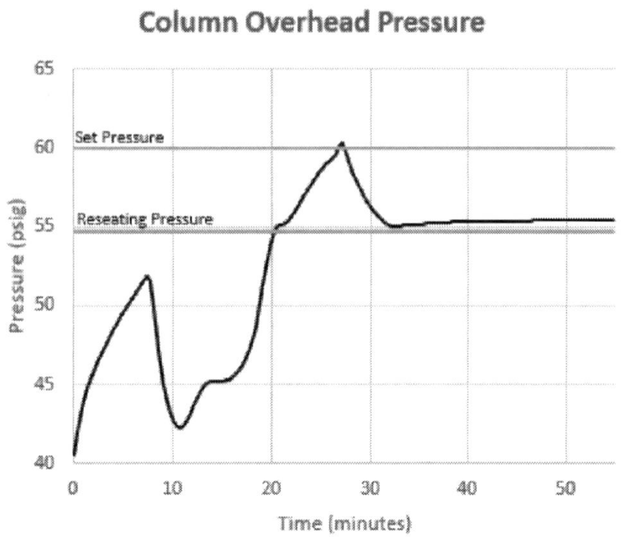

Figure 9. Dynamic simulation column overhead pressure over time

3.2 Top Tower Reflux Failure

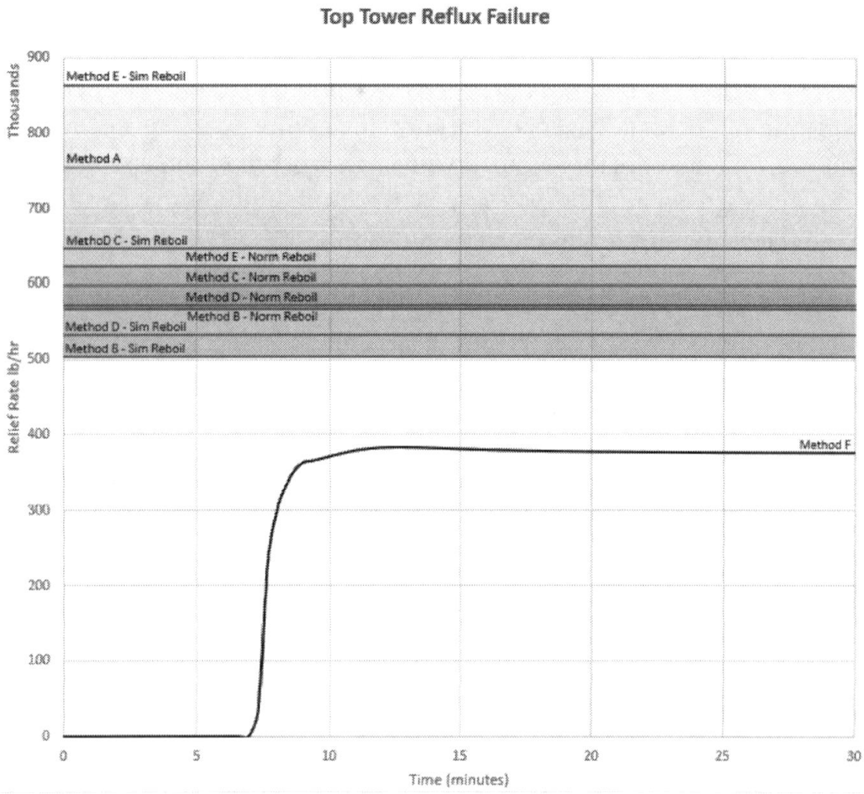

Figure 10. Estimated relief rates for top tower reflux failure scenario

The graph above shows the predicted relief rates for the closed outlet scenario for each of the 10 methods evaluated. The simulated reboiler methods are shown in red, and the normal reboiler duty methods are shown in blue. As with the closed outlets scenario, the dynamic simulation predicts a lower required relief rate, and the unbalanced heat method predicts a significantly larger rate. The difference between the various options for the simulated reboiler and normal reboiler duty methods is less stratified than the closed outlets scenario, with a variance of 40% for the simulated reboiler cases and only 9% for the normal reboiler duty cases. The cause of this variance for the simulated reboiler cases is due to a large impact on the heat input by the simulation methods.

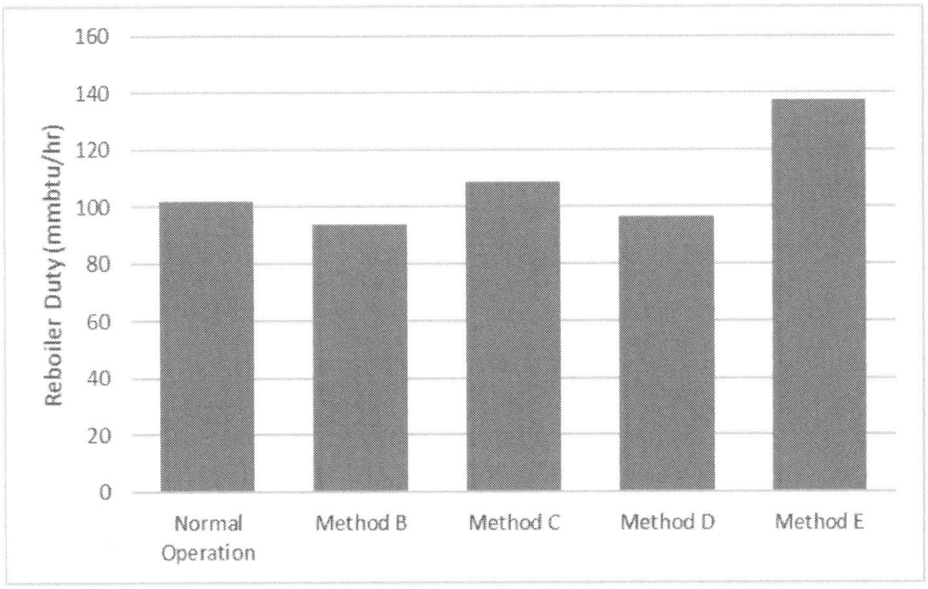

Figure 11. Variance of calculated reboiler duties based on the simulation method

As the number of simulated trays decreases, the quality of separation is lowered with more light components remaining in the cooler bottoms liquid, and the simulated reboiler duty increases as shown above in figure 11. The single flash drum case (method E) was found to have a reboiler duty of nearly 40% higher than the normal operation duty, while the remaining methods aligned more closely with the normal value.

3.3 Power Failure

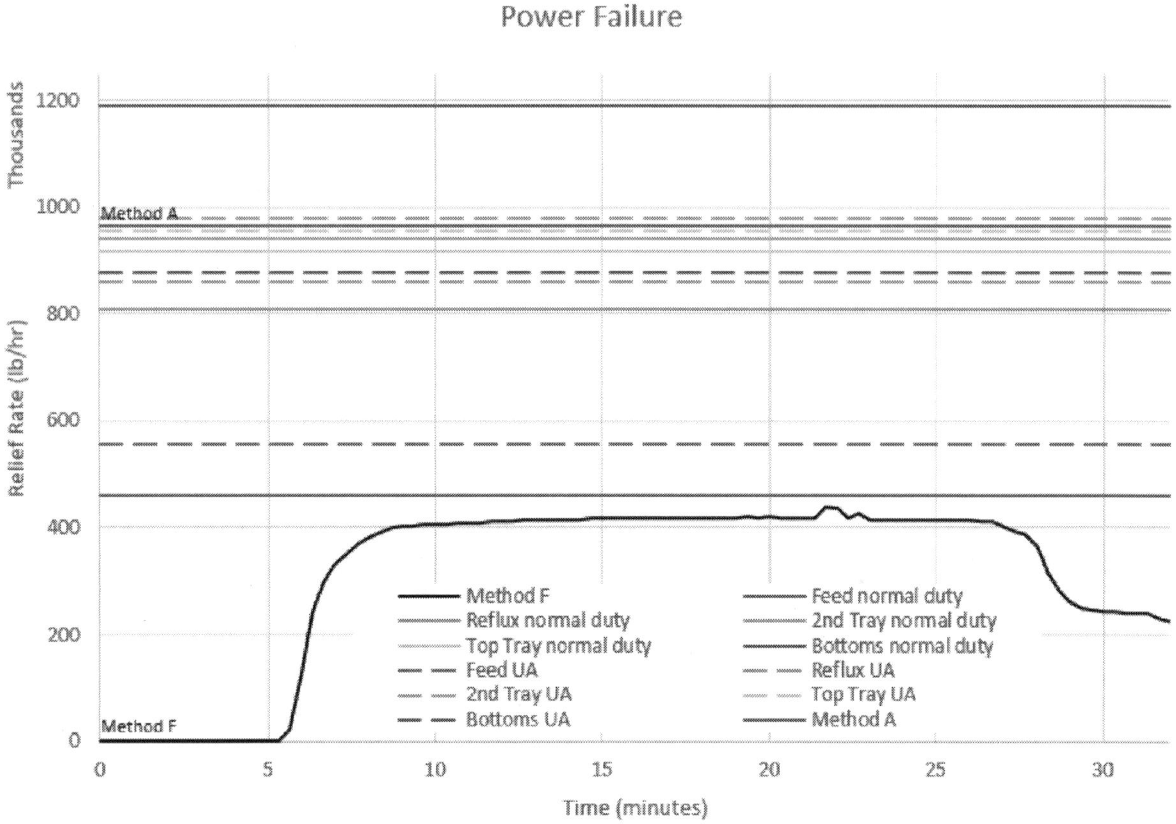

Figure 12. Estimated relief rates for power failure scenario

The graph above shows the estimated relief rates for the power failure scenario for the 12 methods evaluated. The steady state methods using the simulated reboiler with clean UA are shown with dotted lines. The most conservative approach to estimation of the boil up relief rate was found to be bottoms liquid with the normal reboiler duty applied. Modelling the system with a dynamic simulation resulted in a significantly lower estimated rate, by over 60%. Of the simulated steady state methods, the feed liquid with the normal reboiler duty most closely matched with the results of the dynamic simulation.

The unbalanced heat method estimates a required relief rate of nearly 1,000,000 lb/hr and matches closely with the top and second tray liquid with reboiler UA cases. For the simulated reboiler cases, every liquid used apart from the bottoms resulted in a duty exceeding 100% of the normal duty, and as a result the normal duty was used. Due to this, the bottoms liquid cases were the only liquid evaluated in which the normal duty method resulted in a larger relief rate than the simulated reboiler method.

4 Conclusions

The estimated required relief for the closed outlets and top tower reflux failure scenarios were closely matched among the three methods by using a series of flash drums in place of a column unit operation for the steady-state simulation. Increasing the number of flash drums used in this method increased separation and decreased the overhead vapor rate, resulting in lower estimated relief rates. Using a more simplified flash drum model can be completed quickly but does not match with the system when benchmarked at normal conditions, resulting in variances for the results. The unbalanced heat input method estimated conservative required relief rates for all scenarios. The dynamic simulation (method F) estimated lower required relief rates than other methods but required significantly more time and information to develop. As a result of this, dynamic simulations may be better suited for critical and/or complicated systems. For the total power failure scenario, the most conservative approach to estimation of the boil up relief rate was found to be bottoms liquid with the normal reboiler duty applied. As boil up methodologies are typically defined by project guidelines, there may not be an option to evaluate the scenario differently. The potential benefit of using a more detailed dynamic simulation should be evaluated, comparing the downsides and added expense necessary to create and validate the simulation to any potential mitigations required by using a more conservative method, such as significant piping rework. These results are not meant to be prescriptive for future systems, but to provide additional information so that the most fitting approach can be used given the requirements and constraints of the project, and show the flexibility of evaluation methods and their impacts on a system.

617540 Is Your SIF Trying to Do Too Much?

Abstract:

The number of inputs and outputs to a safety instrumented function (as well as how they vote) affects the probability of failure on demand (PFD) and the SIL (safety integrity level), assuming the test interval remains the same. The larger the number of inputs and outputs, the higher the PFD and potentially the lower the SIL.

What if the SIF includes all the sensors and final elements that are involved in any trip of a large processing unit like a heater, a reformer, or a distillation column? What if the SIF includes the actions to shutdown upstream units that feed the unit being tripped? What if the SIF includes the actions to shutdown downstream units that feed the unit being tripped? When SIL verification is done for an SIF as described here, the result may be that the **target PFD and SIL cannot be achieved**. The temptation may be to add redundancy in sensors and final elements or to reduce the proof test interval in an attempt to reduce the calculated PFD. Frustration may abound as capital and operating costs rise steeply.

This paper shows how to use the principles of LOPA (layer of protection analysis) and the information in the PHA (process hazard analysis) to split up the massive SIF into smaller SIFs that more manageable. The smaller SIF need include only the sensors, logic solver(s), and final elements that detect and prevent a specific scenario (one cause leading to one consequence). The approach makes sure all the smaller SIFs can protect against all the scenarios that the massive SIF was intended to prevent. Trips of upstream and downstream units are considered as orderly shut-down actions. If needed, trips of upstream and downstream units are analyzed as small SIFs, as well.

With reasonable size SIFs, there is an opportunity to design the SIF with a reasonable number of sensors and final elements, and a reasonably long proof test interval.

Examples will be included

618701 Determination of the Best Vent Stack Design for High-Pressure Hydrocarbon Blowdown Using CFD Modelling – a Case Study

Abstract:

Risk assessment plays a vital role in the project development to identify the hazard, assess the risk, and provide the safety mitigation measures to reduce the risk to As Low As Reasonable Practice (ALARP). This paper presented an industrial case study of low-temperature hydrocarbon gas depressurization from high-pressure process equipment. The incident happened during the plant's start-up after a few days of shutdown for some maintenance work. An investigation was conducted and found the depressurization incident was due to an inactive battery in the Uninterruptible Power Supply (UPS) used to back-up Safety Integrated System (SIS). The depressurization has caused some of the hydrocarbon releases returned into the test hall that suggests an improvement to the vent stack's existing design to prevent re-entry of flammable gas in depressurization due to emergency. In this study, a CFD modeling and consequence analysis was performed for a cryogenic pilot plant to identify the most suitable vent stack design for a safe high-pressure hydrocarbon venting to the atmospheric. The analysis was conducted for four different vent stack designs. From the CFD simulation for methane and CO_2, the existing vent stack design (U-shape) resulted in the highest gases trapped in the testing hall due to poor vent design leading to unsatisfactory gas dispersion during high-pressure gas blowdown. The L-shape vent stack design (design #3) is selected from the study because it had a good gas dispersion in the atmosphere, and the methane and CO_2 gas trapped inside the building were less than 5% LEL and 1%, respectively, below the safety threshold limit.

621793 Sponsored Technology Workshop - Reactive System ERS Design Basics & the Importance of Two-Phase Flow - Fauske & Associates

Abstract:

ERS design becomes more complex when runaway reactions and two-phase flow are considered, but these are often the dictating situations resulting in the greatest emergency relief requirements. It is important to evaluate these worst-case situations to ensure proper prevention or mitigation. This workshop will cover reactive system relief system design basics and highlight the importance of two-phase flow. Example reactive system ERS design problems will be setup and performed using FERST powered by CHEMCAD.

SPRING21
+17TH GCPS
A Joint AIChE and CCPS Meeting

Importance of Safety Critical Devices and Emergency Shutdown Systemsin a Process Plant

Presenter/Principal Author: Sheikh Adeel Ahmad
Affiliation: Fatimafert Limited
Address: Lahore, Pakistan
Email address: adeel.ahmad@fatima-group.com

Other Author(s): Umair Siddiq
Affiliation(s): Fatimafert Limited
Address (es): Lahore, Pakistan
Email address (es): Umair.siddiq@fatima-group.com

Prepared for Presentation at
American Institute of Chemical Engineers
2020Spring Meeting and 17thGlobal Congress on Process Safety
Virtual
April 18 - 22, 2021

Importance of Safety Critical Devices and Emergency Shutdown Systems in a Process Plant

Presenter/Principal Author: Sheikh Adeel Ahmad
Affiliation: Fatimafert Limited
Address: Lahore, Pakistan
Email address: adeel.ahmad@fatima-group.com

Other Author(s): Umair Siddiq Umair.siddiq@fatima-group.com
Affiliation(s): Fatimafert Limited

Keywords: Asset Integrity and Reliability, Emergency Management and Hazards Identification and Risk Analysis

Abstract

Engineered Safety Systems are always a critical part of any hazardousprocess plant and should always align with the design intentto avoid a catastrophic failure of the plant that can cause multiple fatalities and lost time injuries. The maintainability of the safetysystems, especially emergency shutdown systems, is always given a high priority for safe operation and controlled shut down of theprocess plants.

Fatimafert Limited is an Ammonia-Urea Complex situated in Lahore, Pakistan. The Ammonia and Urea plant is a high pressure and hightemperature technology that deals with highly flammable and toxicchemicals at its manufacturing facility. The plant is considered a Highly Hazardous Process and requires special focus on the maintenance and reliability of the equipment and devices to maintain safe operation ofthe plant and avoid loss of containment.

It is always a challenge to maintain and operate a highly hazardousplant of this vintage (1971). As part of Company's roadmap forProcess Safety Management (PSM) implementation at the site, acomprehensive review of the existing procedures and practices wasinitiated with special focus on safety critical devices and equipment and their reliability.

This paper will cover the following major activities that were performed to improve the inherent safety of the process plant:

- Preparation of a list of Safety Critical Equipment and Devices of allplants
- Development of a comprehensive Preventive Maintenance (PM) Plansof all Safety Critical equipment and devices.
- Development of a regime for functional testing of all availableemergency shutdown securities and their logics for availability duringextreme emergencies.

- Gap assessment of old vintage plant with latest plants for trip securities and emergency shutdown systems.
- A study for revamp of Emergency shutdown system of Ammonia plant to reduce reliance on human intervention for emergency handling.
- Development of detailed protocols for replacing these Safety Critical devices with alternate control measures for safe plantoperation.
- A comprehensive review of all the securities available in the designand their in-service status; putting them back in service throughrelevant HSE forums.
- Implementation of these systems made the facility safer and reduced plant downtime by avoiding malfunctions / false trips.

1.0 Introduction

Fatimafert limited is an ammonia urea manufacturing complex located in Lahore, Pakistan and was commissioned in 1971 with a name plate capacity of 1100 Metric Tons per Day (MTPD) prilled urea. The plant was revamped in 1991 and capacity was enhanced to 1350 MTPD prilled urea. The ammonia manufacturing unit was designed by MW Kellogg and involves highly toxic and flammable gases at high temperature and pressure mainly due to steam methane reforming for hydrogen production at elevated operating conditions and operation of synthesis loop above 1800 psig for ammonia formation. Since ammonia urea manufacturing plants comes with inherent hazardous substances, therefore the vintage plants are more prone to risk and may result in catastrophic failure if inadequate safety systems exist. The safe and reliable operation of the plant andmitigation of risk associated with normal operation and emergency scenario has always been given utmost priority in Fatimafert. Consequently, special consideration was given to the safety system especially emergency shutdown system and safety critical system as part ofimplementation of company's roadmap of process safety management.

2.0 Safety Critical Devices / Equipment

To ensure safe and reliable plant operation, critical equipment were segregated from non-critical to evaluate the process risks, control measures, additional measures for risk mitigation and keeping them in healthy state during all operational scenarios including plant shutdown period.

Safety criticaldevices / equipment are defined as "Devices, Equipment or Systems whose failure can result in, allow or contribute to the release of or exposure to sufficient quantities of hazardous substances or their energy (fires, explosions, etc.) that can result in serious injuries or irreversible health effects, significant property damage or significant environmental impact or adversely affect response to such failures as opposed to other equipment failures which are unlikely to result in more than minor injuries e.g. medical treatment or first aid case." This includes:

- Any plant **equipment containing hazardous material** in sufficient quantity (e-g vessels; Pipelines etc.)

- **Devices** that prevent the loss of containment of hazardous material (e-g **interlocks** etc.)
- **Equipment to respond or mitigate** the loss of containment (e.g fire hydrants etc.)

Safety critical is applicable to both High hazard process (HHP) and Low hazard operation (LHO).

2.1 High Hazard Process (HHP)

It is any activity of manufacturing, handling, storing or using substances that, when released or ignited, have significant potential for death or irreversible human health effects, significant property or environmental damage or off-site impacts due to acute toxicity, flammability, explosivity, corrosivity, thermal instability, latent heat or compression.

2.2 Low Hazard Operation (LHO)

It is any activity of manufacturing, handling, storing or using substances that, when released or ignited, have lower potential for death or irreversible human health effects, significant property or environmental damage or off-site impacts due to toxicity, mechanical hazards (including stored energy), asphyxiation (suffocation potential).

2.3 Methodology for Development of Safety Critical List

As a baseline of the Process Safety Management system, Safety Critical items (devices and equipment) should be developed and the Preventive Maintenance (PM) plans and Quality Assurance program should be updated.

The following methodology was adopted for the development of safety critical list after constituting a dedicated teamcomprising of process and operation engineers.

1. All highly hazardous chemicals present at each unit that exceed their allowable threshold quantities, as defined by the OSHA and EPA were listed.

2. Boundary limits of each unit were determined based upon latest revised process and instrumentation diagrams (P&IDs), unit plot-plan and discussions with plant personnel.

3. Selection of nodes of each section was carried out taking into consideration the following criteria:

   i. Remotely operated emergency block valves
   ii. Remotely Operable Control Valves
   iii. Check valves etc.

4. Each P&ID was highlighted to indicate the Equipment / Line which is to be included on the Safety Critical Equipment List.
5. Quantification of hazardous components in each section was done based on the P&IDs and

isometrics and compared with the unit's list of highly hazardous chemicals and flammable materials.

6. Baseline Process Hazard Analysis (FMEA/HAZOP) of each loop was reviewed to determine whether failure of specific Electrical and Instrumentation (E&I) loop or control circuits/devices would result in loss of process controlor would provide false information that could result in the loss of process control.

7. Each critical component having a driving force (either electric or steam) for failure was evaluated for inclusion in Safety Critical equipment list in case of loss of process control or a release of hazardous material.

8. Safety Critical Equipment / device lists were approved from relevant forums; these lists were then shared with all relevantpersonnel for the development of Mechanical Integrity and Quality Assurance plans.

2.3.1 Identification of Hazardous Chemicals

1. Each hazardous chemical, quantity in area, concentrations, hazardous event was listed and applicable chemicals were identified on the site "safety critical hazardous substances".
2. The following "Chemical classification index" table was developed based on OSHA guidelines and used.

Table 1. Criteria for Hazardous Chemicals

Chemical	Criteria	Reference
Anhydrous Ammonia (NH_3)	At or above 10,000 lb. (4,545 kg)	OSHA (CFR 1910.119)
Ammonia solution >20%	***>20%, At or above 20,000 lb.***	EPA (550-B-15-001)
Hydrogen (H)	At or above 10,000 lb. (4,545 kg)	OSHA (CFR 1910.119)
Methane (NG)	At or above 10,000 lb. (4,545 kg)	EPA (550-B-15-001)
Carbon Monoxide (CO)	***> 200kg***	CCPS List of Chemicals Hazardous for Humans
Sulfuric Acid (H_2SO_4)	***>65%, At or above 5,000 lb.***	OSHA (CFR 1910.119)
Hydrochloric Acid (HCl)	***>33%, At or above 5,000 lb.***	EPA (550-B-15-001)
Sodium Hydroxide (NaOH)	***>49%, At or above 1,000 lb.***	EPA (550-B-15-001)
Diesel Oil	***Category IV flammable liquid At or above 25,000 lb.***	OSHA (CFR 1910.106), DEPC (2012/18/EU)
Gasoline	***At or above 25,000 lb.***	DEPC (2012/18/EU)
Kerosene	***At or above 25,000 lb.***	DEPC (2012/18/EU)
Nitric Acid	***>60%, At or above 5,000 lb.***	EPA (550-B-15-001)
Noxes (NO_x)	***NO : At or above 250 lb. NO2 : At or above 250 lb.***	OSHA (CFR 1910.119)
Ammonium Nitrate (AN)	***Node containing all quantities of***	DEPC (2012/18/EU)

	AN with conc. ≥50%	
Nitro Phosphate (NP) Melt	***Node containing all quantities of AN with conc. ≥50%***	DEPC (2012/18/EU)
Steam	***All Steam at & above 18 kg/cm2***	As per criticality of Service
Hydrazine	***At or above 15,000 lb.***	EPA (550-B-15-001)
Ammonia Carbamate	***Ammonia>20%, At or above 20,000 lb.***	EPA (550-B-15-001)

2.4 Preventive Maintenance Plans of Safety Critical (SC) Devices & Equipment:

The SC devices and equipment were reviewed by relevant Maintenance personnel and history of failures / malfunction was referred. The PM plans of these were then prepared / revised based on the criticality of service as highlighted by the SC development team. The intent was to control the hazard and avoid loss of containment.

The PM plans were approved from the relevant Sub HSE forums (Sub HSE Mechnical and Sub HSE E&I) and then issued for implementation. The PM plans are stewarded by Operations and any deviation is not preferred. In case the PM of any SC equipment / device is not possible, Executive Waiver protocol is followed and thorough risk evaluation and consequence analysis is performed. The waiver is approved from Plant Manager for a limited time only.

3.0 Emergency Shut Down (ESD) Systems:

As the plant was designed and constructed in late 1960's, it lacked the number of trip securities and instrumentation for control and auto tripping of the equipment in case of deviations. In case of emergencies, most part of the plants was manually shut down initiated by the DCS and field operators that resulted in more reliance on human interaction. Due to shortage of natural gas supply in the country, the plant remained in shut down condition especially after 2010 resulting in high turnover of personnel. The gas supply situation improved after 2016, staffingwas improved and special focus was put on the training of personnel to maintain the required skill level and competency index.

Due to lean manpower and high turnover, such critical and hazardous plants especially Ammonia plant become vulnerable for catastrophic failure, so a study was initiated for improving the emergency shutdown systems.

3.1 Benchmarking with Best Engineering Practices and Development of Systems:

Detailed review of the existing system was conducted and special focus was put on the following:

- Review of instrumentation / securities on latest Ammonia plants was carried out and a comparison was made with the existing securities present at our plant.
- Cause and Effect matrix for existing plant was also reviewed and updated.
- Existing securities / relays testing procedure was developed and a regime was developed for all critical loop testing in Turnarounds. After multiple tests, a database will be

developed and the testing regime will be modified based on the Mean Time Between Failures (MTBF) of each security.

- Loop testing of all trip securities was carried out and observations were recorded; 02 securities out of 75 tested were highlighted as faulty; both relating to fuel system of Ammonia plant.
- Out of service securities were put back in service after detailed review.

3.2 Gap Analysis and Project Study:

Initial gap analysis of the system was carried out and a detailed study was initiated to improve the ESD in line with requirements of IEC 61511 and IEC61508 due to following concerns:

- Currently Safe shutdown in case of upset is ensured through few relay-based loops & standard operating procedures (SOPs)
- Multiple failures (Steam Turbo Generator STG and CO_2 Absorber) in 2018 due to absence of ESD
- ESD can bring the plant to a safe state without human intervention
- Depletion of experienced manpower can result in major failure during emergency shut down

The gap analysis revealed:

- Major gaps in existing ESD logics such as no integrated furnace protection
- Incomplete logics – 75% less input/outputs (I/Os) as compared with latest plants.
- Each existing logic contains a number of human dependent actions
- Gaps in existing ESD hardware – existing Solonoid valve (SOVs), tripping philosophy

3.3 Infrastructure Study:

Detailed study of existing hardware was also carried out to determine the scope of work of the project. The benchmark of latest plants was used for comparison and hardware requirements were evaluated that comprised of:

1. Transmitters
2. SOVs
3. Control Valves

Based on these, the cost of hardware was calculated and a business case has been prepared for approval from the Management.

4.0 Development of Procedure for Handling Bypassing SC Devices:

After development of SC Equipment / devices lists, awareness on the importance and risks associated with their failures were communicated to all tiers of the organization. Consequence analysis of each was also shared as per relevance to apprise each individual about the significance of the health of these components of a hazardous plant.

Procedures were also developed for removing any of the Safety Critical Equipment / Device that included detailed risk analysis, control measures and additional control measures required during the bypass period. Strong vigilance by Operations, Process and Inspection personnel is also ensured during such events. No deviation from the defined procedure was allowed due to the criticality of the SC equipment / devices and hazardous nature of Ammonia – Urea plants. The procedures were shared within the organization along with detailed training sessions for Operations, Maintenance and Process personnel. Necessary forms like Out of service, taking back in service and waiver were also developed and implemented.

5.0 **Conclusion:**

The study of Safety Critical Equipment / Devices and Emergency Shut down Systems carried out as part of our journey to PSM excellence revealed that:

- The detailed working for development of SC Equipment / Devices lists is mandatory for containment of hazardous process chemicals
- Health of SC equipment / devices should be regularly assessed through a logical PM regime
- Bypassing of SC equipment / devices should be avoided; in case it is required, proper bypass procedure should be developed and followed.
- Emergency Shut Down systems should be updated throughout the life cycle especially for hazardous plants like Ammonia – Urea complex
- Latest technologies and the failure incident worldwide should be critically reviewed with reference to the existing facilities being operated in order to avoid similar one at other sites.

6.0 **References:**

- Standard no. 1910.119 "Pròcess safety management of highly hazardous chemicals" Occupational safety and health administration ,https://www.osha.gov/laws-regs/regulations/standardnumber/1910/1910.119 (accessed Jan. 21, 2021)

- "Consolidated List of Chemicals Subject to the Emergency Planning and Community Right-to-Know Act (EPCRA), Comprehensive Environmental Response, Compensation and Liability Act (CERCLA) and Section 112(r) of the Clean Air Act," Environmental Protection Agency

- https://www.epa.gov/sites/production/files/2015-03/documents/list_of_lists.pdf (accessed Jan. 21, 2021)

- John C. Wincek, Croda, Inc., 8 Croda Way, Mill Hall, PA 17751, Is all Safety-Critical Equipment Critical to Safety? IChemE SYMPOSIUM SERIES NO 159

SPRING21
+17TH GCPS
A Joint AIChE and CCPS Meeting

Design and Full-Scale Testing of a Blast-Resistant Roller Shutter Door

John Puryear
ABS Consulting
140 Heimer Rd. Suite 300 San Antonio, TX 78232
jpuryear@absconsulting.com

Darrell Barker
dbarker@absconsulting.com

Jerry Collinsworth
jcollinsworth@absconsulting.com

Michael Wong
Gliderol Doors (S) Pte Ltd
86 International Rd. Singapore
mw@Gliderol.com.sg

Leow Chyan
leowchyan@Gliderol.com.sg

US Patent Pending – USPTO Pub No US/2021/0047881 A1

Prepared for Presentation at
American Institute of Chemical Engineers
2021 Spring Meeting and 17th Global Congress on Process Safety Virtual
April 18 - 22, 2021

Design and Full-Scale Testing of a Blast-Resistant Roller Shutter Door

John Puryear
ABS Consulting
140 Heimer Rd. Suite 300 San Antonio, TX 78232
jpuryear@absconsulting.com

Darrell Barker
dbarker@absconsulting.com

Jerry Collinsworth
jcollinsworth@absconsulting.com

Michael Wong
Gliderol Doors (S) Pte Ltd
86 International Rd. Singapore
mw@Gliderol.com.sg

Leow Chyan
leowchyan@Gliderol.com.sg

Keywords: roller shutter door, blast loading, finite element analysis, shock tube testing

Abstract

Conventional roller shutter doors can become hazardous debris when subjected to blast loading. To mitigate this hazard, Gliderol and ABS Consulting are developing a blast-resistant roller shutter door. The two-phase development includes finite element analysis (FEA) to design the door system and shock tube testing to validate the design. Phase 1 included analysis of concept and performance testing of full-scale door components to gather data necessary to develop an FEA model of the entire door. Phase 2 entails detailed design development, analysis and validation of the complete door system.

In this paper, details of the Phase 2 design and validation of the first phase blast testing are discussed. Design blast loads and structural performance requirements are provided. Shock tube testing on full-scale components in Phase 1 is described, and the validation of the FEA modeling techniques against the Phase 1 testing is presented. Finally, the FEA design for the full door system is discussed.

1 Introduction

Blast overpressure caused by deflagration of energetic materials is a hazard at many oil, gas and chemical (OGC) sites. Also common at OGC sites are roller shutter doors (RSDs) composed of light-gauge steel elements. These conventional RSDs are often inadequate to resist blast overpressures that may develop at a site. The construction of conventional RSD is such that the elements composing them can hazardous debris for occupants inside an OGC building.

Mitigation of this debris hazard requires that the resistance of the RSD be increased and that elements be connected to mitigate the possibility of becoming debris. To meet these requirements, Gliderol and ABS Consulting are developing a blast-resistant RSD. The goal is to significantly increase the resistance of a conventional RSD to blast and connect the elements of the RSD in a way that mitigates the debris hazard.

The two-phase development of the blast-resistant RSD includes finite element analysis (FEA) to design the door system and shock tube testing to validate the design. The first phase, which was completed spring of 2020, included analysis of concept and performance testing of door components to gather data necessary to develop an FEA model of the entire door. The second phase entails detailed design development, analysis and validation of the full-scale door system.

2 Hazard Characterization

2.1 *Air Blast*

Review of a representative specification for blast design in a chemical processing facility suggested that an RSD which prevents debris hazards for blast impulses in the range of 1100 kPa-ms (160 psi-ms) would provide significant mitigation for vapor cloud explosion hazards. The pneumatic shock tube in which the RSDs are being tested is able to reach a comparable impulse with an associated peak pressure of 125 kPa (18 psi) while accommodating a full-sized RSD. This pressure-time and impulse-time history, shown in Figure 1, served as the design basis for developing the RSD.

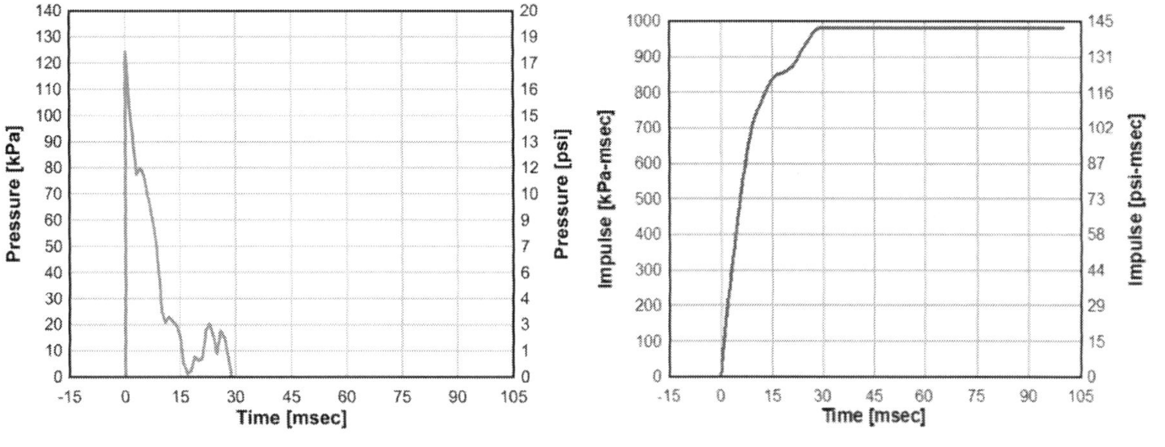

Figure 1. Test Case Applied Blast Load

2.2 Debris

The primary hazard posed by an RSD is that it becomes high-speed debris when it fails in response to a blast load. Much of the RSD is light-gauge steel curtains by design for ease of operation. The blast overpressure applied to the presented area of the curtain adds up to a large total force. Because the mass of the curtainwall is small relative to its presented area, the curtainwall accelerates at a high rate and can travel at high speed into an occupied space. This hazard is illustrated with stills from high-speed photography in Figure 2.

Figure 2. Debris from RSD Failure

3 Roller Shutter Door Design

3.1 Design Considerations

From the hazard characterization, the design goals are to resist the blast overpressure and to prevent the components of the RSD from becoming debris. The approach adopted to develop the blast-resistant roller shutter door was to reinforce an existing hurricane-rated door design because it included heavy-gauge curtains and connections.

Because operability requires that the RSD remains lightweight, the design cannot rely on increasing the mass of the RSD significantly, a common approach to blast-resistant design. Furthermore, maintaining the economy of the RSD precludes significantly increasing the strength of the connecting elements, and overly-strong connecting elements have the additional disadvantage of attracting high loads and thereby failing in a brittle manner.

3.2 Design Strategy

From all these considerations, the design strategy became allowing the RSD to plastically deform and progressively arresting the moving components. The Phase 1 testing and FEA modeling demonstrated the viability of a novel connection for attaching steel rods that reinforce the curtains to a larger supporting rod [1].

Building on the Phase 1 developments, the door designs consist of curtains supported by horizontal connector rods and vertical wire rope which are unrestrained at the bottom of the door. Under blast loading, curtains will deform and transfer load to the vertical wire rope which are restrained at the top by the roller drum. The bottom of the door may be displaced substantially upward in response to the blast load, either unrestrained for lower loads or tethered for higher loads.

4 Shock Tube Testing

To understand the response of the RSD to blast loading, the Phase 1 development entailed shock tube testing of full-scale curtains reinforced with rods and the novel connection of the rods to a supporting element. Measurements from those tests are the basis for validating the FEA model discussed in Section 5. The validated FEA model is being used to design the complete RSD system, which will be validated with additional shock tube testing in Phase 2.

4.1 Test Loads

Blast loads were applied using the pneumatic shock tube shown in Figure 3. This device uses a sudden burst of compressed air to create a blast pulse, which travels down the tube and is applied to the test specimen which is secured to the end of the tube. The blast load creates a specified positive blast pressure and impulse on the test specimens. Negative phase blast loads may occur, but the system is not designed to produce a negative pulse typical of a high explosive load. The applied blast loading was selected for each specimen from a set of pre-recorded calibration shots based on the anticipated response of the test article; the histories shown in Figure 1 are an example loading.

Figure 3. Shock Tube and Test Fixture

4.2 Instrumentation

High-speed video was used to capture response of the test articles. The cameras were placed in various positions to capture specific information about the response of the components. Posttest photography included documentation of failure modes for the components. Blast pressure measurements for the load applied to the test articles were captured by piezoelectric gauges collected by digital oscilloscope.

4.3 *Test Measurements*

The tested door panel systems were all nominally 3340 mm (131.5 in) tall and 3600 mm (142.0 in) wide. Panels were secured to the shock tube by semi-rigid vertical supports as shown in Figure 3 above. An example response of the test article to blast loading is shown in Figure 4. The FEA model was validated on the basis of peak inbound deflection. The loadings and deflections for the applicable test cases are shown in Table 1.

Figure 4. Response of Phase 1 Test Article to Shock Tube Loading

Table 1. Max Deflection Phase 1 Test Articles

Test ID	Pressure [kPa] (psi)	Impulse [kPa-ms] (psi-ms)	Max Deflection [mm] (in)	Time at Max [ms]
A	101 (14.6)	544 (78.9)	621 (24.4)	35
B	123 (17.8)	664 (96.3)	613 (24.1)	33

5 FEA Model - Validation

Prior to continuing with design of the full-scale RSD, FEM models were developed that numerically replicated the shock tube tests. LS-PrePost and LS-DYNA were used, respectively, to generate model geometry and perform the calculations. The results from these calculations were compared to the test measurements, and this comparison served to validate the FEA models.

5.1 *Geometry*

The geometry for validation is shown in Figure 5. Symmetry was used to reduce the number of elements, and the figure renders the geometry about the symmetry axes.

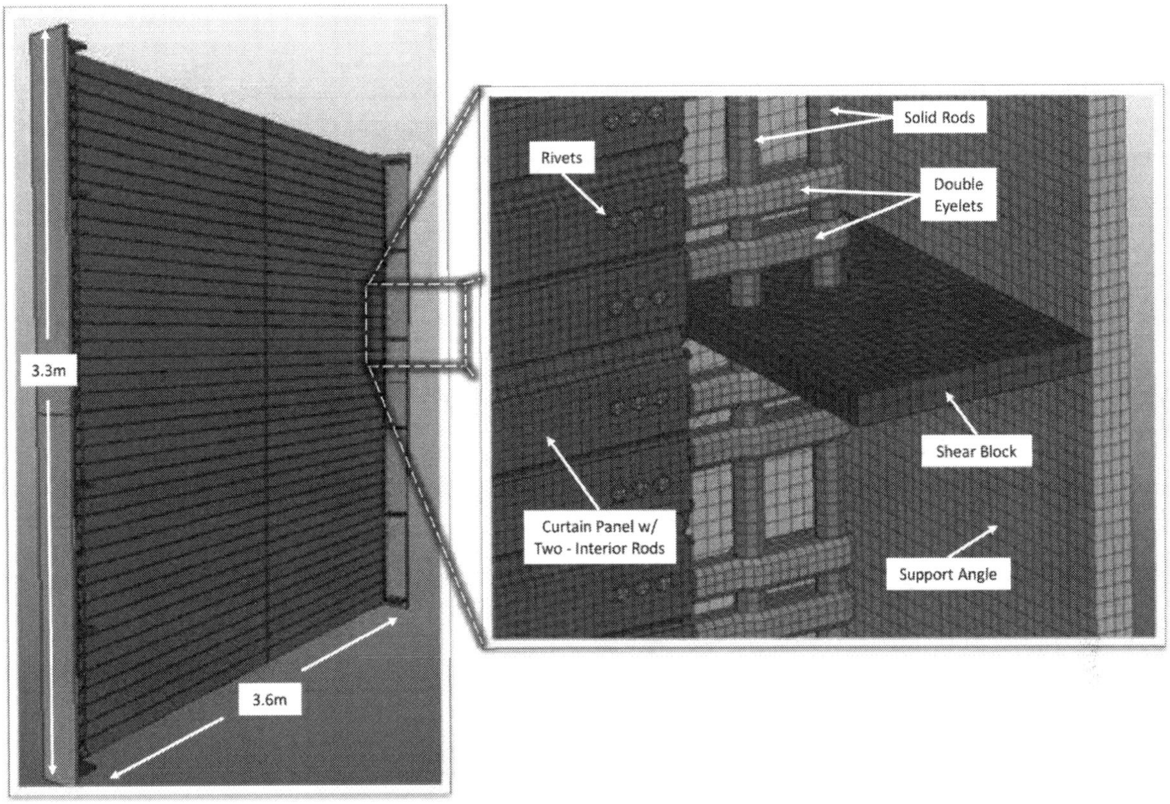

Figure 5. Validation Model Geometry

5.2 Element Formulation

All light gauge steel components were modeled as thin-shell elements (elform type -16 in LS-DYNA), which is a fully integrated shell element. The support rod and threaded eyes were modeled as solid elements (elform 1) which is a constant stress solid element. Through-rods and rivets were modeled as Hughes-Liu beam elements with cross section integration (elform 1).

5.3 Material Models

Several different types of steel were used in the model to represent the RSD components. Exact material constants used for modeling for the specified materials were not available. As a result, surrogate material grades with the properties as close to the specified grades as possible were used.

A ductile damage and failure model, the "generalized incremental stress-state dependent damage model" (GISSMO) damage model, was used in the FEA. GISSMO is a general treatment of damage and fracture in metals. Its main feature is a fracture locus (or surface) defined as equivalent plastic strain to failure (ε_f) as a function of triaxiality (η) and load angle parameter (ξ). The use of GISSMO in place of more primitive methods of predicting failure, such as a constant limiting effective plastic strain, permitted more efficient design of the RSD by reducing conservativism in the failure prediction.

5.4 Validation: Calculation versus Test Results

The qualitative agreement between the FEA results and testing are illustrated in Figure 6, which compares deformed shapes. Quantitative comparison of the calculated and measured deflections is shown in Table 2. The agreement between the calculation and the measurement is good, ~10% for the two cases. This result gives confidence that the modeling approach is sufficiently accurate for design.

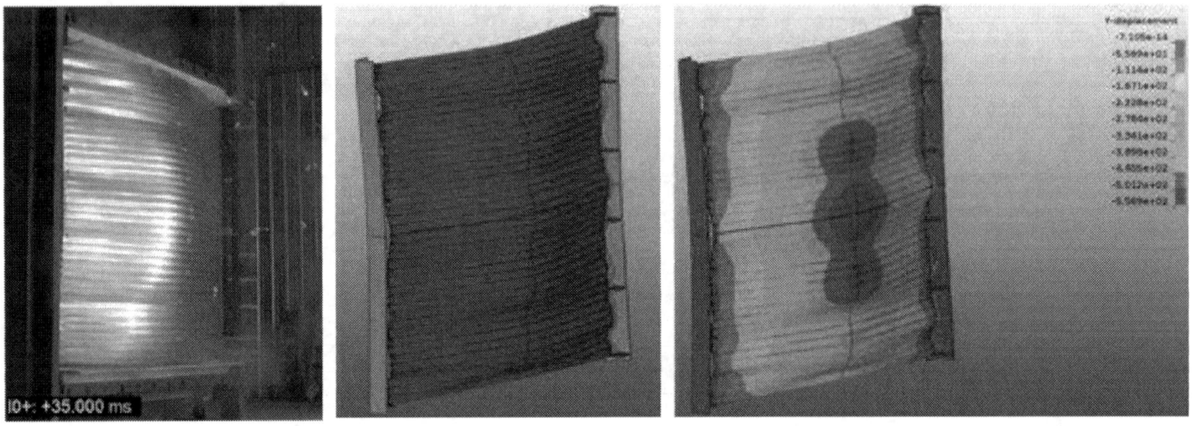

Figure 6. Deformed Shapes – Test versus Calculation

Table 2. Max Deflection Phase 1 Test Articles

Test ID	Test Measurement			FEA Model			Model Deflection vs. Measurement - % Difference
	Max Deflection [mm]	Max Deflection [in]	Time at Max [ms]	Max Deflection [mm]	Max Deflection [in]	Time at Max [ms]	
A	621	24.4	35	554	21.8	29.2	11.4%
B	613	24.1	33	675	26.6	28	9.6%

6 FEA Model – Design

6.1 Geometry

The validated FEA models were modified to examine practical design configurations. The geometry for the Phase 1 testing was a simplification a functioning RSD for the purpose of isolating the structural behavior of critical components. For the Phase 2 FEA design, components necessary for RSD operation are being added. These include the drum around which the curtains are wound, supporting structure, wire rope for supporting the rods inside the curtain and a shoe at the end of the wire rope. The ½ symmetry FEA model used for design is shown in Figure 7.

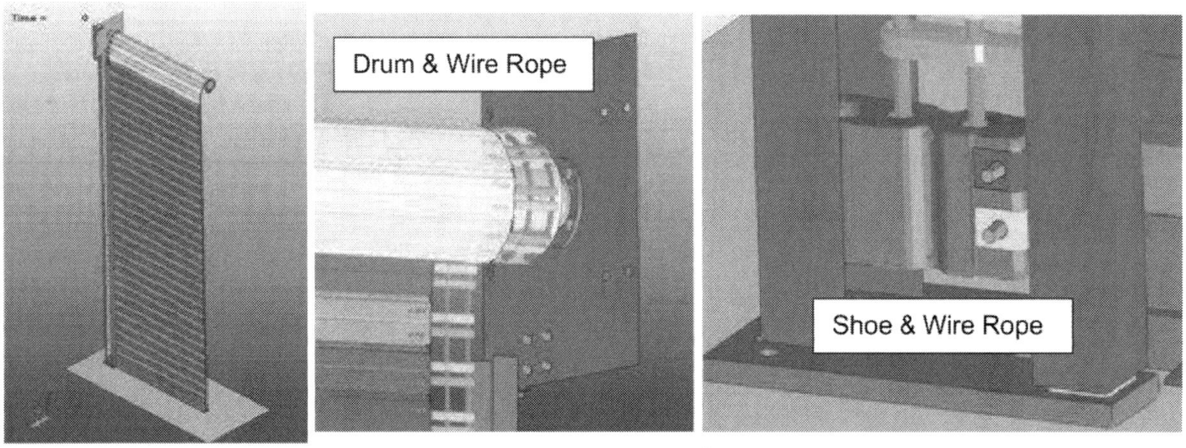

Figure 7. Design Model Geometry

6.2 Simulation Results

The detailed FEA modeling identified weak points in the design which controlled response of the RSD under blast loads. As component failures were observed in the simulations, the design and model were changed to reduce or eliminate unwanted failure modes and increase the energy absorbing capability of the door. This iterative procedure, while lengthy due to simulation run time, was substantially more efficient than physical testing and allowed the design team to establish confidence in the performance during testing. This approach also significantly reduced the number of tests necessary to validate the design. Moreover, this validated engineering model can now be used to develop designs for other blast load levels and door opening sizes.

Two performance levels were considered for design. For Performance Level 1 (PL1), the shoe of the RSD is permitted to displace up and into the occupied space, but the amount of displacement is controlled by a tether at the RSD's base. The requirement for PL1 is that components of the RSD not detach and become debris.

For PL2, the shoe of the RSD is restrained laterally and out-of-plane so that it does not move upward or inward into the occupied space. The RSD will be moderately damaged and remain substantially intact. Accordingly, the limiting load for PL2 is considerably less than the load for PL1.

Stills of the response of the response at PL1 and PL2 are shown respectively in Figure 8 and Figure 9. The limiting load for PL1 was equal to the pressure-time history shown in Figure 1, whereas PL2 is 25% of the PL1 load.

The tether for PL1 is included as a constraint with a birth time, i.e., at a specific time the constraint activates and arrests the shoe. The drum pays out the wire rope and supported curtains to limit the force applied to the structure supporting the drum. For PL2, the constraint is present from the beginning, representing a non-responding connection for the shoe. The force histories on the shoe determined by these constrains will be the design basis for the tether cross section, base connection and anchorage in the Phase 2 test program.

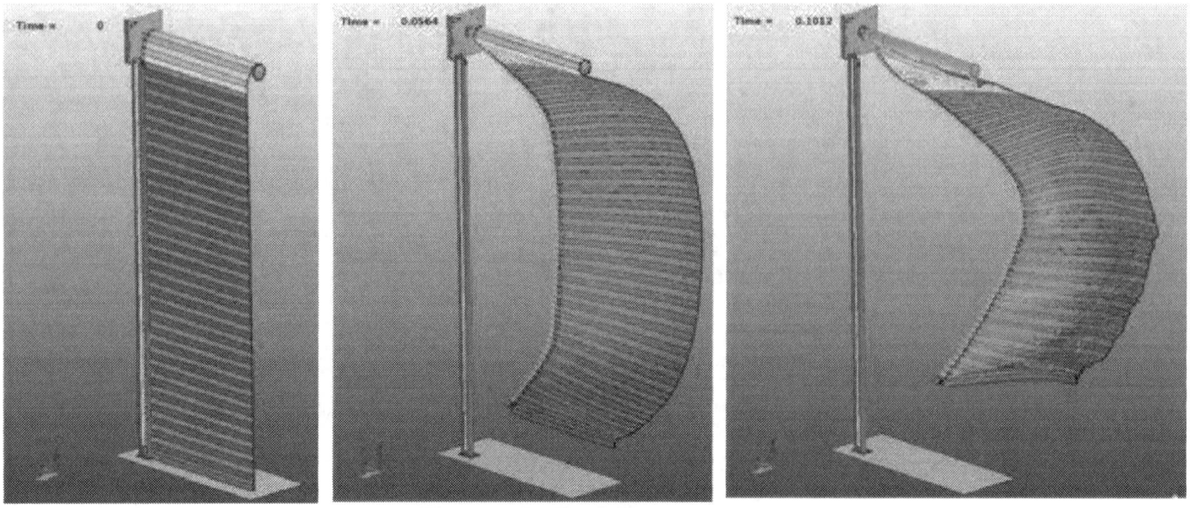

Figure 8. PL1 Response

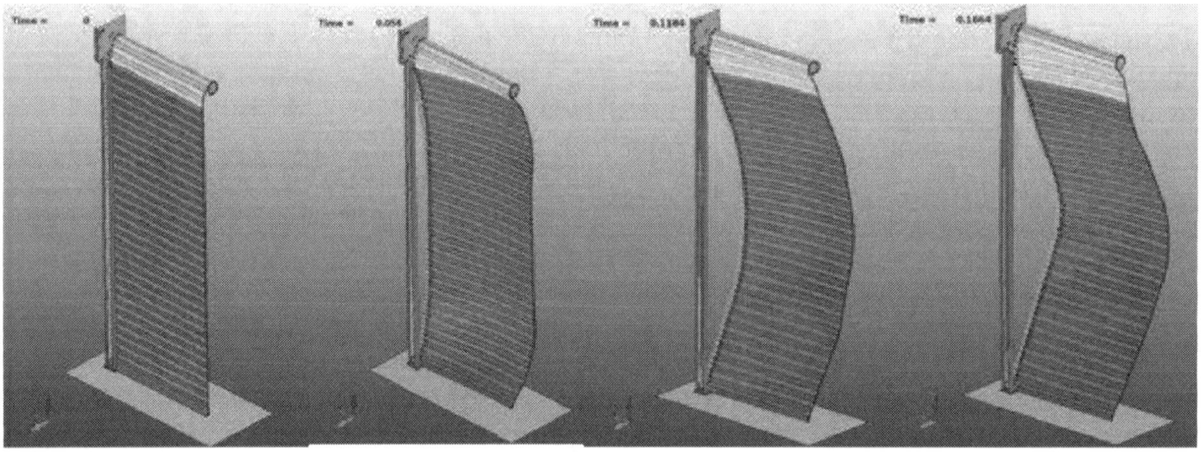

Figure 9. PL2 Response

7 Conclusion

The FEA modeling compared very well with the Phase 1 component testing. This gave the development team confidence that the physics of the highly deforming RSD could be accurately predicted and used for design. Detailed modeling for the Phase 2 design allowed the team to identify failure modes early in the design process and develop and viable configuration to validation testing. The upcoming Phase 2 testing in the ABS Consulting shock tube will validate the FEA model design for the entire RSD system.

8 References

[1] D. Barker, J. Puryear, J. Collinsworth, M. Wong and L. Chyan, "Development of a Blast-Resistant Roller Shutter Door," in *Mary Kay O'Connor Process Safety Center*, College Station, TX, 2019 (Presentation Only).

SPRING21
+17TH GCPS
A Joint AIChE and CCPS Meeting

Gone with the wind: Ventjet is a new engineering model for atmospheric dispersion

Derek Miller
Air Products
7201 Hamilton Boulevard, Allentown, PA 18195, USA
millerd3@airproducts.com

Elena Vyazmina
Air Liquide Research and Development
Innovation Campus Paris, 1, chemin de la Porte des Loges, 78354 Jouy en Josas, France
elena.vyazima@airliquide.com

Amy Shen
Air Products
Bldg. 88, Lane 887, Zuchongzhi Rd, Pudong District, Shanghai, China
shenc2@airproducts.com

Elizabeth Lutostansky
Air Products
7201 Hamilton Boulevard, Allentown, PA 18195, USA
lutostem@airproducts.com

Prepared for Presentation at
American Institute of Chemical Engineers
2021 Spring Meeting and 17th Global Congress on Process Safety
Virtual
April 18 - 22, 2021

in papers or printed in its publications

Gone with the wind: Ventjet is a new engineering model for atmospheric dispersion

Derek Miller
Air Products
7201 Hamilton Boulevard, Allentown, PA 18195, USA
millerd3@airproducts.com

Elena Vyazmina
Air Liquide Research and Development
Innovation Campus Paris, 1, chemin de la Porte des Loges, 78354 Jouy en Josas, France
elena.vyazima@airliquide.com

Amy Shen
Air Products
Bldg. 88, Lane 887, Zuchongzhi Rd, Pudong District, Shanghai, China
shenc2@airproducts.com

Elizabeth Lutostansky
Air Products
7201 Hamilton Boulevard, Allentown, PA 18195, USA
lutostem@airproducts.com

Keywords: gas dispersion model, gas venting, wind tunnel data, field test data, accidental release data, CFD modeling, effective source, ground effects

Abstract

Modeling of atmospheric dispersion of flammable and toxic gases is very common in industry. There is a large spectrum of different models form Gaussian, Integral up to Computational Fluid Dynamics (CFD) models. However, these simple engineering models (Gaussian, Integral) do not always correctly estimate the dispersion of gases with a density higher than air. The purpose of this paper is to present an alternative/new model Ventjet dedicated to atmospheric dispersion of various gases including light gases (hydrogen, methane, etc), neutral gases (nitrogen, oxygen) and heavy gases (CO_2, LNG etc). Ventjet takes into account various weather condition profiles namely neutral D and stable F. The effect of the ground is also integrated into Ventjet. Ventjet can be used for various releases including vertical, angled and even into-the-wind. Ventjet has been broadly validated versus multiple measurements from wind tunnel and field experiments as well as CFD simulations.

1 Nomenclature

symbol	units	description
A	m^2	Top hat cross sectional area
c	Mol/mol	Top hat molar concentration
$C_{centerline}$	Mol/mol	Centerline molar concentration
C_{target}	Mol/mol	Target molar concentration for dispersion model
C_m	kg/kg	Top hat mass fraction
C_d	-	Orifice discharge coefficient
C_{drag}	-	Atmospheric drag on cloud
C_p	J/kg K	Gas heat capacity
C_R	-	Ratio $C_{centerline}$/ c = 1.7
E	J/kg	Energy = Enthalpy + kinetic energy
E	Kg/s	Air entrainment
$C_{rosswind}$	Kg/s	Crosswind air entrainment
E_{jet}	Kg/s	Jet air entrainment
$E_{passive}$	Kg/s	Passive air entrainment
g	m/s^2	Gravity, 9.81 m/s2
H_{eff}	m	Cloud effective height = $z_{tophat\ upper}$. Used to determine grounded cloud Richardson number Ri_*
l_y	m	Passive sideways mixing distance
l_z	m	Passive vertical mixing distance
m_w	Kgmol/kg	Molecular weight
m	Kg/s	Cloud mass rate
M	Kg m/s^2	Cloud momentum
M_x	Kg m/s^2	Cloud horizontal momentum

M_z	Kg s /m^2	Cloud vertical momentum
P	m	Top hat cloud wetted perimeter
R	m	Top hat radius
Ri_*		Cloud Richardson number, used to determine grounded cloud lift off
R_s	m	Radius of circular top hat cloud (=R away from ground)
R_0	J/kgmol K	Gas constant, 8.314
r	m	Cloud radius to concentration C_{target}
s	m	Distance along cloud centerline
T	K	Release temperature
T_a	K	Ambient air temperature
u	m/s	Top hat cloud velocity
u_a	m/s	Wind velocity at centerline elevation
U_{10}	m/s	Ind velocity at reference height of 10m
W_{gnd}	m	Top hat cloud ground contact radius
x	m	Distance in horizontal direction (wind direction) of cloud centerline
z	m	Distance in vertical direction of cloud centerline
z'	m	Upward distance from ground to centerline perpendicular to cloud direction
z_0	m	Surface roughness, default = 0.18m
z_{ref}	m	Elevation for reference wind, 10m
$z_{tophat\ upper}$	m	Elevation of top of tophat cloud
$z_{tophat\ lower}$	m	Elevation of bottom of tophat cloud, greater than or equal to 0m
α_1	-	Jet entrainment coefficient, 0.08

α_2	-	Crosswind entrainment coefficient, 0.5
α_3	-	Passive entrainment coefficient, 1.0
ε	-	Passive turbulence parameter
γ	-	C_p/C_v heat capacity ratio, 1.4 for diatomic gases like O_2, N_2, H_2, CO, 1.3 for larger molecules like CO_2, CH_4
ϕ	radians	Centerline angle to horizontal in the downstream direction
ρ	Kg/m^3	Top hat density
ρ_a	Kg/m^3	Ambient air density

2 Introduction

2.1 Overview

Arguably, the current De Facto industry standard is the DNV-GL PHAST [1] program. The Industrial Gas industry, for example, has used PHAST for many years, not only for performing risk analyses, but also for design purposes such as the sizing and location of vents for atmospheric gases such as oxygen, nitrogen, and argon. The Industrial Gas industry has experienced several "near misses" at sites operating air separation units (ASUs), where either high or low oxygen levels were measured at operating areas surroundings the ASU. As a result, Air Products and Air Liquide initiated a joint program to develop a new engineering model for gas dispersion. The work is summarized in this paper.

The joint program has identified several issues with current models. For each issue, relatively simple enhancements have been developed and incorporated into a new model called Ventjet.

Ventjet has been validated against a wide range of test data (wind tunnel and full-scale field experiments), CFD results and actual incidents and near misses.

The new enhancements presented in this paper impact gas dispersion modelling for many applications, not just the industrial gas industry.

A summary comparison between Ventjet and PHAST versions 6.54-8.2 is provided in Table 2 at the end of this paper.

The primary intention of this paper is to highlight important issues with available models today, provide a comprehensive set of test data and CFD modeling results and present a relatively straightforward, robust model that works well across the whole dataset. The authors recognize that there may be alternative modeling approaches that can be developed in the future that work just as well or perhaps better and that more experimental work is required for testing, validation, and refinement.

During the Ventjet model development, the authors have been in discussion with DNV-GL, who have provided valuable input. In parallel with the development of Ventjet presented in this paper, DNV-GL have developed their own different approach to address issues identified with cross wind entrainment. DNV-GL have incorporated their new approach into the most recent release PHAST 8.4 as a non-default option. This paper does not describe the DNV-GL approach, but does include results for a beta version of PHAST 8.4 with the new crosswind entrainment option applied, as well as results for previous versions of PHAST.

2.2 4.0 Real example of typical gas dispersion issue in the industrial gas industry

An ASU cryogenically separates air into oxygen, nitrogen and argon products. Such a plant has a significant number of vents that continuously or intermittently discharge nitrogen/argon rich or oxygen rich streams to atmosphere. Such vents must be located with adequate safety distances from areas where operators might be, especially nearby elevated platforms, such as on the cryogenic distillation cold box, or control rooms. At the design stage of the plant, dispersion calculations are performed to determine safe separation distances, usually using PHAST (versions 6.54-8.2), or using separation distance tables generated using PHAST.

The European Industrial Gas Association (EIGA) publishes a Standard on this subject [2], that is currently undergoing revision, in the light of the work presented in this paper.

When modeling either nitrogen/argon or oxygen releases, the dispersion model does not take into account the nitrogen/argon and oxygen already in the air when determining concentrations. Consequently, the dispersion model is run based on excess nitrogen/argon or oxygen. For example, if releasing nitrogen in air, with a conservative safe lower limit of 19.5% total oxygen (80.5% total nitrogen) net within the surroundings, this is equivalent to an excess nitrogen of 7.1 %. If releasing oxygen in the air, with a conservative safe upper limit of 25% total oxygen, this is equivalent to an excess oxygen level of 5.2 %. APPENDIX C provides the relevant conversions.

It should be noted that not all separation distances can be set using "simple" engineering models, since the effects of nearby structures can cause effects such as down-drafting, which can only be predicted with accuracy using advanced CFD models. This subject is outside the scope of this current work.

ASU plant operators usually wear personal monitors that will detect either low oxygen or elevated oxygen. Consequently, potentially hazardous atmospheres are detected, and personnel warned if within.

In the example provided here, Air Liquide ASU plant personnel measured low oxygen (17%), by personal monitor,16m downwind of the main ASU waste vent (almost pure nitrogen) on a cold box platform at about the same level [3], see Figure 1. Figure 1 shows simulation results for FLACS [4] (a commercial CFD code dedicated to modeling of dispersion and explosion and fire for risk assessment), Ventjet and PHAST[1] respectively, all run to 18.9% excess N_2 (red isopleth) and 7% excess N_2 (blue isopleth) (this equates to 17% and 19.5% total O_2 respectively). PHAST 6.7 fails to predict that the measured concentration reaches the platform, whereas both FLACS and Ventjet predict at least this level. For this type of vent, it is generally found that PHAST (versions 6.54-8.2) predicts horizontal distances to given concentrations that are about half those predicted by FLACS or Ventjet. PHAST 8.4 (beta version with new cross wind entrainment method) performs much better than PHAST 6.7.

For modeling of this example, the following conditions have been used: Molecular weight 28, release rate 110 kg/s, release temperature 295 K, Vent diameter 3.55m, elevation 18m, vertical release, ambient air temperature 303.2 K, 9 m/s wind @ 10m, D stability, surface roughness 0.18m

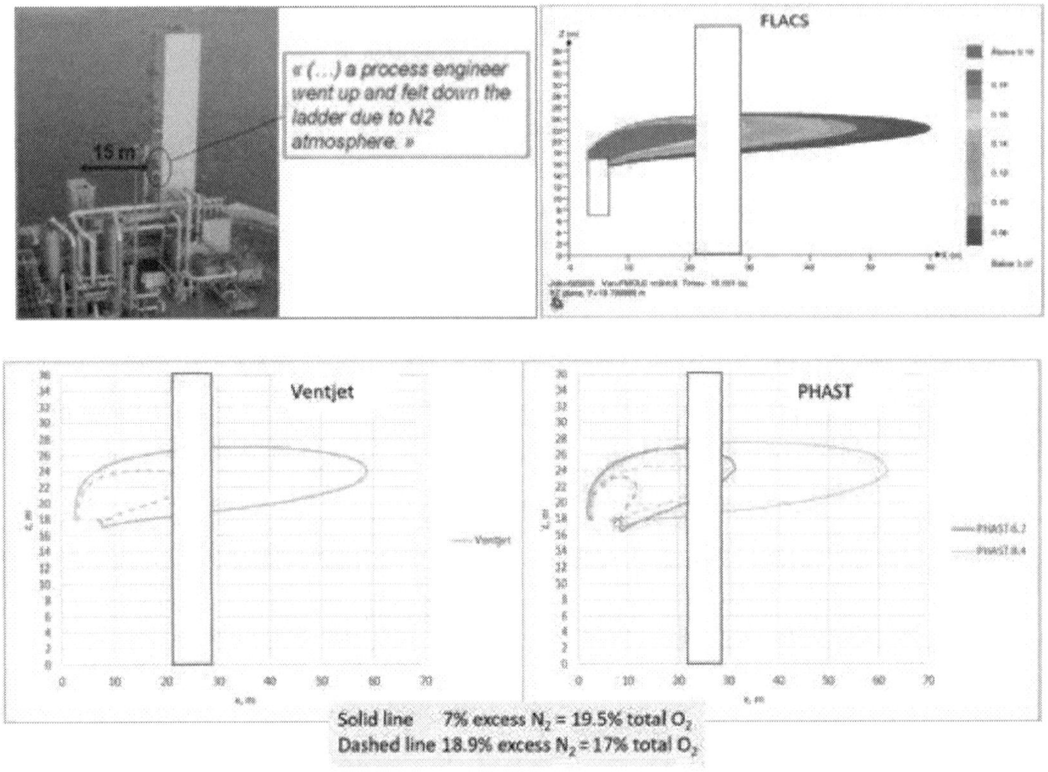

Figure 1. Incident at an Air Liquide ASU plant, an engineer felt down the ladder due to too high N_2 concentration in the ambient air. N_2 was released through a chimney 15m away from the platform. Top left frame shows the ASU plant, top right frame demonstrates the CFD (FLACS) modeling of the incident, Ventjet and PHAST simulations of waste nitrogen dispersion correspond to middle left and bottom right frames respectively.

2.3 Simple history of atmospheric dispersion modeling

Table 1 presents a very simplified overview of some of the key model developments in the history of atmospheric dispersion modeling.

Table 1. Summary of key developments in gas dispersion modeling

Development	Time period	References
Gas Jet dispersion model	1950's – 1960's	Morton Taylor Turner [5], Ricou Spalding[6]
Integral (Jet, cross wind, passive dispersion) model	1970's	Ooms[7]
Integral model with dense gas ground dispersion	1980's	DEGADIS[8]
Choked flow dispersion effective source model	1980's	Birch[9] [10]
Integral model with two phase dispersion	1980's	HGSYSTEMS[11]
Fully integrated model with cloud & pool, etc.	1990's to present	PHAST[1]

3 Ventjet Model Description

3.1 Model basis

Where possible, Ventjet uses the same equations as PHAST [1] (versions 6.54-8.2). It only deviates where an issue has been identified and an alternative solution has been developed. These alternative solutions either come from the Ooms [7] model or are original developments as part of this work. The main differences between Ventjet and PHAST (versions 6.54-8.2) are summarized in Table 2.

Ventjet is a "simple" top-hat gas dispersion model using ideal gas equations. A top-hat model assumes a uniform concentration, velocity, and density across the cloud, which is assumed to be radially symmetrical (in other words a circular cross section). All entrainment and momentum calculations are performed on this top-hat profile. Onto the top-hat profile is subsequently superimposed a standard Gaussian concentration profile, used to predict the centerline concentration and the width of the cloud to a specified concentration

As the cloud approaches the ground, part of the cloud contacts the ground, which changes the shape of the cloud and the "wetted" perimeter available for air entrainment. A simple approach is used to model the ground impact on cloud shape, described later in the paper.

A fundamental assumption is that the whole cloud is at atmospheric pressure, this assumption is also made by other models such as PHAST (versions 6.54-8.2). This is a correct assumption where the initial source nozzle / hole condition is non-choked. Where the source is choked, the nozzle condition is above atmospheric pressure and Ventjet adopts the conventional approach of using an "effective source" which is at atmospheric pressure. There are several different methods for determining this effective source condition, which is also called the pseudo, expanded, or depressurized condition. Discussion and conclusion on the most appropriate method for Ventjet is included in this paper.

Horizontal gas jets passing near to the ground can be pulled down towards the ground (even if buoyant) due to different phenomena including the Coanda effect [31]. A new simple modification to the vertical momentum balance is presented that addresses this issue reasonably well. As far as the authors are aware, this is the first time that the ground effect is modelled in an engineering model dedicated to atmospheric dispersion. This paper includes validation against CFD and published data.

It is found that the Ventjet model works well for releases into the wind, with the only required changes being to modify the geometry terms, to accommodate the cloud potentially traveling in any direction in the vertical plane parallel to the wind. Note: there is no experimental test data available on this subject, that can be found by the authors, but this paper provides a summary of CFD modeling that supports the proposed approach

The primary reference sources for the Ventjet equations are PHAST (versions 6.54-8.2) and the Ooms model. Each equation presented in the paper is clearly labelled as to its source or labeled as "this work" if an original development.

Ventjet is designed to be effective across a wide range of gas release conditions:

- low to high pressure (sub-sonic and sonic)
- low to high temperature
- buoyant to dense gases
- vertical, angled and horizontal releases with wind and into the wind
- airborne (in free space) clouds and those that reach the ground

The current version of Ventjet is not effective for:

- Two phase releases
- Dense gases that travel along the ground for extended distances

Ventjet is not yet validated for concentrations below 0.01%

3.2 Mass Balance and air entrainment

3.2.1 Cloud overview

Figure 2 is a simple sketch of a gas cloud dispersing from a hole, which could be a purpose designed vent, or an accidental hole in a vessel or pipe. The gas dispersion source term is the initial atmospheric pressure condition of the cloud at the point of release and is defined by its mol wt., mass release rate or initial pressure, initial temperature, hole diameter & discharge coefficient and release orientation. The gas is released into an ambient air environment that is defined by temperature, reference wind velocity @ 10m above the ground, weather stability class and ground surface roughness. Three dimensions are defined: the horizontal downstream x axis, the vertical z axis, and the cloud centerline s axis. As the gas expands from the release orifice, it experiences jet air entrainment (driven by the difference between the jet velocity and the velocity of the ambient air), crosswind air entrainment (driven by the wind velocity) and passive air entrainment (driven by atmospheric turbulence). All three entrainment terms are active simultaneously. Entrainment of air has two effects on the cloud – it dilutes it and it impacts its momentum and hence direction. Initially in this paper, it will be assumed that the release is low pressure sub-sonic. The appropriate method for addressing high pressure choked flow releases will be addressed later in the paper. It will also initially be assumed that the release is in the direction of the wind, releases into the wind will be addressed later in the paper.

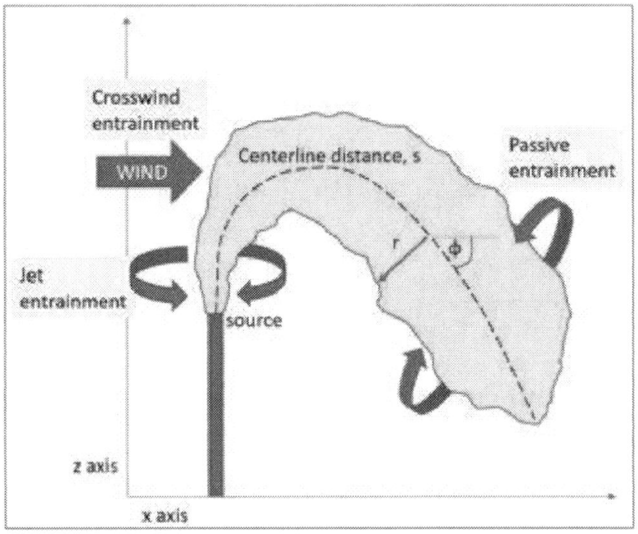

Figure 2. Simple sketch of the main features of a dispersing cloud model.

3.2.2 Mass balance

Equation 1 and 2 describes the basic mass balance equation where air entrainment dilutes the cloud. The mass balance is established incrementally along the direction of the cloud centerline (s direction)

Mass Balance [1][7] $$\frac{dm}{ds} = E$$ Equation 1

Entrainment[1][7] $$E = E_{jet} + E_{crosswind} + E_{passive}$$ Equation 2

3.2.3 Jet entrainment

Equation 3. The equation was originally derived by Morton Taylor Turner in the 1950's and is still used in many dispersion models today. Entrainment is proportional to the relative velocity of the jet to the wind, in the centerline direction.

Jet entrainment [1][5][7][8][11] $E_{jet} = \alpha_1 \, P \, \rho_a \, |u - u_a \cos(\Phi)|$ Equation 3

3.2.4 Cross wind entrainment

The equation was originally described by Ooms [7]. There is a key difference between this equation and the one used in PHAST (versions 6.54-8.2), namely the presence of the cos (Φ) term, included in Ooms and Degadis but subsequently eliminated in HGSYSTEMS and PHAST. It is not intuitively obvious why the cos(Φ) term should be there and the explanation for this cos (Φ) term by Ooms is that it was required to match the data available at that time. The same has proved true in the current work presented here. This term effectively reduces the initial cross wind entrainment for vertical releases (for example, it drives it to zero when the cloud is vertical). The improved estimate of initial cross wind entrainment dilution is critical to determine safe distances from vents which have been under-estimated by PHAST (versions 6.54-8.2) by a factor of about 2. To model falling (dense) clouds, it is found effective in Ventjet to set the crosswind entrainment to zero in such circumstances, even though it is not intuitively obvious why this should be the case. This will be shown later in the paper.

Cross wind entrainment [7][8] $E_{crosswind} = \alpha_2 \, P \, \rho_a \, u_a |\sin(\Phi)\cos(\Phi)|$ Equation 4

If $\Phi < 0$ [this work] $E_{crosswind} = 0$

3.2.5 Passive entrainment

Equation 5. Passive entrainment was not well understood when the Ooms paper was published. It is better understood now. All the sub-equations required are provided in APPENDIX A. All equations are from PHAST (version 6.54-8.2). Note that the Equation 5 used in Ventjet is a modified version of the one from PHAST. As the falling cloud approaches the ground, PHAST (version 6.54-8.2) phases out passive entrainment and phases in dense gas entrainment. Since Ventjet does not include a dense gas entrainment term, the simple approach of not phasing out passive entrainment in Ventjet seems to work reasonably well.

Passive entrainment [1][this work] $E_{passive} = \alpha_3 \, \pi \, \rho_a \, \varepsilon^{1/3} \, (l_y^{\frac{4}{3}} + l_z^{\frac{4}{3}})$ Equation 5

Passive entrainment PHAST [1] $E_{passive} = \left(1 - \frac{W_{gnd}}{R}\right) \alpha_3 \, \pi \, \rho_a \, \varepsilon^{1/3} \, (l_y^{\frac{4}{3}} + l_z^{\frac{4}{3}})$

3.3 Momentum balance

There is an interaction between the use of an atmospheric drag term, C_{drag} in the two momentum equations and the selection of the crosswind entrainment equation discussed above. For example, PHAST (versions 6.54-8.2) uses the same momentum equations, but sets this drag term to zero. Since the Ooms crosswind entrainment equation is selected for Ventjet, the Ooms $C_{drag} = 0.3$ value is also used.

3.3.1 Horizontal momentum

Equation 6. Cloud horizontal momentum is changed as air is entrained with a velocity equal to the wind velocity (wind is horizontal)

Momentum (horizontal) [7] $\frac{dM_x}{ds} = E\ U_a + C_{drag}\left(\frac{P}{2}\right)\rho_a\ u_a^2\ |\sin(\Phi)^3|$ Equation 6

3.3.2 Vertical momentum

Equation 7. Cloud vertical momentum is changed due to gravity acting on the difference between the cloud density and the surrounding air density (buoyancy), plus drag. The final term on the right-hand side is a new addition, developed as part of this work, to address the impact of the ground in pulling the cloud downwards discussed later in this paper, it is only active if the lower edge of the top hat cloud has reached the ground.

Momentum (vertical) [7] [this work]

$$\frac{dM_z}{ds} = -g\ (\rho - \rho_a)A + if(\Phi > 0, -1,1)C_{drag}\left(\frac{P}{2}\right)\rho_a\ u_a^2\ \sin(\Phi)^2\ \cos(\Phi)\quad - if\ (z > z_{min}, 1,0) * 2\ W_{gnd}\ 0.5\ \rho\ u_x^2$$ Equation 7

3.4 Cloud velocity and trajectory

Momentum is mass flow rate times velocity, so if momentum and mass flow rate are known, local cloud horizontal and vertical velocity (u_x and u_z) can be determined, and hence the cloud centerline velocity is found by Equation 8.

Centerline velocity $u^2 = {u_x}^2 + u_z^2, u_x = \frac{M_x}{m}, u_z = \frac{M_z}{m}$ Equation 8

The cloud trajectory angle Φ is determined from the horizontal and vertical velocities, equation 9.

Trajectory $\tan(\Phi) = \frac{u_z}{u_x}$ Equation 9

The incremental centerline step sizes in the horizontal and vertical directions are determined by Equation 10.

Centerline step sizes $dx = u_x \sin(\Phi)dt\,, dz = u_z \cos\,(\Phi)dt$ Equation 10

3.5 Cloud geometry and physical properties

3.5.1 Airborne (in free space) cloud away from ground

The cloud cross section is assumed to be circular and of simple top-hat form (same concentration, velocity and density across cloud radius). At any centerline distance, s, the cloud radius, R is found by Equation 11 since all other terms are known. The cloud perimeter, P is required to determine air entrainment and is found using equation 12, $W_{gnd} = 0$ if cloud is not impacted by the ground. W_{gnd} is the grounded cloud contact radius or, put another way, the half width of the cloud in contact with the ground.

Mass Rate [7][1] $\boldsymbol{m = \pi R^2\ u\ \rho}$ Equation 11

3.5.2 Cloud near ground

Figure 3 shows how the ground impacts the cloud. This is a simple approach that does not take into account cloud sideways spreading. The key parameter to determine is the wetted perimeter, P that is still exposed to the ambient air for entrainment. The grounded cloud radius must be such that the cloud area is correct to achieve mass balance, consequently a relationship is needed between R and z'. The new Equation 14 achieves this and is based on a curve fit of calculated values of R/R_s vs z'/R_S, where R_s is the radius of a circular cloud, not impacted by the ground. For grounding and grounded clouds, it is important to determine the appropriate wind velocity, for purposes of entrainment estimation. Ventjet uses the same concept as PHAST (versions 6.54-8.2), to use the elevation of the Centroid (center of mass) of the cloud. For clouds in the in the air, this will be the same as the cloud centerline, but for clouds near the ground it will be at a higher elevation (resulting in a higher wind speed). The elevation of the centroid in Ventjet is simply taken as half-way between the lower and upper elevation of the top hat cloud, Equation 15. For a grounded cloud, z_{tophat} lower = 0.

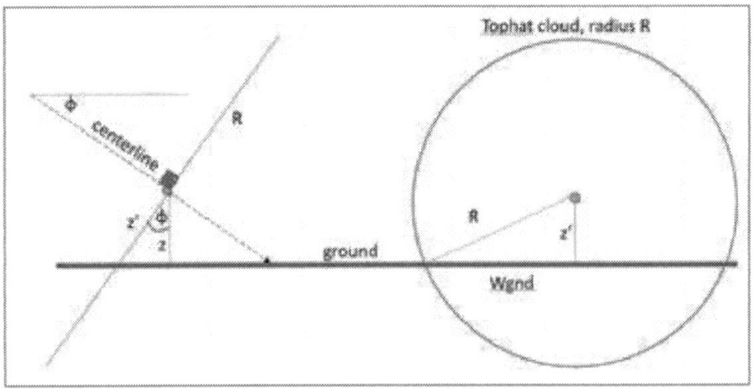

Figure 3. Cloud geometry for a cloud interacting with the ground.

Wetted perimeter [1] $$P = R\,\{2\pi - 2\,\text{asin}\,(\frac{W_{gnd}}{R})\}$$ Equation 12

Grounded cloud contact radius [1] $$W_{gnd} = \left(R^2 - z'^2\right)^{0.5} \quad , z' = \frac{z}{\cos(\Phi)}$$ Equation 13

Grounded cloud equivalent radius [this work]

$$R = R_s\,\{\,0.2701(\frac{z'}{R_s})^2 - 0.695\left(\frac{z'}{R_s}\right) + 1.4188\}\ , R_s = (\frac{A}{\pi})^{1/2}$$ Equation 14

Cloud Centroid elevation [this work] $$z_{centroid} = \frac{z_{tophat\,upper} + z_{tophat\,lower}}{2}$$ Equation 15

3.5.3 Cloud density and mol fraction

Top-hat density is determined by simple mixing Equation 16, based on the mass fraction and the atmospheric densities of the releasing gas and the ambient air. The top-hat molar concentration is also determined by simple mixing Equation 17.

Density [1][7] $$\frac{1}{\rho} = \frac{C_m}{\rho_0} + \frac{1-C_m}{\rho_a} \quad , C_m = \frac{m_o}{m}$$ Equation 16

Mol concentration [1][7] $$C = \frac{m_o}{m_w} / \{\frac{m_0}{m_w} + \frac{(m-m_0)}{m_{wa}}\}$$ Equation 17

3.5.4 Gaussian concentration distribution across cloud

As previously stated, all calculations on the cloud are based on the top-hat assumption. It is obvious that the cloud concentration profile is not actually of this form and so a means is required to convert from a top hat to a "real" concentration profile. This is achieved through the assumption of a standard Gaussian profile and all that is then required to be known is the ratio of the centerline concentration to the top-hat concentration, C_R. Equation 18. Ventjet assumes the same constant ratio as PHAST (C_R=1.7). So, the centerline concentration is determined from the top-hat concentration and Equation 19 is used to determine the radius, r of the cloud to any given concentration of interest, C_{target}. Derivation of this equation is provided in APPENDIX E. Note: the 1.7 factor is applied differently in Ventjet (where it applied all the way along the centerline) than it is in PHAST (where it is only incorporated in a modified entrainment constant for jet entrainment, not cross wind or passive entrainment constants).

Centerline concentration [1] $$\frac{C_{centerline}}{C} = C_R = 1.7$$ Equation 18

Radius to concentration of interest [this work]

$$r\left(C_{target}\right) = R\left(-\ln\left(\frac{C_{target}}{C_{centerline}}\right) * \left(\frac{1}{(C_R)}\right)\right\}^{0.5}$$ Equation 19

3.6 Grounded cloud lift-off

A cloud traveling along the ground will not necessarily lift off if it becomes buoyant. In order to lift-off, the cloud will need enough buoyancy to overcome surface boundary layer turbulent forces. Ventjet follows PHAST (versions 6.54-8.2) and uses the Briggs lift-off criteria Equation 20, which uses the cloud Richardson number, Ri_*, based on the grounded tophat cloud height, H_{eff}.

Cloud lift off criteria [1] $$Ri_* = \frac{g\,(\rho - \rho_a)\,H_{eff}}{\rho_a\, u_*^{\,2}} < -20$$ Equation 20

3.7 Wind profile

Wind speed increases with elevation (zero at the ground). It is conventional to define a wind speed at a reference elevation (z_{ref}) usually at 10m. Wind speeds at other elevations are then determined based on Pasquill atmospheric stability classes and surface roughness (z_o). Surface roughness is a function of the nature of upwind ground level obstacles. Ventjet default z_o= 0.18m, the same as PHAST (versions 6.54-8.2). Equations for wind speed are from PHAST (versions 6.54-8.2) and are provided in APPENDIX A. Wind speed is determined at the centroid elevation z centroid, which for a cloud in the air is the same as z centerline.

4 High pressure choke flow release effective source determination

4.1 Available methods

It is important to establish the correct atmospheric pressure effective source condition, so that the downstream cloud air entrainment (function of jet momentum) and buoyancy (function of jet density relative to ambient air density) are modelled correctly by the dispersion model. There are a number of available published methods, but there are two that are well established and in common usage: conservation of Mass, Momentum and Enthalpy used in Birch '87 [10] and conservation of Mass, Momentum and Energy generally used in PHAST (versions 6.54-8.2) [1]. The two methods predict the same initial momentum, but very different initial temperatures and consequently different downstream buoyancy. Let us look at these two methods…

4.2 Birch '87 method

Conservation of Mass, Momentum, Enthalpy. Subscript 1 is for initial condition, Subscript 2 is for the nozzle (elevated pressure) condition, subscript 3 is for the effective (atmospheric pressure) condition. Note: Birch '87 is based on ideal gas, so conservation of enthalpy is equivalent to conservation of temperature

Conservation of mass [10] $$\rho_3\, u_3\, A_3 = \rho_2\, u_2\, A_2 \qquad \text{Equation 21}$$

Conservation of momentum [10]

$$\rho_3\, {u_3}^2\, A_3 = \rho_2\, {u_2}^2\, A_2\, Cd^2 + A_2(P_2 - P_3) \qquad \text{Equation 22}$$

Combining 21 & 22 [10] $$u_3 = u_2\, Cd + \frac{(P_2 - P_3)}{\rho_2\, u_2\, Cd} \qquad \text{Equation 23}$$

Conservation of enthalpy (ideal gas) [10] $$T_3 = T_1 \qquad \text{Equation 24}$$

4.3 PHAST (versions 6.54-8.2) method

Conservation of Mass, Momentum and Energy. Equations 21, 22 and 23 are the same. The difference is that Equation 24 is replaced by conservation of energy, Equation 25 and 26. Note that PHAST uses real gas properties, not ideal gas, so Energy, $E = H + 0.5\ u^2$. For ideal gas $H = C_pT$

Conservation of energy (ideal gas) [1]

$$C_pT_3 + 0.5\, {u_3}^2 = C_pT_1 + 0.5\, {u_1}^2 \qquad \text{Equation 25}$$

For almost all cases the initial velocity $u_1 << u_3$, and if constant C_p is assumed (ideal gas)

$$T_3 = T_1 - \frac{0.5}{C_p}\, {u_3}^2\, ,\; C_p = \frac{\gamma}{\gamma - 1}\left(\frac{R_o}{m_w}\right) \qquad \text{Equation 26}$$

4.4 Comparison of methods

If we compare equation 25 with Equation 24, we can see the difference between conservation of Enthalpy vs. Energy. For conservation of Energy, since the effective source velocity u3 for choked flow releases is very high (usually super-sonic) this means that the effective source temperature will be very low, based on Equation 25, but for conservation of Enthalpy, the effective source temperature will be the same as the initial temperature of the process based on Equation 24 (at least for ideal gas, for real gases the differences are usually slight)). At this point we need to consider the nature of the downstream dispersion model. Engineering models (PHAST, Ventjet etc.) are based on conservation of Enthalpy, not Energy. In such a model, as the very high velocity releasing gas mixes with air, if initially cold, it will certainly start to warm due to simple mixing with ambient air, but it will not correctly account for the warming as the gas slows down and kinetic energy is converted back into enthalpy / temperature. One might expect such a dispersion model to predict too cold downstream temperatures, if combined with an effective source model based on conservation of Energy, even though it might accurately predict the very initial temperature downstream of the nozzle (accurate prediction of jet properties close to the nozzle is rarely important in gas dispersion modeling, whereas prediction of conditions downstream is always important). The prediction of the effective source condition does not need to be thermodynamically correct; it simply needs to be the condition that results in the correct downstream dispersion modeling, which is why it is sometimes called the pseudo source condition. It is interesting to note that both Birch '87 and PHAST (versions 6.54-8.2) will predict the same jet momentum, even though densities will be different. This suggests that initial jet entrainment will be the same, it will only be downstream, where buoyancy effects become important, that the differences will be seen.

4.5 Comparison with experimental data and CFD (Computational Fluid Dynamic) modeling

See Figure 4. A simple experimental test at an Air Separation Unit was performed by Air Products as part of this work, to measure temperatures downstream of a choked flow release. Air, at about 100 psig, was released horizontally though a ¾" nozzle and downstream centerline temperatures were measured using a simple handheld temperature probe. Due to the force of the escaping jet, it was not possible to measure temperatures very close to the nozzle. The releasing air initial temperature was very slightly below ambient. Measured jet temperatures were all between the initial temperature and the ambient temperature. A CFD model of the release was done using FLUENT [12]. FLUENT uses ideal gas equations to model the complex thermodynamics of a high-pressure gas as it passes through a series of shock waves until it reaches the atmospheric pressure condition. The FLUENT results match the actual measured results well and confirm that the initial condition downstream of the shock waves is indeed quite cold (PHAST version 6.54-8.2 also match this condition well), but that the gas rapidly warms as kinetic energy converts to enthalpy and that the prediction of downstream temperatures is best predicted using Birch '87. Conclusion: Conservation of mass, momentum and Enthalpy is one of the most appropriate method to use to predict the effective source condition, when used with an atmospheric dispersion model that conserves Enthalpy, not Energy, such as Ventjet.

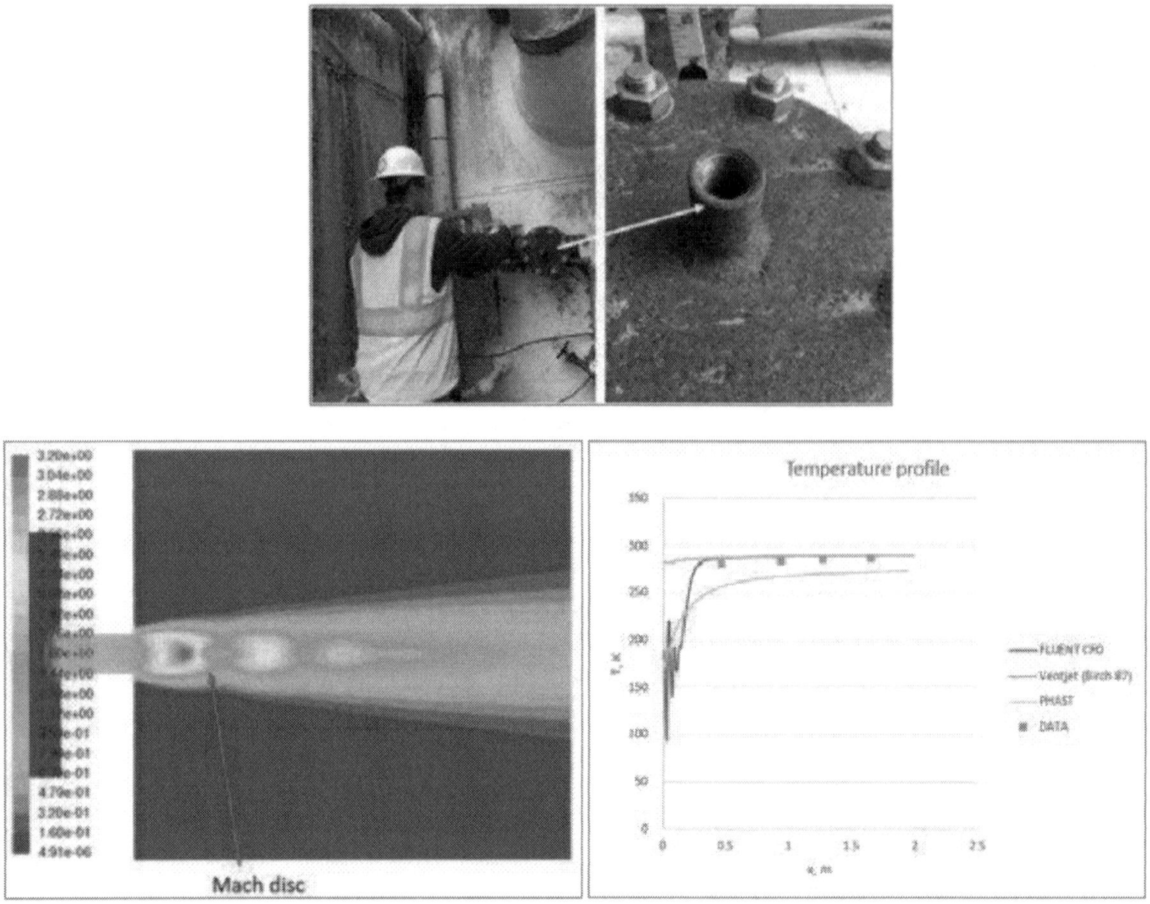

Figure 4. Experiments of Air Products: measurements (top frame), CFD simulations (bottom left frame) and Ventjet & PHAST 8.2 modeling results (bottom right frame).

5 Selection of entrainment constants

5.1 Jet entrainment

Two choices have been evaluated for the entrainment constant α_1 in Equation 3. Ooms uses a value of 0.057, whereas PHAST (versions 6.54-8.2) uses a value of 0.08, modified to reflect the Gaussian profile used in PHAST, which is different than the top hat profile used in the original references and by Ventjet. As described by DNV-GL [1]· the 0.08 value is consistent with a jet expansion half angle of 9.1 degrees, measured by early work by Ricou Spalding. This value is used extensively in simple jet dispersion models. Ventjet has been run against several high-pressure gas jet release tests, dominated by jet entrainment, to establish the most appropriate value for α_1.

5.1.1 Birch [9] Natural Gas tests

Birch performed a series of natural gas vertical release tests at different pressures, with a 2.7mm nozzle, vertical release, no wind. Birch presented the results in terms of 1/c (inverse of molar concentration) against z/d (vertical distance / nozzle diameter), which proves to be a good way of assessing the choice jet entrainment constant. Results using Ventjet clearly show that $\alpha_1 = 0.08$ provides the best fit (see Figure 5). PHAST 8.2 gives the same results as Ventjet ($\alpha_1 = 0.08$).

Modeling set-up: *mol wt. 17, release pressure 1.14 bara (C_d=0.8) and 3.5 bara (C_d=0.85), release temperature 288 K, Vent diameter 2.7mm, elevation 0m, vertical release, ambient air temperature 288 K, 0 m/s wind @ 10m, F stability, surface roughness 0.18m.*

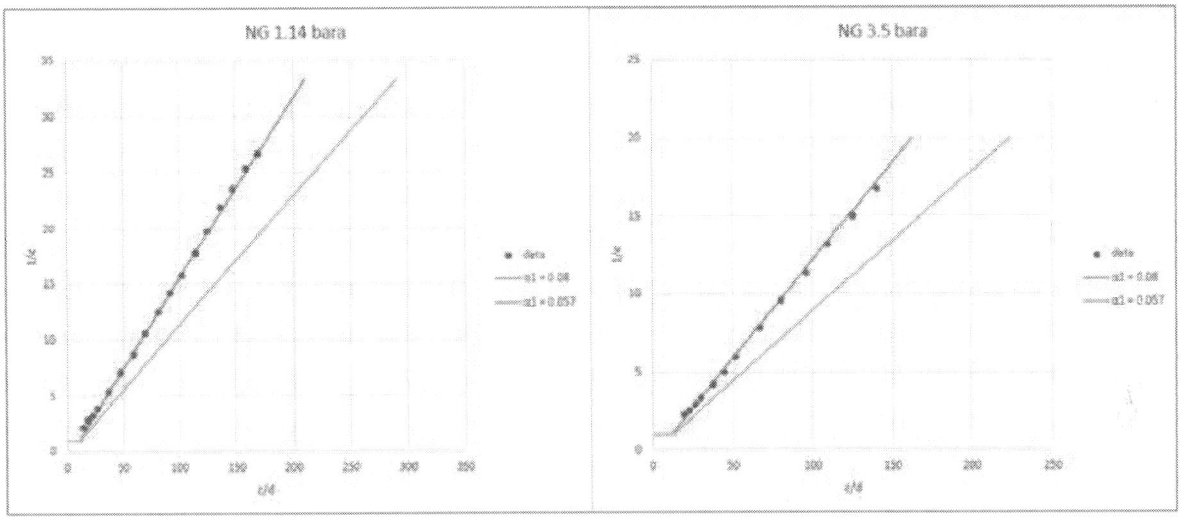

Figure 5. Birch [9] Natural Gas tests vs modeling results of Ventjet: red line corresponds to fitting constant $\alpha_1 = 0.08$, whereas green line corresponds to fitting constant $\alpha_1 = 0.057$.

5.1.2 Molkov [13]

Molkov [13] provides a summary of a wide range of previously published high pressure hydrogen jet tests. Ventjet simulation of all these test results vs. measured concentration test data is provided in FIGURE 6 and again shows a better match using Ventjet with $\alpha_1 = 0.08$.

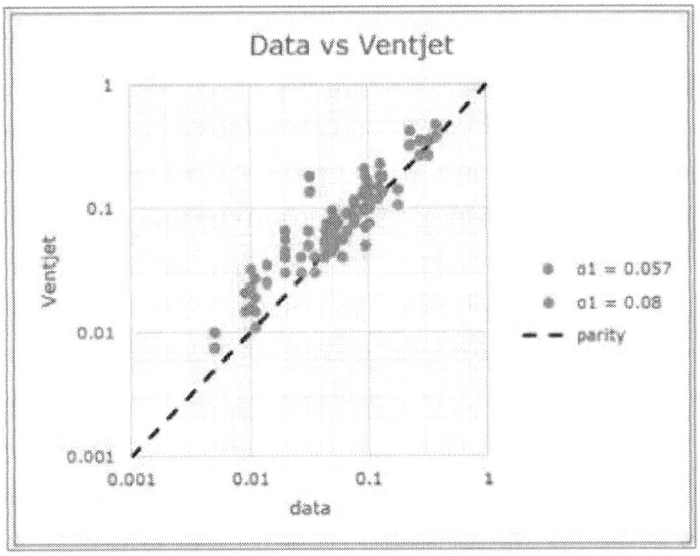

Figure 6. Ventjet simulation results vs. measured concentration test data summarized by Molkov [13].

5.1.3 Daubech [14] H_2 tests

Daubech [14] H_2 tests (see figure 7). A series of high-pressure horizontal releases were performed, with downstream concentrations measured at the release elevation. Wind was reported as "very low" (the experimental setup is located inside an ancient stone pit). Release rates were carefully measured at the release orifice as well as release temperature and pressure. Experiments were reproduced three times for each measurement point to obtain reliable experimental data. FLACS [4] modeling has shown that for the chosen locations of the sample points, the jet is still horizontal and so these readings are in fact the centerline readings… at further out locations this would not have been true, due to buoyancy and ground effects discussed later in this paper where these test results are discussed in more detail. FIGURE 7 shows the comparison between the data, FLACS simulations [15] and Ventjet modeling results. Ventjet with α_1 = 0.08 provides the best fit. Modeling set-up: *mol wt. 2, release rate 0.25 kg/s, release temperature 293.2 K, Vent diameter 0.012m, elevation 1.5m, horizontal release, ambient air temperature 283.2 K, 0 m/s wind @ 10m, F stability, surface roughness 0.18m*

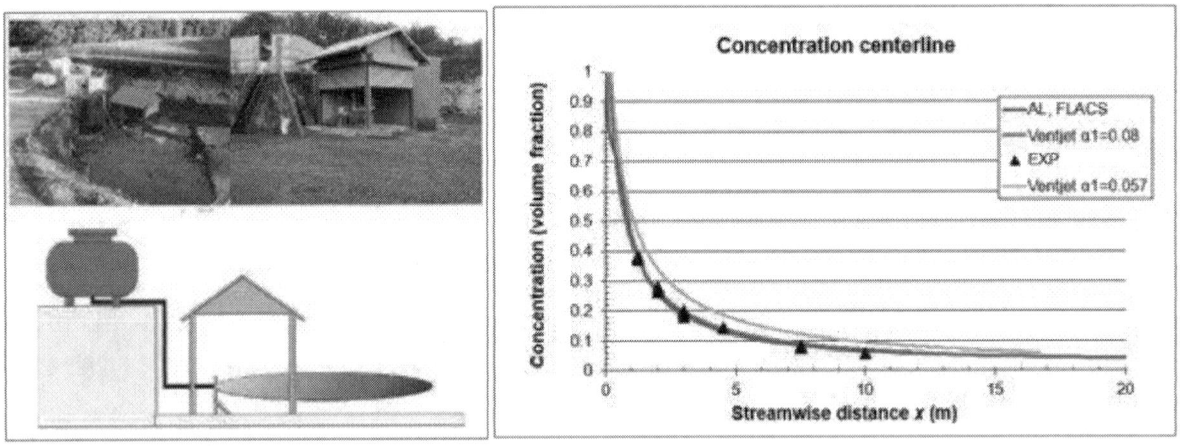

Figure 7. Left frame corresponds to the description of the experimental facility of Daubech [14, 15]: the upper picture shows the experimental rig (including the gas reservoir storage and the release point), whereas the lower picture shows a sketch of the same experimental rig. The right frame show the comparison between the experimental data, FLACS and Ventjet modeling results.

5.1.4 Roberts[16] H_2 tests

Shell and the UK HSL performed a series of high-pressure horizontal hydrogen release tests. Many were hampered by crosswind effects, but some good data was obtained where release and wind direction coincided relatively well. Weather conditions were not reported, although wind reported as "low". Release rates were calculated rather than measured directly. Concentrations were measured at the same elevation as the release and so may not accurately reflect true centerline concentrations, especially at locations further from the source. Ventjet was run against one set of data: Run #6, 119 bara, 3mm nozzle, see figure 8. Results show a quite good match, again with $\alpha_1 = 0.08$ performing best. Modeling set-up: *mol wt. 2, release rate 0.053 kg/s, release temperature 288 K, Vent diameter 3mm, elevation 1.5m, horizontal release, ambient air temperature 288 K, 1.5 m/s wind @ 10m, F stability, surface roughness 0.18m*

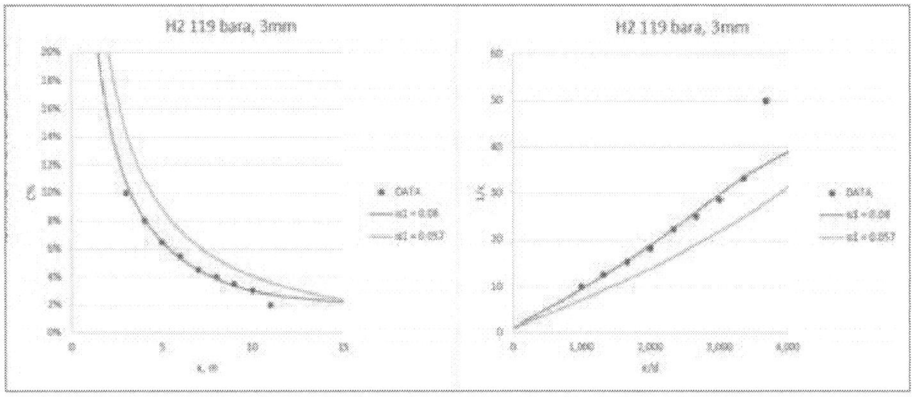

Figure 8. Comparison between the experimental data of Roberts [16] and Ventjet modeling results.

5.2 *Cross wind entrainment*

Ventjet adopts the Ooms value for $\alpha_2 = 0.5$ without modification.

5.3 *Passive entrainment*

Ventjet adopts the PHAST value for $\alpha_3 = 1$ without modification

6 Jet interaction with ground

When a gas jet passes close to a solid surface it will get drawn towards that surface. The jet interacts with the surface and this can be simply modeled assuming that the velocity difference between the jet and the ground (0 m/s) causes the pressure on the lower side of the cloud to be lower than on the upper surface of the jet, therefore drawing the jet downwards. This effect can occur, even if the jet is buoyant. Once the jet attached to the surface, if that surface is convex, the jet may stay attached to that surface, but if the surface is flat (like the ground), the downward force is not maintained and the cloud will lift off again, if buoyant. This effect is easily demonstrated using CFD codes such as FLACS.

This complicated phenomenon is modelled in a simple way in Ventjet by applying an additional downward force to the vertical momentum equation 7 equal to the Bernoulli pressure reduction $0.5\ \rho\ u_x^2$ (top hat conditions, u_x is horizontal component of cloud velocity) acting over the width of the jet in contact with the ground, 2 W_{gnd}. Once the jet centerline is at the ground, this force is set to zero and buoyancy effects may then cause the cloud to lift again. The effect only impacts the cloud trajectory if the cloud is close enough to the ground so that $W_{gnd} > 0$. Note: the approach adopted here might be suspected to somewhat "overdo" the ground effect impact on elevated, diluted clouds at some distance from the ground, as the equivalent top hat dimension gets large, although the approach is seen to be quite effective when compared to the currently available wind tunnel tests and CFD presented in this paper.

The high-pressure hydrogen horizontal tests described above, Daubech [14,15] H_2 tests, demonstrate the impact of the ground. Tests were performed at two release elevations of 1.5m and 0.75m. FLACS runs match the experimental data well and clearly demonstrate the ground effect, especially for the 0.75m elevation release. Fortunately, the data collected happens to be in the near region of the cloud, where the trajectory is still horizontal and so the centerline concentration comparison between data, FLACS and Ventjet presented above is valid.

Comparison of centerline concentrations data vs FLACS vs Ventjet. See Figure 9. Results show an excellent fit between FLACS and data. Ventjet does a good job matching the 1.5m elevation data, but overestimates entrainment a little bit for the 0.75m elevation release.

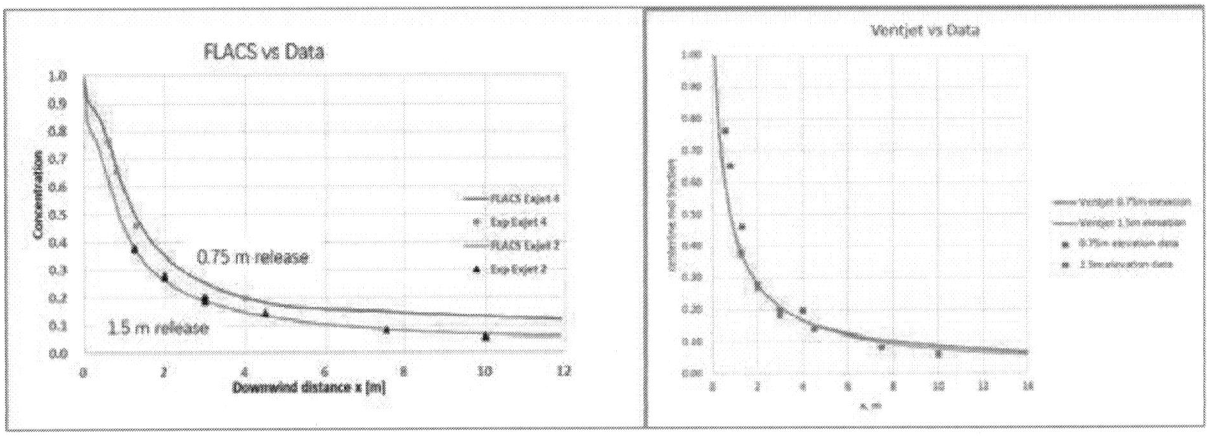

Figure 9. Comparison between the experimental data of Daubech [14,15] and FLACS & Ventjet modeling results.

Cloud profiles. See Figure 10-12. The FLACS runs show the buoyant hydrogen jet being pulled down to the ground, followed by it lifting off again. The highly simplified approach incorporated into Ventjet does a quite good job in matching the cloud profiles from FLACS. Examination of the FLACS vertical cloud slices shows the general fall and rise of the cloud and illustrates why a simple top hat model such as Ventjet cannot be expected to model such complex phenomena with a high degree of accuracy.

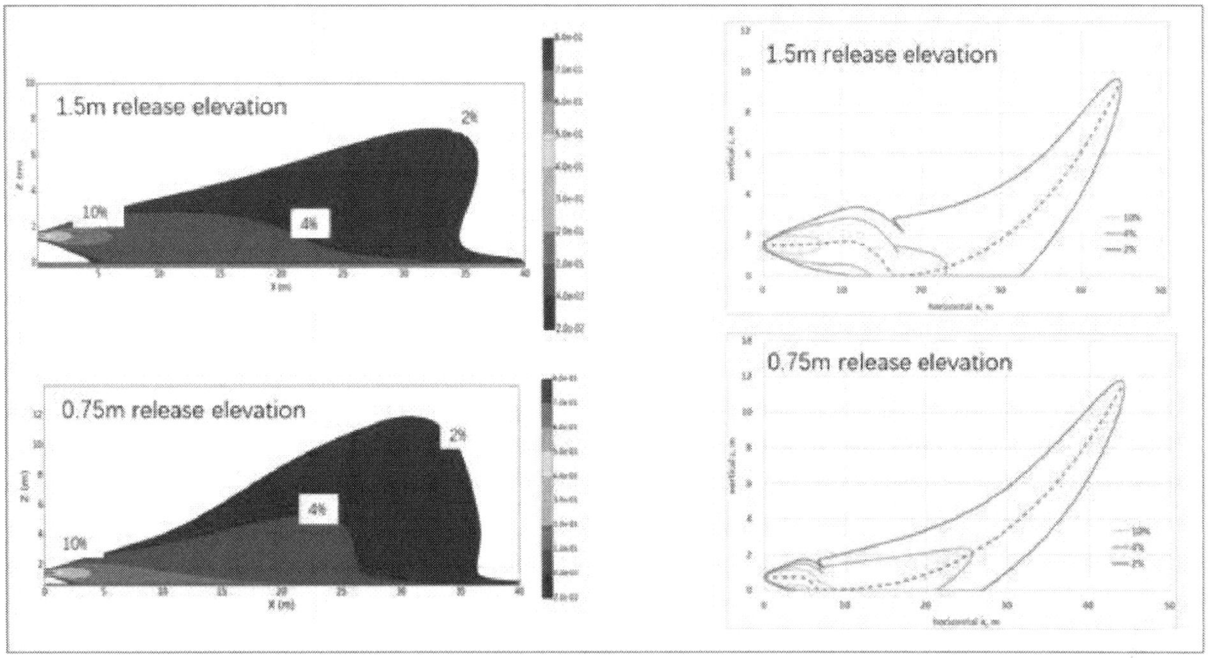

Figure 10. Comparison between FLACS & Ventjet modeling results.

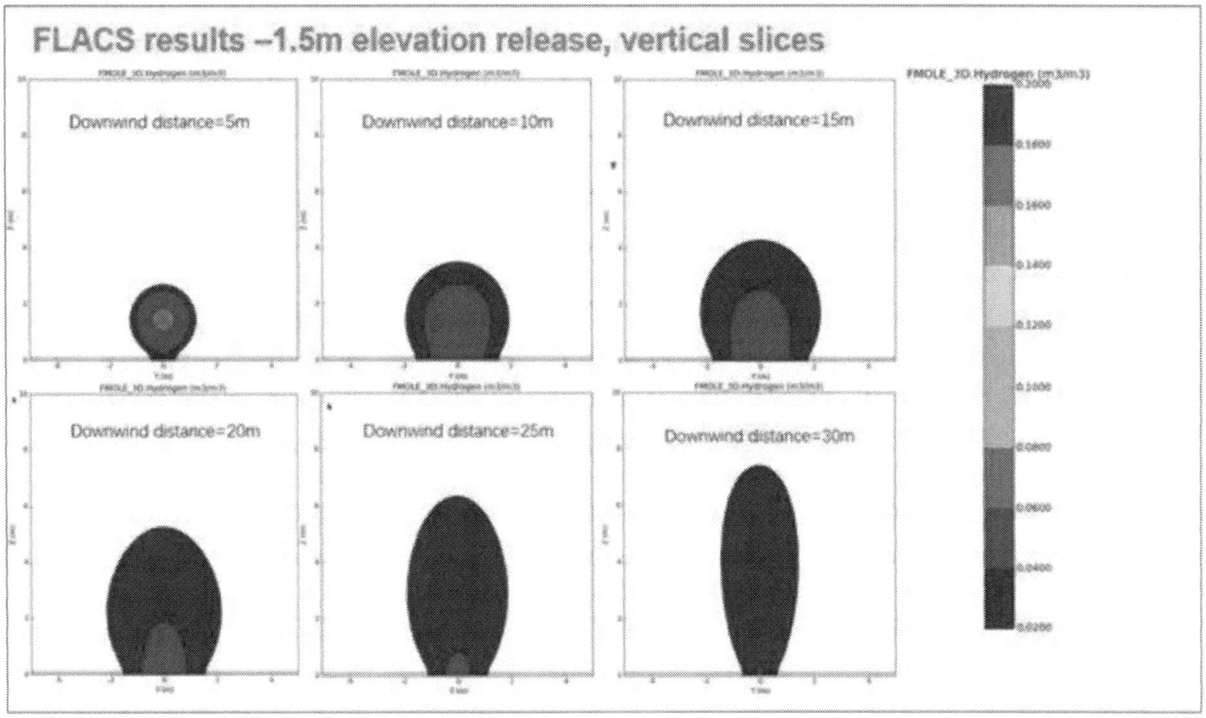

Figure 11. Concentration distributions at various distance from the source for release at 1.5m from the ground: FLACS modeling results.

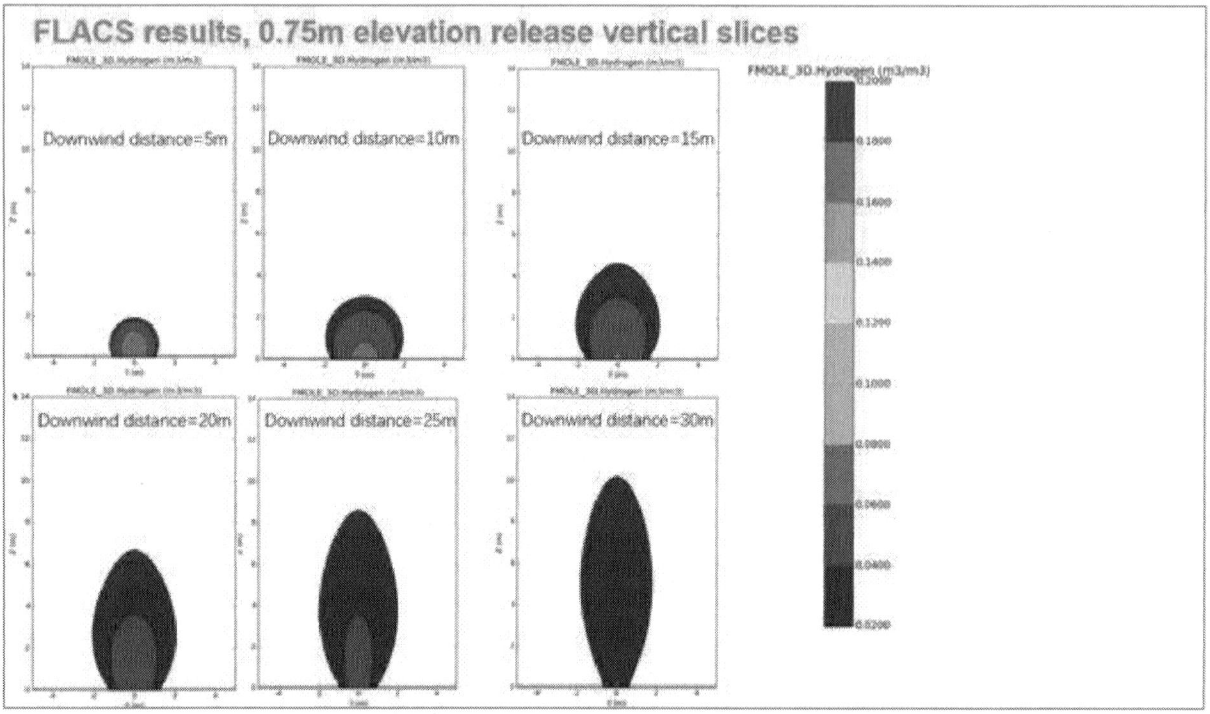

Figure 12. Concentration distributions at various distance from the source for release at 0.75m from the ground: FLACS modeling results.

Additional FLACS modeling of ambient temperature 2" sub-sonic methane horizontal releases (294K methane and air temperature, F stability 1.9 m/s, Surface roughness 0.18m) has also been performed to further test the validity of the ground effect approach for modeling the ground effect, across a range of velocities at 10 m/s and 226 m/s (Mach 0.02 and 0.5), figures 13 and 14. The impact of the ground be clearly seen and Ventjet performs quite well, predicting when the cloud will be grounded and when it will not, as well as travel distances, with acceptable accuracy, when compared to FLACS.

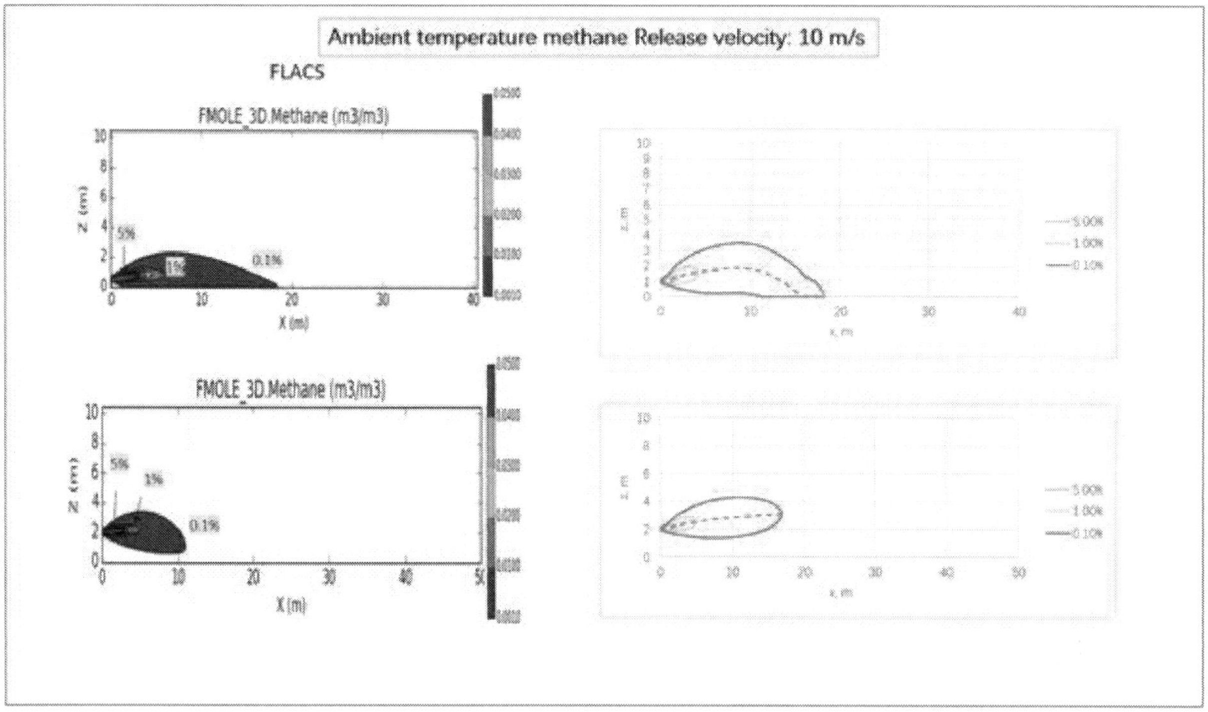

Figure 13. Concentration distributions at various distance from the source for methane release velocity of 10m/s: FLACS (left) and Ventjet (right) modeling results

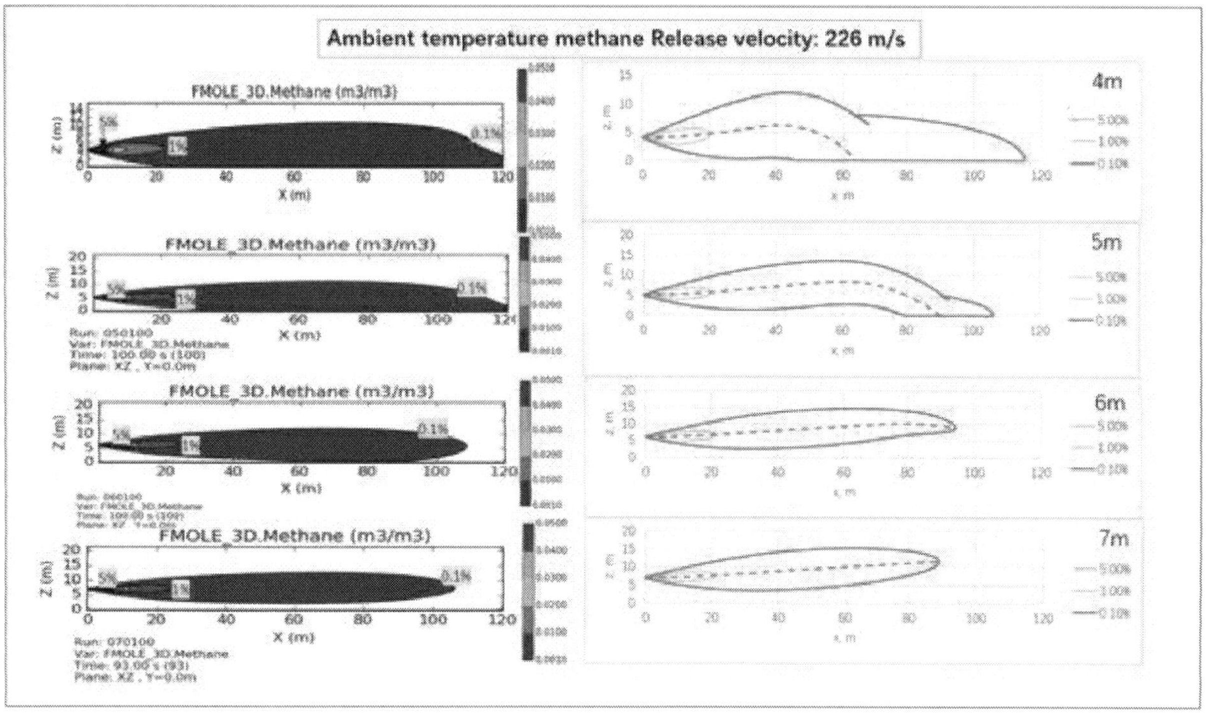

Figure 14. Concentration distributions at various distance from the source for methane release velocity of 226m/s: FLACS (left) and Ventjet (right) modeling results.

Experimental data on the ground effect is available from Desilts [17] who presents test data showing the impact of adjacent flat plates on the dispersion of hydrogen jets. Results are presented in the form of distance to LFL (4%) as a function of the distance, h, between the plates and the release nozzle. Results are compared to the results of CFD modeling. In figure 15 these results are compared against Ventjet predictions for three different test supply pressures and nozzle diameter. Results show that Ventjet performs as well as CFD and better for cases where the plate is close to the jet and the ground effect is greater. There is some offset in the distances predicted (by both CFD and Ventjet) vs data which is unexplained and the same offset can be observed in comparisons with results also included in the paper for gas jets away from any surfaces (not presented here). Overall, although these are relatively small-scale tests, they provide some validation of the approach to the ground effect developed for Ventjet.

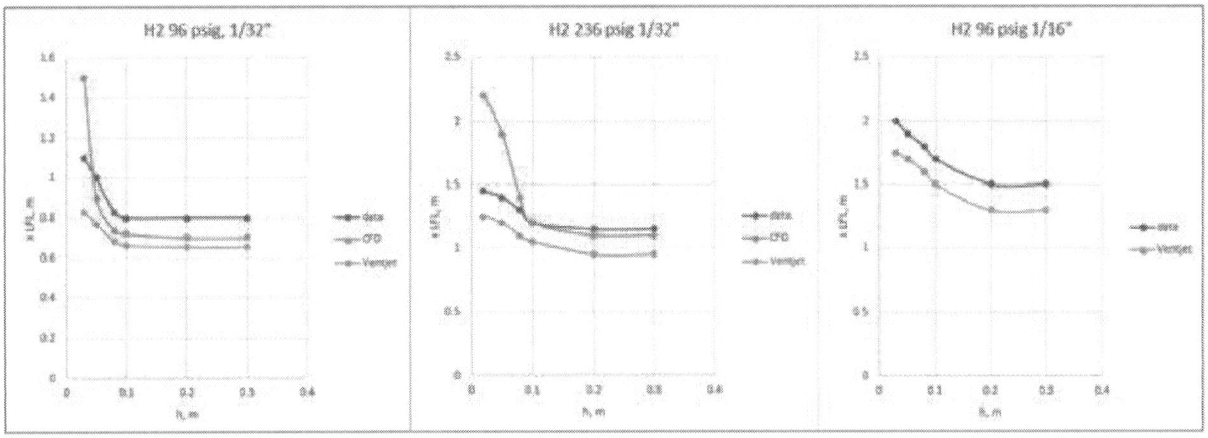

Figure 15. Concentration distributions versus distances h for hydrogen releases: experimental data and modeling results from FLACS and Ventjet.

From the above analysis it can be observed that the ground effect can affect even very buoyant gases such as hydrogen, and that these effects can be modeled with quite good accuracy by FLACS. In addition, simple modifications to engineering models like Ventjet can be quite effective in making predictions. Including the ground in the dispersion model is important when modeling releases near the ground, since the cloud will stay near the ground longer and possibly travel further overall than if this effect is not included – for example, it is standard industry practice to model horizontal releases at elevations between 1m – 3m when performing onsite and offsite Quantified Risk Assessments (QRAs).

7 Step by step method for solving dispersion equations

7.1 Set up problem

Define initial release conditions (mol wt., release rate or pressure, temperature, γ, C_d) and atmospheric conditions (temperature, wind speed at 10m, surface roughness). The choice of C_d is up to the user and depends on whether non-choked or choked flow and whether the release is from a vent pipe or an accidentally generated hole. Unless otherwise stated in the paper a simple $C_d = 1$ is used.

7.2 1.1Set a concentration of interest

Set a concentration of interest, C_{target}, the mol fraction that defines the extent of the cloud of interest.

7.3 Determine release condition

In order to establish the correct source condition for the gas dispersion, the upstream pressure must be determined. This can either be set by the user, or the program can be set up to accept a mass rate input, in which case the upstream pressure must be iterated until the correct mass rate is calculated. All the required (ideal gas) equations for non-choked and choked flow are provided in APPENDIX B. If the flow is non-choked, then then the nozzle condition is equal to the dispersion source condition. If the flow is choked, then the equations 23 and 24 are used to determine the effective source condition.

7.4 Dispersion calculations

Dispersion calculations are performed on the top-hat model equations along the cloud centerline s axis. It is convenient to set a maximum centerline distance, s_{max} and divide this up into equal lengths – 1000 steps works quite well. The program can be set up to run once through to s_{max}, or more usefully, the program runs through an automatic convergence loop until the concentration at s_{max} is equal to C_{target}.

7.5 Cloud width

At each incremental step along the centerline, the radial distance to C_{target} is determined using the Gaussian distribution Equation 15. The x, y values for the upper and lower limit of the cloud at each centerline distance are determined by this radius and the orientation of the cloud at this location, φ.

8 Model versus experimental data

8.1 Model Validation vs. wind tunnel tests

8.1.1 Shatzmann [18] wind tunnel tests

Experiments were performed in a wind tunnel, where a small-scale boundary layer was established in the 16 m long section of the tunnel. The test section was 4 m long. Roughness elements were distributed around the bottom of the boundary layer establishment section. Only neutral stability boundary layer without vertical temperature gradient were investigated. In the lower part of the boundary layer the mean velocity can be fitted by a logarithmic flow profile. Different boundary layers with different roughness and velocity were generated. For more detailed information see,

Ref. [18],[19] and [20]. Mixtures composed of Sulphur hexafluoride (SF_6), carbon dioxide (CO_2) and ethane (C_2H_6) in various proportions were used in experiments. Releases were done at various heights and angles. For some releases, the concentration was also measured close to the ground. Results were presented in the non- dimensional form, hence certain similarity requirements must be used to transfer experimental data from small wind-tunnel scale to "real world" industrial scale, see Appendix D for the detailed description. Results are compared to Ventjet and some of the results were compared to DISPAL, an in-house CFD code developed by Air Liquide for atmospheric dispersion modeling.

Shatzmann T4. Modeling set-up: *mol wt. 48.1, release rate 1.18kg/s, release temperature 291.7 K, Vent diameter 0.0477m, elevation 1.5m, vertical release, ambient air temperature 291.7 K, 13.76 m/s wind @ 10m, D stability, surface roughness 0.0111m.* Results are compared to Ventjet, PHAST 6.54, PHAST 8.4 (beta)and DISPAL, see fig 16.

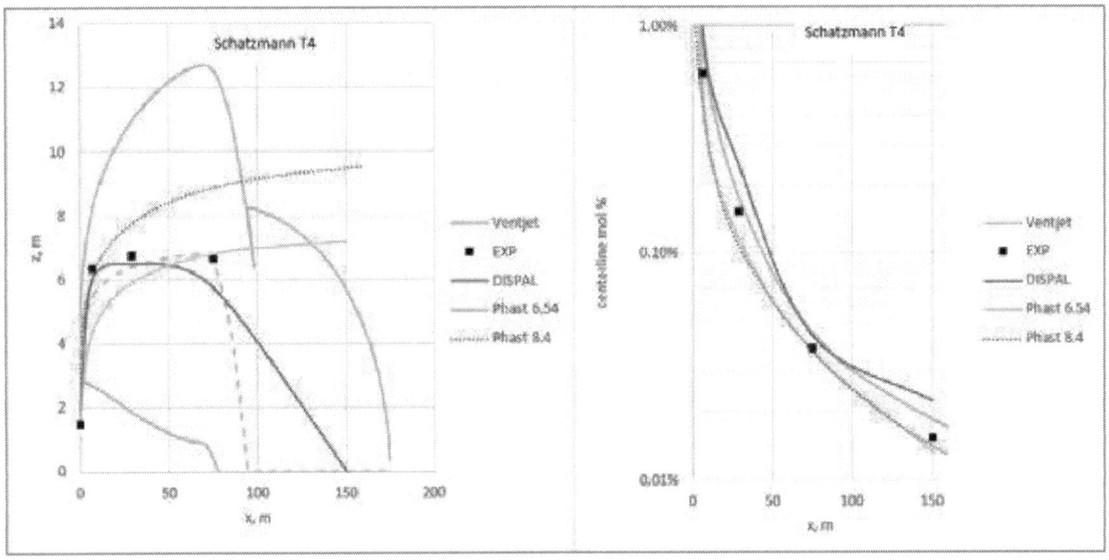

Figure 16. Comparison between the experimental data of Shatzmann [18] T4, DISPAL (an in-house CFD code of Air Liquide [21]), PHAST 6.54, PHAST 8.4 (beta) and Ventjet modeling results: the left frame corresponds to the jet trajectory, the right frame corresponds to the concentration decay on the centerline versus the distance from the release.

Shatzmann T10. Modeling set-up: *mol wt. 66.6, release rate 58kg/s, release temperature 291.7 K, Vent diameter 1.27m, elevation 8.5m, vertical release, ambient air temperature 291.7 K, 6.7 m/s wind @ 10m, D stability, surface roughness 0.037m. The concentration was also measured at 1m above the ground.* Results are compared to Ventjet, PHAST 6.54, PHAST 8.4 (beta) and DISPAL, see fig 17. This comparison shows a good agreement between experimental data, Ventjet and DISPAL for the jet trajectory and for the concentration decay along the centerline. However, PHAST 6.54 demonstrates a different trajectory. The concentration close to the ground is in reasonable agreement between DISPAL, Ventjet and experimental measurements, whereas the PHAST 6.54 underestimates it by a factor of three. PHAST 8.4 (beta) does better predicting the trajectory, but still overpredicts dilution as the cloud approaches the ground.

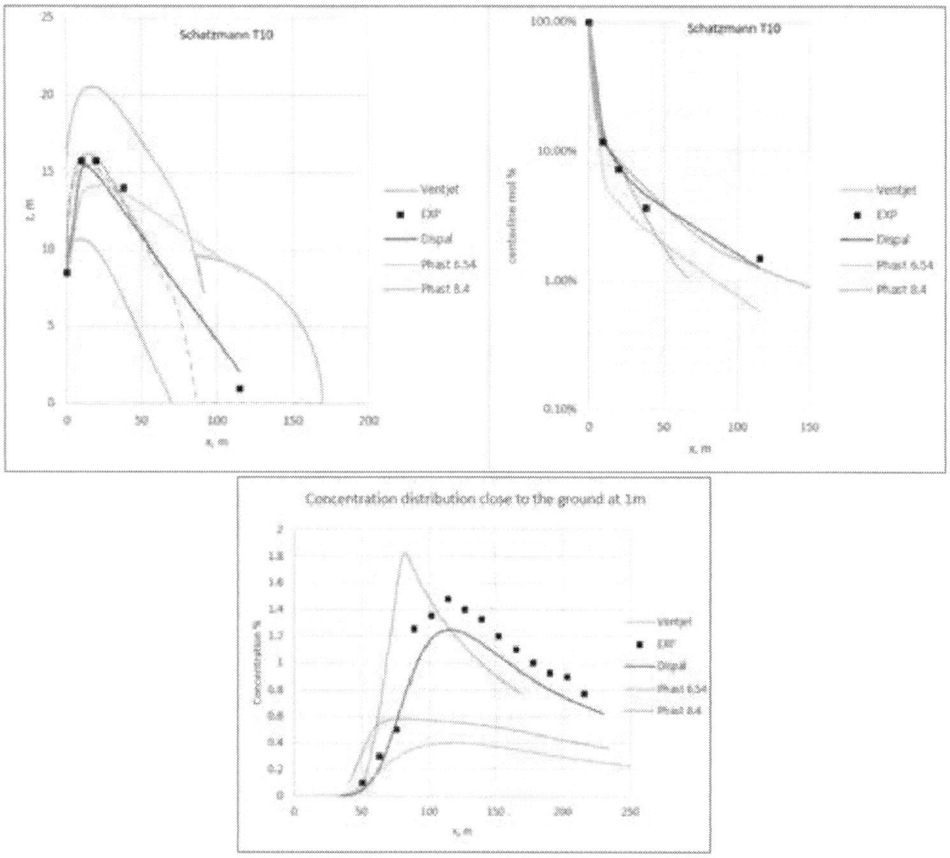

Figure 17. Comparison between the experimental data of Shatzmann [18] T10, DISPAL (an in-house CFD code of Air Liquide [21], PHAST 6.54, PHAST 8.4 (beta) and Ventjet modeling results : on the top - the left frame corresponds to the jet trajectory, the right frame corresponds to the concentration decay on the centerline versus the distance from the release, on the bottom the frame corresponds to the concentration distribution around the ground at 1m.

Shatzmann T11. Modeling set-up: *mol wt. 66.6, release rate 58kg/s, release temperature 291.7 K, Vent diameter 1.27m, elevation 8.5m, vertical release, ambient air temperature 291.7 K, 13.4 m/s wind @ 10m, D stability, surface roughness 0.037m.* Results are compared to Ventjet, PHAST 6.54 and DISPAL, see fig 18. This comparison shows a reasonable agreement between experimental data, Ventjet and DISPAL for the jet trajectory and for the concentration decay along the centerline. However, PHAST 6.54 demonstrates a different trajectory that does not reach the ground. PHAST 8.4 (beta) performs better than PHAST 6.54 but still fails to predict that the cloud reaches the ground.

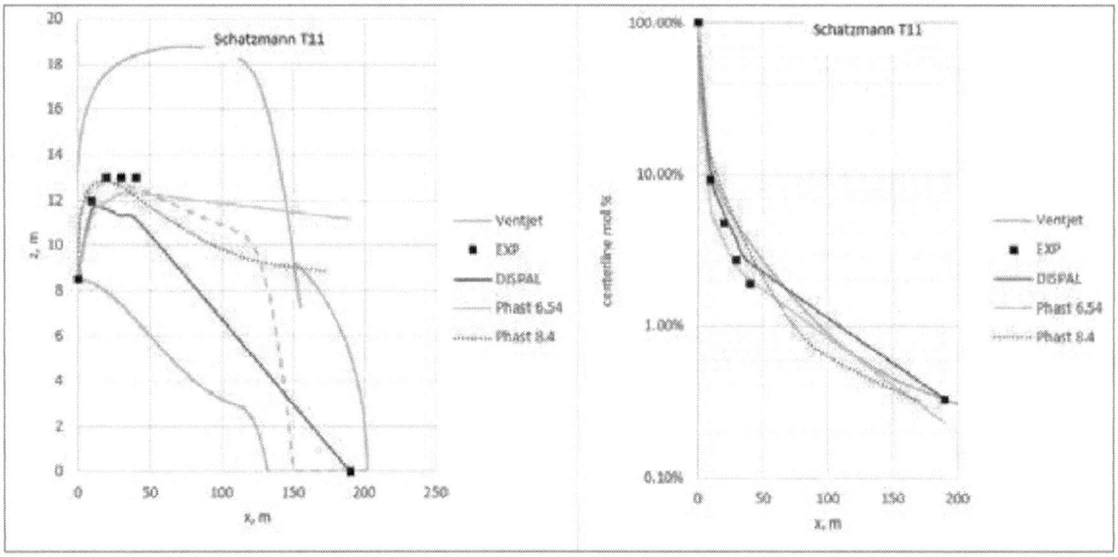

Figure 18. Comparison between the experimental data of Shatzmann [18] T11, DISPAL (an in-house CFD code of Air Liquide [21], PHAST 6.54, PHAST 8.4 (beta) and Ventjet modeling results: the left frame corresponds to the jet trajectory, the right frame corresponds to the concentration decay on the centerline versus the distance from the release.

Shatzmann T13. Modeling set-up: *mol wt. 141.4, release rate 212.4kg/s, release temperature 291.7 K, Vent diameter 1.27m, elevation 8.5m, vertical release, ambient air temperature 291.7 K, 11.5 m/s wind @ 10m, D stability, surface roughness 0.037m.* Results are compared to Ventjet, PHAST 6.54, PHAST 8.4 (beta) and DISPAL, see fig 19. This comparison shows a good agreement between experimental data, Ventjet and DISPAL for the jet trajectory and for the concentration decay along the centerline. However, PHAST 6.54 demonstrates significant differences in the trajectory, compared to the experimental investigation. PHAST 8.4 (beta) performs better than PHAST 6.54 but still fails to predict that the cloud touches the ground.

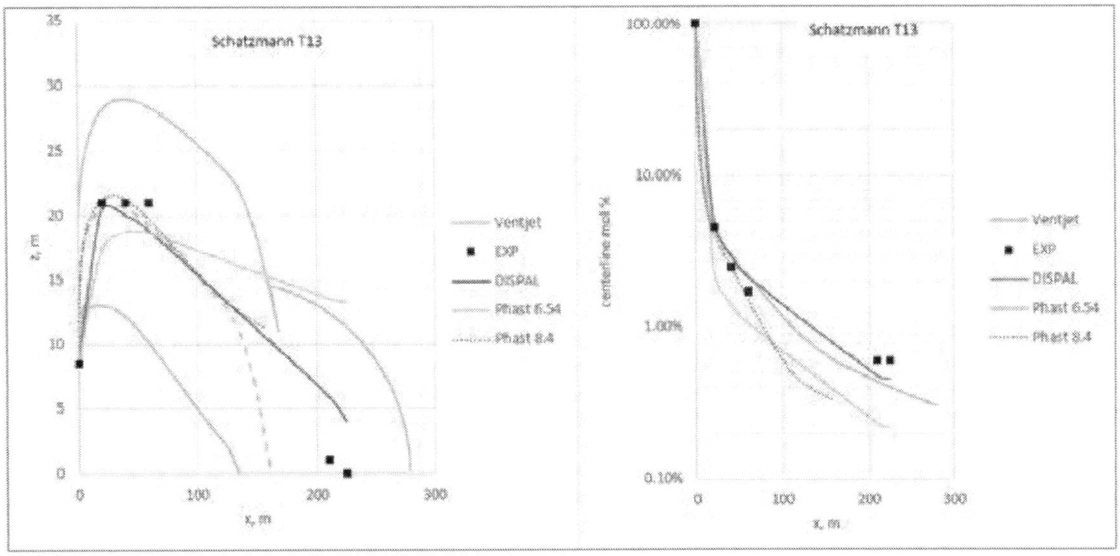

Figure 19. Comparison between the experimental data of Shatzmann [18] T13, DISPAL (an in-house CFD code of Air Liquide [21], PHAST 6.54, PHAST 8.4 (beta) and Ventjet modeling results: the left frame corresponds to the jet trajectory, the right frame corresponds to the concentration decay on the centerline versus the distance from the release.

8.1.2 Donat [19] wind tunnel tests.

Experiments of Donat [19, 20] were performed in the same wind tunnel as Shatzmann, however for different experimental configurations (release velocities, ambient velocities, stack height and diameter, angle of the release and roughness). For the upscaling up to the industrial scale the same approach is followed as for Shatzmann experiment (see appendix D). *The concentration was also measured at 0,58m above the ground.* Results are compared to Ventjet and some of the results were compared to DISPAL, an in-house CFD code developed by Air Liquide for atmospheric dispersion modeling.

Donat #1. Modeling set-up: *mol wt. 45.2, release rate 21.7 kg/s, release temperature 291.7 K, Vent diameter 0.58m, elevation 7m, vertical release, ambient air temperature 291.7 K, 8.88 m/s wind @ 10m, D stability, surface roughness 0.008m.* Results are compared to Ventjet, PHAST 6.54, PHAST 8.4 (beta) and DISPAL, see fig 20. This comparison shows a reasonable agreement between experimental data Ventjet and DISPAL for the jet trajectory and for the concentration decay along the centerline. However, PHAST 6.54 demonstrates a different trajectory. The concentration close to the ground is in reasonable agreement between DISPAL, Ventjet and the experiment, whereas PHAST 6.54 underestimates the concentration. PHAST 8.4 (beta) performs better than PHAST 6.54 but still fails to predict that the cloud reaches the ground.

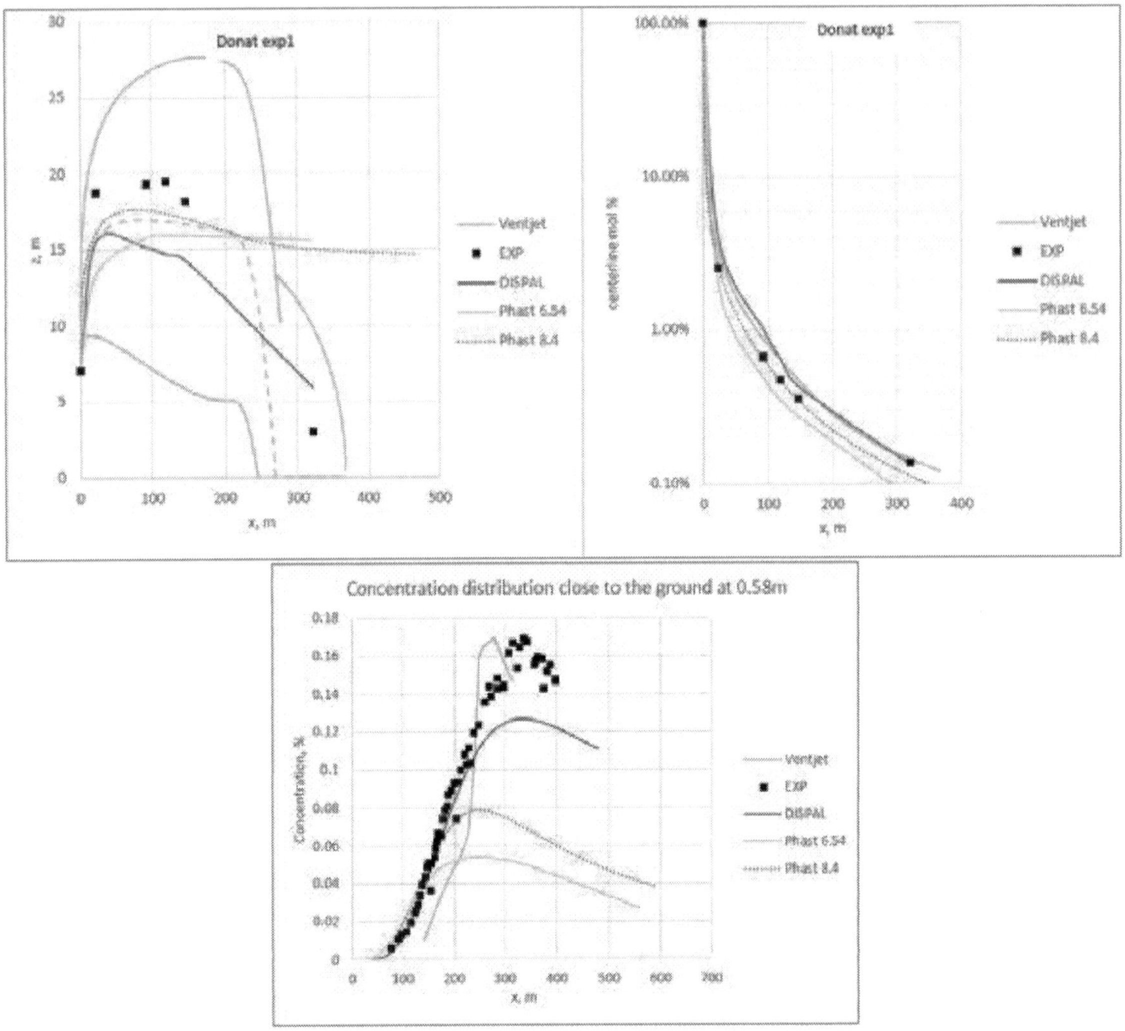

Figure 20. Comparison between the experimental data of Donat [19] #1, DISPAL (an in-house CFD code of Air Liquide [21], PHAST 6.54, PHAST 8.4 (beta) and Ventjet modeling results: results: on the top - the left frame corresponds to the jet trajectory, the right frame corresponds to the concentration decay on the centerline versus the distance from the release, the bottom frame corresponds to the concentration distribution around the ground at 0.58m.

Donat # 3: Modeling set up: *Mol wt. 66.6, release rate 8.17 kg/s, release temperature 291.7K, Vent diameter 0,58m, elevation 3.9m, vertical release, ambient air temperature 291.7K, 10.15m/s wind @ 10m, D stability, surface roughness 0.008m.* Results are compared to PHAST 8.4 (beta) and Ventjet, see fig 21. This comparison shows a good agreement between experimental data and Ventjet for the jet trajectory and for the concentration decay along the centerline. PHAST 8.4 (beta) performs well in the near field but fails to predict that the cloud reaches the ground.

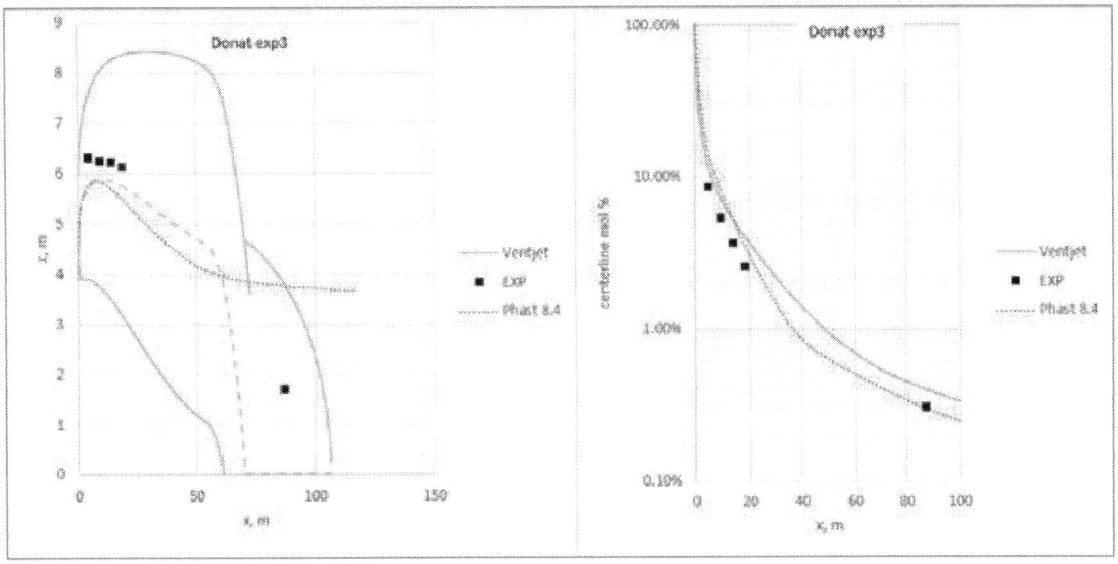

Figure 21. Comparison between the experimental data of Donat [19] #3, PHAST 8.4 (beta) and Ventjet modeling results: the left frame corresponds to the jet trajectory; the right frame corresponds to the concentration decay on the centerline versus the distance from the release.

Donat # 4: Modeling setup: *Mol wt. 66.6, release rate 8.17 kg/s, release temperature 291.7K, Vent diameter 0,58m, elevation 3.9m, vertical release, ambient air temperature 291.7K, 11.5m/s wind @ 10m, D stability, surface roughness 0.19m.* Results are compared to PHAST 8.4 (beta) and Ventjet, see fig 22. This comparison shows a reasonable agreement between experimental data, PHAST 8.4 (beta) and Ventjet for the jet trajectory and a good agreement for the concentration decay along the centerline.

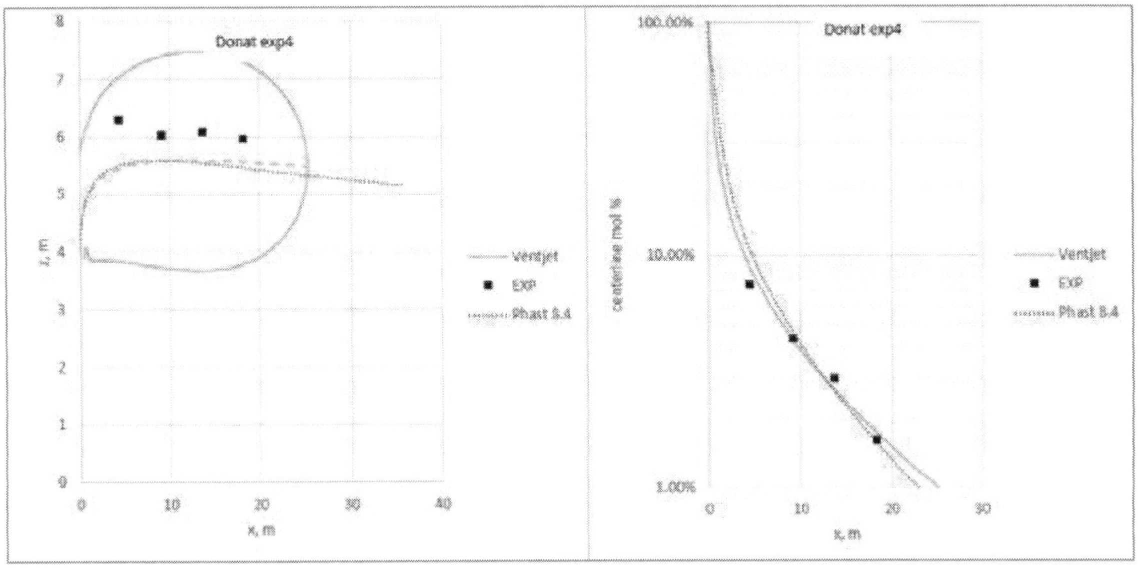

Figure 22. Comparison between the experimental data of Donat [19] #4, PHAST 8.4 (beta) and Ventjet modeling results: the left frame corresponds to the jet trajectory; the right frame corresponds to the concentration decay on the centerline versus the distance from the release.

Donat # 5: Modeling set up: *Mol wt. 141.4, release rate 29.9 kg/s, release temperature 291.7K, Vent diameter 0,58m, elevation 3.9m, vertical release, ambient air temperature 291.7K, 8.75m/s wind @ 10m, D stability, surface roughness 0.008m.* Results are compared to PHAST 8.4 (beta) and Ventjet, see fig 23. This comparison shows a reasonable agreement between the experimental data and Ventjet for the jet trajectory and a good agreement for the concentration decay along the centerline. PHAST 8.4 (beta) performs quite well in the near field, but fails to predict that the cloud reaches the ground.

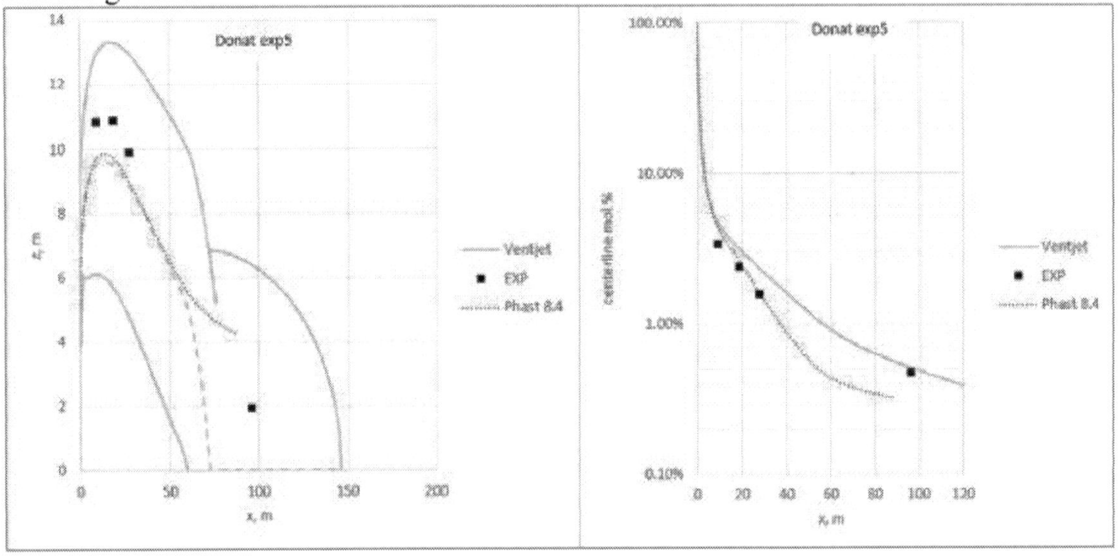

Figure 23. Comparison between the experimental data of Donat [19] #5, PHAST 8.4 (beta) and Ventjet modeling results: the left frame corresponds to the jet trajectory; the right frame corresponds to the concentration decay on the centerline versus the distance from the release.

Donat #10 (45 degree release) Modeling set-up: *mol wt. 66.6, release rate 45.4 kg/s, release temperature 291.7 K, Vent diameter 0.8m, elevation 12m, 45 degree release, ambient air temperature 291.7 K, 15.71 m/s wind @ 10m, D stability, surface roughness 0.07m.* Results are compared to PHAST 8.4 (beta) and Ventjet, see fig 24. This comparison shows a good agreement between experimental data, Ventjet and DISPAL for the jet trajectory and for the concentration decay along the centerline. PHAST 8.4 (beta) fails to predict that the cloud reaches the ground.

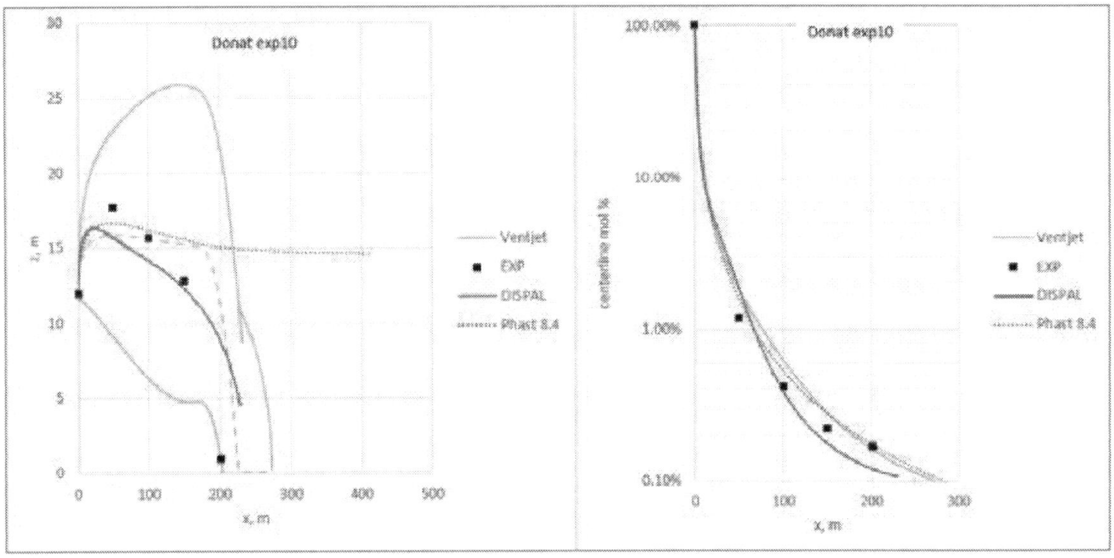

Figure 24. Comparison between the experimental data of Donat [19] #10, DISPAL (an in-house CFD code of Air Liquide [21] PHAST 8.4 (beta) and Ventjet modeling results: the left frame corresponds to the jet trajectory; the right frame corresponds to the concentration decay on the centerline versus the distance from the release.

Donat # 11 : Modeling set up: *Mol wt. 66.6, release rate 13.1 kg/s, release temperature 291.7 K, Vent diameter 0,58m, elevation 8.7m, vertical release, ambient air temperature 291.7K, 9.22m/s wind @ 10m, D stability, surface roughness 0.07m.* Results are compared to PHAST 8.4 (beta) and Ventjet, see fig 25. This comparison shows a good agreement between experimental data and Ventjet for the jet trajectory and for the concentration decay along the centerline. PHAST 8.4 (beta) performs equally well, perhaps better than Ventjet in this particular case.

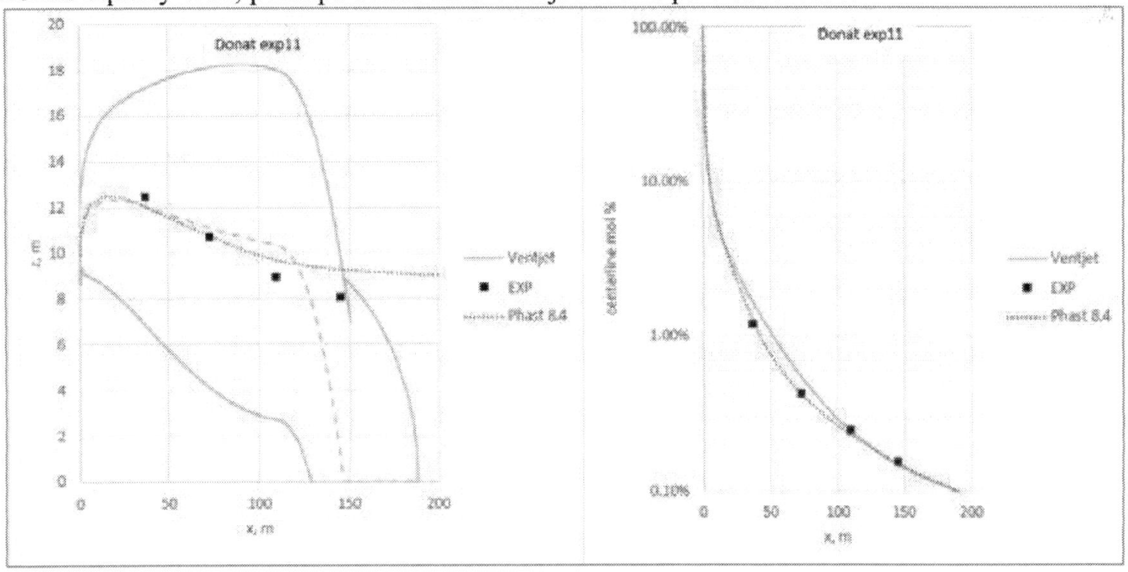

Figure 25. Comparison between the experimental data of Donat [19] #11, PHAST 8.4 (beta) and Ventjet modeling results: the left frame corresponds to the jet

trajectory; the right frame corresponds to the concentration decay on the centerline versus the distance from the release.

Donat # 19: Modeling setup: *Mol wt. 66.6, release rate 177.31 kg/s, release temperature K, Vent diameter 0,58m, elevation 17.4m, horizontal release, ambient air temperature 291.7K, 14.73m/s wind @ 10m, D stability, surface roughness 0.07m.* Results are compared to PHAST 8.4 (beta) and Ventjet, see fig 26. This comparison shows a reasonable agreement with experimental data of Ventjet for the jet trajectory and a close agreement for the concentration decay along the centerline. PHAST 8.4 (beta) performs well in near field but fails to predict that the cloud touches the ground.

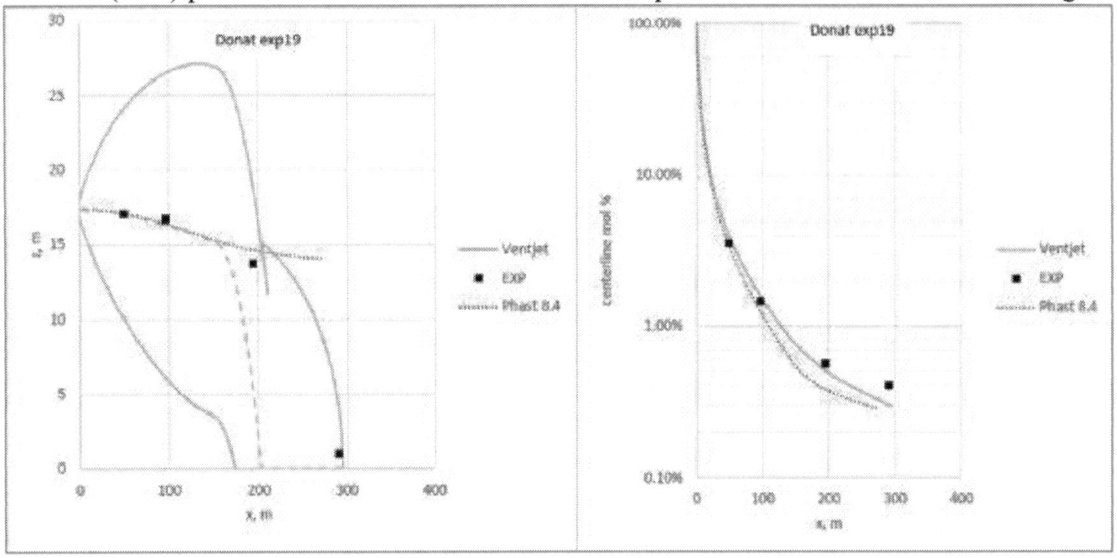

Figure 26. Comparison between the experimental data of Donat [19] #19, PHAST 8.4 (beta) and Ventjet modeling results: the left frame corresponds to the jet trajectory; the right frame corresponds to the concentration decay on the centerline versus the distance from the release.

Donat # 21. Modeling set-up: *mol wt. 66.6, release rate 31.8 kg/s, release temperature 291.7 K, Vent diameter 0.3m, elevation 9m, vertical release, ambient air temperature 291.7 K, 11.15 m/s wind @ 10m, D stability, surface roughness 0.07m.* Results are compared to PHAST 8.4 (beta) and Ventjet, see fig 27. This comparison shows a close agreement with the experimental data, Ventjet and DISPAL results for the jet trajectory and the concentration decay along the centerline. PHAST 8.4 (beta) performs less well than Ventjet, but still acceptable.

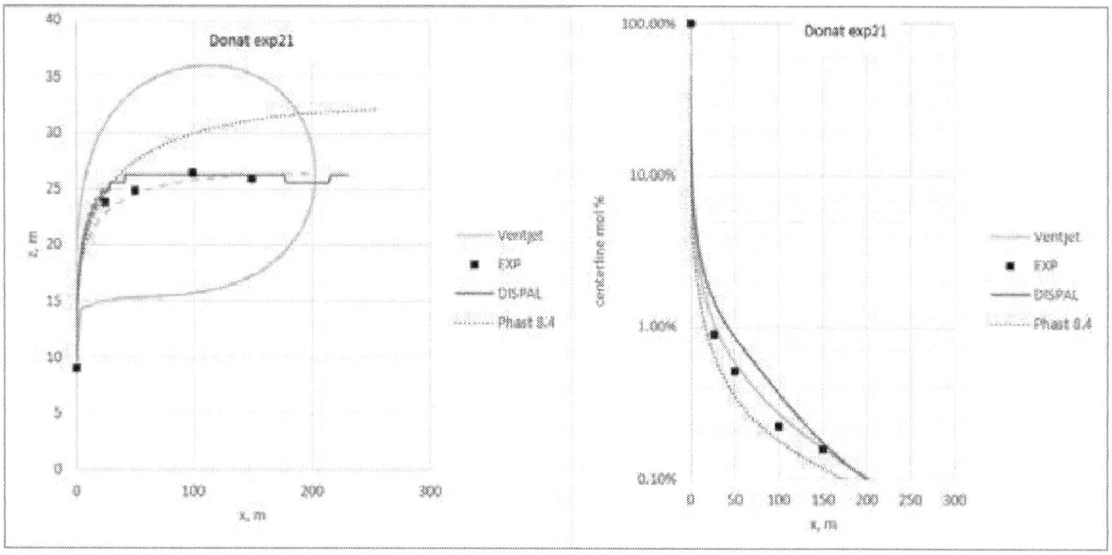

Figure 27. Comparison between the experimental data of Donat [19] #21, DISPAL (an in-house CFD code of Air Liquide [21], PHAST 8.4 (beta) and Ventjet modeling results: the left frame corresponds to the jet trajectory; the right frame corresponds to the concentration decay on the centerline versus the distance from the release.

Donat # 44: Modeling set up*: Mol wt. 66.6, release rate 52.49 kg/s, release temperature K, Vent diameter 0,58m, elevation 8.7m, horizontal release, ambient air temperature 291.7K, 18.29m/s wind @ 10m, D stability, surface roughness 0.008m.* Results are compared to PHAST 8.4 (beta) and Ventjet, see fig 28. This comparison shows a quit good agreement between Ventjet and experimental data for the jet trajectory and the concentration decay along the centerline. PHAST 8.4 (beta) performs well in near field but fails to predict that the cloud touches the ground.

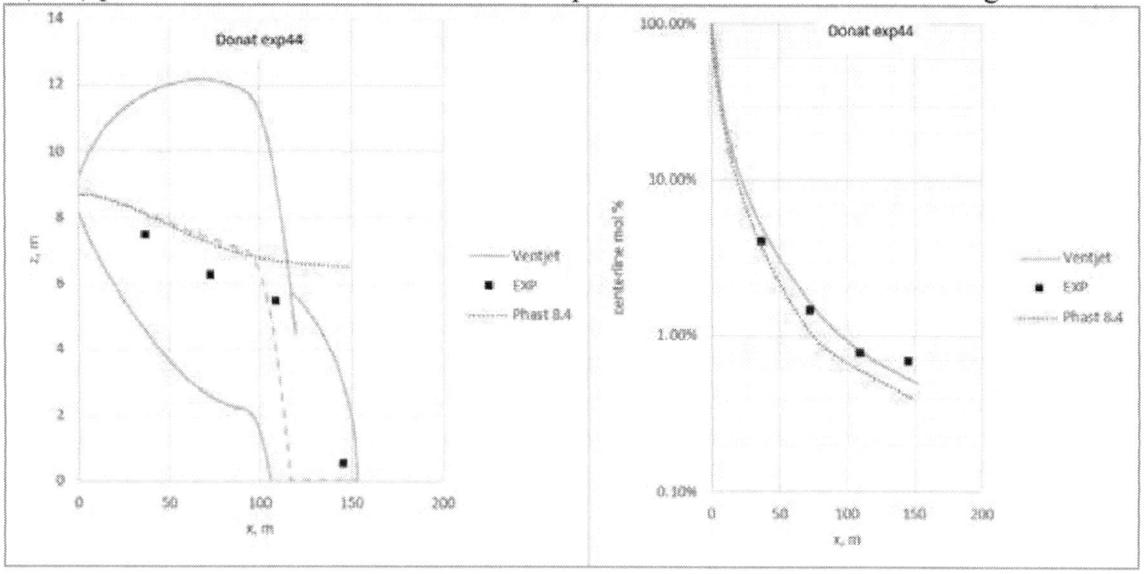

Figure 28. Comparison between the experimental data of Donat [19] #44, PHAST 8.4 (beta) and Ventjet modeling results: the left frame corresponds to the jet

trajectory; the right frame corresponds to the concentration decay on the centerline versus the distance from the release.

8.1.3 Vidali [22] wind tunnel tests.

For these experiments, the large wind-tunnel of the Laboratoire de Mecanique de Fluide et de Acoustique of the Ecole Centrale de Lyon (France) was used. The size of the working sections is 14 m, the width is 3.7 m. The ceiling is adjustable to better control longitudinal pressure gradients. Experiments were done for a turbulent neutral boundary layer. The release was done via a metallic pipe of internal diameter 0.012 m and 0.075 m height. A mix of carbon dioxide (CO_2), air and ethane (C_2H_6) is used in the experiment. Concentration measurements are performed with a Flame Ionization Detector (FID). Results are compared to PHAST 8.4 (beta) and Ventjet, see fig 29. Ventjet performs quite well, PHAST 8.4 (beta) performs quite well in near field but fails to predict that the cloud touches the ground.
92.5% CO_2, 6% air, 1.5% C_2H_6 release. *Mol wt. 42.94, release rate 47.85 kg/s, release temperature 293K, Vent diameter 1.2m, elevation 7.6m, vertical release, ambient air temperature 293.2 K, 10.73 m/s wind @ 10m, D stability, surface roughness 0.0113m.*

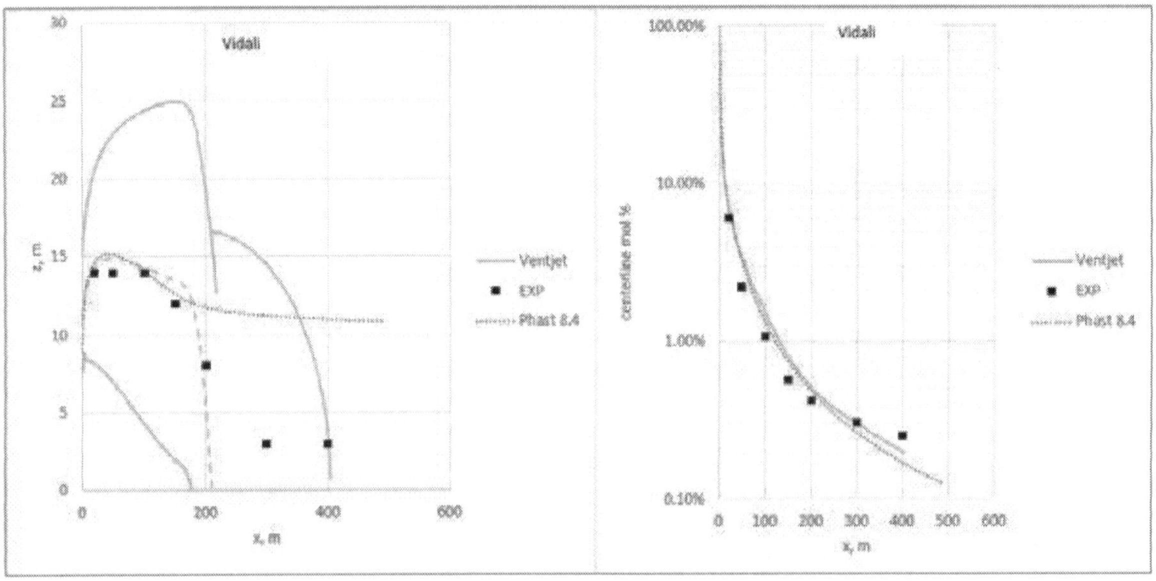

Figure 29. Top frame corresponds to the experimental facility of Vidali [22]. Comparison between the experimental data of Vidali [22] PHAST 8.4 (beta) and Ventjet modeling results: the left bottom frame corresponds to the jet trajectory; the right bottom frame corresponds to the concentration decay on the centerline versus the distance from the release.

8.2 Model validation vs. field experiments

8.2.1 CRIGEN field tests [23,24].

Middle-scale experiments were performed in field conditions on the site close to the LNG terminal of Montoir de Bretagne, France, see fig 30. For the dispersion LNG vapor releases of temperatures between -110°C and -150°C and controlled flow rate from 2000 to 4000 kg/h. Experiments were performed in May-June 2013 for stable and neutral atmospheric conditions. The concentration of LNG vapor and O_2 was measured using probes (see fig 30) attached to masts as well as the associated temperature, measured by infrared camera. In most tests the vapors of LNG containing 94.82% of CH_4, 5% C_2H_6, 0.14% C_3H_8 and 0.04% N_2 was used. Results are compared to PHAST 8.4 (beta) and Ventjet, see fig 31 and 32. PHAST 8.4 (beta) and Ventjet demonstrate a good agreement with the experiment for the concentration decay for both releases (centerline trajectory data not available)

Figure 30. Experimental facility of CRIGEN[23,24].

CRIGEN horizontal 6" release: Modeling set up*: Mol wt. 16.79, release rate 0.568 kg/s, release temperature 150.95K, Vent diameter 0.1524m, elevation 5.65m, horizontal release, ambient air temperature ~283K, 1.55m/s wind @ 10m, D stability, surface roughness 0.05m*

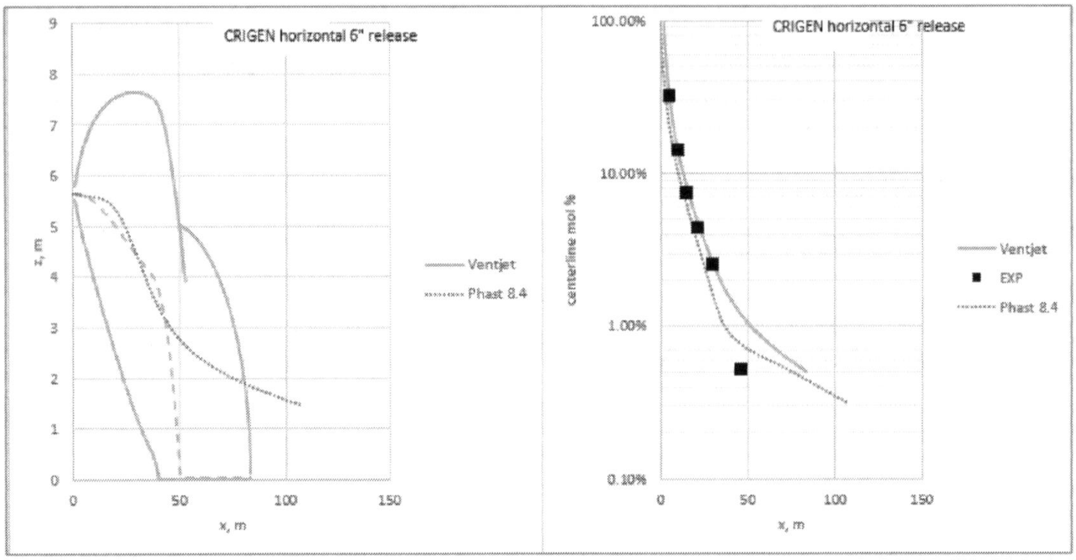

Figure 31. Comparison between the experimental data of CRIGEN[23,24] PHAST 8.4 (beta) and Ventjet modeling results for 6" horizontal release: the left frame corresponds to the jet trajectory; the right frame corresponds to the concentration decay on the centerline versus the distance from the release.

CRIGEN vertical 6" release: Modeling set up: *Mol wt. 16.79, release rate 0,433 kg/s, release temperature 150.95K, Vent diameter 0.1524m, elevation 4.65m, vertical release, ambient air temperature ~281K, 2.36m/s wind @ 10m, D stability, surface roughness 0.01m*

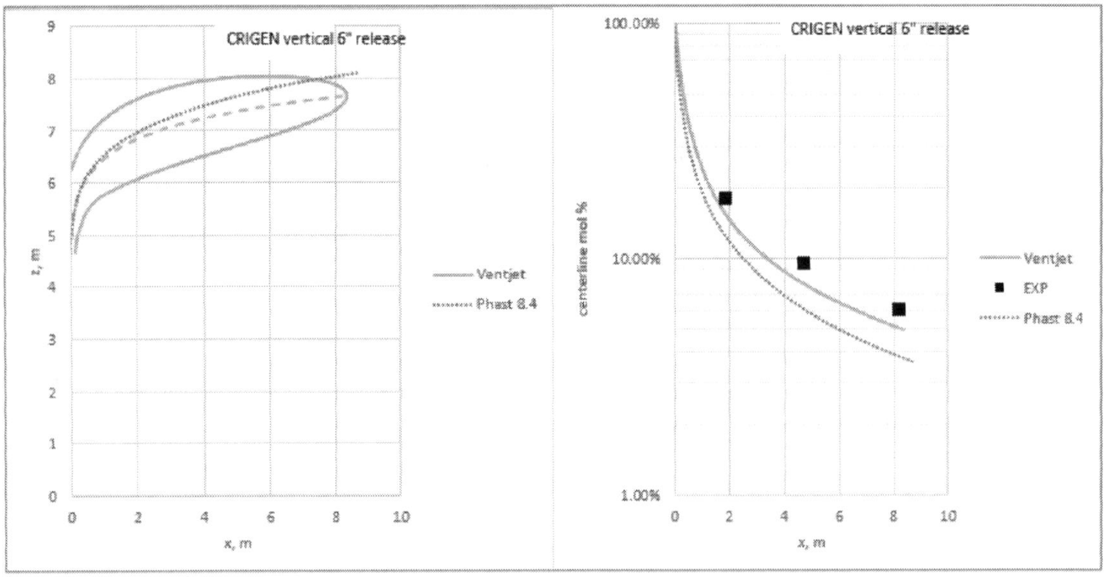

Figure 32. Comparison between the experimental data of CRIGEN[23,24] and Ventjet modeling results for 6" vertical release: the left frame corresponds to the jet trajectory; the right frame corresponds to the concentration decay on the centerline versus the distance from the release.

8.2.2 Kit Fox Experiment [25-28].

Middle-scale experiments were performed in field conditions for the dispersion of CO_2 on a flat desert land covered by three types of artificial roughness elements, see fig 15.

Figure 33. Experimental facility of Kit Fox[25-28].

In this experiment, three different types of artificial roughness elements were used (see Fig. 33 and 34):

- The inner region is a 39m x 85m ERP (Equivalent Roughness Pattern) area, corresponding to a matrix of obstacles 2.4 m high and 2.4 m large, 6.1 m apart in lateral direction and 8.5 m in longitudinal direction. It is suggested to use an equivalent roughness = 0.12-0.24 m for dispersion modeling;
- The inner region of 120m x 314m corresponds to URA area (Uniform Roughness Pattern), corresponding to obstacles 0.2 m high and 0.8 m large, 2.4 m apart in lateral direction and 2.4 m in longitudinal direction. It is suggested to use an equivalent roughness = 0.01-0.02 m for dispersion modeling;
- The outer region is a desert area with a suggested equivalent roughness of 0.0002 m.

Instantaneous concentration sensors are located on four arrays at distances of 25m, 50m, 100m and 225m downwards from the source 0.5m above the ground. CO_2 was released from a 1.5m x 1.5m approximately 1m above ground level located in the ERP area.

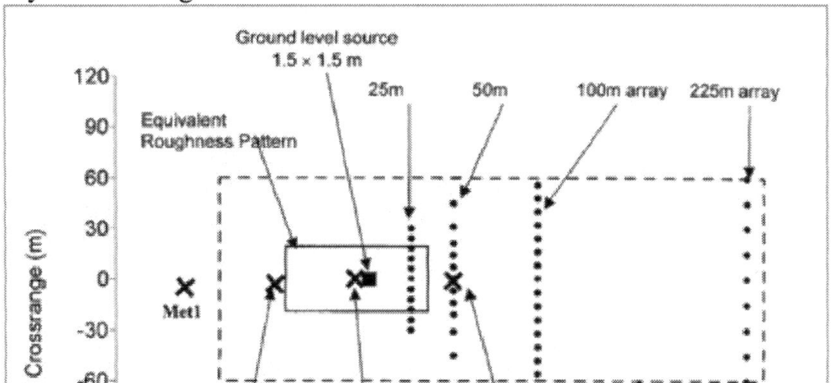

Figure 34. Scheme of the experimental facility of Kit Fox[25-28].

In the chosen experiment, the flow rate of the CO_2 release was set to 3.77 kg/s, the release duration is 180s, the release temperature was 296.76K, see reference [28]

In this experiment the following atmospheric conditions were observed:

- The wind direction was almost parallel downrange direction (fig. 34)
- The atmosphere during the experiment was stable, corresponding to the F 5.3 Pasquill stability class (the velocity 5.3m/s was measured at 10 m height)
- Temperature at the ground level estimated as T = 301.77K
- Temperature at 10m above the ground T = 306.5 K
- Monin-Obukhov length was assumed to be 8 m, see reference [28].

Results are compared to Ventjet modeling, see fig 35. Ventjet results are in close agreement with the experimental measurements. It is important to note that this is a complex test to model. Ventjet has been modified for the purposes of this analysis to have two zones of different roughness and to allow the specification of a fixed Monin-Obukov length L_{mo} (see appendix A), with the specified value of 8m, is based on the upstream roughness of 0.0002m, as provided in [28]. L_{mo} impacts passive entrainment. Figures 35 plot shows that with these modifications, Ventjet performs well. Ventjet can also be run, allowing L_{mo} to vary, according to local roughness and by using a fixed surface roughness, as shown in the Figure 36, which shows the impact of the surface roughness on the dispersion. These Ventjet results suggest that passive entrainment is based more on upwind roughness rather than local roughness, whereas wind speed near the ground is more based on surface roughness at the specific location in the cloud. This finding can help in understanding of gas dispersion in general and highlights the importance of correct surface roughness selection. It is interesting that Ventjet performs well, given that this is an initially dense gas on the ground, which suggests that the cloud is acting more like a neutrally buoyant cloud than a true dense gas, at least after initial dilution.

Modeling set up: *Mol wt. 44.01, release rate 3,77 kg/s, release temperature 296.76K, Vent diameter 1.69m, elevation 1m, vertical release, ambient air temperature 306.5K, 5.3m/s wind @ 10m, F stability L_{mo}=8m, inner surface roughness 0.12m, outer surface roughness 0.01m, inner - outer transition distance 85m.*

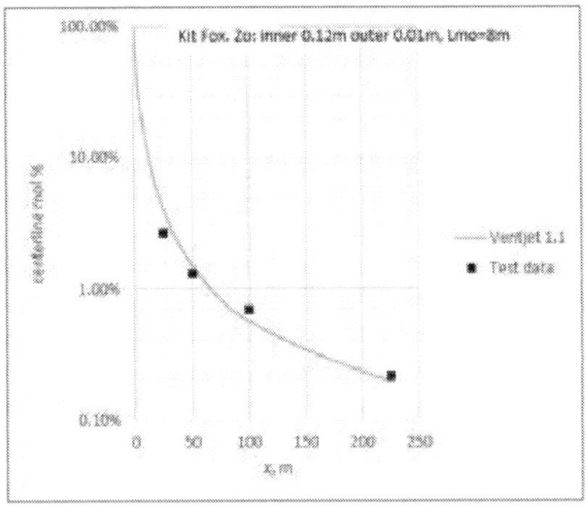

Figure 35. Comparison between the experimental data of Kit Fox[25-28] and Ventjet modeling results

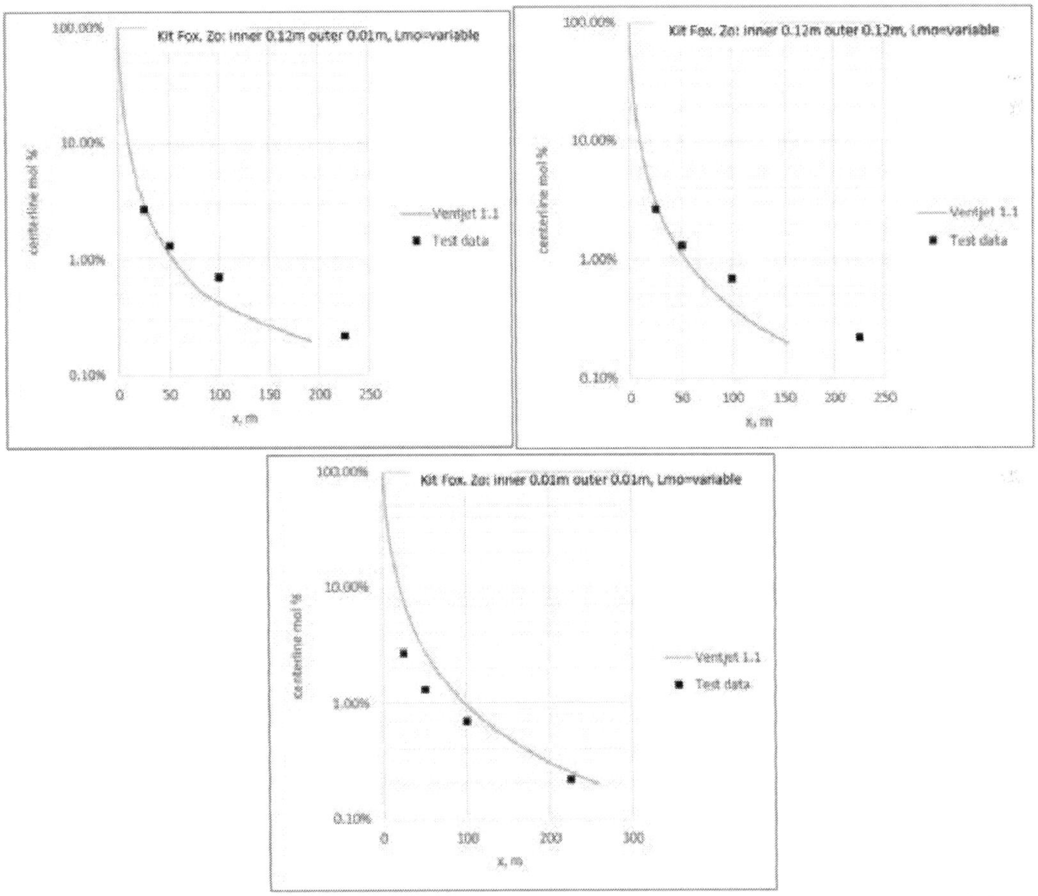

Figure 36. Comparison between the experimental data of Kit Fox[25-28] and Ventjet modeling results – sensitivities

8.3 Model validation vs. industrial scale accidental releases

8.3.1 Air Products Air Separation unit oxygen vent.

Air Products Air Separation unit oxygen vent, high oxygen (25% - 26% total O_2) detected at local operating platform 12m downwind of plant vent. 26% total oxygen is equivalent to 6.5% excess O_2 in air. Model down to 6.5%. All conditions are rough estimates. Ventjet does a good job matching reported readings, PHAST 8.4 (beta) also performs well, see figure 37. Note: previous versions of PHAST underpredict distances for this release by a factor of about 2 (not shown). *Mol wt. 32, release rate 2.95 kg/s, release temperature 288K, Vent diameter 0.6m, elevation 12m, vertical release, ambient air temperature 293 K, 5 m/s wind @ 10m, D stability, surface roughness 0.18m.*

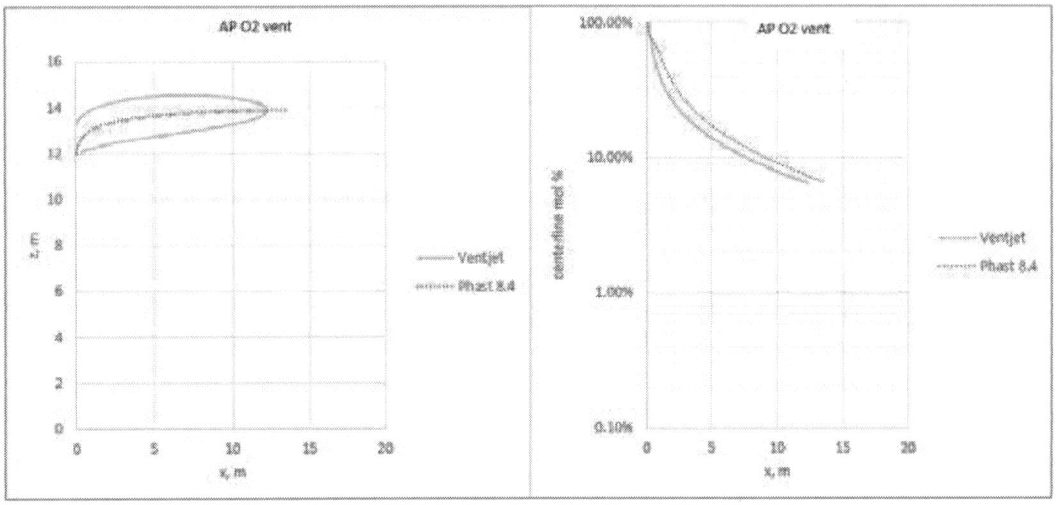

Figure 37. Comparison between PHAST8.4 (beta) and Ventjet modeling results for Air Products Air Separation unit oxygen vent.

8.3.2 LNG tank relief valve cold vent [29]

LNG tank relief valve cold vent [29], hydrocarbons detected at grade on personal flammable gas detector at about 200m downwind. Information provided at FLUG meeting 2012 (Bergen). FLACS and Ventjet model the dispersion down to 0.5% concentration, see figure 38. All conditions are rough estimates. FLACS is run with a wind speed of 0.8 m/s. Ventjet matches FLACS quite well, with only a minor adjustment in wind speed from 0.8 m/s to 0.5 m/s (note that FLACS and Ventjet / PHAST 8.4 (beta) have slightly different wind speed vs. elevation profile equations). PHAST 8.4 (beta), also run at 0.5 m/s wind speed, does reasonably well. *Mol wt. 17, release rate 10.16 kg/s, release temperature 145K, Vent diameter 0.25m, elevation 60m, horizontal release, ambient air temperature 299 K, 0.8 m/s for FLACS (0.5 m/s for Ventjet and PHAST 8.4 (beta)) wind @ 10m, F stability, surface roughness 0.18m.*

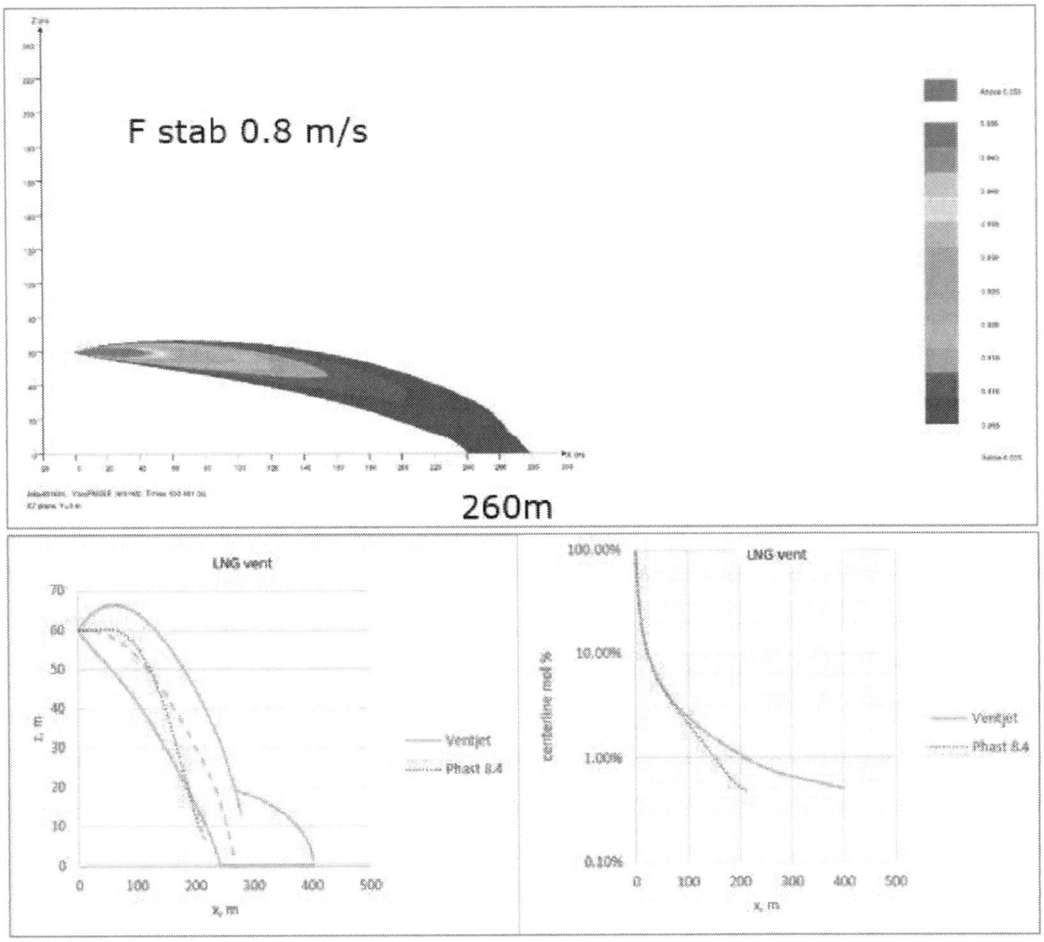

Figure 38. Top frame shows FLACS simulations results for LNG tank relief valve cold vent, down to 0.5% concentration. On the bottom results from Ventjet and PHAST 8.4 (beta) simulations for LNG tank relief valve cold vent, down to 0.5% concentration are presented.

9 Extension of model to "into the wind" releases

Prediction of how gases disperse when released into the wind is important when sizing and locating horizontal vents on structures, since potentially hazardous clouds can blow back to areas where personnel may be located.

The authors of this paper have not been able to find any experimental data on "into the wind" releases. Since FLACS has proved successful in matching all the available "with the wind" release data (wind tunnel, field tests and real-world incidents), it has been decided to simply compare Ventjet against FLACS.

Three different release scenarios have been selected for this comparison: buoyant gas, neutral density gas and heavy gas.

The only modifications done to Ventjet are those geometry changes required to correctly model entrainment and momentum over the full range of angles from 0 to 360 degrees rather than the more limited -90 to + 90 degrees. No changes beyond this were made. The method for making these changes is described in APPENDIX F.

Expected dispersion patterns into the wind. The expected flow patterns for the three release scenarios are shown in Figure 39. Positive buoyancy releases are expected to behave reasonably close to regular downwind dispersing jets and Ventjet should be able to model these reasonably well. Neutrally buoyant releases are likely to be the most challenging, since the jet will turn over on itself, with the jet center rolling out towards the outside of the turning jet, in addition at the point where the jet turns there will be high sideways velocities and associated mixing / entrainment (higher at higher wind speeds) but Ventjet will actually predict zero velocity at this point where it reverses into the wind. Negatively buoyant releases will head down towards the ground and it can be expected that the mixing / entrainment due to interaction between the turning cloud and the ground may be more than Ventjet would predict… it should also be remembered that Ventjet is not set up to model dense gases traveling along the ground and will be conservative once this occurs. Overall it can be anticipated that Ventjet will somewhat underpredict mixing due to cloud turning, with the most marked difference being for neutrally buoyant clouds.

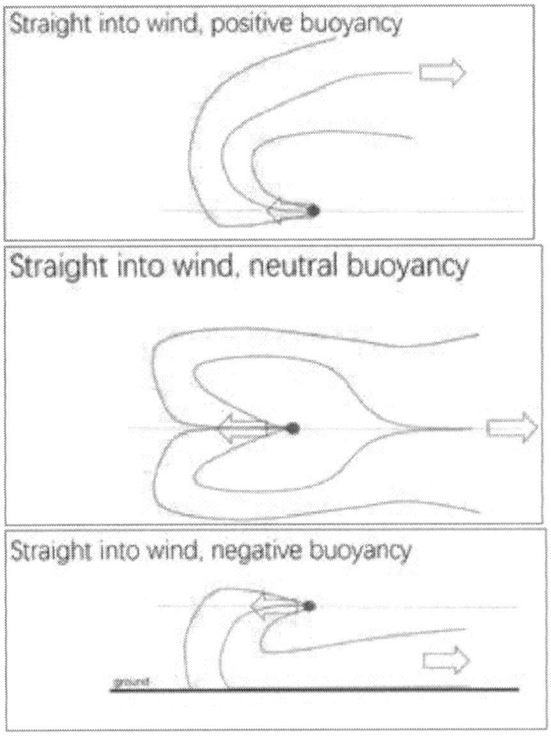

Figure 39. Sketches of Ventjet into wind model for light, neutral and heavy gases respectively.

9.1 Buoyant Gas comparison

Results are shown in Figure 40 &41. Ventjet performs reasonably well. Once the cloud turns into the wind, Ventjet matches FLACS well for F 1.5 and somewhat overpredicts downwind distances at lower concentrations for D 5. FLACS appears to show a slight drop in elevation at lower concentrations for D 5, which cannot easily be explained for a buoyant gas. Modeling set-up: *mol wt. 45.2, release rate 21.7 kg/s, release temperature 291.7 K, Vent diameter 0.58m, elevation 7m, vertical release, ambient air temperature 291.7 K, 8.88 m/s wind @ 10m, D stability, surface roughness 0.008m*

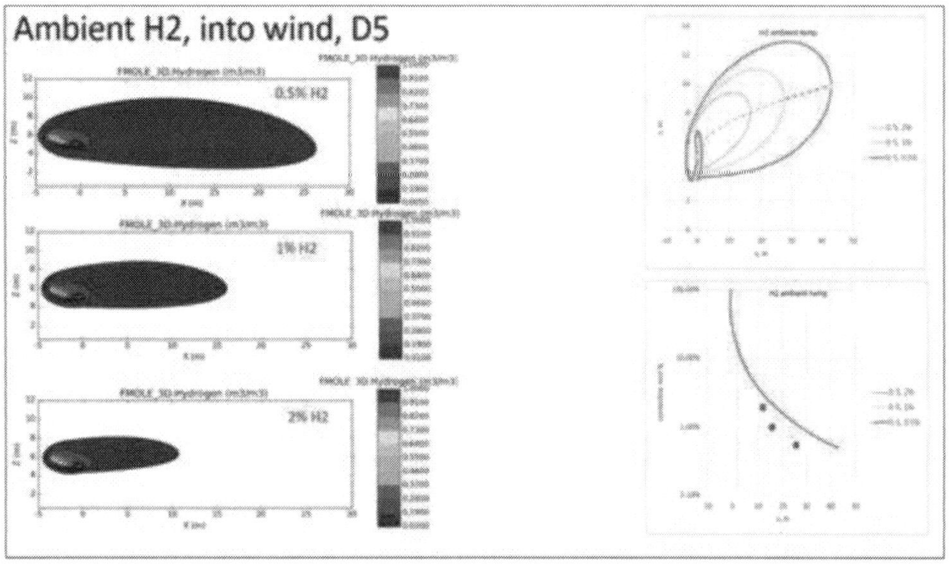

Figure 40. FLACS and Ventjet simulations results for light gas dispersion into wind D5 weather conditions

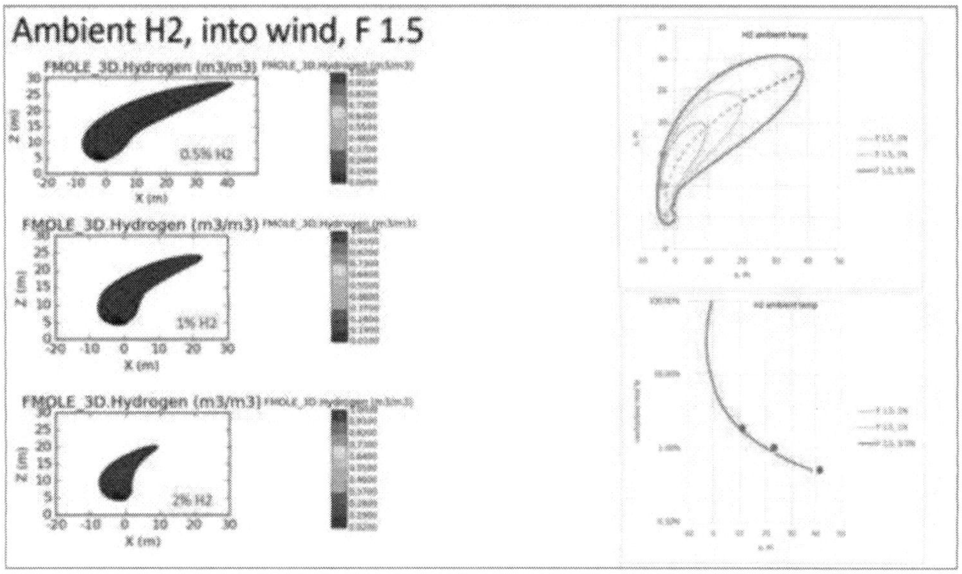

Figure 41. FLACS and Ventjet simulations results for light gas dispersion into wind, F1.5 weather conditions.

9.2 *Neutral density Gas comparison*

Ambient temperature nitrogen. Results are shown in Figure 42 & 43. Ventjet performs better than might reasonably be expected, given the complexities of this release. Once the cloud turns, Ventjet predicts downwind distances about the same as FLACS for F 1.5 and about double for D 5. Modeling set-up: *mol wt. 28, release rate 0.6 kg/s, release temperature 288 K, Vent diameter 0.2m, elevation 5m, horizontal into wind release, ambient air temperature 288 K, 5 m/s wind @ 10m, D stability, and also 1.5 m/s, F stability, surface roughness 0.18m*

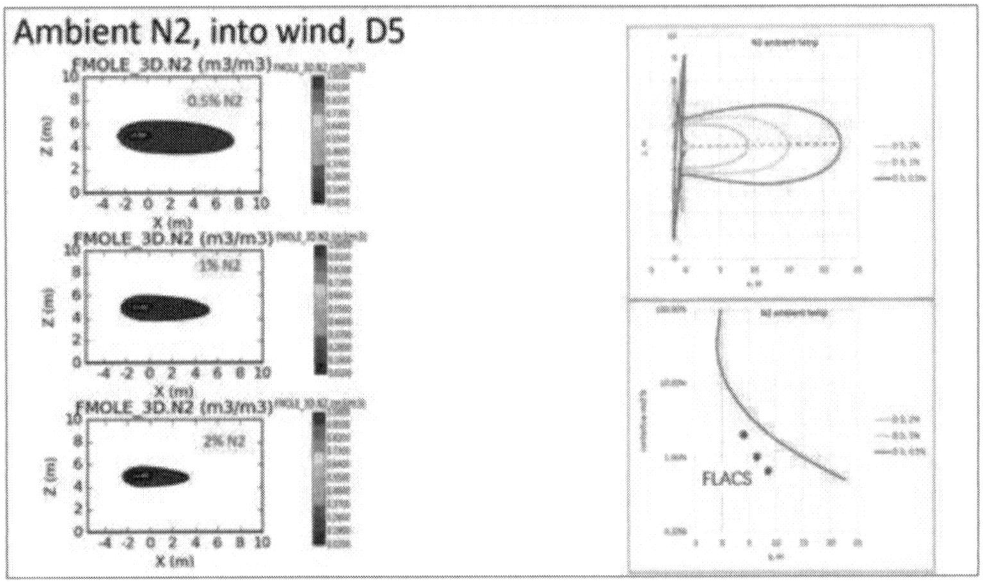

Figure 42. FLACS and Ventjet simulations results for neutral gas dispersion into wind, D5 weather conditions.

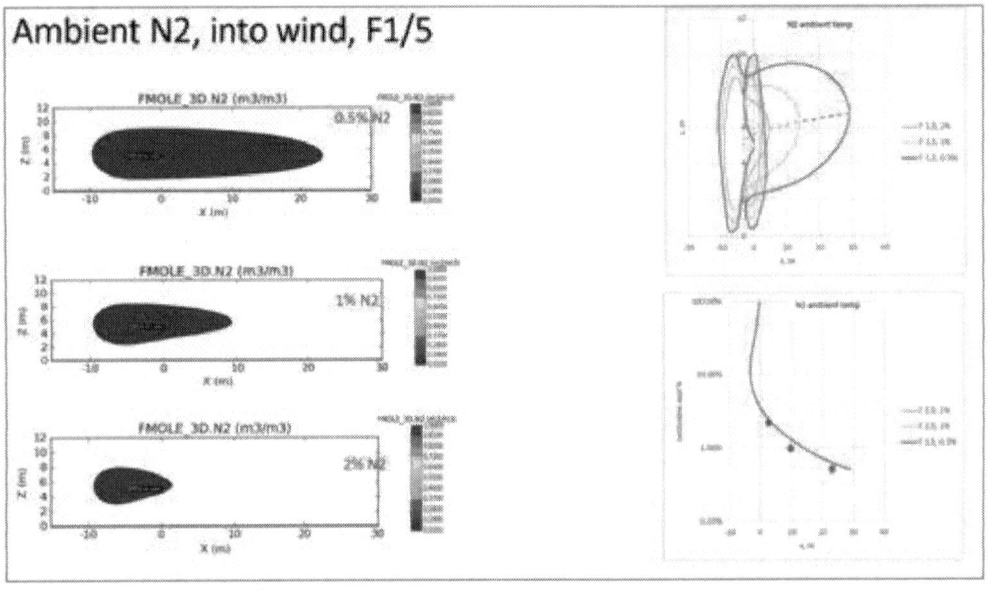

Figure 43. FLACS and Ventjet simulations results for neutral gas dispersion into wind, F1.5 weather conditions.

9.3 Dense Gas comparison. Cold oxygen

Results are shown in Figure 44 & 45. Again, Ventjet matches FLACS quite well, bearing in mind that Ventjet will probably be conservative once the dense gas reaches the ground, the best way to compare results is by looking at the concentration and location of cloud touchdown. For D 5, Ventjet predicts touchdown at about 4% and -3m, whereas FLACS predicts touchdown at about 3% and -6m. For F 1.5, Ventjet predicts 3% at -22m, FLACS 4% at -25m. Modeling set-up: *mol wt. 32, release rate 5 kg/s, release temperature 80.5 K, Vent diameter 0.1m, elevation 5m, horizontal into wind release, ambient air temperature 288 K, 5 m/s wind @ 10m, D stability, and also 1.5 m/s, F stability, surface roughness 0.18m*

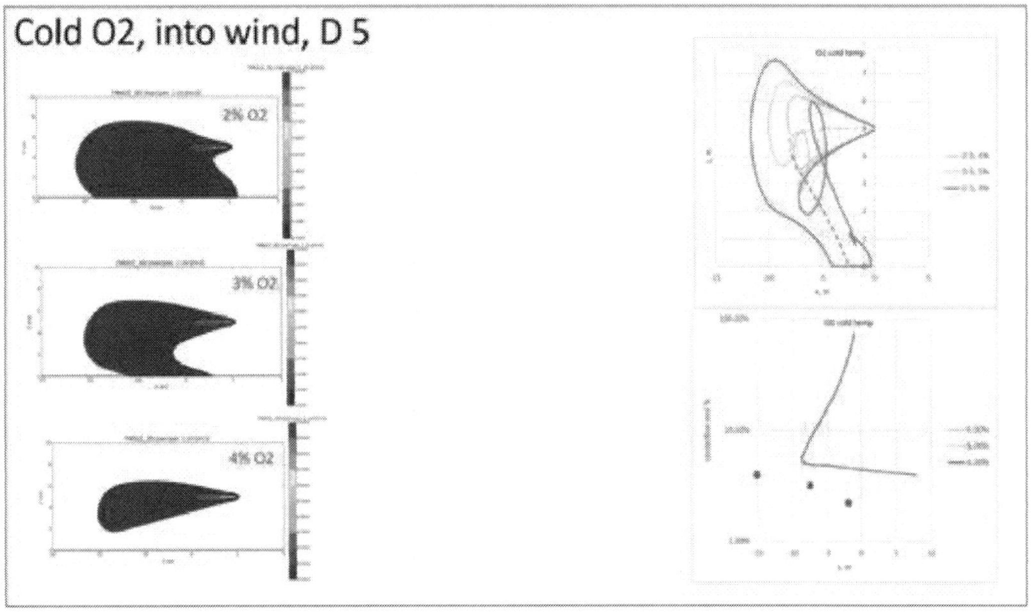

Figure 44. FLACS and Ventjet simulations results for heavy gas dispersion into wind, D5 weather conditions.

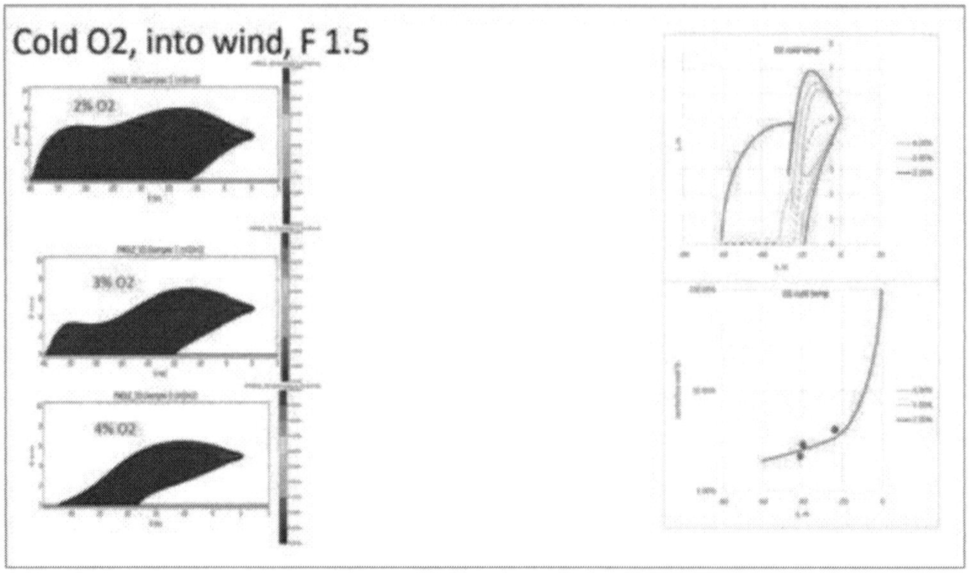

Figure 45. FLACS and Ventjet simulations results for heavy gas dispersion into wind, F1.5 weather conditions.

Overall, Ventjet does quite well at predicting the dispersion of into the wind releases, compared to FLACS. In fact, for the F 1.5 cases, Ventjet is close FLACS. For the higher wind speed D 5 cases, Ventjet somewhat underpredicts the additional air entrainment as the cloud turns and, consequently, predicts somewhat longer downwind distances than FLACS, which is, at least, a bit conservative.

10 Low velocity releases

Releases with velocities equal (isokinetic) or lower (hypokinetic) than the ambient wind velocities are also of high importance for industrial applications.

For the comparison the experiments performed in the wind tunnel of Laboratoire de Mecanique de Fluide et de Acoustique of the Ecole Centrale de Lyon (France) was used [32, 33]. The size of the working sections is 14 m, the width is 3.7 m. The ceiling is adjustable to better control longitudinal pressure gradients. Experiments were done for a turbulent neutral boundary layer. The release was done via a L-shape metallic pipe (directed downstream) of internal diameter 0.003 m and 0.006m & 0.152 m and 0.05 m height. Ethane (C_2H_6) is used in the experiment since its density is similar to air. Results are compared to PHAST 8.4 (beta) and Ventjet.
ES3-isokinetic. *Mol wt. 29, release rate* 0,073 *kg/s, release T= 291.7K, Vent diameter 0,15m, elevation 7.6m, horizontal release, ambient air temperature 291.7K, 3,635 m/s wind @ 10m, D stability, surface roughness 0.0055m*. This comparison shows a reasonable agreement between the experimental data, Ventjet and Phast 8.4 (beta) for the jet trajectory, see Figure 46. Phast 8.4 (beta) correctly estimates the concentration decay along the centerline, whereas Ventjet overestimates the concentration, giving somewhat conservative results.

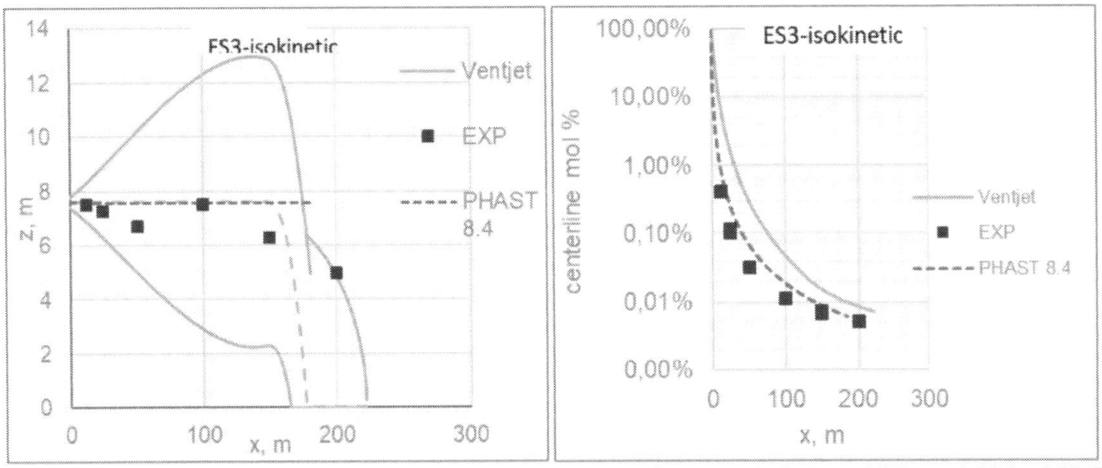

Figure 46. Comparison between the experimental data of Nironi [32, 33] ES3-isokinetic, PHAST 8.4 (beta) and Ventjet modeling results: the left frame corresponds to the jet trajectory; the right frame corresponds to the concentration decay on the centerline versus the distance from the release.

ES6-isokinetic. *Mol wt. 29, release rate* 0,29 *kg/s, release T= 291.7K, Vent diameter 0,3m, elevation 7.6m, horizontal release, ambient air temperature 291.7K, 3,635 m/s wind @ 10m, D stability, surface roughness 0.0055m.* This comparison shows a reasonable agreement between the experimental data Ventjet and Phast 8.4 (beta) for the jet trajectory, see Figure 47. Phast 8.4 (beta) correctly estimates the concentration decay along the centerline, whereas Ventjet overestimates the concentration, giving conservative results.

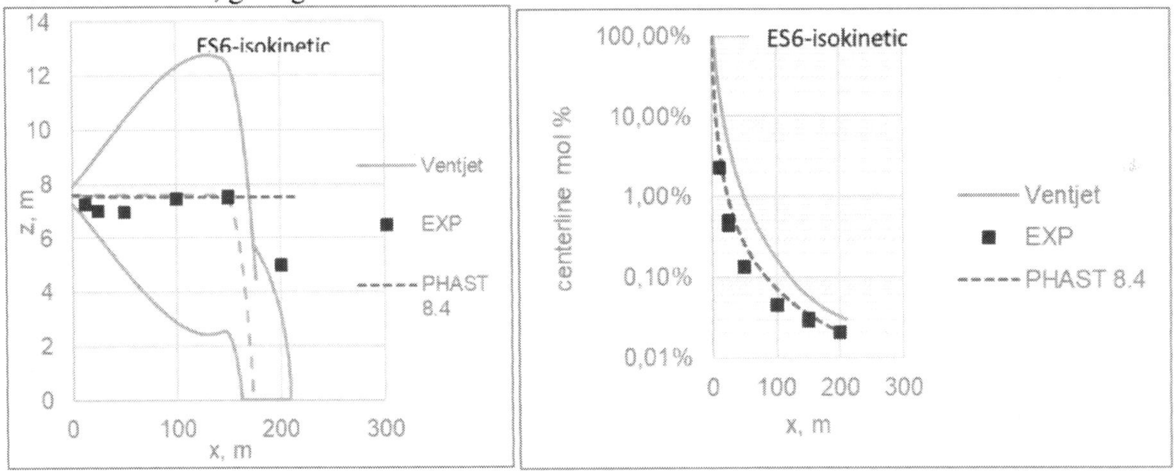

Figure 47. Comparison between the experimental data of Nironi [32, 33] ES6-isokinetic, PHAST 8.4 (beta) and Ventjet modeling results: the left frame corresponds to the jet trajectory; the right frame corresponds to the concentration decay on the centerline versus the distance from the release.

LLS-isokinetic. *Mol wt. 29, release rate* 0,0563 *kg/s, release T= 291.7K, Vent diameter 0,15m, elevation 2.5m, horizontal release, ambient air temperature 291.7K, 3,635 m/s wind @ 10m, D stability, surface roughness 0.0055m.* This comparison shows a reasonable agreement between the experimental data Ventjet and Phast 8.4 (beta) for the jet trajectory (experimentally the concertation is very hard to measure exactly on the ground), see Figure 48. Phast 8.4 (beta) and Ventjet correctly estimate the concentration decay along the centerline.

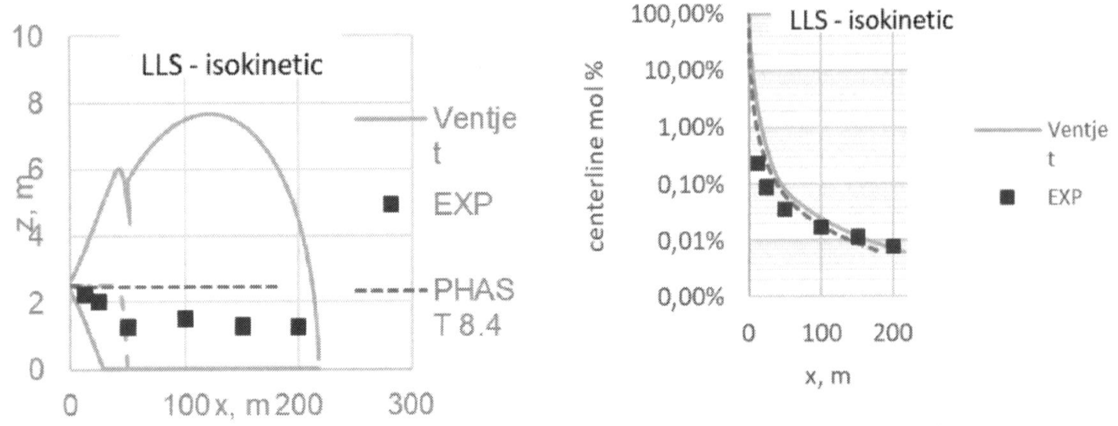

Figure 48. Comparison between the experimental data of Nironi [32, 33] LLS-isokinetic, PHAST 8.4 (beta) and Ventjet modeling results: the left frame corresponds to the jet trajectory; the right frame corresponds to the concentration decay on the centerline versus the distance from the release.

ES3-hypokinetic. *Mol wt. 29, release rate* 0,00218 *kg/s, release T= 291.7K, Vent diameter 0,15m, elevation 7.6m, horizontal release, ambient air temperature 291.7K, 3,635 m/s wind @ 10m, D stability, surface roughness 0.0055m.* This comparison shows a reasonable agreement between the experimental data Ventjet and Phast 8.4 (beta) for the jet trajectory, see Figure 49. Ventjet correctly estimates the concentration decay along the centerline, whereas Phast 8.4 (beta) significantly underestimates the concentration.

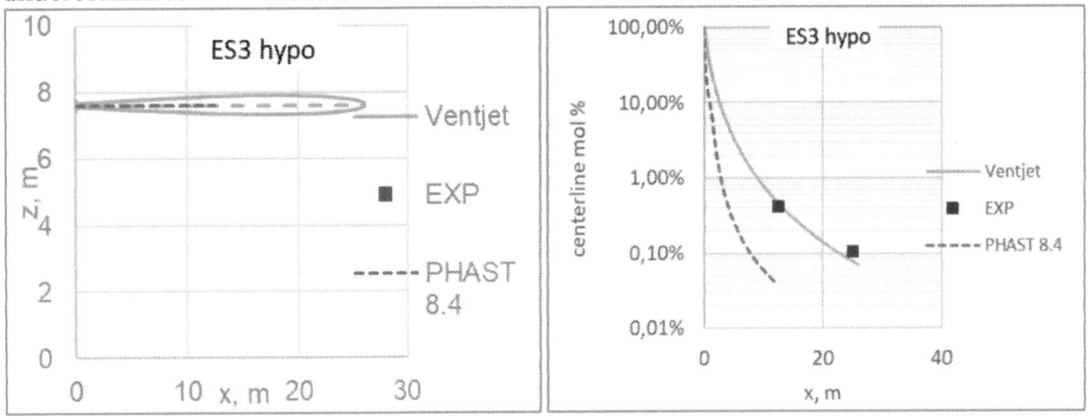

Figure 49. Comparison between the experimental data of Nironi [32, 33] ES3-hypokinetic, PHAST 8.4 (beta) and Ventjet modeling results: the left frame corresponds to the jet trajectory; the right frame corresponds to the concentration decay on the centerline versus the distance from the release.

ES6-hypokinetic. *Mol wt. 29, release rate* 0,29 *kg/s, release T= 291.7K, Vent diameter 0,3m, elevation 7.6m, horizontal release, ambient air temperature 291.7K, 3,635 m/s wind @ 10m, D stability, surface roughness 0.0055m.* This comparison shows a reasonable agreement between the experimental data Ventjet and Phast 8.4 (beta) for the jet trajectory, see fig 50. This comparison shows a reasonable agreement between the experimental data Ventjet and Phast 8.4 (beta) for the jet trajectory, see Figure 49. Ventjet correctly estimates the concentration decay along the centerline, whereas Phast 8.4 (beta) significantly underestimates the concentration.

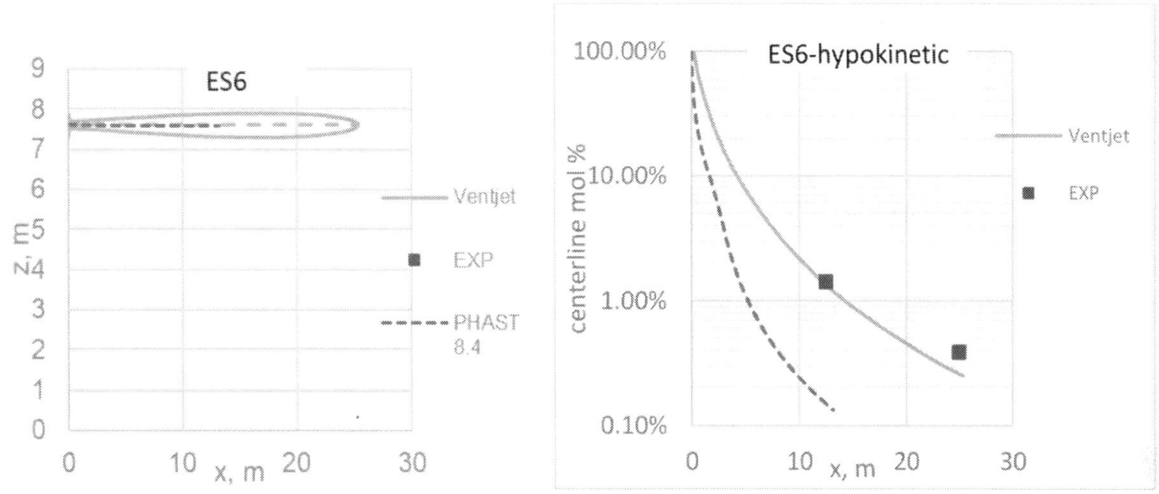

Figure 50. Comparison between the experimental data of Nironi [32, 33] ES6-hypokinetic, PHAST 8.4 (beta) and Ventjet modeling results: the left frame corresponds to the jet trajectory; the right frame corresponds to the concentration decay on the centerline versus the distance from the release.

11 Summary comparison between Ventjet and PHAST

Ventjet is based on PHAST version 6.54-8.2 with changes described in this paper. Table 2 provides a simple summary of the main differences between Ventjet and PHAST version 6.54-8.2, with comments.

Table 2. Summary comparison between Ventjet and PHAST version 6.54-8.2

	PHAST version 6.54-8.2	Ventjet	Comments
Scope	Full integrated model for gas and two-phase, including aerosol, rainout and pool	Gas only	
Thermodynamics	Rigorous	Ideal gas	
Equation solution	Direct solution of gaussian cross wind cloud concentration profile	Solution of tophat cloud followed by superimposed gaussian concentration profile with centerline concentration 1.7 times tophat average	

Jet entrainment	Applied to gaussian model. Entrainment constant adjusted to allow for gaussian profile vs tophat profile	Applied to tophat model	There is some confusion in the literature about whether published entrainment constants are based on gaussian or tophat models
Cross wind entrainment	Equally applied to be rising, horizontal and falling plumes. Entrainment constant NOT adjusted to allow for gaussian profile vs tophat profile	Only applied to rising / horizontal clouds. Applied to top-hat model	
Passive entrainment	Near field and far field. Applied to gaussian model. Entrainment constant NOT adjusted to allow for gaussian profile vs tophat profile	Near field only, applied to tophat model	
Dense gas entrainment	Included	Not included	
Choked flow expanded source	Conservation mass, momentum, energy	Conservation mass, momentum enthalpy	For choked flow releases PHAST will predict colder releases than Ventjet. To use Conservation energy would require that the gas dispersion model also conserves energy, which PHAST does not (both dispersion models conserve enthalpy)
Ground impact force	Included	Not included	For falling clouds, the ground will tend to convert vertical momentum into horizontal momentum (gas goes sideways) which has the effect of reducing the vertical velocity as the cloud approaches the ground – a "cushioning" effect that reduces the ability of the cloud to touchdown

Ground effect	Not included	Included	Engineering models for gas dispersion in the public domain do not include the ground effect. The approach of using Bernoulli to determine down force appears to be novel, but is well validated against a number of CFD models and the limited available test data
Releases into the wind	Not included	Included	Most engineering models do not allow releases into the wind. Beyond ensuring the correct geometries, no changes were required to Ventjet and results match CFD quite well in most cases
Low velocity releases	Not recommended to use	Included	Ventjet gives accurate results for releases with velocity lower than the ambient wind speed and somewhat conservative results for the cases where the release velocity is equal to the wind speed

12 Conclusions

A new engineering model for gas dispersion, Ventjet, has been developed and is presented in this paper.

The model is based on PHAST (versions 6.54-8.2), but with method enhancements to crosswind entrainment, choked flow effective source, the inclusion of ground effect and the ability to model releases into the wind

The model has been validated against a wide range of test data (wind tunnel and field scale), CFD and actual incidents.

The enhancements described in this paper can be incorporated into more comprehensive models that include two -phase releases and dense gas ground dispersion.

In the future, other modeling approaches may prove to be equally or more effective than those presented in this paper. It is hoped that the quite comprehensive dataset provided in this paper will assist in validation of any other such approaches that may be developed.

For releases that are not impacted by the ground, the latest version PHAST 8.4 (with new DNV-GL cross wind entrainment method) shows significantly better agreement with experimental data than previous versions of PHAST and compares quite well with Ventjet, even though different approaches to resolving crosswind entrainment issues have been taken. For releases that are impacted by the ground, PHAST 8.4 does not perform as well as Ventjet, since no ground effect is yet included. Ventjet gives good results for cases, where the release velocity is much lower than the ambient wind speed and somewhat conservative results where the release velocity is about the same as the wind speed.

13 Acknowledgements

The authors gratefully thank M. Schatzmann, C. Vidali, M. Marro and P. Salizzoni for multiple discussions and the technical support during this work. The authors gratefully acknowledge DNV for the fruitful exchange.

14 References

[1] DNV-GL, Unified Dispersion Model Theory, Supporting documentation for PHAST software package, 2014, Available at https://www.dnv.com/software/services/phast/index.html Accessed on April 12, 2021.

[2] Safe Location of Oxygen and Inert Gas Vents. EIGA Document 154/16, 2016, Available at https://www.eiga.eu/publications/eiga-documents/doc-15416-safe-location-of-oxygen-and-inert-gas-vents/ Accessed on April 12, 2021.

[3] W.Yip and O.Ettighoffer, Air Gas Venting, 15 CGA Safety and Reliability of Industrial Gases, Equipment and Facilities Seminar, Tampa, Florida, USA, 20-21 October, 2015.

[4] FLACS User's Manual, GexCon, Bergen, continuously updated, Available at https://www3.gexcon.com/files/manual/flacs/pdf/flacs-users-manual.pdf Accessed on April 12, 2021.

[5] B.R. Morton, G.I. Taylor and J.S. Turner, Turbulent Gravitational Convection from Maintained and Instantaneous Sources, Proc. Roy. Soc, 1950.

[6] F.P. Ricou and D.B. Spalding, Measurements of Entrainment By Axisymmetrical Turbulent Jets,. J. Fluid Mech, 1961.

[7] G. Ooms, A New Method For The Calculation Of The Plume Path Of Gases Emitted By A Stack,. Atmos. Env., 1972.

[8] T. Spicer and J. Havens, A Dispersion Model For Elevated Dense Gas Jet Releases, Volume 1, EPA-450 / 4 – 88-006a, 1988.

[9] A.D. Birch, D.R. Brown, M.G. Dodson and F. Swaffield, The structure and concentration decay of high-pressure jets of natural gas, Combust. Sci. and Tech., 36, pp. 249–26, 1984.

[10] A. D. Birch, D.J. Hughes and F. Swaffield, Velocity decay of high pressure jets, Combust. Sci. and Tech., 52, pp. 161–171, 1987.

[11] HGSYSTEM Technical Reference Manual. Chapter 5. Plume or jet models in HGSYSTEM, 1990, Available at www.HGSYSTEM.com Accessed on April 12, 2021.

[12] ANYSY® FLUENT®, Version 16.0, 2014 ANSYS, Inc., Available at https://www.ansys.com/products/fluids/ansys-fluent Accessed on April 12, 2021.

[13] V. Molkov, Fundamentals of Hydrogen Safety Engineering I, 2012, Available at Bookboon.com Accessed on April 12, 2021.

[14] J. Daubech, J. Hebrard, S. Jallais, E. Vyazmina, D. Jamois and F. Verbecke, Un-ignited and ignited high pressure hydrogen releases: Concentration - turbulence mapping and overpressure effects, Journal of Loss Prevention in the Process Industries, 36 (2015) 439-446, DOI:10.1016/j.jlp.2015.05.013.

[15] S. Jallais, E. Vyazmina, D. Miller and J. K.Thomas, Hydrogen Jet Vapor Cloud Explosion: A Model for Predicting Blast Size and Application to Risk Assessment, Process Safety Progress, Vol 37, No 3 (2018) 397-410, 2018. DOI:10.1002/prs.11965.

[16] P.T. Roberts, L.C. Shirvill, T.A. Roberts, C.J. Butler and M. Royle, Dispersion of Hydrogen from high-pressure sources, Symposium Series no. 151, 2006, Available at https://www.icheme.org/media/9814/xix-paper-26.pdf Accessed on April 12, 2021.

[17] S. Desilets, S. Core, G. Nadeau, B. Benard and A. Tchouvelev, Experimental Results and Comparison with Simulated Data of a Low Pressure Hydrogen Jet, Available at https://h2tools.org/sites/default/files/2019-08/Experimental%20Results%20and%20Comparison%20with%20Simulated%20Data%20of%20a%20Low%20Pressure%20Hydrogen%20Jet.pdf Accessed on April 12, 2021.

[18] M. Shatzmann, W.H. Snyder, and R. E. Lawson Jr., Experiments with heavy gas jets in laminar and turbulent cross-flows. Atmospheric Environment, Vol. 27A, No. 7, pp. 1105-1116, 1993.

[19] J. Donat and M. Shatzmann, Wind tunnel experiments of single-phase heavy gas jets released under various angles into turbulent cross flow. J of Wind Engineering and Industrial Aerodynamics Vol. 83, pp. 361-370, 1999.

[20] J. Donat, Windkanalexperimente zur Ausbreitung von Schwergasstrahlen, Dissertation, Berichte aus dem Zentrum fuK r Meeres- und Klimaforschung der Universitat Hamburg, Reihe A: Meteorologie, 24, 1996.

[21] E. Vyazmina and M. Till, Comparison of DISPAL and PHAST Computations with Experimental Data for the dispersion of a heavy gas from a stack, R&D Report CRCD_MatApp_13/046, Air Liquide.

[22] C. Vidali, M. Marro, L. Gostiaux, H Correia, S. Jaillais, D. Houssin, E. Vyazmina, P. Salizzoni, "Dispersion of heavy gas and passive scalar releases", The 15th International Conference on Wind Engineering, Beijing, China; September 1-6, 2019.

[23] P. Quillatre, Relevance of the current modelling methods for the prediction of LNG vapour dispersion and development to be carried on, HAZARDS 27, Symposium Series no. 162, 2017.

[24] J.Sail, V. Blanchetière and A. Giacosa, LNG vapor dispersion test campaign to qualify spectral cameras and dispersion models, GASTECH Singapore, 2015.

[25] S.R. Hanna, J.C. Chang and D.G. Strimaitis, Hazardous gas model evaluation with field observations, Atmospheric Environment, Vol.27A, No.15, pp.2265-2285, 1993.

[26] S.R. Hanna and J.C. Chang, Use of the Kit Fox eld data to analyze dense gas dispersion modeling issues, Atmospheric Environment, 2001, 35: p 2231-2242.

[27] S.R. Hanna, O.R. Hansen and S. Dharmavaram, FLACS CFD air quality model performance evaluation with Kit Fox, MUST, Prairie Grass, and EMU observations, Atmospheric Environment, 2004. 38: p. 4675–4687.

[28] B. Truchot, J. M. Lacome, A. Albergel, P. Armand, X. Boesflug, N. Boisson N., P. Calot, B. Carissimo, A. Carrau, A. Champassith, B. Chardin, J. Commanay, A. Daubarede, C. Desut Dhaultfoeuille, L. Donnat, G. Dussin, B. Geniaut, M. Gossin, M. Ichard, S. Jallais, C. Lapierre, M. Muller, M. Nibart, C. Olry, E. Pacard, S. Peruccio, J.L. Revel, L. Soulhac, V. Soumoy, C. Souprayen, A. Stassinopoulos, M. Till, A. Tripathi, B. N. Vazquez, E. Vyazmina, Guide de Bonnes Pratiques pour la réalisation de modélisations 3D pour des scénarios de dispersion atmosphérique en situation accidentelle : Rapport de synthèse des travaux du Groupe de Travail National, INERIS report DRA-15-148997-06852A, 2016, Available at https://aida.ineris.fr/sites/default/files/gesdoc/86009/Guide_Bonnes_Pratiques.pdf Accessed on April 12, 2021.

[29] M. Torre, Differences observed when modeling an atmospheric dispersion with non-collinear and collinear wind direction and leak direction, FLACS User Group (FLUG) May meeting, 2012.

[30] W. H. Snyder, Guideline for fluid modeling of atmospheric diffusion. Report No. EPA-600/8-81-009, U.S. Environmental Protection Agency, Research Triangle Park, NC, USA, 1981.

[31] https://en.wikipedia.org/wiki/Coand%C4%83_effect

[32] C. Nironi, P. Salizzoni, M. Marro, P. Mejean, N. Grosjean and L. Soulhac, Dispersion of a Passive Scalar Fluctuating Plume in a Turbulent Boundary Layer. Part I: Velocity and Concentration Measurements, Boundary-Layer Meteorol (2015) 156:415–446, DOI 10.1007/s10546-015-0040-x

[33] C. Nironi, Concentration fluctuations of a passive scalar in a turbulent boundary layer, PhD thesis, ECL, 2013, Available at https://tel.archives-ouvertes.fr/tel-00964852 Accessed on April 12, 2021.

15 APPENDIX

15.1 APPENDIX A: Equations for wind speed as function of height and for passive air entrainment. The wind speed is a function of the elevation of the cloud centroid.

For convenient, this is simply referred to as z in this Appendix.

15.1.1 Wind speed

Equations are based on the concept of a friction velocity, u_*

Friction velocity $$u_* = k \frac{u_{10}}{\ln\left\{\frac{10+z_0}{z_0}\right\}} - \Psi(z_{ref}/L)$$

Where, k = Von Karman const = 0.4, u_{10} is the specified wind speed at the reference height, z_{ref} of 10m, z_o is the surface roughness, m

$\Psi\left(\frac{zref}{L}\right) = 0$ D stability

$\Psi\left(\frac{zref}{L}\right) = -2\left(\frac{zref}{L}\right)$ F stability

Monin-Obukov length $$L = a\, z_o{}^b$$

Where

a =-425, b= 0.375 for D stability

a = 26, b= 0.17 for F stability

From the friction velocity the actual wind profile is determined as follows

Wind velocity at height z $$\boldsymbol{u_a(z) = \frac{u_*}{k} \ln\left\{\frac{z+z_o}{z_o}\right\} - \Psi(z/L)}$$

$\Psi\left(\frac{z}{L}\right) = 0$ D stability

$\Psi\left(\frac{z}{L}\right) = -2\left(\frac{z}{L}\right)$ F stability

15.1.2 Passive entrainment

$$Z = \frac{(z+z_0)}{L}$$

$l_y(Z) = l_z(Z) = E(Z) = 1$ D stability

$l_y(Z) = \frac{1}{1+0.1Z}, l_z(Z) = \frac{1}{E(Z)}$, E(Z) = 1 + 4Z F stability

$$l_y = \min\{R, 0.88\,(z+z_o)l_y(Z)\}$$

$$l_z = \min\{R, 0.88\,(z+z_o)l_z(Z)\}$$

$$\varepsilon = E\,(Z)\,\frac{u_*{}^3}{(k(z+z_o))}$$

Passive entrainment from main text of paper $$E_{passive} = \alpha_3\, \pi\, \rho_a\, \varepsilon^{1/3}\, (l_y^{\frac{4}{3}} + l_z^{\frac{4}{3}})$$

15.2 *APPENDIX B: Non-choked and choked flow equations (ideal gas)*

Equations represent isentropic expansion from initial to nozzle throat condition, with conservation of energy.

P0, T0, ρ0 are source condtiions	
Choked flow @ orifice throat	
temperature	Tc = T0*(2/(γ+1))
pressure	P = P0*(Tc/T0)^(γ/(γ-1))
sonic velocity	a=(γ R T/mw)^0.5
Mach	M=1
velocity	u = M* a
density	ρ = mw P/(R T)
mass flux	G = u* ρ
throat area	A = (π/4) d^2
mass flow	mdot = G* A * Cd
Non Choked flow @ orifice throat	
temperature	T = mw P / (R* ρ)
pressure	P = Patmos
sonic velocity	a=(γ R T/mw)^0.5
Mach	M=u/a
velocity	u = (2 Cp (T0 - T))^0.5
density	ρ = ρ0 *(P/P0)^(1/γ)
mass flux	G = u* ρ
throat area	A = (π/4) d^2
mass flow	mdot = G* A * Cd
heat capacity	Cp = (R/mw)*(γ/(γ-1)
choked or non choked?	if choked P >= Patmos then choked

15.3 *APPENDIX C Excess O_2 and Excess N_2*

When releasing either nitrogen or oxygen into air, the dispersion model does not account for the nitrogen or oxygen already in the air when determining concentrations. It is conventional to use the concept of excess nitrogen or oxygen, which is what the dispersion model does calculate.

Example 1: oxygen is released into air. The safe upper limit is 23.5% total oxygen, what is the excess O_2 limit that the dispersion program should be run to, in order to establish the limit of the hazardous cloud? Answer: 3.3% O_2

Example 2: nitrogen is released into air. The safe lower limit is 19.5% total oxygen, what is the excess N_2 limit that the dispersion program should be run to in order to establish the limit of the hazardous cloud? Answer: 7.1% N_2

EXCESS OXYGEN		
O2 + x (0.21 O2 + 0.79 N2)	total O2 frac = (1 + 0.21x)/ (1+ X)	excess O2 = 1/(1+x)
EXCESS NITROGEN		
N2 + x(0.21O2+0.79N2)	total O2 frac = (0.21x/(1+x)	excess N2= 1/(1+x)

EXCESS NITROGEN	
total O2	excess N2
0	1.000
0.14	0.333
0.16	0.238
0.18	0.143
0.19	0.095
0.195	0.071
0.2	0.048

EXCESS OXYGEN	
total O2	excess O2
1.0000	1.000
0.4000	0.241
0.3000	0.115
0.2500	0.052
0.2350	0.033

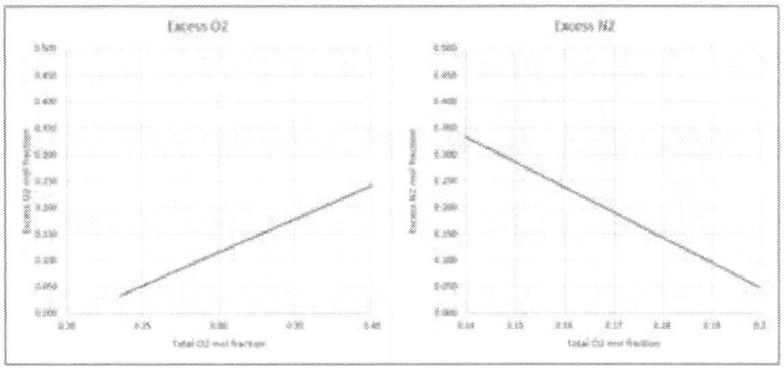

15.4 APPENDIX D : Similarity analysis used In Shatzmann [18] and Donat [19,20] experiments

To scale up from wind scale up to the real industrial scale a similarity dimensional analysis was suggested by Shatzmann [18], see figure D. Here the sketch of a continuous, steady-state jet release is given. For dimensional analysis following parameters are considered: the elevation of the release point from the ground (chimney height) h_s , diameter of the release (the diameter of the chimney) D_s, angle the release (chimney inclination) β_s (can be vertical, horizontal or at 45°), velocity $U_s$of the release, its density ρ_s and viscosity μ_s, C_s is the source concentration. Other important parameters considered by Shatzmann [18] are the properties of the ambient wind: wind velocity at the height of the release point $U_a(h_s)$, ambient density $\rho_a$and viscosity μ_a.

The local dilution C Shatzmann [18] characterized as

$$\frac{C}{C_s} = f\left(\frac{x}{D_s}, \frac{y}{D_s}, \frac{z}{D_s}, \frac{h_s}{D_s}, \frac{\delta}{D_s}, \beta_s, \frac{\rho_s}{\rho_a}, Fr, \frac{U_s}{U_a(h_s)}\right) \quad \text{Equation D1}$$

where x, y, z are the downstream, lateral and vertical directions x, y, z, respectively, δ corresponds to the height of the boundary layer, Froude number $Fr = \left(\frac{(U_s - U_a(h_s)\cos(\beta_s))^2 \rho_s}{g D_s(\rho_s - \rho_a)}\right)^{0.5}$.

In Shatzmann [18] and Donat [19,20] experiments all similarity parameters were set to prototype real scale values with the exception of the two Reynolds numbers($Re_s = \frac{\rho_s D_s U_s}{\mu_s}$ and $Re_a = \frac{\rho_a D_s U_a(h_s)}{\mu_a}$.). However, they have only marginal influence on the dilutions since they were kept above critical values [30] ($Re_s > 2300$). The ambient Reynolds number Re_a defines the ambient flow around the stack and how the released jet is downwashed. Downwash occurs only at velocity ratios $\frac{U_s}{U_a(h_s)} < 1$, which is not the case of Schatzmann on Donat experiments. Hence, the effect of ambient Reynolds number does not considerably influence the local dilution.

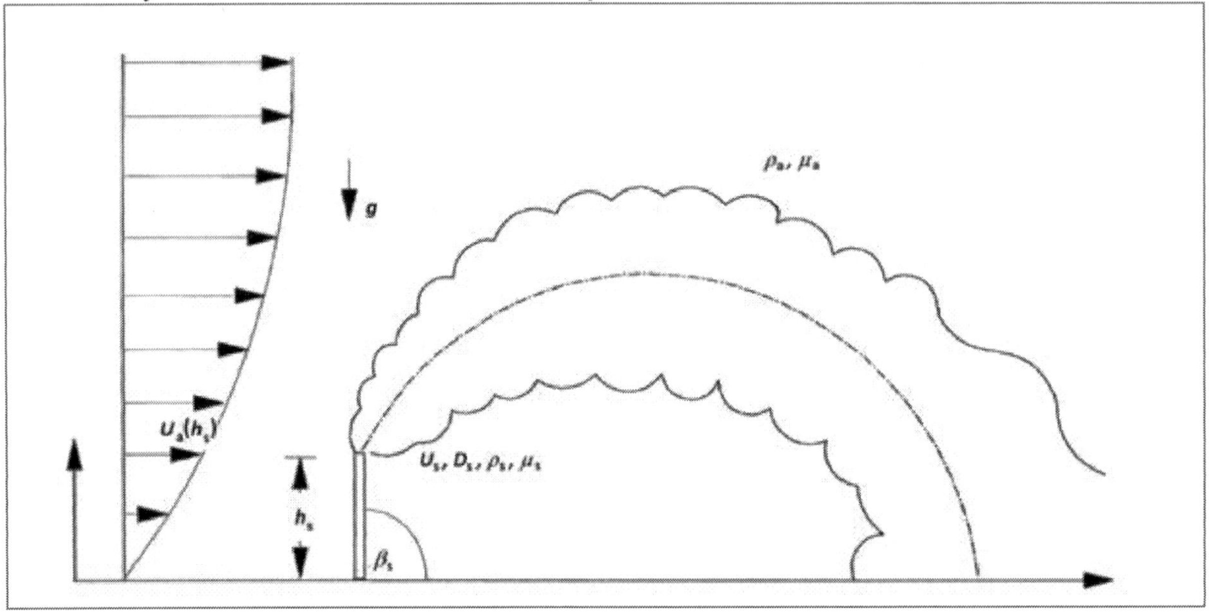

Figure D. Heavy gas jet discharged into a boundary layer. Sketch adapted from Shatzmann [18].

For scaling up to the industrial applications following movement parameters should be kept as the experimental values as in the equation 20.

For instance, multiplying the release diameter by 100, the elevation of the release point h_s also should be multiplied by 100 to keep the same ration in equation D1. The density of the ambient gas ρ_a and the density of the released gas ρ_s can be kept unchanged (as the experimental data) as well as viscosities (μ_s and μ_a). Velocities of the released gas and ambient flow can be found from Fr number and the ratio $\frac{U_s}{U_a(h_s)}$, hence the real scale velocities (of the release and wind velocity) will be multiplied by factor 10 (not by 100!).

15.5 APPENDIX E : Derivation of radial Gaussian concentration profile

Ventjet is a top hat model that assumes constant radial concentration, density and velocity at any given centerline distance from the release point. All calculations done to predict trajectory and entrainment dilution are based on this assumption. In order to determine the centerline concentration and the radial distance to any concentration of interest, a Gaussian profile is superimposed.

At any given distance along the centerline, the top hat concentration and cloud radius are known. If we assume a constant ratio between the centerline concentration and the top hat concentration, what is the radial distance to any target concentration?

Standard Gaussian profile equation: $\frac{C(r)}{C_o} = exp\left(\frac{-0.5\, r^2}{\sigma^2}\right)$ Equation (i)

C(r) = concentration at radial distance r. C_o = centerline concentration, σ = standard deviation

C = top hat mol concentration, R = top hat radius, C_{target} = target concentration.

Mols in top hat cloud cross section per unit length along centerline:

$$m_{TH} = \pi R^2 C \qquad \text{Equation (ii)}$$

Mols in Gaussian cloud cross section per unit length along centerline:

$$m_G = \int_0^{\infty} C_o\, 2\pi r\, exp\left(\frac{0.5\, r^2}{\sigma^2}\right) dr \qquad \text{Equation (iii)}$$

Substitute $u = r^2$, so du = 2 rdr

$$m_G = Co\, \pi \int_0^{\infty} exp\left(\frac{0.5\, u}{\sigma^2}\right) du \qquad \text{Equation (iv)}$$

So, $m_G = 2\pi\, C_o\, \sigma^2$ Equation (v)

By mass balance $m_{TH} = m_G$ (Equations (ii) and (v)) $\pi R^2 C = 2\pi C_o \sigma^2$

$$\sigma^2 = 0.5R^2 \frac{C}{C_0} = 0.5R^2 \frac{1}{C_R} \qquad \text{Equation (vi)}$$

Where $\frac{C_{centerline}}{C} = C_R = 1.7$ Main paper Equation 17

Inserting into Equation (i) and rearranging, and replacing C(r) with C_{target}, we get

$$r\left(C_{target}\right) = R\left(-\ln\left(\frac{C_{target}}{C_{centerline}}\right) * \left(\frac{1}{(C_R)}\right)\right\}^{0.5} \qquad \text{Main paper Equation 18}$$

15.6 APPENDIX F: Ventjet modifications required to accommodate releases into the wind

Certain modifications are required to the treatment of the angles found in the equations for jet entrainment (Equation 3), cross wind entrainment (Equation4) , angled elevation (Equation 13), vertical momentum (Equation 7), horizontal momentum (Equation 6), trajectory horizontal and vertical incremental step size (Equation 10), to correctly accommodate a jet that can now travel in any direction within the 2D vertical plane parallel to the wind.

The first action is to convert the trajectory angle Φ from a range of -90 degrees (straight down) through + 90 degrees (straight up) to an anticlockwise angle starting a 0 degrees horizontal with wind, all the way through 180 degrees horizontal, into the wind and round to 360 degrees horizontal, into the wind again.

The next step is to transform this angle Φ to the angle β. In each of the impacted equations, Φ is replaced by β according to the table (note angles are in radians, not degrees), and the sign is either positive or negative according to the value of Φ

ϕ	β	Jet entrain ment	Cross wind entrainment	Angled elevation	Vertical momentum	Horizontal momentum	Trajectory vertical step increment, dz	Trajectory horizontal step increment, dx
0 < ϕ < π/2	ϕ	- Cos β	Sin β cos β	Cos β	- Sin β^2 cos β	Sin β^2 sin β	Sin β	Cos β
π/2 < ϕ < π	π - ϕ	+Cos β	Sin β cos β	Cos β	Sin β^2 cos β	Sin β^2 sin β	Sin β	- Cos β
π < ϕ < 3π/2	ϕ - π	+Cos β	Sin β cos β	Cos β	- Sin β^2 cos β	Sin β^2 sin β	- Sin β	- Cos β
3π/2< ϕ < 2π	2π - ϕ	- Cos β	Sin β cos β	Cos β	Sin β^2 cos β	Sin β^2 sin β	- Sin β	Cos β

Example :

Jet entrainment [5][7][8][11][1] $E_{jet} = \alpha_1 \, P \, \rho_a \, |u - u_a \cos(\Phi)|$ Equation 3

0< Φ < π/2, β= Φ $E_{jet} = \alpha_1 \, P \, \rho_a \, |u - u_a \cos(\beta)|$

π/2< Φ < π, β= π-Φ $E_{jet} = \alpha_1 \, P \, \rho_a \, |u + u_a \cos(\beta)|$

π < Φ < 3π/2, β= Φ-π $E_{jet} = \alpha_1 \, P \, \rho_a \, |u + u_a \cos(\beta)|$

3π/2 < Φ < 2π, β= 2π-Φ $E_{jet} = \alpha_1 \, P \, \rho_a \, |u - u_a \cos(\beta)|$

Concepts for Toxic Facility Siting

Chris Buchwald, PE
ExxonMobil Research and Engineering
22777 Springwoods Village Parkway
Spring, TX 77389
Chris.R.Buchwald@ExxonMobil.com

Prepared for Presentation at
American Institute of Chemical Engineers
2020 Spring Meeting and 17th Global Congress on Process Safety
Virtual
April 18 - 22, 2021

Concepts for Toxic Facility Siting

Chris Buchwald, PE
ExxonMobil Research and Engineering
22777 Springwoods Village Parkway
Spring, TX 77389
Chris.R.Buchwald@ExxonMobil.com

Keywords: "Toxic Facility Siting", "Facility Siting"

Abstract

Developing a protocol for a toxic facility siting program requires an underlying framework that embodies four concepts. These concepts provide the principles for establishing the basis for using an engineering approach to prove a toxic refuge is safe for the credible release scenario versus a qualitative approach for designing a toxic refuge.

This paper develops the four concepts and highlights how they contribute to the effectiveness of the overall program. The concepts are:

1) The importance of considering the aggregation of risk on a buildings location as well as the individual risk
2) The significance of a centerline dispersion curve and the relationship to the tightness of a safe haven, shelter-in-place building, or toxic refuge building
3) The understanding of the plume shape (not a circle) of the toxic dispersion and the correlation with emergency response planning
4) The interaction between a toxic refuge tightness, the emergency response plan, and the credible scenario

An effective toxic facility siting protocol will encompass elements of all four of these concepts and in the unexpected case of a toxic release provide the flexibility to react to variations for each of these concepts.

1 Conceptual Overview of Toxic Facility Siting

1.1 Scope

To provide an example framework for completing a toxic facility siting is practicable, auditable to applicable industry standards, and technically sound. While this framework provides one approach, the process of completing a toxic facility siting has many components that interact and affect each other. Hence there are many viable variances and options with how the toxic facility siting study can be completed.

1.2 Definitions

Credible Design Point

The credible design point scenario establishes a basis for evaluating/constructing a toxic refuge shelter and the credible design point scenario includes benefits of safety systems functioning correctly.

Non-Toxic Refuge Building

Buildings that are not a safe haven or a shelter-in-place. This can include administration buildings, maintenance buildings, and/or other buildings.

Refuge-of-Opportunity

This is a shelter of last resort that may be used by occupants to provide a level of isolation between themselves and the ongoing dispersion. An example could be a motor vehicle cabin area.

Safe Haven

A building that has a specific engineered design level of protection. A safe haven has a design basis that uses an external contaminant concentration and duration to determine the exposure level for occupants inside the building based on infiltration. Safe havens can allow typical control functions to continue as occupants required to stay can bring a facility to a safer state. Safe havens can include elements of pressurization, supplied breathing air, door airlocks, scrubbing systems, and duct isolation controls. A safe haven has periodic testing to assure that the design features are in place and functioning to achieve the design basis.

Shelter-In-Place

A building that gets additional isolation by the occupants closing doors, closing windows, and shutting off heating, ventilation, and air conditioning systems. A building designated as shelter-in-place can also include supplies to allow the doors and windows to be taped to limit leakage around openings.

Toxic Refuge

A class of building that has been designated as a safe haven or a shelter-in-place.

1.3 Concepts

The process of determining toxic facility siting is linked by three concepts that work interactively with each other: 1-Emergency Response Plan, 2-Toxic Refuge Design Basis, and 3-Scenario Design Basis. Each of these affect the other two such that toxic facility siting is an iterative approach as the concepts evolve and develop over time. See Figure 1.

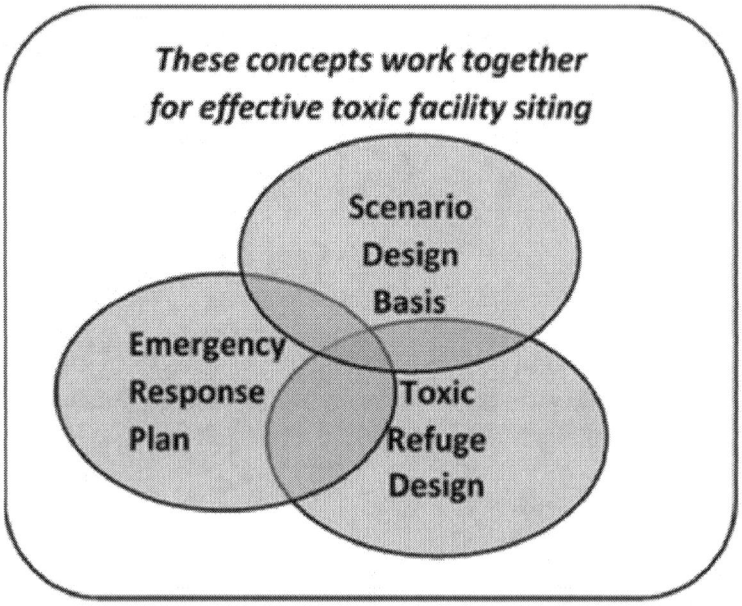

Figure 1: Three Connected Facility Siting Concepts

As all three concepts work together there is no order or weighting of importance to anyone of the three. To be effective all three must be in place and functioning.

In applying this approach, there is a center line concentration dispersion concept in that the closer a building is located to a release source the tighter the building must be to provide protection for occupants. This is illustrated in Figure 2.

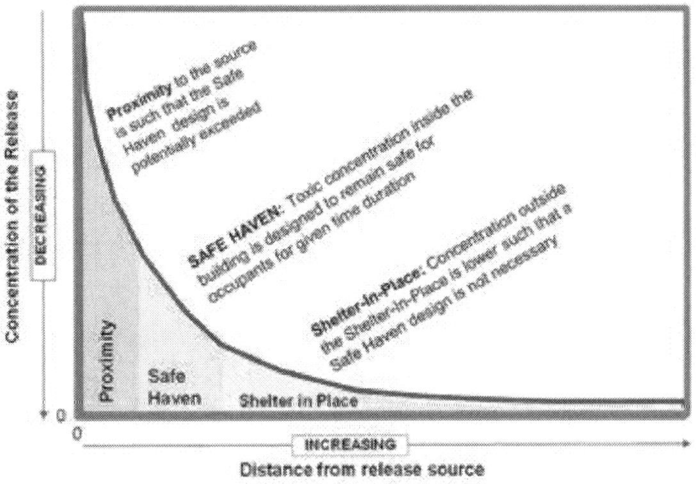

Figure 2: Center Line Concentration and Toxic Refuge Protection

At locations far from the source, shelter-in-place is adequate (and can be for longer durations of time). However as the outside concentration increases there are closer locations which may be overwhelmed unless the building is designed against infiltration (like a safe haven). The distance scale and the shape (how fast the concentration drops) is dependent on the actual release scenarios. For many smaller release scenarios the concentration may drop off such that a safe haven design is not needed. Conversely depending on the toxicity of the chemical and the proximity to the source, there could be larger releases with a center line concentration that could potentially overwhelm a tight safe haven design. For these cases there is a proximity zone where the outside concentration is too close to the source for the building. In this approach the credible design point scenario establishes the basis for the toxic refuge design to protect the occupants and the emergency response plan uses the credible design point scenario to determine the extent and magnitude of evacuation, if necessary, and protection. For instance, if the ongoing event is larger than the credible design point scenario then Emergency Response Plan (ERP) would call for evacuation of the toxic refuge. If the ongoing event is smaller than the credible design point scenario then the emergency response plan would not evacuate the toxic refuge. In the same manner the credible design point scenario can be used to determine the magnitude of evacuation zones and the extent for community shelter-in-place instructions. The actual shape and the various zones under the curve will adjust based on the actual scenario under consideration.

In addressing the various options available there is a hierarchy of preference. While the hierarchy is listed below there are many factors to consider in each review, and the hierarchy below does not replace engineering. Also, a site may not have all the line items listed in the hierarchy; for instance, if evacuation can be proven to be safely effective then there may not be a need for a safe haven.

1st Hierarchy Level is Evacuation - The purpose of evacuation is to place distance between the ongoing event and potential occupants of buildings. When evacuation is safely completed then people are out of the potential hazard zone. Evacuation must be carefully planned as dispersion plumes can have varying shapes. Rarely are dispersion plumes a full circle, so evacuation can be used to safely move people who are not in the immediate dispersion plume path but whom may be in the dispersion potential hazard zone. See Figure 3.

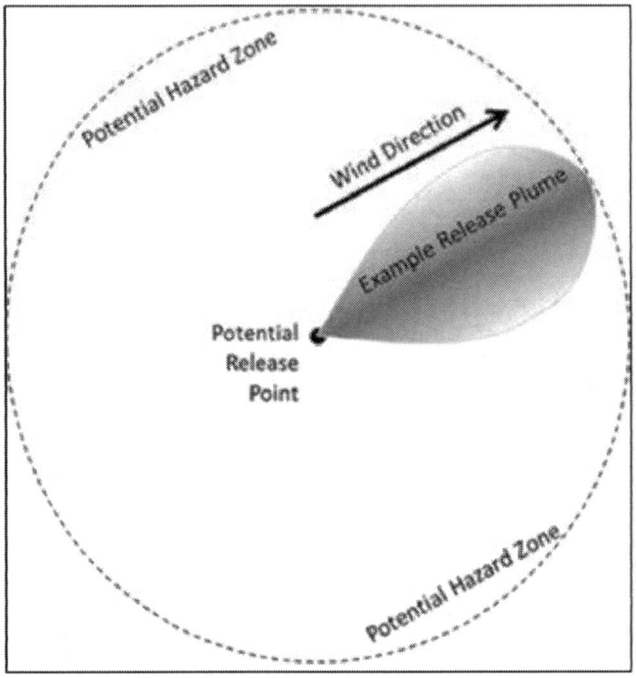

Figure 3: Relation between the Dispersion Plume and Potential Hazard Zone

With this in mind care should be exercised in evacuation to assure that people do not cross into a path of potential harm. Evacuation is more than a simple statement (evacuate) and needs to be an evaluation process (completed in advance when emergency response plan is developed) as part of a process of understanding the toxic risk. The emergency response plan should pre-determine how and when to evacuate so that occupants whom may be safer inside a toxic release shelter are not placed at greater risk by evacuating into a potential release plume.

2nd Hierarchy Level is Safe Haven – When occupants must remain within the dispersion release plume, then the safe haven concept can be employed. This approach keeps occupants safe such that they may shelter or safely bring plant conditions to a safer state. For longer duration events the emergency response plan may need the option of safely egressing people from the safe haven.

3rd Hierarchy Level is Shelter-In-Place - This option is most viable at lower centerline dispersion concentrations. At these lower levels a shelter-in-place approach can provide longer durations of protection.

4th Hierarchy Level is Non-Toxic Refuge building - This option can be useful if shelter-in-place is not available. In multi-story buildings (such as a large administration building) additional protection may be afforded by simply going to higher floors to get above lower concentrated layers of heavier than air dispersions.

5th Hierarchy Level is Refuge of Opportunity – This is used as a last resort when no other levels of protection are available. This level could be a closed motor vehicle cabin that is not running. This is the ability for an individual to create a barrier between themselves and the release that is occurring. Care and caution should be taken to not start the vehicle or induce any spark as the toxic dispersion may also include flammables, and a spark could ignite the vapor cloud.

1.4 Scenario Development

Scenario Development: Overview

Scenario Development is the one of three parts that must work together for effective toxic chemical management. The other two parts are the emergency response plan and the toxic refuge design.

When managing a process with toxic chemicals there are many possible scenarios that could lead to a potential release occurring. However, only one of those scenarios establishes the design criteria for a toxic release shelter and the associated emergency response process. What is necessary is that a site determines a credible design release basis and that the emergency response plan and toxic release shelter design accommodate the credible design point.

When managing toxics there are often multiple independent layers of protection that are in place which minimizes the frequency of larger releases. As such typically the larger release or the worst case release is not the controlling scenario. This worst case scenario is often at a much lower risk due to the lower frequency such that a smaller size scenario may be the controlling credible risk at a higher frequency. There is a cumulative effect of all the independent protection layers in place and possible release scenarios that lends to managing the risk for a credible design point scenario versus a worst case scenario.

There are many ways that a company/site can determine the credible design point scenario. The following list gives a few example approaches that can be used individually or combined together to develop the credible scenario:

1. Maximum consequence at given frequency (such as frequency of event to number of fatalities curves).
2. Fixed consequence at determined frequency (individual risk curves).
3. Probability of release (a mass leak rate to include some level of all the possible releases that could occur ie: 75%, 95%, or some other level).
4. A preselected condition (a given release case such as a ½", ¾", or some other leak size).
5. Other design cases established by the company/site.

The basis of the above approach used by the company/site should be written and documented. Once the credible scenario is determined a centerline concentration can be calculated using a dispersion model.

Scenario Development: Tools

Some tools that can be helpful in selecting the credible design point scenario are listed here.

1. Quantified Risk Analysis (QRA). As many scenarios are provided it may be possible to determine scenarios with higher consequences above a certain frequency level. May also look at multiple scenarios and determine which ones are controlling at differing locations.
2. Layer of Protection Analysis (LOPA) approach. This identifies the initiating event and the protection layers in place.
3. Hazard Identification Methodologies. Differing hazard identification methods may be helpful in identifying technology specific scenarios that could be used as a credible design point case.
4. Other approaches. Companies may use other methods to identify controlling scenarios.

Scenario Development: Deliverables

The scenario development deliverables are listed below.

1. Development of the credible design point scenario that is auditable to applicable industry standards for the toxic refuge design.
2. Development of the credible design point scenario basis for emergency response plan development.

1.5 Emergency Response Plan

Emergency Response Plan: Overview

The emergency response plan is one of three parts that must work together for effective toxic chemical management (the other two parts are the Scenario Development and the Toxic Refuge Design).

Based on the distance of a toxic dispersion plume, the emergency response may need to cover both offsite response and onsite response conditions. Additionally when the site has the potential for offsite impacts from toxic dispersion plumes, the emergency response plan may need a prepared messaging statement for community notifications for both incidents involving the toxic material and for those incidents that may occur at the site but not involving toxics.

The emergency response plan should use the credible design basis and a determination of the actual release to make a decision if community shelter-in-place or community evacuation is the appropriate course of action. This same logic addresses onsite toxic refuge evacuation decisions. If the credible design basis event (or less severe) is occurring and all the safety systems are functioning the emergency response can tolerate people remaining inside toxic refuge locations. However, if the ongoing event escalates beyond the credible design basis scenario, then evacuation may be necessary for building occupants. As such the people occupying the decision making roles in the incident command system must be aware of the basis of the credible design scenario and a recognition of whether the safety systems are working as designed or not.

The emergency response plan uses the credible design point scenario to determine (as a clear decision point) if the actual event is worse than design point or not (ie: two branches) for occupied buildings:

1. If safe haven needs evacuation or can remain in place for the event duration
2. If the shelter-in-place zone needs evacuation or can remain in place for the event duration.

Emergency Response Plan: Implementation

An effective emergency response plan will include training on toxic emergency response which will include:

1. Drills involving credible design point scenario releases or smaller,
2. Drills on evacuation of toxic refuge for larger than credible design point scenario,
3. Drills on effectiveness of placing toxic refuges into their safe mode of operation.

Emergency Response Plan: Deliverable

1. Development of an emergency response plan based on releases larger than the credible design point scenario and smaller than the credible design point scenario.

1.6 Toxic Refuge Design

Toxic Refuge Design: Overview

Toxic refuge design is one of three parts that must work together for effective toxic chemical management. The other two parts are the scenario development and the emergency response plan.

Toxic refuges can see many potential scenarios. As safe havens can be of varying designs, including fresh air supplies for occupants and scrubbing systems for air ingress points, engineering is involved in determining the suitability of the safe havens response for a given external release condition. Shelter-in-place buildings can also be of varying designs and include various tightness levels. For both types of buildings the credible design point scenario is used to establish the auditable, to applicable industry standards, toxic refuge design basis. This is the measurable external concentration used to determine that the interior of the safe haven or the shelter-in-place buildings remains within safe exposure levels for the duration of the event.

A key difference between safe havens and shelter-in-place buildings is that a safe haven building has an engineered design air change per hour rate for the building (or portion of the building) while a shelter-in-place building uses a reasonable air change per hour value for the style of building.

As there are many different dispersion models that can calculate the concentration at the toxic refuge shelter, the model selected as well as the dispersion assumption basis used for the analysis should be documented. This dispersion analysis provides the external concentration at the toxic refuge location buildings air intake so that an interior concentration calculation over time can be completed. The interior concentration level selected is determined by the company's risk acceptance criteria.

Toxic Refuge Design: Implementation

Effective implementation of the toxic refuge needs to include a verification of the engineering assumptions used in the calculations. For safe havens this includes assuring the process design assumptions are still valid on the unit that established the credible design scenario and periodically validating the building tightness integrity is maintained (air change per hour leak test, ducting damper seals, door/window seals, wall piercings, and so on).

Safe havens also need occupant training on the process to put the building into the "safe mode" of operation.

Toxic Refuge Design: Deliverables

1. A toxic refuge that is designed to protect occupants from external hazards based on the credible design point scenario.
2. A toxic refuge design that works in conjunction with the emergency response plan.
3. A toxic refuge design that is auditable to applicable industry standards.

2 Application of this Process

When dealing with toxic chemicals the reference concentration and the duration of the exposure should be considered together as a dosage. The dosage in turn can be used to establish a criteria for the scenario evaluations. This dosage approach applies to both outside or inside the building exposures.

2.1 Determination of the endpoint

Selection of the reference concentration involves understanding the tolerable dosage exposures for the specific toxic chemical and the scenario that may lead to the exposure. When making this determination there are published dosages that are associated with pre-determined exposure times such as acute exposure guideline levels (AEGL), emergency response planning guidelines (ERPG), immediate dangerous to life or health (IDLH), short term exposure levels (STEL) and other 3rd party published endpoints, as well as calculated dosages that utilize a probit to estimate a probability of lethality. As such for any release that occurs, there is an infinite seriatim of values that can span orders of magnitudes when selecting a reference concentration. Figure 4 arranges the seriatim for commonly available reference concentrations (see sources) for an example chemical and arranges them on a typical centerline dispersion graph. Visually there is a range of endpoints that encompasses three orders of magnitude. Selection of the endpoint is a company specific determination and it may vary based on the scenario under review. As this graphic is a single example for a specific release scenario of one toxic chemical, the actual distances will vary based on each scenario and chemical reviewed. The underlying center line concentration curve will be different for each scenario (smaller leaks will not go as far as larger leaks thus for the same endpoint the distance reached will be closer for smaller leaks and farther for larger leaks).

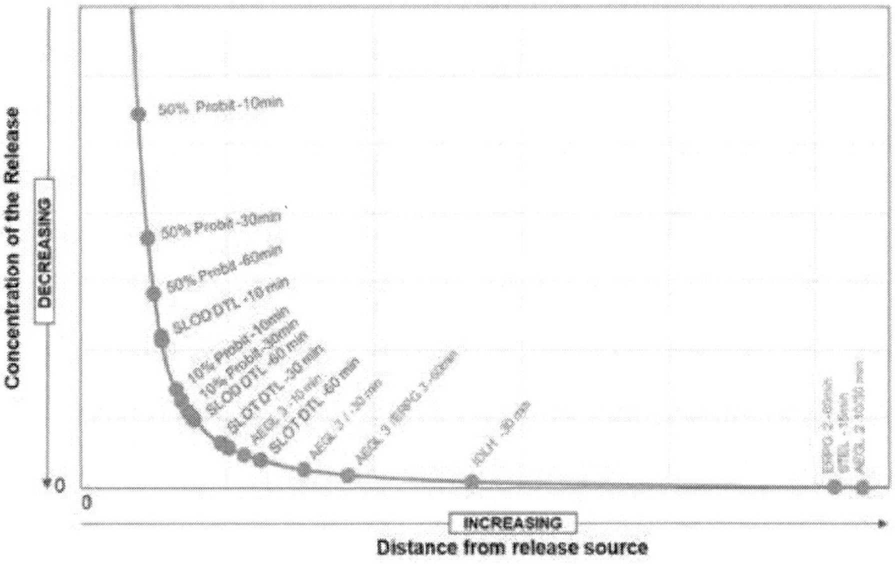

Figure 4: Example Center Line Dispersion Curve and Toxicity Endpoints

Source references for the data in this graphic can be found in the public domain (from references 1, 2, 3, and 4).

The selection of the reference concentration does not have to be limited to a single point. For instance a higher reference concentration may be tolerated at a lower frequency and/or a lower reference concentration may be used for a higher frequency scenario. In this manner multiple endpoints may be selected and evaluated.

2.2 Occupied Building Evaluation

Once the endpoint(s) is (are) selected the building evaluation can be completed. Dispersion analysis is used to determine the external concentration at the building of concern, and then a building tightness calculation can be used to determine the time it takes for the outside concentration to ingress to the inside of the building. Published literature is available to perform the building leak in calculations (such as reference 5).

These calculations rely on air change rates which can be measured via tracer gas tests (see reference 6).

In general the tighter the building then the longer time it takes for infiltration to occur. So safe haven buildings are tighter than shelter-in-place buildings.

Consideration for building factors can affect the performance of the toxic refuge. These can be items such as:

- Time factor – how long it takes to implement the safe mode during an emergency,
- Mechanical factor – the integrity of the building envelope and the maintenance of the buildings engineered features
- Effectiveness factor – the number of times the external door is opened during the event

Applying this basis, each toxic refuge has a "safe time" that the building will protect occupants from the selected reference concentration. This process is repeated for each scenario that can potentially impact the location where the building is sited.

The dispersion calculation for each scenario provides the centerline concentration that can occur at the buildings location. The footprint of the dispersion is also important to incorporate the plume effects of the release together with the wind rose for impacting the building.

2.3 Safe Time Determination

The safe time determination requires knowledge of the duration of the scenario. As many acute toxics systems have multiple independent layers of protection, these processes have a means to "safe-park" or isolate/control toxic inventories/pressures expediently which shortens the toxic dosage at the impacted location. While there may be groupings of scenarios that have similar times/dispersions, this dosage is fundamentally unique for each scenario/weather condition.

This safe time determination can also include building mitigation systems such as separate tighter rooms within the affected building, building chemical filtration systems, separate exterior entry door vestibules, pressurization from clean air sources, or other engineering solutions.

If the scenario duration time is less than the building leak in time, then the building has met its design criteria (such as the safe haven in Figure 5).

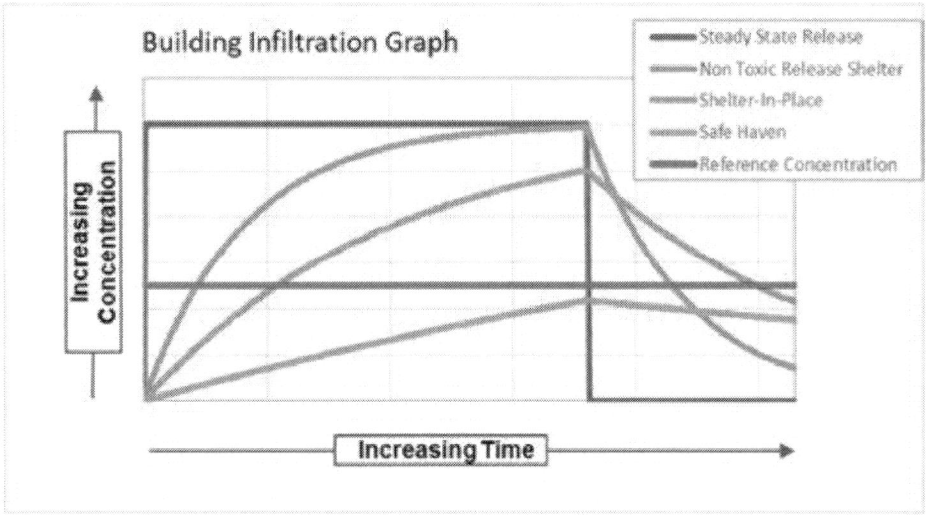

Figure 5: Example infiltration rates for buildings

2.4 Safety Margin Consideration

As with any analysis, there are many assumptions that are made throughout the process (availability of safety systems, and so on) and there are many uncertainties in the analysis itself (dispersions and frequency numbers). When comparing the building safe time to the scenario duration time, it is a convenient location in the process to introduce a safety factor. For instance, buildings that have leak times just a little bit longer (versus a lot longer) than the scenario duration time, it may be prudent to look at the sensitivity in the assumptions and data for those buildings. This essentially introduces a safety margin (described in Figure 6) for toxic refuge results that are close to the limits.

Buildings that are potentially at risk

These are buildings that the safe time is shorter than the event duration.

Buildings that may consider a safety margin

These are buildings that are sufficient for the controlling scenario (Safe Time is longer than Event Duration Time) however they are within the safety margin

Buildings with no need for a safety margin

These are buildings that are sufficient for the controlling scenario (Safe Time is longer than Event Duration Time with a safety margin)

Figure 6: Safety Margin Concept applied to Toxic Refuges

2.5 Toxic Facility Siting Risk Considerations

For many acute toxics, the selected reference concentrations are much lower (ie: 1-1,000 ppm) than typical hydrocarbon flammable limits (~10,000 to 20,000 ppm). Thus using the centerline concept this means that the toxic dispersion will have a farther reach and a wider footprint. In this context, the toxic scenario plumes have a larger impact zone than a comparable flammable plume. And the toxic plume may impact more locations within the same wind rose direction than a flammable plume, which infers that there could be potentially larger aggregated risk zones.

In addition to evaluating the impact on the building (which can impact multiple people within a building), it can also be beneficial to evaluate the impact footprint of a release in the direction that may have more than a single building impact. This requires an analysis of both the wind rose and dispersion footprint. The results of this can be important elements in developing emergency response plans.

Risk can also be analyzed geographically on a building location. For instance a location near the source release may be impacted by many different sized scenarios, whereas a location farther away may only be impacted by larger scenarios.

It is important that the facility siting analysis identifies risk aggregation in addition to individual risk as this can be a requirement of industry standards (see reference 7).

3 Iterative Nature of the Process

The process identified in this framework is iterative in that limitations on a toxic refuge design or emergency response can involve altering the credible design point scenario and re-evaluating the outcomes. There is a balance between the scenario selection, the building design, and the emergency response actions. For instance a scenario that is too large may overwhelm the building design or not allow enough time for occupants to put the building into the safe mode. Conversely a scenario too small may trigger evacuation needlessly to the point that occupants are desensitized to the actions.

Initial responses by building occupants to implement the safe mode will take a fixed amount of time (even if automated) and emergency response also will take some amount of time to establish the incident command system and receive the initial leak evaluation. These times and each building design may require iteration within this framework to establish a viable and practicable solution.

4 Summary

There are many variations to this process that can have a solid engineering basis. While the approach outlined in this paper is one method it is not meant to eliminate or disfavor other technical approaches.

In managing the effects of toxic release there is recognition that emergency response, toxic refuge design, and credible design point scenario all work in conjunction with each other and are interactive with each other. Changes to one will affect the other two. This process can involve the shape of the plume and the concept of the relation to the toxic refuge on a centerline dispersion curve. As such an effective toxic facility siting analysis can be an iterative process that requires multiple disciplines working together to achieve an optimal result.

Applying the approach in this paper has the following benefits:

1. Is a science and data based approach
2. Provides an engineering basis that is determines the toxic concentration and duration that is safe for the occupants
3. Provides an auditable to applicable industry standards scenario, building design, and emergency response criteria

These concepts provided in this paper can also be used to provide a technical basis for showing measurable impacts of potential changes (ie: projects to increase the safe time or alter emergency response). When compared to a base case it can provide the basis for a cost to benefit analysis.

5 References

/1/ Chemical Center for Process Safety Guidelines for Chemical Process Quantitative Risk Analysis, 2nd edition 2000, Center for Chemical Process Safety, American Institute of Chemical Engineers, New York, NY, 2000.

/2/ United Kingdom Health & Safety Executive Toxicity Level of Chemicals, Available at https://www.HSE.gov.uk/chemicals/haztox.htm Accessed on January 18, 2021.

/3/ United States Environmental Protection Agency, Available at https://www.EPA.gov/aegl Accessed on January 18, 2021.

/4/ United States Center for Disease Control, Available at https://www.CDC.gov Accessed on January 18, 2021.

/5/ Mannan, S, "Lee's Loss Prevention in the Process Industries", 4th Edition 2012, Equation 15.51.3.

/6/ American Society for Testing Materials, "Standard Test Method for Determining Air Change in a Single Zone by Means of Tracer Dilution." E741-11, Nov 2017.

/7/ American Petroleum Institute Recommended Practice 752, "Management of Hazards Associated with Location of Process Plant Permanent Buildings", third edition December 2009.

SPRING21
+17TH GCPS
A Joint AIChE and CCPS Meeting

Facility Siting - Negotiating the Minefield of "Maximum Credible Event"

Kristen Graham
Baker Engineering and Risk Consultants, Inc. (BakerRisk®)
11011 Richmond Avenue Suite 700, Houston, TX 77042
KGraham@Bakerrisk.com

J. Kelly Thomas
BakerRisk
3330 Oakwell Court Suite 100, San Antonio, TX 78218
KThomas@Bakerrisk.com

Prepared for Presentation at
American Institute of Chemical Engineers
2020 Spring Meeting and 17th Global Congress on Process Safety
Virtual
April 18 - 22, 2021

Facility Siting - Negotiating the Minefield of "Maximum Credible Event"

Kristen Graham
Baker Engineering and Risk Consultants, Inc. (BakerRisk®)
11011 Richmond Avenue Suite 700, Houston, TX 77042
KGraham@Bakerrisk.com

J. Kelly Thomas
BakerRisk
3330 Oakwell Court Suite 100, San Antonio, TX 78218
KThomas@Bakerrisk.com

Keywords: Flammability, Thermal Radiation, Explosions, Toxics, Risk, Safety, Facility Layout, Facility Siting, Managing Process Safety, Design Criteria

Abstract

The purpose of this paper is to provide a case study that compares the results from a consequence-based facility siting study (FSS) limited to a specified release size as the maximum credible event (MCE) versus a risk-based study with a range of release sizes and associated frequencies. The results are intended to highlight potential differences in the building siting decisions made using a consequence-based approach with a limited MCE versus those made with a risk-based approach.

Most recommended practices and other FSS guidance documents advise evaluating MCEs for an FSS. However, there can be uncertainty in MCE selection and the values used in industry span a considerable range. Arguments for limiting the MCE in a consequence-based FSS to a 2-inch release include that most accidental releases (i.e., > 95%) are smaller and that a 2-inch release size is the largest that provides actionable information upon which siting decisions can be made. However, a number of well-known incidents in industry involve releases larger than 2-inches, which introduces uncertainty as to a reasonable upper-bound for an MCE. The use of a risk-based FSS approach effectively eliminates the decision regarding release size since this approach aggregates the full range of operational scenarios and release sizes along with the associated likelihood of failure. Based on the MCE scenario(s) employed and the criteria selected, consequence-based and risk-based FSS evaluations for a given site may lead to very different conclusions as to which buildings are considered safe and what types of mitigation actions should be taken.

1 Introduction

Recommended practices such as API RP 752 [1] and RP 753 [2] advise evaluating maximum credible events (MCEs) as the basis of a facility siting study (FSS). However, there is uncertainty across the industry as to what release size constitutes an MCE for the purposes of an FSS. Consequence-based studies are sometimes limited to a 2-inch release on the basis that most accidental releases (i.e., > 95%) are 2-inches or smaller [3]. However, a number of well-known incidents in industry involve releases significantly larger than 2-inches, which introduces uncertainty as to a reasonable upper-bound for an MCE. Examples of such events include the 1989 Phillips Chemical Company explosion in Pasadena, Texas, which gave rise to the OSHA PSM "Process Safety Management of Highly Hazardous Chemicals" regulations, and the BP Texas City Refinery explosion in 2005, which gave rise to API RP 753 for portable buildings [4].

The occurrence of incidents involving large release sizes (i.e., greater than 2-inches) may prompt evaluation of such releases as MCEs. One option is to consider full bore releases, or up to 6-inches in recognition that the release rate associated with even larger release sizes generally decreases rapidly. However, larger releases assessed on a consequence-basis may provide results that would suggest impractical or extremely costly mitigation. Many companies considering larger releases (i.e., in recognition of actual incident history within industry) have therefore moved to a risk-based approach, which allows evaluation of small, medium, and large events as well as the likelihood of occurrence. The risk-based approach effectively eliminates consideration of MCE release size since it incorporates likelihood into the full range of operational and release scenarios.

The purpose of this paper is to explore the differences in siting decisions based on a consequence-based FSS for release sizes up to 2-inches and for full bore (i.e., up to 6-inches) and a risk-based FSS. Two examples are presented to illustrate the conclusions that may be drawn depending on the analysis type and hole size chosen. The examples are intended to represent a "typical" small gas fractionation plant and a "typical" ammonia distribution terminal. It is acknowledged that other facility types, operations, and layouts could lead to different conclusions regarding consequence- and risk-based FSS analyses, but these examples are illustrative.

2 Risk and Consequence Criteria

As with the variation across industry in terms of MCE selection for FSS analysis, there is variation in the criteria or thresholds established for occupied buildings. Just as the company must select the MCE basis for an FSS based on corporate requirements and/or industry best practice, they must also set building risk criteria. The criteria provided in Table 1 are representative of those used for consequence- and risk-based FSS analyses, and hence were also used for the purposes of examples provided in this paper.

Table 1. Risk and Consequence Criteria for Buildings

Hazard Type	Consequence-Based Study	Risk-Based Study
Explosion	Building Damage Level < 2.5 (i.e. widespread building damage)	Building Societal risk **per hazard type** • Lower threshold: 1E-5 fatality/year • 1E-5 to 1E-3 ALARP Region • Upper threshold: 1E-3 fatality/year
Toxic	<1% lethality for escape	
Flash Fire (flammable)	< LFL	
Thermal (jet or pool fire)	<1% lethality for escape	

If a predicted consequence exceeds the corresponding criteria (e.g., higher building damage or percent lethality than the criteria), then mitigation would typically be applied to reduce the predicted consequence to an acceptable level. Risk criteria can be applied to facilities, process units, buildings, workgroups, and/or offsite populations. API RP 752 recommends both individual and societal risk criteria, however this paper only invokes building societal risk criteria for the purposes of illustration. Risk criteria can also be applied per hazard type or as an overall criterion. For ease of comparison, this paper uses a risk criteria per building per hazard type. Risk criteria are often applied with two sets of thresholds, an upper threshold and lower threshold, as shown in Figure 1. If risk exceeds the upper threshold, the risk is considered intolerable and mitigation is required. If risk is below the lower threshold, the risk is considered broadly acceptable and no action is required. If risk falls between the two thresholds, referred to as the ALARP (as low as reasonably practicable) region, mitigations should be sought and implemented if practicable. Obviously, the use of different consequence or risk criteria would lead to different conclusions and mitigation recommendations [5].

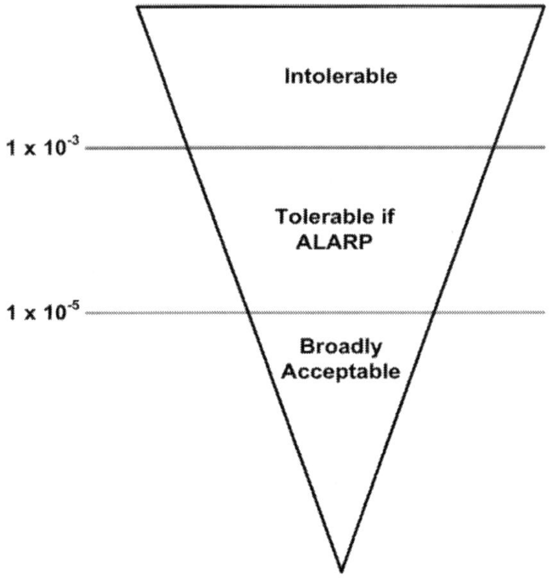

Figure 1. Risk Criteria [6]

3 Modeling Methodology

The methodology shown on the following page was used in conjunction with the SafeSite$_{3G}$© and QRATool© software tools for consequence and risk calculations, respectively, for the purposes of this case study. Similar results would be expected with other widely accepted consequence/risk methodologies and software tools.

Release scenarios were selected by reviewing process flow diagrams (PFDs) to define a representative set of hazard sources.

- A range of release sizes was considered for each identified hazard source.
- Hole diameters of 0.5-inch and 2-inch for consequence FSS.
- Full-bore rupture up to 6-inch was added to 0.5-inch and 2-inch for risk FSS.

Dispersion of the material from releases was modeled taking into account thermodynamic properties, flashing, rainout, pooling, evaporation, and gas dispersion.

- Flammable impacts were assessed for "open" and "closed" buildings.
- Vulnerabilities were predicted for each toxic scenario as personnel evacuate or shelter-in-place (SIP) using dose calculations and probit expressions and can credit mitigations such as HVAC isolation and PPE.

Jet and pool fire consequences were assessed using industry appropriate fire models.

- Vulnerabilities were predicted for each fire scenario as personnel evacuate or shelter-in-place (SIP) using dose calculations and probit expressions.

Vapor cloud explosions were modeled using the Baker-Strehlow-Tang methodology to determine blast loads on buildings. Building damage levels were predicted for each onsite building using the proprietary Building Evaluation and Screening Tool (BEAST©).

- Vulnerability values were predicted for each onsite building for each blast scenario assessed based on the building construction type.

To convert vulnerabilities to risk results:

- Vulnerabilities were converted to predicted number of fatalities based on occupancy data for each building.
- Frequencies were assigned to each release based on estimated pipe length, quantities of other process equipment, and failure rates based on industry experience.
- Scenario frequency was calculated by multiplying the release frequency by probabilities for weather conditions, release/wind direction, ignition timing, and the time of day.
- Frequency and consequence results were combined to define risk for each scenario and the overall risk was calculated by adding the risk of each scenario assessed.

4 Case Studies

Two examples are presented to illustrate the conclusions that may be drawn depending on the analysis type and hole size chosen. The examples are intended to represent a "typical" small gas fractionation plant and a "typical" ammonia distribution terminal. These examples do not represent actual facilities. The results from these examples are not intended to provide recommended mitigation based on facility type alone.

4.1 Case Study 1: Gas Plant

Case study 1 represents a small gas plant including fractionation and NGL storage. Figure 2 shows an overview of the facility including areas of congestion/confinement along with the locations of release sources, buildings, and ignition sources. Table 2 gives the building structural models and occupancy. Table 3 provides the release sources and conditions modelled.

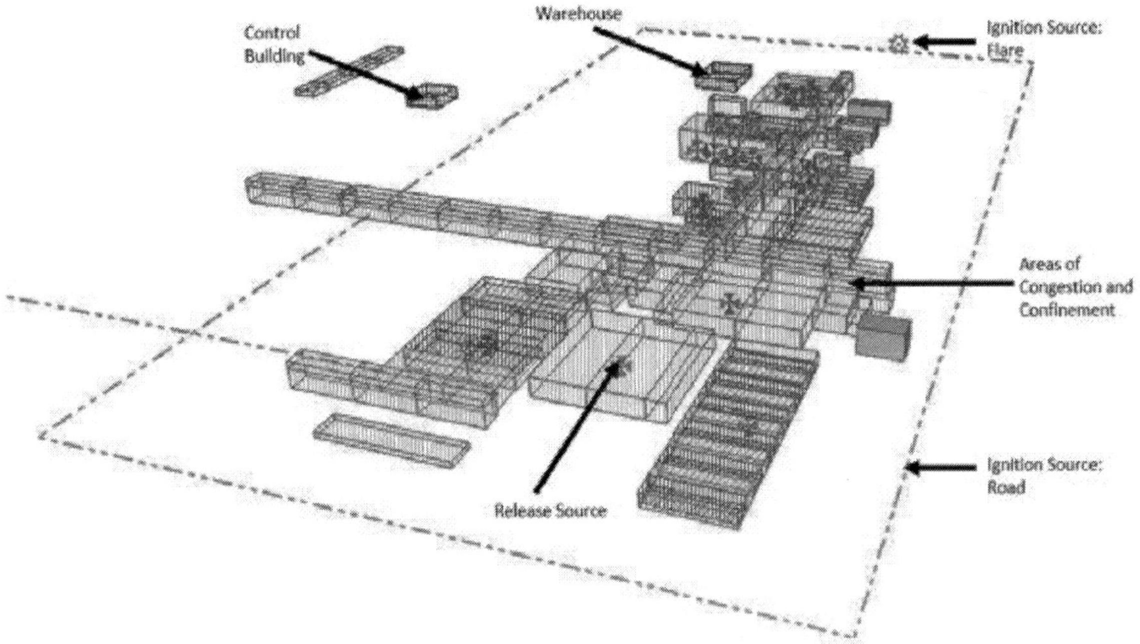

Figure 2. Facility Siting Model (Case 1)

Table 2. Buildings Modelled (Case 1)

Building	Construction Type	Day Occupancy	Night/Weekend Occupancy
Control Building	Wood Trailer	5	2
Warehouse	Pre-Engineered Building- Large, High Bay	2	0

Table 3. Release Sources Modelled (Case 1)

Source	Description	Material	Temperature (F)	Pressure (psig)	State	Max Size (in dia)
01-ScrubberOvd	Inlet gas scrubber overhead to amine filter/separator	C1/C2/C3/N2 73/15/10/2	50	1,000	Gas	6
02-ScrubberBtm	Inlet gas scrubber pumps to condensate stabilization trains	C1/C2/C3/C4/C5 37/21/22/14/6	50	1,000	Liquid	6
03-AmineCont	Amine contactor overhead to scrubber to dehy inlet gas filter separator	C1/C2/C3/C4/N2 74/13/6/5/2	100	1,000	Gas	6
04-InletFilter	Inlet gas filter separator to dehydrators to cryo gas/gas exchanger	C1/C2/C3/C4/N2 72/13/8/5/2	120	955	Gas	6
05-RegenComp	Regeneration gas compressor to gas heater	C1/C2/C3/C4/N2 72/13/8/5/2	125	1,000	Gas	6
06-RegenDehy	Regeneration dehydrator to cooler	C1/C2/C3/C4/N2 72/13/8/5/2	525	980	Gas	6
07-RecycleExch	Recycle gas exchanger to demethanizer	C1/N2 97/3	-60	725	Gas	6
08-GasExch	Gas/gas exchanger to cold separator	C1/C2/C3/C4/N2 72/13/8/5/2	-30	950	Liquid	6
09-ColdSepBtm	Cold separator bottoms to demethanizer	C1/C2/C3/C4/N2 52/20/16/11/1	-30	950	Liquid	6
10-ColdSeptOvd	Cold separator overhead through expander to demethanizer	C1/C2/C3/N2 85/9/3/3	-30	950	Gas	6
11-DemethBtm	Demethanizer bottoms to bottoms pumps	C2/C3/C4/C5 52/30/12/6	65	270	Liquid	6
12-DemethOvd	Demethanizer overhead through exchangers to compressor	C1/N2 97/3	-165	270	Gas	6
13-Comp	Compressor discharge to cooler	C1/N2 97/3	140	300	Gas	6
14-Expander	Expander/compressor discharge cooler to residue gas compression	C1/N2 97/3	120	290	Gas	6
15-DemethPump	Demethanizer bottoms pumps to cryo NGL pumps	C2/C3/C4/C5 52/30/12/6	65	615	Liquid	6
16-RefridgeComp	Refrigeration suction scrubber to refrigeration compressors	C2/C3/C4 2/96/2	1	20	Gas	6
17-CompDisch	Refrigeration compressor discharge to surge vessel	C2/C3/C4 2/96/2	180	245	Gas	6
18-Economizer	Refrigerant surge vessel through economizer to gas chiller	C2/C3/C4 2/96/2	115	235	Liquid	6
19-NGLPumps	Cryo NGL pump discharge	C2/C3/C4/C5 52/30/12/6	70	1,000	Liquid	6
20-ResidueComp	Residue gas compressor discharge	C1/N2 97/3	120	1,200	Gas	6
21-StabilizerComp	Condensate stabilizer overhead to compressors	C1/C2/C3/C4 53/27/15/5	60	230	Gas	6
22-DrumBtm	Condensate flash drum bottoms through deethanizer to condensate stabilizer	C2/C3/C4/C5/C6 9/32/28/15/16	185	280	Liquid	6
23-StabilizerBtm	Condensate stabilizer tower bottoms to oil storage tanks	C4/C5/C6 3/30/67	350	240	Liquid	4
24-StabilizerOvd	Condensate stabilizer tower overhead to LP gas header	C2/C3/C4/C5 31/48/19/2	140	230	Gas	6
25-RefluxPump	Condensate stabilizer reflux pumps to tower and NGL bullets	C2/C3/C4/C5 11/42/37/10	135	230	Liquid	4.
26-NGLStorage	NGL storage vessels	C2/C3/C4/C5 11/42/37/10	80	130	Liquid	6

4.2 Case Study 1 Results

Contours plots were generated to provide consequence results for releases up to 2-inches and for full bore releases (i.e., up to 6-inches). Each release case is rotated in 16 directions, with the results for all release cases and directions combined to form a composite contour. Figure 3 shows the flammable gas concentration contours to the endpoints of the ½ of the lower flammability limit (LFL), the LFL, and the upper flammability limit (UFL). Figure 4 provides thermal radiation contours to endpoints of 4, 12.5, and 37.5 kW/m^2. Figure 5 shows the VCE blast overpressure contours to thresholds of 0.6, 0.9, 3, 5, and 10 psig. No toxic scenarios were evaluated for this facility.

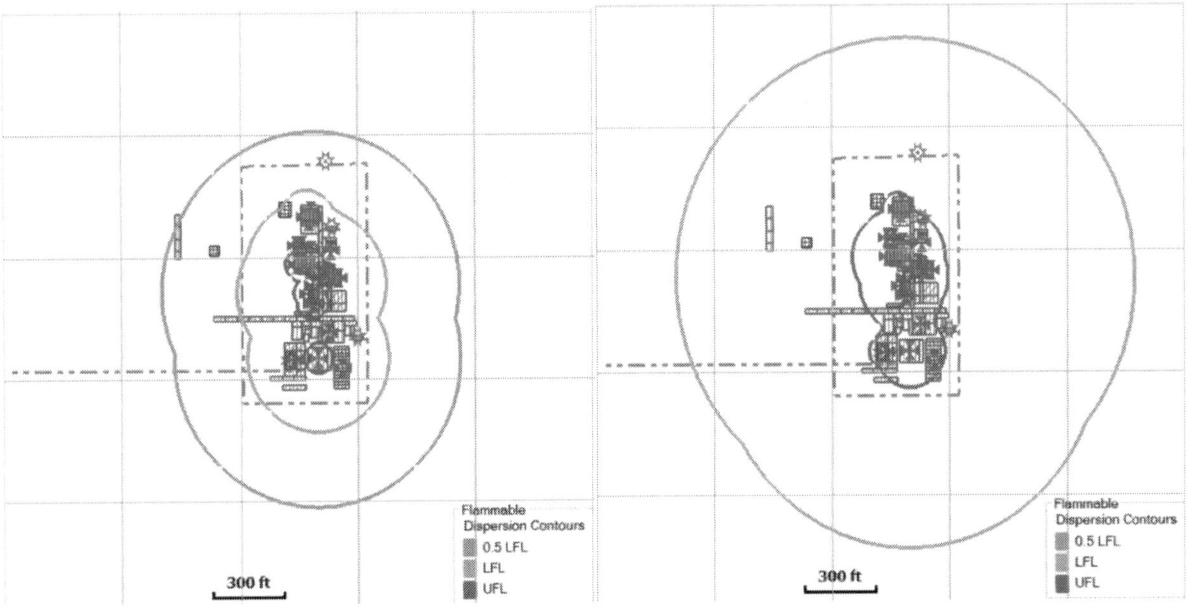

Figure 3. Flammable Contours for Releases up to 2" (left) and Full-Bore up to 6" (right)

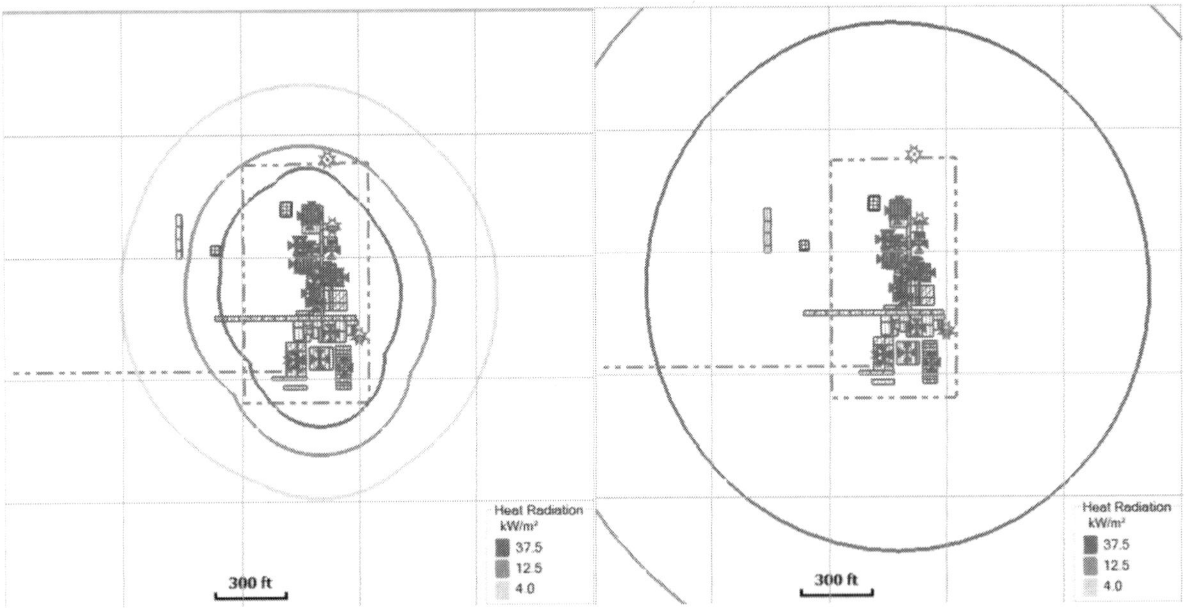

Figure 4. Thermal Radiation Contours for Releases up to 2" (left) and Full-Bore up to 6" (right)

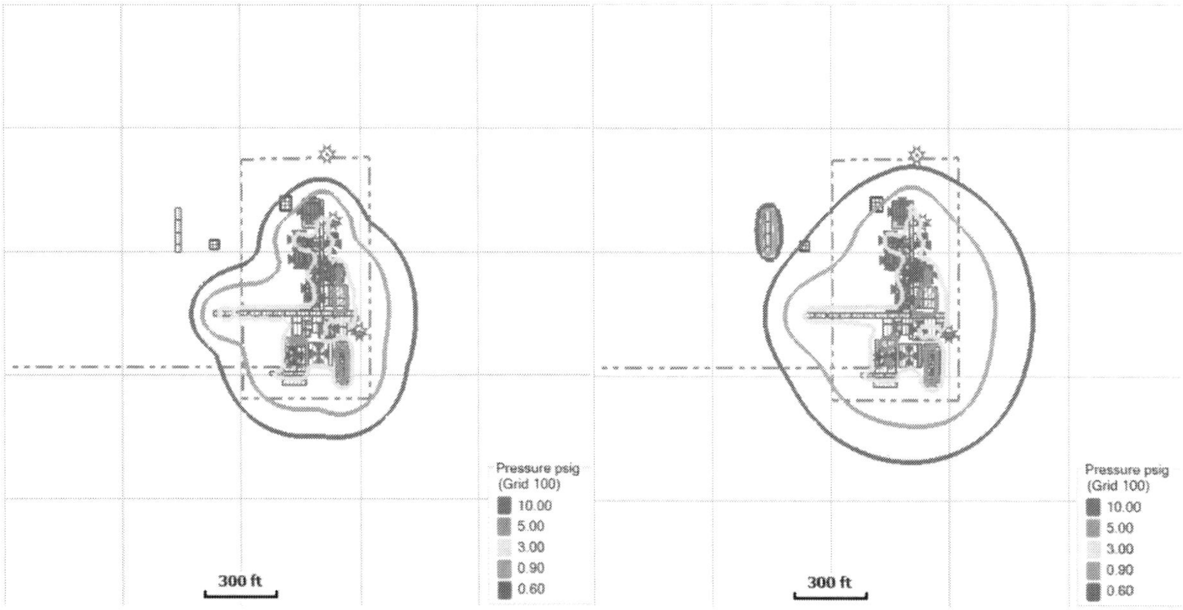

Figure 5. Overpressure Contours for Releases up to 2" (left) and Full-Bore up to 6" (right)

The consequence results summary table is shown in Table 4. The Warehouse exceeds the explosion criteria (BDL>2.5), the flammable criteria (>LFL), and the thermal criteria (>1% lethality) for both 2-inch and full-bore releases. The Control Building does not exceed any consequence criteria for releases up to 2-inches but exceeds the thermal and flammable criteria for full-bore releases. The risk results summary is shown in Table 5. The Warehouse risk is in the ALARP region (between the high and low risk thresholds) for explosion, flash fires, and jet fires. The Control Building is in the ALARP region for flash fires and jet fires and below the lower threshold for explosion risk. Table 6 shows a comparison for each building and hazard type describing whether they meet or exceed the consequence criteria for 2-inch and full-bore releases along with the comparison against the risk criteria.

Table 4. Building Consequence Results (Case 1)

Building: Section	Maximum Impacts								
	Building Damage Level			Flammable Gas Conc.			Fire Lethality		
	0.5 in	2 in	Full Bore	0.5 in	2 in	FB	0.5 in	2 in	Full Bore
Control Building			2		>1/2 LFL	>LFL			100%
Warehouse		3	3		>LFL	>LFL		100%	100%

Table 5. Building Societal Risk (Case 1)

Building	Building Societal Risk (fatalities /year)			
	Explosion	Flash Fire	Jet Fire	Total
Control Building	3.0E-10	1.1E-5	**7.1E-4**	7.1E-4
Warehouse	1.6E-5	2.5E-5	**1.6E-4**	1.8E-4
Totals	**1.6E-5**	**3.6E-5**	**8.7E-4**	**8.9E-4**
	2%	**4%**	**94%**	

Table 6. Consequence and Risk Results Comparison (Case 1)

Hazard	Study Type	Control Building		Warehouse	
		Threshold	Comment	Threshold	Comment
Blast	2" Consq.	Meets Criteria	No action	Exceeds Criteria	Structural Upgrades, Relocate
	6" Consq.	Meets Criteria	No action	Exceeds Criteria	Structural Upgrades, Relocate
	Risk	Broadly Acceptable	No action	Tolerable if ALARP	Consider Upgrades, Relocation
Flammable	2" Consq.	Meets Criteria	No action	Exceeds Criteria	LEL Detection, HVAC Shutdowns
	6" Consq.	Exceeds Criteria	LEL Detection, HVAC Shutdowns	Exceeds Criteria	LEL Detection, HVAC Shutdowns
	Risk	Tolerable if ALARP	LEL Detection, HVAC Shutdowns	Tolerable if ALARP	LEL Detection, HVAC Shutdowns
Thermal	2" Consq.	Meets Criteria	No action	Exceeds Criteria	Thermal Shielding, Relocation
	6" Consq.	Exceeds Criteria	Thermal Shielding, Relocation	Exceeds Criteria	Thermal Shielding, Relocation
	Risk	Tolerable if ALARP	Consider Thermal Shielding, Relocation	Tolerable if ALARP	Consider Thermal Shielding, Relocation

For the Control Building, the selection of the hole size makes a difference in whether the building exceeds the consequence criteria for flammable and fire scenarios. While the building is far enough from the process to not be impacted by the flammable clouds or thermal radiation resulting from 2-inch releases, a larger release could exceed acceptance criteria. Considering only 2-inch releases would imply the building location is adequately far from the process. Conversely, considering larger releases may require the building to be moved (which may not be possible) or upgraded (which may be expensive).

The risk analysis for the Control Building shows that the building falls within the ALARP region. This risk analysis allows the operator to not only understand how likely the events are to occur but also to make business decisions with regards to mitigation under ALARP considerations. The risk analysis also helps the operator prioritize thermal risk over flammable risk since the thermal risk is more than an order of magnitude larger than the flammable risk; this prioritization is not obvious based on the consequence analysis results.

For the Warehouse, the building is close enough to the process that the consequence results are identical whether a 2-inch or full-bore release is considered. In this case, the building exceeds the consequence criteria for all three hazard types and would require upgrades to mitigate each hazard type, which may be costly enough that relocating the building is a better option.

The risk analysis for the Warehouse shows that the building is within the ALARP region for each hazard type. Instead of mitigation being definitively required, an ALARP analysis may be performed to determine what level of mitigation should be implemented. The risk analysis also helps the operator prioritize thermal risk for this building. Last, the risk analysis allows mitigation prioritization for the Warehouse vs. Control Building; the Control Building mitigations should be prioritized as it is subject to higher risk.

While many of the results are in alignment for risk-based versus consequence-based analysis, a 2-inch consequence-based study alone would not have identified that the majority of risk was due to thermal risk at the Control Building.

4.3 Case Study 2: Ammonia Terminal

Case study 2 represents an ammonia terminal including ammonia storage and loading/unloading. Figure 6 shows an overview of the facility including areas of congestion/confinement along with the locations of release sources, buildings, and ignition sources. Table 8 gives the building structural models and occupancy. Table 7 provides the release sources and conditions modelled.

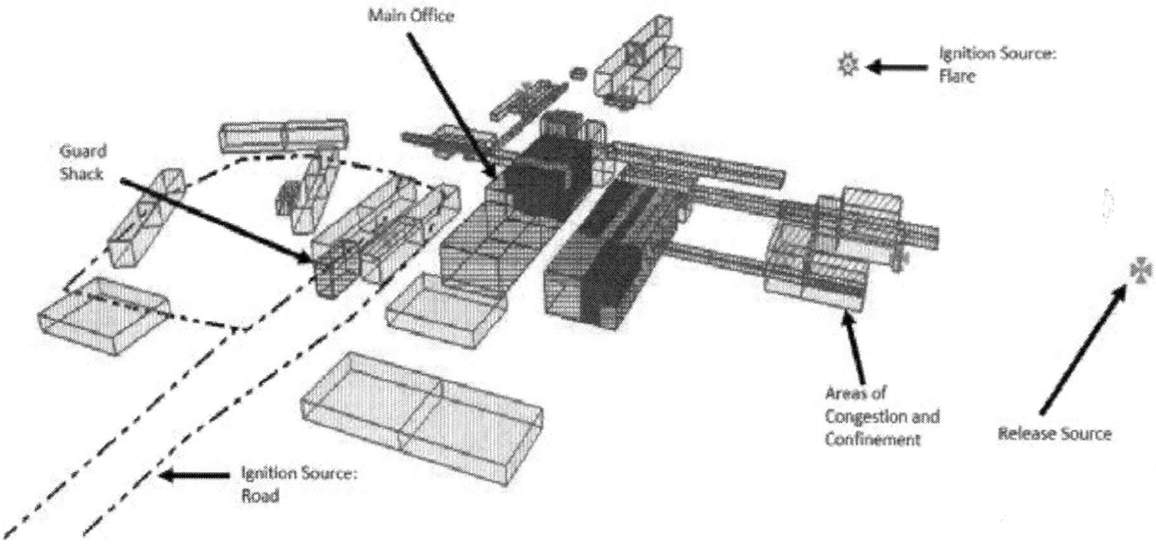

Figure 6. Facility Siting Model (Case 2)

Table 7. Buildings Modelled (Case 2)

Building	Construction Type	Shelter in Place	Air Exchange Rate	Egress Delay	PPE Reliability	Day Occupancy	Night/ Weekend Occupancy
Guard Shack	Wood Trailer	No	1 ACH @ 5mph	2 min	0.95	2	1
Main Office	Pre-Engineered Building- Large, High Bay	No	1 ACH @ 5mph	2 min	0.95	5	0

Table 8. Release Sources Modelled (Case 2, all releases are ammonia)

Source	Description	Temperature (F)	Pressure (psig)	State	Max Size (in dia)
01-Bullet	Ammonia Storage Bullet	60	250	Liquid	6
02-ColdPumps	Cold Pumps from storage through heater to Truck Loading	-30	250	Liquid	6
03-TruckLoad	Truck Loading racks	30	200	Liquid	2
04-Storage	Storage Tank to Desuperheater	-10	1	Gas	6
05-Desuperheater	Desuperheater overheads to Compressor	-25	1	Gas	6
06-CompStg1	1st STG. Compressor to interstage cooler	200	40	Gas	6
07-Interstage	Interstage separator overheads to 2nd STG. Compressor	30	40	Gas	6
08-CompStg2	2nd STG. Compressor outlet to condenser	270	200	Gas	4
09-Condenser	Condenser to Receiver	95	200	Liquid	6
10-RecieverBtm	Receiver bottoms to Interstage drum	95	200	Liquid	2
11-DrumBtm	Drum bottoms to Interstage drum	-10	120	Liquid	1
12-Desuperheater	Desuperheater overheads to Compressor	-25	1	Gas	6
13-CompStg1	1st STG. Compressor to interstage cooler	200	40	Gas	4
14-Interstage	Interstage separator overheads to 2nd STG. Compressor	30	40	Gas	6
15-CompStg2	2nd STG. Compressor outlet to condenser	270	200	Gas	4
16-Condenser	Condenser to Receiver	95	200	Liquid	4
17-RecieverOvh	Receiver overheads to Purger	110	200	Gas	1
18-HotPump	Hot pump from Storage Bullet to Truck loading	60	200	Liquid	4
19-Metering	Pipeline Meter Skid to Ammonia Storage Bullet	60	250	Liquid	4
20- TruckLoad	Truck Loading racks	30	200	Liquid	2

4.4 Case Study 2 Results

As with the Case 1 results, contours plots were generated to provide consequence results for releases up to 2-inches and for full bore releases (i.e., up to 6-inches). Figure 7 shows the flammable gas concentration contours to the endpoints of the ½ LFL, LFL, and UFL. Figure 8 shows the thermal radiation contours to endpoints of 4, 12.5, and 37.5 kW/m^2. Figure 9 provides toxic gas concentration contours to the endpoints of 1%, 10%, and 90% lethality for a 10-minute exposure. Figure 10 shows the VCE blast overpressure contours to thresholds of 0.6, 0.9, 3, 5, and 10 psig.

The consequence results summary table is shown in Table 9. The Guard Shack and Main Office both exceed the flammable criteria (>LFL), the thermal criteria (>1% lethality), and toxic criteria (>1% lethality) for both 2-inch and full-bore releases. The risk results summary is shown in Table 10. The Guard Shack risk is above the toxic risk criteria but below the lower criteria for the other hazard types. The Main Office is in the ALARP region for explosion and toxic risk and below the lower criteria for the other hazard types. Table 11 shows a comparison for each building and hazard type describing whether they meet or exceed the consequence criteria for 2-inch and full-bore releases along with the comparison against the risk criteria

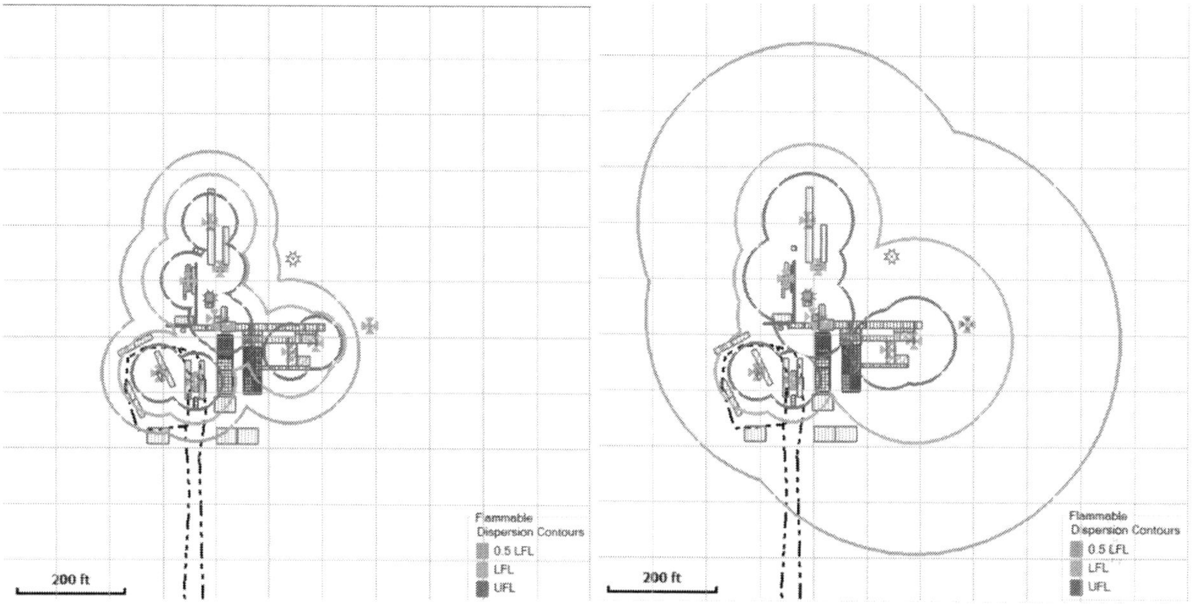

Figure 7. Flammable Contours for Releases up to 2" (left) and Full-Bore up to 6" (right)

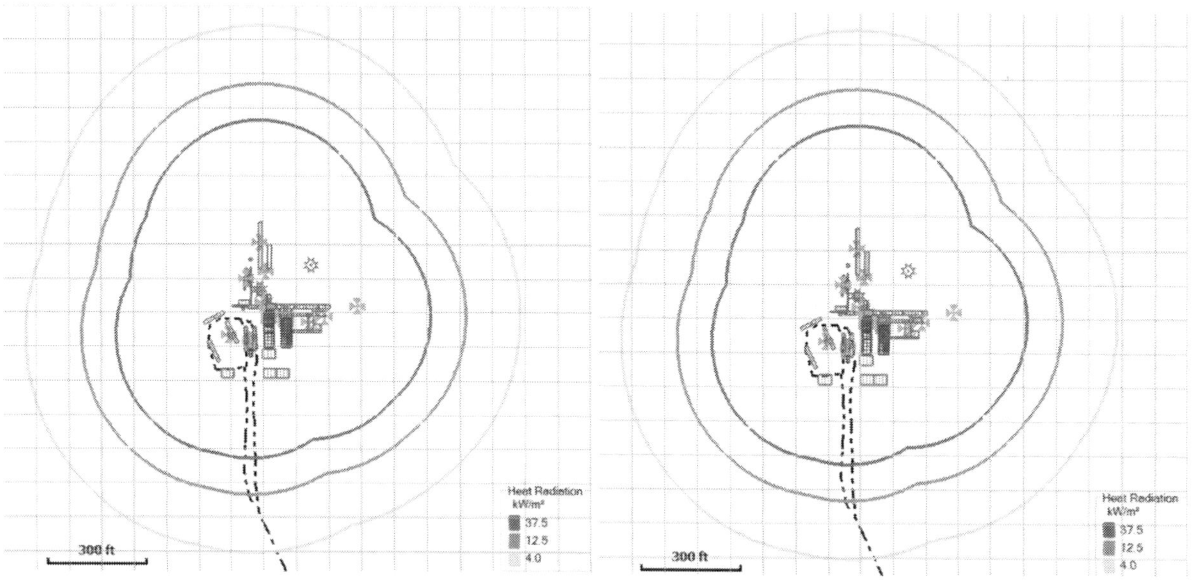

Figure 8. Thermal Radiation Contours for Releases up to 2" (left) and Full-Bore up to 6" (right)

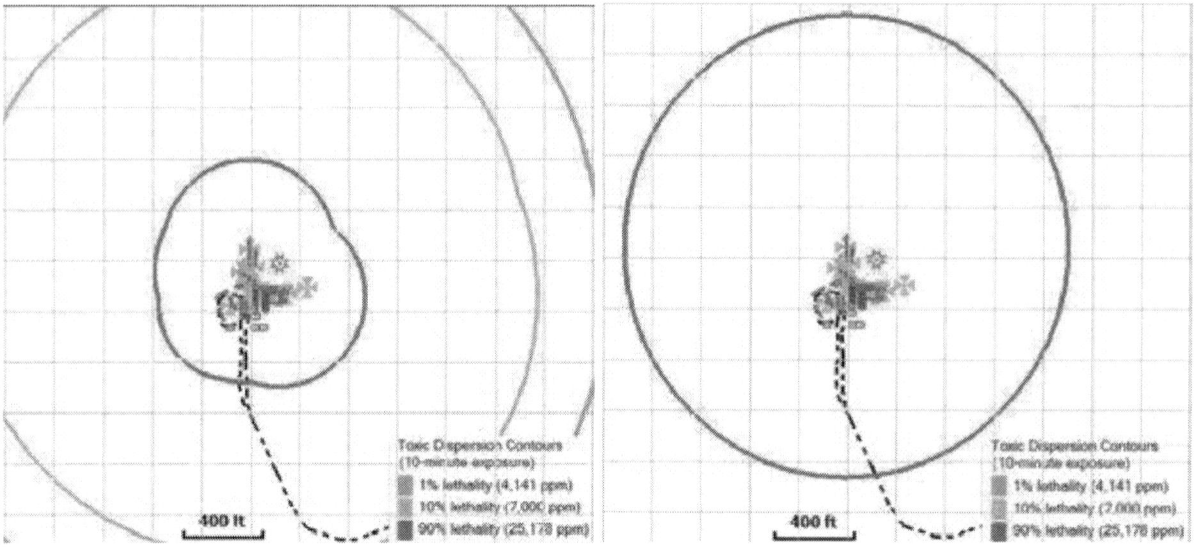

Figure 9. Toxic Contours for Releases up to 2" (left) and Full-Bore up to 6" (right)

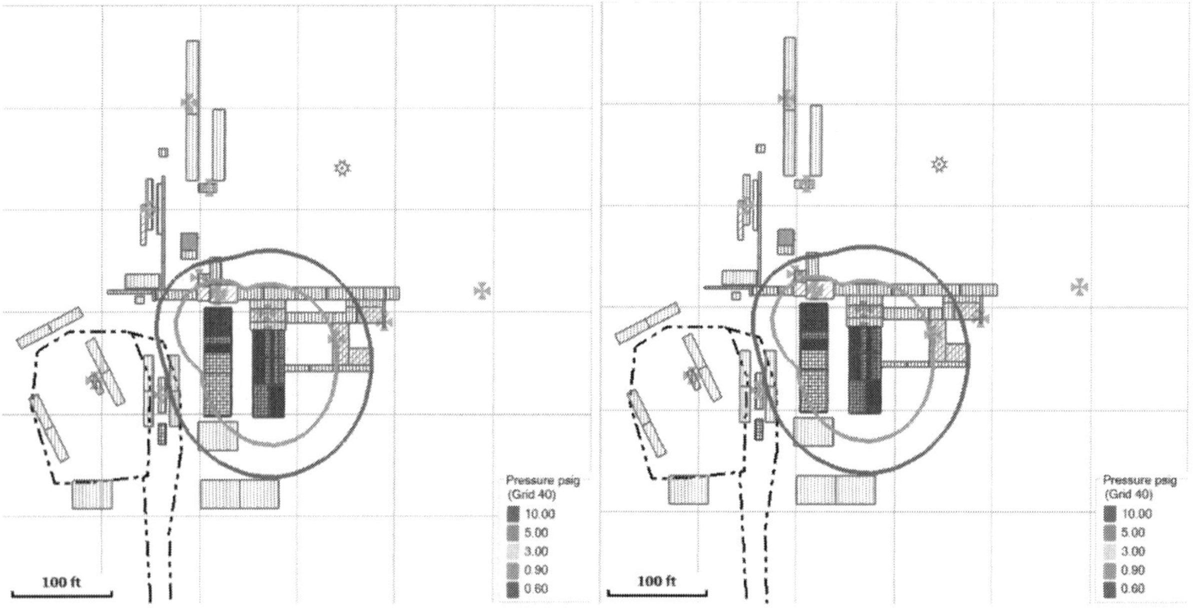

Figure 10. Overpressure Contours for Releases up to 2" (left) and Full-Bore up to 6" (right)

Table 9. Building Consequence Results (Case 2)

Building	Maximum Impacts											
	Building Damage Level			Flammable Gas Conc.			Fire Lethality			Toxic Lethality-NH_3		
	0.5 in	2 in	Full Bore	0.5 in	2 in	Full Bore	0.5 in	2 in	Full Bore	0.5 in	2 in	Full Bore
Guard Shack		2	2	>1/2 LFL	>UFL	>UFL		100%	100%	2%	78%	78%
Main Office		2	2	>1/2 LFL	>UFL	>UFL		100%	100%	1%	37%	37%

Table 10. Building Societal Risk (Case 2)

Building/Area	Building Societal Risk (fatalities /year)				
	Explosion	Flash Fire	Toxic	Jet Fire	Total
Guard Shack	5.2E-9	4.4E-7	**1.7E-3**	3.3E-6	1.7E-3
Main Office	2.5E-5	5.3E-7	**8.2E-4**	7.8E-6	8.5E-4
Totals	**2.5E-5**	**9.7E-7**	**2.5E-3**	**1.1E-5**	**2.6E-3**
	1%	**<0.1%**	**99%**	**<0.1%**	

Table 11. Consequence and Risk Results Comparison (Case 2)

Hazard	Study Type	Guard Shack		Main Office	
		Threshold	Comment	Threshold	Comment
Blast	2" Consq.	Meets criteria	No Action	Meets criteria	No Action
	6" Consq.	Meets criteria	No Action	Meets criteria	No Action
	Risk	Broadly Acceptable	No Action	Tolerable if ALARP	Consider Upgrades, Relocation
Flammable	2" Consq.	Exceeds criteria	LEL Detection, HVAC Shutdowns	Exceeds criteria	LEL Detection, HVAC Shutdowns
	6" Consq.	Exceeds criteria	LEL Detection, HVAC Shutdowns	Exceeds criteria	LEL Detection, HVAC Shutdowns
	Risk	Broadly Acceptable	No Action	Broadly Acceptable	No Action
Toxic	2" Consq.	Exceeds criteria	Tighten building, PPE, HVAC Shutdown	Exceeds criteria	Tighten building, PPE, HVAC Shutdown
	6" Consq.	Exceeds criteria	Tighten building, PPE, HVAC Shutdown	Exceeds criteria	Tighten building, PPE, HVAC Shutdown
	Risk	Intolerable	Tighten building, PPE, HVAC Shutdown	Tolerable if ALARP	Tighten building, PPE, HVAC Shutdown
Thermal	2" Consq.	Exceeds criteria	Thermal Shielding, Relocation	Exceeds criteria	Thermal Shielding, Relocation
	6" Consq.	Exceeds criteria	Thermal Shielding, Relocation	Exceeds criteria	Thermal Shielding, Relocation
	Risk	Broadly Acceptable	No Action	Broadly Acceptable	No Action

For the ammonia terminal, the maximum building impacts are the same whether a 2-inch or full bore release size was used since, relative to Case 1, the facility is smaller and the buildings are much closer to the release scenarios.

For the Guard Shack, the consequence-based study shows that mitigations should be implemented to address flammable, thermal, and toxic risk or the building should be relocated farther from the hazards. Based on the small footprint of the facility and the requirement for security personnel to be located onsite, it may not be possible to move the building. Alternatively, the building may be upgraded, although this may be costly.

The risk analysis for the Guard Shack shows that the building is in the intolerable region for toxic risk and the acceptable region for thermal and flammable hazards. Not only does this help the operator prioritize which risk to mitigate first, but also shows the mitigating thermal and flammable risk is not required due to the low frequency associated with those events. Equally importantly, risk analysis can help the operator prioritize what types of toxic risk reduction measures (e.g., PPE, tightening the building, HVAC shutdown interlocks, providing air cylinders for supplied breathing air for sheltering during a release, etc.). The risk analysis allows each mitigation option combination to be evaluated so that the optimal strategy can be selected based on risk reduction.

Similar to the Guard Shack, the Main Office consequence-based study shows that mitigations should be implemented to address flammable, thermal, and toxic risk or the building should be relocated farther from the hazards. Relocating the Office may be an option depending on how much of the surrounding area is owned by the company; however, the hazard contours indicate that the building would need to be relatively far removed from the facility to be outside of the consequence thresholds.

The risk analysis for the Main Office shows that the building is within the ALARP region for toxic and explosion hazards and in the acceptable region for thermal and flammable risk. Again, this helps the operator prioritize toxic risk as it is more than an order of magnitude higher than the other hazard types. It also shows that thermal and flammable risk are not required due to the low likelihood of occurrence. The risk analysis would also help the operator prioritize mitigations for the Guard Shack vs. Main Office. In this case the Guard Shack mitigations should be prioritized as the building is subject to higher risk.

While the results align for the 2-inch and full-bore release consequence analyses, the risk analysis points to mitigating toxic risk as the top priority as it comprises 99% of the risk onsite. For toxic risk specifically, the risk analysis will help quantify the amount of risk reduction required for the occupants of both buildings as well as providing a platform for comparing the effectiveness of different types of toxic mitigations.

5 Conclusions

The case studies in this paper provide a comparison between the results and mitigation decisions based a consequence-based FSS (i.e., with a limited release size MCE) and those based on a risk-based FSS incorporating a range of release sizes and associated frequencies.

The difference in consequence results for a 2-inch vs. 6-inch MCE in Case Study 2 (ammonia terminal) is negligible based on the small footprint of the facility, the small diameter piping/equipment in the facility, and the location of the buildings with respect to the process. The selection of the FSS MCE release size for some facilities therefore may not change the results and mitigation decisions, and an MCE representative of historical incidents (e.g., full-bore rupture) can be used. However, Case Study 1 (gas plant) shows that for the Control Room, which is located relatively far from the process, the 2-inch MCE does not adequately represent larger release scenarios that could impact the building. It therefore could be concluded that a building is safe based on a 2-inch MCE. Alternatively, evaluating larger MCE releases on a consequence basis may suggest costly building relocations or upgrades, which may not actually be required based on the low likelihood of the events.

In both case studies the facilities evaluated were small enough that the 2-inch MCEs covered most of the onsite congested/confined volumes. However, for larger facilities, it is expected that 2-inch MCEs would not reach all onsite congested/confined volumes, but that larger releases would. A consequence-based 2-inch MCE would therefore fail to capture the VCE vulnerabilities in such cases. Furthermore, releases in excess of 2-inches are often required for the resulting flammable cloud to reach off-site congestion, such as neighboring industrial facilities and/or vegetation at the site boundary. For example, 2-inch releases from process units may not reach wooded areas at the site boundary, in which case the explosion vulnerability of a building sited away from the process and close to the site boundary could be significantly underestimated. Hence, while a 2-inch MCE may be adequate for some facilities, it may fail to identify significant vulnerabilities and risks posed by larger releases representative of past industry incidents.

For the case studies presented in this paper, many of the mitigation decisions arising from the consequence- and risk-based analyses are in alignment. For example, in Case Study 1 the Warehouse exceeds the consequence threshold for each hazard type (i.e., mitigation required) and the risk results also indicate that cost-effective mitigation should be sought for each hazard type. However, the risk analysis also allows prioritization of jet fire mitigation over flash fire and explosion mitigation, as well as allowing the prioritization of building risk mitigation based on relative risk. In addition, a risk-based FSS provides the opportunity to determine whether a mitigation measure is cost-effective (i.e., risk is ALARP). Conversely, a consequence-based FSS does not provide the information to determine which buildings and hazards to prioritize for mitigation or whether a proposed mitigation action is cost effective.

However, in other cases, consequence-based and risk-based FSS analyses can lead to very different conclusions. For example, a 2-inch consequence-based FSS for the Control Building in Case Study 1 shows that the building is not impacted. Conversely, the risk-based FSS shows the building is in the ALARP region and risk mitigation should be sought for thermal and flammable hazards. In fact, ~75% of the risk for the facility is thermal risk to the Control Building, which is entirely missed by the 2-inch consequence-based study. A consequence-based approach may work for some facilities and building locations, but may miss the highest risks in other cases.

A risk analysis not only supports risk mitigation prioritization, but can also provide insight on the effectiveness of different risk mitigation measures. Toxic releases exceed the consequence threshold for both buildings in Case Study 2 and toxic risk dominates both buildings; with either approach a decision must still be made as to what level of toxic risk mitigation is adequate. With a consequence-based study, the operator may choose to implement the mitigations they have at other facilities without consideration for how effective those measures will be. A risk analysis allows comparison between different types and combinations of mitigation (e.g., PPE, increasing building tightness, HVAC isolation, etc.) and selection of the optimal and most cost-effective mitigation option.

In addition to helping prioritize mitigations, risk analysis also provides additional mitigation options. Consequence-based analysis can only credit mitigation that reduces consequences (e.g., using less hazardous materials, reducing congestion, upgrading building, etc.). Risk based studies can credit these types of mitigation as well as those that reduce initiating event frequency (e.g., double-seal or seal-less pumps, double walled pipes, etc.) or outcome likelihood (e.g., rapid detection and isolation) [7].

The recommended practices and guidance documents for facility siting allow the facility owner to determine the appropriate hole size for the MCE and decide whether a consequence or risk basis should be used. While the results of a 2-inch MCE consequence-based, larger release MCE consequence-based, and risk-based analysis may be similar sometimes, they can also lead to very different building siting conclusions. The case studies provided in this paper illustrate instances where the 2-inch MCE consequence-based study, while covering the most likely consequences, completely missed the less frequent but higher potential consequence events. The 6-inch MCE consequence-based studies cover these lower-frequency high-consequence events but may result in spending significant capital mitigating events that are extremely unlikely. The risk-based approach allows risk-based decisions to be made using a full range of release sizes and to prioritize mitigation on the highest risk hazards and buildings.

6 References

[1] API (2009) "Management of Hazards Associated with Location of Process Plant Permanent Buildings," American Petroleum Institute Recommended Practice 752 (API RP 752), 3rd Ed., API, Washington D.C.

[2] API (2007) "Management of Hazards Associated with Location of Process Plant Portable Buildings," American Petroleum Institute Recommended Practice 753 (API RP 753), 1st Ed., API, Washington D.C.

[3] Marx, J. and Nicotra, A. *Is a Two-Inch Hole [Size] Adequate for a Siting Study?* Prepared for the 12th Global Congress on Process Safety 50th Loss Prevention Symposium. April 2016.

[4] Grounds, C, Moosemiller, M, Thomas, JK. Historical overview of facility siting. Proc Safety Prog. 2020; 39:e12185. https://doi.org/10.1002/prs.12185

[5] Mander, TJ, Sarrack, A, Diakow, P, Bruce-Black, J. Current state of the practice for facility siting studies. Proc Safety Prog. 2020; 39:e12181. https://doi.org/10.1002/prs.12181

[6] Reducing Risk, Protecting People, HSE's Decision-Making Process, paragraph 130-132.

[7] Ashiofu, A., et al., "Leveraging Facility Siting to Optimize Mitigation Decisions," prepared for Process Safety Progress magazine. Published September 2020 issue.

619228 Leveraging CFD to Effectively Evaluate Sensitivity to Ignition Location When Designing for Explosion Loads

Abstract:

Vapor cloud explosions are among the most dangerous high consequence events that can occur at petrochemical facilities. As the size and complexity of facilities increase, designs must consider the potential adverse effects associated with vapor cloud explosions in large congested areas. When evaluating facility siting issues such as building locations or blast ratings, it is critical to understand the variables that may affect the results—otherwise there is a risk of producing suboptimal results that could impact the safety of those at the facility.

Among the factors that are generally not well studied during facility siting evaluations is the sensitivity of explosion pressures to ignition location. Depending on the size of cloud and particular arrangement of congestion, there can be significant differences in the pressure measured at nearby buildings when varying the ignition position by just a few meters or with changes in elevation. Typical CFD studies consider only a few ignition locations and may only minimally explore the effects of ignition elevation. Further, phenomenological models are not suitable for investigation of the sensitivity to ignition location.

This paper will explain via a series of case studies the importance of evaluating sensitivity to ignition location in more detail. The paper will also demonstrate a novel technique that can be used to visualize the sensitivity of specific targets to ignition location—thus making it easier to communicate results and identify areas of concern.

SPRING21
+17TH GCPS
A Joint AIChE and CCPS Meeting

Minimizing Risk of Unit Trains of Hazmat

Jessica M. Morris, Ph.D.
Enakshi Wikramanyake, Ph.D.
Sisi Song, Ph.D.
Ryan Hart, Ph.D., P.E.
Delmar "Trey" Morrison, Ph.D., P.E., CCPSC, FAIChE
Exponent, Inc.
4580 Weaver Parkway, Suite 100,
Warrenville, Illinois, USA 60555

Prepared for Presentation at
American Institute of Chemical Engineers
2020 Spring Meeting and 17th Global Congress on Process Safety
Virtual
April 18 - 22, 2021

Minimizing Risk of Unit Trains of Hazmat

Jessica M. Morris, Ph.D.
Enakshi Wikramanyake, Ph.D.
Sisi Song, Ph.D.
Ryan Hart, Ph.D., P.E.
Delmar "Trey" Morrison, Ph.D., P.E., CCPSC, FAIChE
Exponent, Inc.
4580 Weaver Parkway, Suite 100,
Warrenville, Illinois, USA 60555

Keywords: Hazard Identification and Risk Analysis, Safety & Health, and Transport Safety

Abstract

In 2020, the CFR was updated to include a rule allowing LNG shipment in DOT-113 class tank cars. Prior to this rulemaking, LNG shipment on railroads was only possible via Special Permit approval resulting in a very limited number of LNG shipments on rail. The recent regulation has simplified the regulatory path to shipping LNG in a series of tank cars by rail as a manifest train and by unit train configuration. As a result, we may expect to experience large growth in LNG shipments via rail in the near future.

This paper undertakes a risk-based analysis to evaluate the risks of blocks of LNG tank cars and unit trains to characterize the risk of large LNG shipments on rail. The risk model for train shipments shall consider: rail tank car derailment, loss of containment, and then fire or explosion for the leaking tank cars. We apply a multinomial probability model to approximate the number of tank cars and the release rate/quantity of LNG multi-car incidents. The results of this study can be used to better understand the scale of risk represented by this emergent shipping activity and identify opportunities for risk reduction through operational limits.

1 Introduction

Until recently Liquid Natural Gas (LNG) shipment on rail was only possible via special permit approval. In 2020, the Pipeline and Hazardous Materials Safety Administration (PHMSA) and Federal Railroad Administration (FRA) updated the Code of Federal Regulations (CFR) 49 CFR to allow LNG shipment in DOT-113 class tank cars. The following sections describe the history behind regulation of LNG transportation on rail and the various containers that can be used for this purpose.

1.1 Background

The recent boom in natural gas production in the United States has led to the rapid growth of natural gas use across multiple industries, including energy and transportation. Natural gas has become a leading fuel source for power generation, as well as transportation in rail, truck, fleet, and marine applications. The increase in demand increases the transportation of LNG across the country resulting in an increase potential in frequency of an accident involving LNG resulting in increased the risk. Historically, transportation records have been relatively free of incidents, and when incidents have occurred, consequences have been relatively minor [1]. For over four decades, LNG has been transported safely by trucks on highways and by marine vessels in the United States. Until recently, Hazardous Materials Regulations (HMR) did not permit the bulk transport of LNG in rail cars.

1.2 LNG Transport on Rail

In 2020, PHMSA and FRA amended the HMR to allow bulk transportation of LNG by rail in specified cryogenic tank cars. 49 CFR § 172, 173, 174, 179 and 180 provide the regulations that must be followed in order to transport LNG by rail. PHMSA and FRA specify that LNG can be transported by rail in DOT-113 rail tank cars, which have a history of safely transporting other cryogenic flammable materials. As stated in 49 CFR, HMR combines packaging design and maintenance, operational controls, package handling, employee training, hazard communication, emergency response information, and security plan requirements to safeguard transportation. Additionally, 49 CFR specifies properties such as operating pressures, tank wall thickness, wall material, and train operational conditions in addition to risk-based framework, derailment comparisons, and safety history. 49 CFR [2] requires that the DOT-113 tank cars used to transport LNG on rail be enhanced with an outer tank that is thicker and made of steel with greater puncture resistance as an additional safety measure.

1.3 Type of LNG Tanks Transported on Rail

Previously, LNG shipment on railroads (via special permit) was primarily transported in T-75 intermodal ISO containers. With the CFR update in 2020, LNG shipment is now authorized by rail in DOT-113 tank rail cars. While both containers transport LNG, the risk associated with the two storage methods has yet to be fully explored. This paper looks at the risk associated with the different container scenarios for LNG that can be transported by rail by comparing unit trains with LNG in T-75 ISOs and DOT-113 tank cars.

The DOT-113 C120W cryogenic liquid tank rail car is designed with an interior pressure vessel to store the cryogenic liquid. This inner tank is surrounded by a sturdy exterior tank with intermediate vacuum space between the two. The interior pressure vessel is constructed with T-304 stainless steel and complies with DOT and AAR regulations [3].

The other option for LNG rail transportation is the UN T-75 intermodal ISO cryogenic portable tank containers (T-75 ISOs). The ISOs are certified against the International Maritime Organization – International Maritime Dangerous Goods Code, Volume 1, which is incorporated into 49 CFR § 178.274. ISO containers are designed to be transported as intermodal freight by railroad, tractor-trailer, and marine vessel, to reduce the frequency of transferring cryogenic liquid between containers during transport from the liquefaction facility to the end user.

Similar to DOT-113 containers, T-75 ISO containers are comprised of an interior pressure vessel for cryogenic liquid, surrounded by a sturdy exterior tank with a vacuum in-between to provide insulation. As a basis for our analysis, we assumed that each ISO has a nominal water capacity of 12,020 gallons with a 90% filling ratio, thus a nominal LNG capacity of 10,818 gallons [4]. A comparison of the design parameters used in this risk analysis comparing the different vessels are listed below in Table 1. Similarly, we assumed that a standard DOT-113 has a nominal LNG capacity of 30,000 gallons.

Table 1. T-75 ISO tank container design parameters.

Parameter	DOT-113	T-75 ISO
MAWP (psig)	90	110
LNG Capacity (gal)	30,000	10,818

In this paper, the overall risk comparing the transportation of LNG via rail in DOT-113 tanks vs. T-75 ISO containers will be explored through a quantitative risk analysis approach.

2 Background

2.1 Risk

Risk is defined as the potential to lose value and is evaluated by taking the product of event likelihood with the event outcome severity. This product is compared to a benchmark risk and deemed as acceptable or unacceptable by the stakeholders. The likelihood of an event can be estimated using experience relating to given equipment in similar service, industry data, or engineering approximations. The consequences of hazards from a potential even can be determined with consequence analysis software. Additional software is able to combine both the consequence and the likelihood to produce the overall risk.

In evaluating risk, two metrics are utilized: (1) Individual Risk (IR) and (2) Societal Risk (SR). Individual Risk is the frequency (yr^{-1}) where an individual with continuous potential exposure may be expected to sustain a serious or fatal injury. For IR surrounding stationary LNG facilities, NFPA 59A identifies three "Zones". Each zone has permitted developments for the potential populations. As the magnitude of the calculated risk increases, the type of occupancy becomes more restrictive. The zones with their permitted developments are provided in Table 2.

Table 2. Quantitative risk criteria for Individual Risk (IR) of fatality around stationary LNG plants as provided by NFPA 59A (2019).

Criterion Annual Frequency (yr^{-1})	Permitted Developments
Zone 1 $IR > 10^{-5}$	All land uses under the control of the plant operator or subject to an approved legal agreement
Zone 2 $3 \times 10^{-7} \leq IR \leq 10^{-5}$	General public areas excluding sensitive establishments*
Zone 3 $IR < 3 \times 10^{-7}$	No restrictions
*Sensitive establishments are institutional facilities that might be difficult to evacuate. Examples include, but are not limited to, schools, daycare facilities, hospitals, nursing homes, jails, and prisons.	

Societal Risk (SR) is one method for evaluating the risk of a given process or operation. Unlike IR, the SR calculation considers the relationship between the cumulative number of potential fatalities (*N*) versus likelihood (*F*) from a series of potential events. Whereas the IR calculation gives insight into the probability of having a fatality, the SR calculation gives the likelihood of a number of potential fatalities as a function of the surrounding population. SR is important for evaluating scenarios with a large potential impact for loss of life, such as train derailments of flammable materials in populated areas.

In this analysis, release scenarios were modeled along the rail line on a per-route mile basis. IR contours cannot be succinctly represented for long routes such as this but are still related to the population level along the line.[1] The highest risk along the mainline will occur at the portion of the track exposed to the highest populations.

2.2 Benchmarking Risk for Transporting LNG (previous studies and literature)

The public perception of LNG risk can be overstated when compared to the risks of other common flammable chemicals. Several studies, discussed below, have been conducted to compare the risk associated with the transportation of flammable chemicals by rail or truck.

The first study analyzed the impact of thermodynamic conditions, such as temperature and pressure, on a loss of containment (LOC) event from a DOT-113 rail car and the impact of the associated consequence due to varying thermodynamic conditions [5]. This study showed that discharge dynamics and distances to hazardous thresholds are a function of the temperature and pressure that

[1] IR is a weak function of population due to the population density effect on the likelihood of ignition model.

the flammable material is stored or released at. This new analysis will be using design conditions specified by the manufacture in Table 1 for simplicity.

Another analysis evaluated the safest method to move LNG: truck, rail, or pipeline [6]. All three of these transport methods have advantages and disadvantages associated with them, including: regulatory requirements, deployment time scales, existing infrastructure, cost, and risk to the public. Is this study, a QRA was conducted to compare the three transportation methods for a hypothetical scenario with an example route of 500 miles and a transport capacity of 430,000 gallons/day of LNG. The analysis revealed the highest risk was associated with pipeline transportation where the lowest risk was for rail transport.

Another study compared the risk of transporting LNG and liquid petroleum gas (LPG) by truck and by rail [7]. Since LPG has a long history of being transported in the U.S and the risks associated by doing so are accepted, it is an appropriate commodity to benchmark LNG transportation risks. LPG is stored at ambient temperature in single-walled pressure vessels. As a result, if a pressurized LPG container develops a leak, the fluid will flow rapidly and will flash, creating a cold dense vapor cloud. Alternatively, LNG is stored cryogenically in double-walled cryogenic tankers. Due to the low storage pressure, a breach in an LNG storage container will result in a lower velocity resulting in an LNG spill that will boil creating a cold, dense vapor cloud. Since the vessel design and storage methods are different, the likelihood of an LNG vessel being involved in a boiling liquid expanding vapor explosion (BLEVE) is lower compared to an LPG vessel. Additionally, the QRA in this study demonstrated that the IR and SR for bulk LNG transportation is similar to or less than that for LPG for both truck and rail transport, providing a good benchmark.

2.3 Hazards Associated with LNG

LNG poses unique hazards relative to other non-pressurized liquid fuels. Comprised of methane, LNG has a flammable range of 5-15%. When stored cryogenically at -260°F, LNG is not pressurized, but must be insulated to avoid excessive boiling due to heat transfer. To avoid excessive pressure buildup under extended duration, portable tanks will vent low volumes of natural gas to the atmosphere via a pressure relief valve.

The cryogenic temperatures of LNG pose unique hazards to rail personnel. Due to a large difference in temperature, the rapid transfer of heat from an object into the cryogenic liquid can cause burns if direct contact with skin occurs or if PPE is inadequate to prevent cold-temperature injury. Additionally, large spills of the liquid onto metal structures can cause embrittlement and fracturing. LNG contains no odorant (unlike residential natural gas supplies), making detection difficult without a flammable gas detector device.

The behavior of a spill of LNG is unique due to the cryogenic temperature of the liquid. For example, a spill of LNG will vaporize rapidly when it contacts ambient air and even faster when in contact with warm solids such as the ground. The cold vapors may condense humid air, causing fog formation and decreased visibility. After vaporization, the denser than air cloud will tend to stay close to the ground as it disperses. The dense vapors can travel great distances without significant dilution, as the mixing with ambient air is limited near the ground, and the vapor will tend to accumulate in low spots or trenches. While all of these hazards are unique to LNG, this study will focus on the hazardous consequences from the ignition of flammable material.

3 Methodology

In this paper, the overall risk comparing the transportation of LNG via rail in DOT-113 tanks vs. T-75 ISO containers was explored. To perform this analysis, a novel approach to determining the probability for the loss of containment per car was performed. The methodology to determine the risk as a function of likelihood and consequence is discussed here.

3.1 Scenarios Analyzed

In this analysis, a standard unit train transporting LNG was assumed. Unit train tank car movements are evaluated along mainline rail track as a function of train speed and exposed population density. The unit train is assumed to travel one trip per day on a mainline track. The return trip with residue tank cars containing cold vapor was not considered in the analysis due to the relatively insignificant risk posed by "empty" cars. This QRA assumed an average daily movement rate of eleven or more tank containers per train. Although more containers may be shipped within the same train, the overall risk is adequately represented by modeling this annual average movement capacity. This study evaluated three rail car / train configurations to consider different rail transportation options:

1. Single DOT-113 tank per train position
2. Single T-75 ISO tank per train position
3. Two T-75 ISO tanks per train position (i.e., two ISOs on an intermodal car)

The risk was calculated for a sample population density along a 1-mile section of track. The per-mile risk results can then be used to estimate the aggregate risk for longer tracks with varied population density along the route. The trains are assumed to carry the same total volume of LNG, 900,000 gallons in each scenario. Based on the capacity of each container and the configuration of the cars, each unit train will have 30, 90, and 45 cars, for one DOT-113 tank per rail car, one T-75 ISO container per rail car, and two T-75 ISO containers per rail car, respectively.

Train movements were evaluated along mainline track at speeds between 25 and 50 mph. The high speed upper threshold value of 50 mph was selected as it is the maximum speed cryogenic materials are allowed to be transported at by 49 CFR [8]. By choosing this upper speed threshold, the risk analysis results will represent upper bound of risk relative to operations at lower train speeds.

3.2 Estimating Accident Rates and LOC Probabilities

The sequence of events leading to a LOC of LNG in the analysis starts with an accident involving the train. From here, the probability of train derailment is determined. If a train is derailed, the probability of the number of LNG containing cars was also determined. LOC can occur from a multitude of release sizes. Each scenario has a specific frequency of occurring that was calculated in this analysis. A sample event tree is shown below in Figure 1. The base assumptions and results are discussed in the subsequent sections.

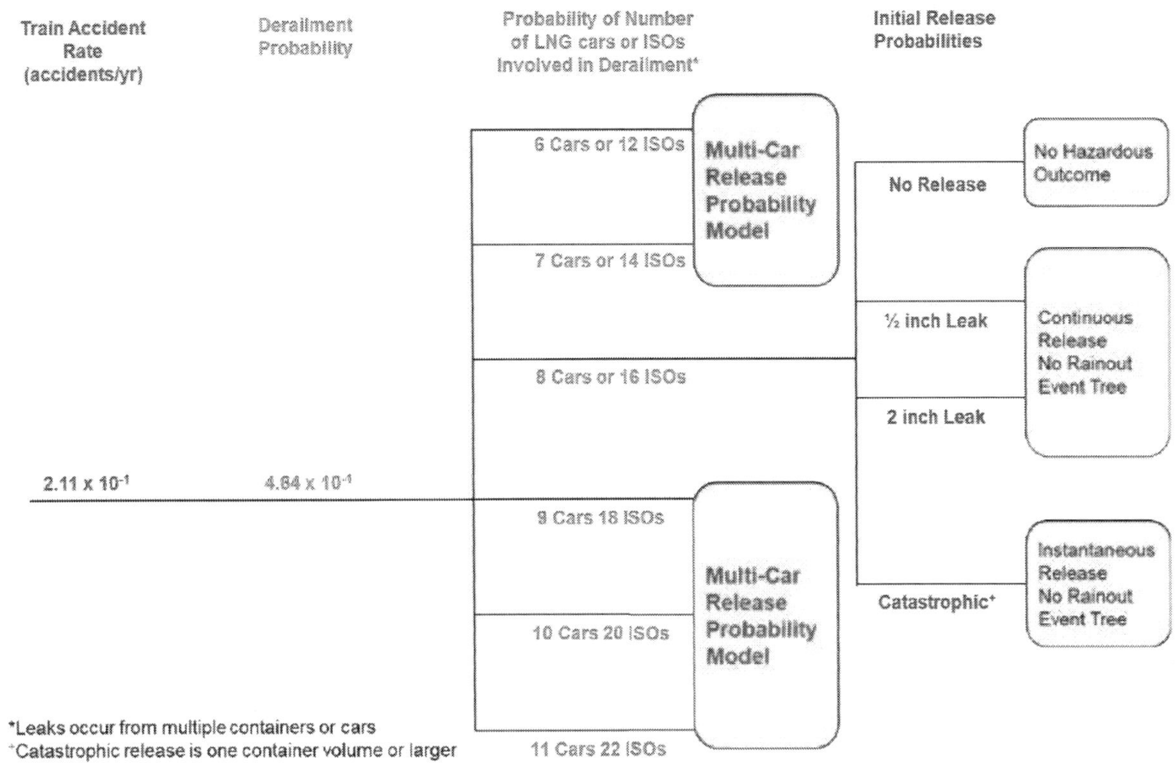

Figure 1. Representative event tree illustrating the relationship between the frequency of train accident and conditional probabilities of subsequent events in the analysis for unit train movements.

3.2.1 Train Accident Rates

LNG shipping by rail is historically uncommon in recent U.S. rail industry history; thus, accident data for the movement of LNG in trains does not exist. In this study, publicly-available data from the Federal Railroad Administration (FRA) [9] was analyzed to estimate generic train accident rates for application to the QRA. For the purposes of this analysis, potential train accidents were only considered to occur along the line of road.

In order to estimate the train accident rate, the FRA industry-wide database for train accidents [10] was first queried for all accident reports during the twenty-year period from January 1, 2000, through December 31, 2019, yielding a total count of 63,502 accidents. The accidents are identified in the database by category and include multiple types of collisions, explosions, fires, other impacts, and other events on mainline tracks, in rail yards, on rail sidings, and industry lines[2], which are consistent with the events necessary to lead to an LOC of LNG from a DOT-113 tank car or T-75 ISO. The FRA data was filtered for all mainline accidents from 2000–2019 (all railroad classes), and

[2] Total number of accidents in Table 4 only includes data on high speed (>25 mph to ≤50 mph) accidents. A small portion of accidents also occur at speeds >50 mph. The sum of all three speed categories will yield the total accident count of 63,502.

the results were analyzed to determine accident frequency and conditional probability of derailment for the less conservative high speed accident derailment range (>25 mph and ≤50 mph), summarized in Table 3.[3] These data were used to determine accident frequency, as will be discussed below.

Table 3. Train accident data from the FRA.

	Statistic	1999–2000
Mainline, Speed > 25 to ≤ 50 mph	Total Accidents	7,093
	Total Derailments	3,290
	% of All Accidents	11.1%
	Probability that Derailment Occurs	46.4%

The raw accident numbers were then divided by train mileage to develop accident frequency estimates as accidents per train mile for the QRA. Operational data tables provided by the FRA were used to determine the total number of mainline[4] train miles for the period from 2000–2019 for all classes of railroad represented in the data [11]. The operational data tables did not subdivide the mainline train miles according to track speed; thus, only the total accident frequency per mainline (non-yard) train mile for all trains could be calculated. This single mainline train accident frequency value was applied to all mainline train movements regardless of train speed. Using the total accident and total mileage values, the accident frequency (on a per train mile basis) was then calculated. The average accident frequency was found to be 2.32×10^{-6} (accidents/train mile) for the mainline travel. The 20-year data was used throughout the analysis due to the relatively large number of data points that provide a larger confidence in the position-in-train derailment probabilities. The results are summarized in Table 4.

Table 4. Train accident rate from FRA data.

	Statistic	2000-2019
Mainline	Total Non-Yard (Mainline) Train Miles	12.95×10^{9}
	Non-Yard Accident Rate (/train mile)	2.32×10^{-6}

The mainline accident frequencies[5] from Table 4 were then multiplied by the total number of annual train miles estimated for the example route (250 miles) to arrive at the yearly accident frequency (accidents per year) used in the calculations for the route on a per mile basis. A summary of the calculated annual accident rate for the example route is provided in Table 5.

[3] Data for train speeds <25 mph and >50 mph is not included.
[4] All "Non-yard" miles were assumed to be mainline miles for the purpose of this analysis.
[5] Note that the terms frequency and rate are used interchangeably.

Table 5. Calculated annual accident frequencies for the example mainline route.

Route	Route Length (track miles)	Estimated Total Annual Route Length (train miles/yr)	Accident Frequency (accident/train mile)	Calculated Annual Accident Frequency (accident/yr)
Example route	250	91,250	2.32×10^{-6}	2.11×10^{-1}

3.2.2 Conditional Probability of Derailment

The train accident values shown above provide an estimate of the frequency that a train accident will occur somewhere along the example route. However, a train accident doesn't necessarily lead to a condition where LOC of an LNG rail car may occur. Therefore, it was assumed that only train accidents leading to the derailment of cars could potentially result in an LOC. The 20-year accident data were analyzed to determine the probability that a train accident will lead to a derailment of any of the DOT-113 or T-75 ISO-carrying cars for mainline movement at speeds from 25 mph up to 50 mph. As listed in Table 3, the calculated derailment probability for high speed mainline movement was 46.4%. These are the probabilities of at least one car being derailed in a train accident; however, this probability does not guarantee that the derailment will involve one or more cars carrying LNG.

As per specified by 49 CFR § 174.85 [12], the LNG containing car may not be nearer than the sixth car from the engine. Hence, the first LNG containing DOT-113 or T-75 ISO-carrying car will start in position six (6) and continued throughout the remainder of the train up to the final car. The historical FRA accident data were analyzed to develop a model for estimating the probability of derailment of a number of individual cars versus position in the train.

The probability of derailment for one or more LNG carrying cars is dependent on the position of the first car derailed in the train, the average number of cars derailed during an accident, and the location of LNG carrying cars in the train. These parameters are expected to be affected by the train speed, which was analyzed using the FRA 20-year accident data.

The FRA 20-year accident data from January 1, 2000, through December 31, 2019, was first filtered to include only mainline accidents. The mainline accidents were then further filtered into mainline accidents with train speeds between 25-50 mph. Next, the accidents were filtered in the database by including only accidents resulting in derailment. The average number of cars derailed in mainline accidents (at speeds of 25-50 mph) was eleven (11).

Based upon the dynamics of a derailment, it was assumed that in an average derailment, the first car would derail plus the immediately following sequence of n-1 cars would derail, where n is the average number of cars derailed. Regarding mainline movements, lower speed derailment accidents involve fewer cars on average than higher speed derailment accidents, making the results from this analysis more conservative.

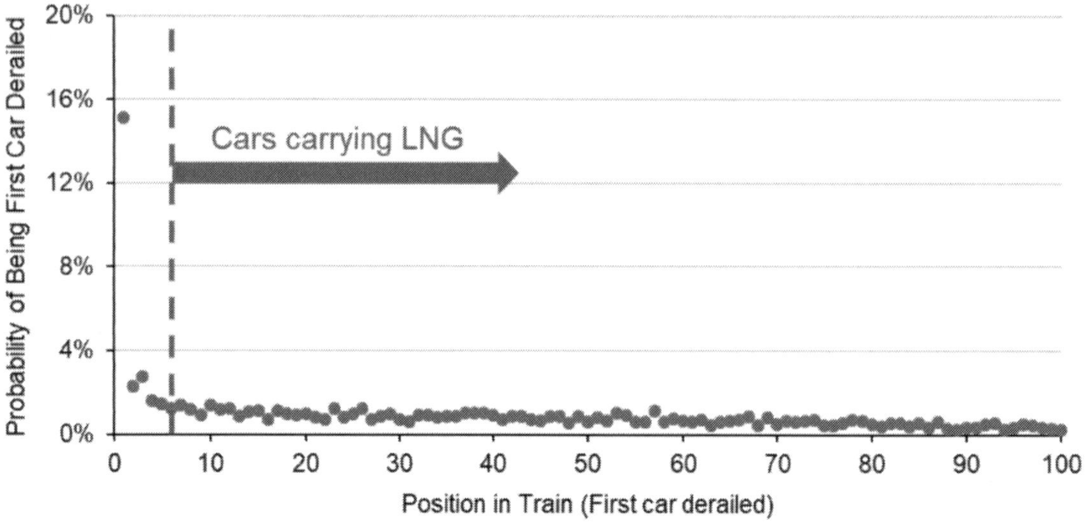

Figure 2. Probability of the first car position-in-train for mainline derailments.

The data reveal that when a train accident results in a derailment, the first car derailed is most often the head car (position 1). The probability that the first car derailed is one of the first ten cars in nearly a third (30%) of all mainline derailments where train speeds are between 25 mph and 50 mph. Similar results are found for the percentage of derailments starting with a car in position 1-10 for the low speed case with 39% for mainline derailments where train speeds are less than 25 mph.

Assuming the train configuration is described by LNG-carrying cars starting at position six (6) and continuing until the end of the train, the probability of having a given number of LNG DOT-113 or T-75 ISO-carrying rail cars involved in the derailment was calculated. Using the average number of involved cars, and the position-in-train derailment probabilities, the probability of having 6–11 cars derailed were calculated. These probabilities are provided in Table 6. Given the accident model, 5 or less cars can only be involved in a derailment if the first car derailed is near the tail end of the train. The nominal probability for such a position of the first car derailed is on the order of 0.5%; thus, the risk of such a derailment is adequately upper bounded by lumping those events into the probability 11 cars derailed.

Table 6. Probability of having 'X' number of LNG T-75 ISO-carrying rail cars involved in a train accident with derailment, by train speed.

'X' Number of Cars Involved in Train Accident with Derailment	Probability of 'X' Number of LNG Cars Involved in Train Accident with Derailment between 25-50 mph
1	*
2	*
3	*
4	*
5	*
6	15.1%
7	2.2%
8	2.8%
9	1.6%
10	1.5%
11	76.8%
*As per specified by 49 CFR § 174.85, the LNG containing car may not be nearer than the sixth car from the locomotive.	

3.3 Multinomial Probability Model

3.3.1 LOC probability

The prior sections showed how accident rate and derailment probability can be estimated for LNG carrying car, however, not all accidents will lead to a LOC of LNG. The specific dynamics of an individual accident will dictate whether and to what extent a LOC may occur. In order to estimate the probability of a LOC event, representative release sizes from a single container with associated conditional probabilities were developed from industry data. Next, a combinatorial probability model was constructed to define the myriad potential permutations of combined releases when multiple containers are involved.

The Pipeline and Hazardous Materials Safety Administration (PHMSA) maintains an online database that provides historical LOC data for rail tank cars, among other transportation vessels [13]. The database complements the FRA database in that the PHMSA database records the inventory of HAZMAT cargo released for each accident; whereas, the FRA database only identifies that an LOC has occurred. The PHMSA database was analyzed in order to estimate the LOC probabilities for the LNG cars. Since the PHMSA database has limited information on accidents related to LNG transport, accidents related to pressure tank car LOC were considered which are a close representation of DOT-113 and T-75 ISO containers.

The PHMSA database listed accident data from 1971 to the present. All rail car data was queried from 1971 to 2020, for incidents including spillage, vapor (gas) dispersion, and no release. The resulting data was then filtered for pressure tank cars only, and incidents where no tank car specification was available were excluded from the analysis. The resulting 5,578 pressure tank car incidents[6] were then sorted by amount released (units are either cubic feet (ft^3) or gallons).

The PHMSA data were grouped into four release volume ranges in order to estimate the probability of a certain leak size. The categories were no release (less than 100 gallons), small release (100 to 1,000 gallons), large release (1,000-30,000 gallons), and catastrophic release (>30,000 gallons) [14] and each volume range was assigned a release scenario based on PHMSA specifications [15]. These volumes were chosen as the PHMSA data appeared to reflect mostly 30,000+ gallon tank cars. The resulting release probabilities for each release category are provided in Table 7.

Table 7. LOC probability from PHMSA pressure tank car incident data and equivalent release scenario for one DOT-113 and one LNG T-75 ISO.

PHMSA Data Analysis			Corresponding PHAST Release Rates r_i	
Quantity Released in gallons	**Probability p_i**	**Release Scenario**	**DOT-113 Release Rate (lb/s)**	**T-75 ISO Release Rate (lb/s)**
≤ 100	0.954	No Release	0	0
100 < x ≤ 1,000	0.015	½-inch Leak	3.66	4.09
1,000 < x ≤ 30,000	0.028	2-inch Leak	58.6	65.5
> 30,000	0.003	Catastrophic	93,279	32,877

3.3.2 Probability analysis and binning

The calculated release probabilities and release rates were then used to find the overall outcome during a derailment scenario. A single derailed container has four possible release outcomes (No Release, ½-inch Leak, 2-inch Leak, Catastrophic Release), corresponding to the rows of Table 7. These four outcomes occur with probabilities p_i and result in release of LNG with rates r_i, $i = 1,2,3,4$. The values of p_i and r_i are given in Table 7 for DOT-113 and T-75 ISO containers.

The aggregate outcome of a derailment scenario involving n containers can be expressed as (x_1, x_2, x_3, x_4) where

x_1 = number of containers experiencing a No Release
x_2 = number of containers experiencing a ½-inch Leak
x_3 = number of containers experiencing a 2-inch Leak
x_4 = number of containers experiencing a Catastrophic Release

Each x_i is an integer with a possible value between 0 and n inclusive, and $\sum_{i=1}^{4} x_i = n$. The aggregate outcome (x_1, x_2, x_3, x_4) describes the number containers with each size of leak, resulting

[6] As of July 9, 2020.

in a combined release rate of $\sum_{i=1}^{4} x_i r_i$ for the n containers. The probability of occurrence of a specific aggregate outcome (x_1, x_2, x_3, x_4) is given by a multinomial distribution [16] with parameters n and p_1, p_2, p_3, p_4:

$$\text{Probability of outcome } (x_1, x_2, x_3, x_4) = \frac{n!}{x_1!\, x_2!\, x_3!\, x_4!} p_1^{x_1} p_2^{x_2} p_3^{x_3} p_4^{x_4}.$$

With $n = 6$, there are a total of 84 unique aggregate outcomes, i.e. unique values of (x_1, x_2, x_3, x_4), each with a probability of occurrence and combined release rate described in the previous section. For example, (5,0,1,0) is a valid aggregate outcome, representing the occurrence of 5 containers with no release and one container with a 2 inch leak, which occurs with probability 0.132 and combined release rate of 58.6 lb/s for DOT-113s and 65.5 lb/s for T-75 ISOs. Figure 3shows a representative case for the 84 probabilities and combined release rates for six T-75 ISOs.

With larger numbers of containers involved, the number of unique aggregate outcomes grows exponentially. For example, with $n = 22$ containers, there are 2,300 unique aggregate outcomes. A procedure to consolidate similar aggregate outcomes is implemented: 1) Choose a representative set of combined release rates that have relatively high probability and are separated from each other in value. In Figure 3, this set is highlighted in red. 2) For outcomes that are not in the representative set, add their probabilities to the probability of the nearest combined release rate in the representative set. 3) Further consolidate release rates and probabilities within the representative set, by taking a weighted average, if necessary, to obtain a reasonable number of representative release rates and probabilities to feed into the next step of QRA.

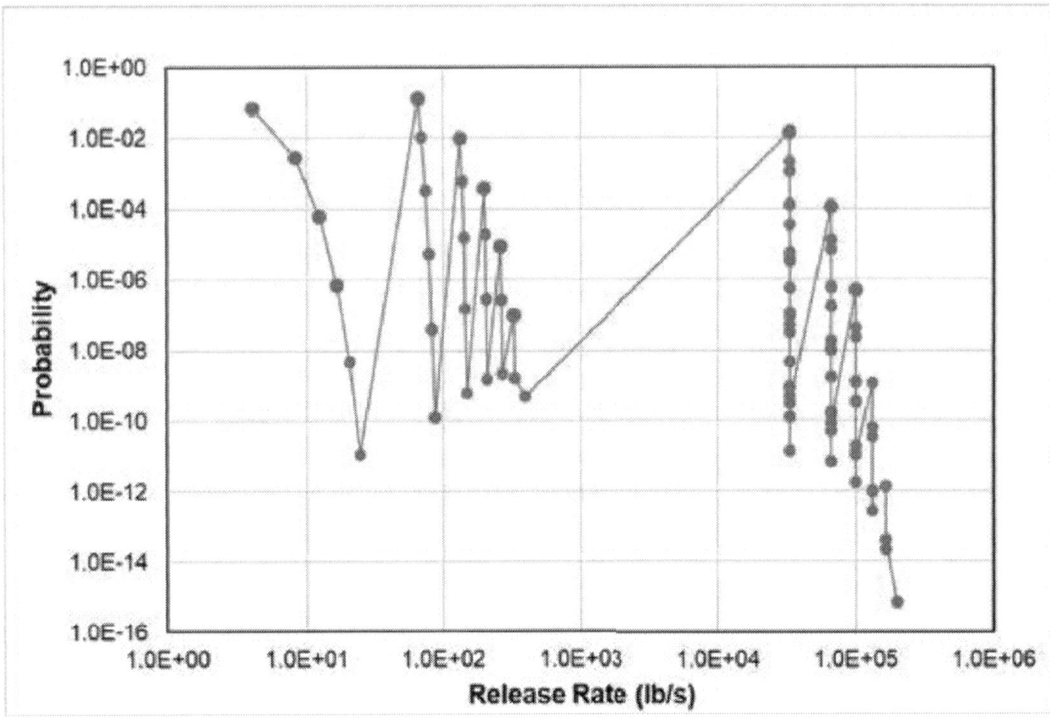

Figure 3. Aggregated outcomes and their probabilities and combined release rates for $n = 6$, representing 6 T-75 ISOs. The blue points show all the probabilities and combined release rates, and the red points show the representative set of combined release rates.

3.4 QRA Model

A quantitate risk assessment combines the likelihood and consequence for a loss of containment event. A QRA model is utilized in this analysis to appropriately take into consideration all parameters that have an effect on the overall risk for the different LNG storage vessels transported via rail.

3.4.1 QRA Software – Phast Risk

The Quantitative Risk Assessment (QRA) performed in this analysis utilized an industry-proven commercial computational model developed and distributed by Det Norske Veritas (DNV) called PHAST Risk. This software takes into consideration the discharge dynamics, dispersion dynamics, consequences, population, weather, and likelihood to quantify the associated risk for each hazardous release scenario.

3.4.2 Probability of Ignition of a Flammable Cloud and Event Trees

For a hazard to occur, a flammable cloud needs to come into contact with an ignition source. In addition to finding an ignition source, the timing that the flammable cloud reaches the ignition source also has an impact in the model. If a flammable release comes into contact with an ignition source immediately upon release, a pool fire and/or jet fire can occur. If a flammable release comes into contact with an ignition source after a delayed period of time, a pool fire, flash fire, and/or explosion may occur. The probability of any of these outcomes occurring (or none at all) is complex and taken into consideration in the PHAST Risk model by the use of event trees (Figure 1). The probability of an individual consequence for a given release depends on whether the release is instantaneous (e.g., catastrophic releases) or continuous (e.g., the other releases considered), the presence of liquid rainout, subsequent pool vaporization, the presence of a persistent liquid pool, and the dispersion behavior of the flammable vapors. For each scenario, PHAST Risk takes into consideration the potential consequences with the event tree and respective probabilities. The probabilities of immediate or delayed ignition are consistent with the guidelines published in the Dutch Purple Book [17] and are thoroughly described in the PHAST Risk documentation [18].

3.4.3 Flammable Effects Exposure for a Population

The population of the surrounding areas have an impact on the overall risk – the larger the surrounding population, the more people exposed to the hazard resulting in increased potential risk. For the societal risk calculations, the QRA model assumes that 90% of the population are indoors and 10% are outdoors. The PHAST Risk model utilizes a grid cell system to calculate the potential fatalities. The probability of fatality is a function of distance. Additional methodology on the PHAST Risk QRA software is provided in Dutch Green Book [19] (and applied to QRA in the Dutch Purple Book [20]). [21]

4 Analysis of Data

In this paper, the overall risk comparing the transportation of LNG via rail in DOT-113 tanks vs. T-75 ISO containers was determined using a QRA model software. In the QRA, the likelihood for a LOC was determined using a novel approach. The likelihood for each LOC event was input into the QRA model along with additional parameters discussed in this section.

4.1 Populations along rail

The population density along the mainline rail route directly affects the risk; thus, an example route with a representative population density was considered in this analysis, assuming the risk with scale proportionally to the increase or decrease in population along the route. The mainline track was evaluated in one-mile long segments to characterize the population density per mile along the example route. The representative population density selected was 1,000 people/mile^2. It must be noted that along this route, the population density can be greater or smaller than the selected representative population density.

4.2 Weather

The ambient air temperature and ground temperature of the route were selected based on guidance for modeling LNG releases in 49 CFR § 193. Default PHAST settings was used where the guidance is open for input (i.e. temperature). The weather conditions used in the analysis are shown below in Table 8.

Table 8. Weather conditions used in analysis

Weather	
Ambient Temperature (°F)	53
Surface Temperature (°F)	53
Relative Humidity	50%
Wind Speed (m/s)	2
Pasquill-Gifford Atmospheric Stability	Category F
Surface Roughness Factor (m)	0.03

4.3 Release Scenarios

For each scenario analyzed in the QRA, the conditions specified for the storage vessels in Table 1 were input into the model (i.e. volume, pressure, saturation temperature, etc). The individual release rates and associated frequencies for combinations of 6 to 11 rail cars for the different scenarios are shown below. These release frequencies per release rate were normalized and input into the model along with total frequency failure for the specified route.

Table 9. Event frequencies for DOT-113 unit train car releases.

Release rate (lb/s)	Release Frequency (/year)	Release rate (lb/s)	Release Frequency (/year)	Release rate (lb/s)	Release Frequency (/year)
6 DOT-113 cars		7 DOT-113 cars		8 DOT-113 cars	
0	4.46E-05	0	4.26E-05	0	4.06E-05
3.80	4.38E-06	3.83	4.91E-06	3.86	5.40E-06
58.6	8.53E-06	58.6	9.59E-06	58.6	1.07E-05
117	6.16E-07	117.	8.35E-07	120	1.14E-06
176	2.36E-08	176	4.01E-08	236	2.30E-09
234	5.19E-10	235	1.18E-09	1 Car	1.39E-06
1 Car	1.05E-06	1 Car	1.22E-06	2 Cars	1.46E-08
2 Cars	7.88E-09	2 Cars	1.10E-08	3 Cars	8.88E-11
3 Cars	3.17E-11	3 Cars	5.55E-11		
Release rate (lb/s)	Release Frequency (/year)	Release rate (lb/s)	Release Frequency (/year)	Release rate (lb/s)	Release Frequency (/year)
9 DOT-113 cars		10 DOT-113 cars		11 DOT-113 cars	
0	3.88E-05	0	3.70E-05	0	3.53E-05
3.89	5.84E-06	3.92	6.22E-06	3.59	6.63E-06
58.6	1.16E-05	58.6	1.25E-05	58.6	1.33E-05
121	1.43E-06	121	1.75E-06	117	1.92E-06
236	4.03E-09	236	6.57E-09	179	1.76E-07
1 Car	1.56E-06	1 Car	1.73E-06	295	4.01E-10
2 Cars	1.88E-08	2 Cars	2.34E-08	1 Car	1.89E-06
3 Cars	1.33E-10	3 Cars	1.89E-10	2 Cars	2.85E-08
				3 Cars	2.59E-10

Table 10. Event frequencies for one T-75 ISO Container per train car on a unit train car.

Release rate (lb/s)	Release Frequency (/year)	Release rate (lb/s)	Release Frequency (/year)	Release rate (lb/s)	Release Frequency (/year)
6 T-75 ISOs		7 T-75 ISOs		8 T-75 ISOs	
0.00	4.46E-05	0.00	4.26E-05	0.00	4.06E-05
4.25	4.38E-06	4.28	4.91E-06	4.32	5.40E-06
65.5	8.50E-06	65.5	9.61E-06	65.5	1.06E-05
131	6.14E-07	131	8.33E-07	135	1.14E-06
197	2.37E-08	197	4.01E-08	264	2.30E-09
262	5.19E-10	263	1.18E-09	1 ISO	1.39E-06
1 ISO	1.05E-06	1 ISO	1.22E-06	2 ISOs	1.47E-08
2 ISOs	7.90E-09	2 ISOs	1.10E-08	3 ISOs	8.85E-11
3 ISOs	3.18E-11	3 ISOs	5.55E-11		

Release rate (lb/s)	Release Frequency (/year)	Release rate (lb/s)	Release Frequency (/year)	Release rate (lb/s)	Release Frequency (/year)
9 T-75 ISOs		10 T-75 ISOs		11 T-75 ISOs	
0.00	3.88E-05	0.00	3.70E-05	0.00	3.53E-05
4.35	5.84E-06	4.38	6.24E-06	4.42	6.60E-06
65.5	1.16E-05	65.5	1.25E-05	65.5	1.33E-05
135.1	1.43E-06	136	1.75E-06	131	1.92E-06
263.9	4.03E-09	264	6.55E-09	200	1.76E-07
1 ISO	1.56E-06	1 ISO	1.73E-06	329	4.01E-10
2 ISOs	1.88E-08	2 ISOs	2.34E-08	1 ISO	1.90E-06
3 ISOs	1.32E-10	3 ISOs	1.89E-10	2 ISOs	2.85E-08
				3 ISOs	2.59E-10

Table 11. Event frequencies for two T-75 ISO containers per train car on a unit train.

Release rate (lb/s)	Release Frequency (/year)	Release rate (lb/s)	Release Frequency (/year)	Release rate (lb/s)	Release Frequency (/year)
12 T-75 ISOs		14 T-75 ISOs		16 T-75 ISOs	
0.00	3.37E-05	0.00	3.06E-05	0.00	2.79E-05
4.45	6.93E-06	4.52	7.48E-06	4.59	7.90E-06
65.5	1.41E-05	65.5	1.54E-05	65.5	1.65E-05
131	2.24E-06	131	2.90E-06	139	4.07E-06
201	2.29E-07	201	3.61E-07	266	4.85E-08
330	6.70E-10	330	1.61E-09	396	1.74E-10
1 ISO	2.06E-06	1 ISO	2.39E-06	1 ISO	2.72E-06
2 ISOs	3.41E-08	2 ISOs	4.68E-08	2 ISOs	6.13E-08
3 ISOs	3.45E-10	3 ISOs	5.68E-10	3 ISOs	8.61E-10
				4 ISOs	8.48E-12
Release rate (lb/s)	Release Frequency (/year)	Release rate (lb/s)	Release Frequency (/year)	Release rate (lb/s)	Release Frequency (/year)
18 T-75 ISOs		20 T-75 ISOs		22 T-75 ISOs	
0.00	2.54E-05	0.00	2.31E-05	0.00	2.10E-05
4.65	8.23E-06	4.72	8.45E-06	4.75	8.60E-06
65.5	1.75E-05	65.5	1.82E-05	65.5	1.88E-05
140	4.95E-06	141	5.87E-06	142	6.81E-06
267	7.75E-08	268	1.16E-07	268	1.67E-07
396	3.82E-10	397	7.55E-10	397	1.38E-09
1 ISO	3.04E-06	1 ISO	3.36E-06	1 ISO	3.67E-06
2 ISOs	7.77E-08	2 ISOs	9.59E-08	2 ISOs	1.16E-07
3 ISOs	1.25E-09	3 ISOs	1.73E-09	3 ISOs	2.33E-09
4 ISOs	1.42E-11	4 ISOs	2.24E-11	4 ISOs	3.36E-11

5 Results and Discussion

In this analysis, the risk was compared for transportation of LNG via rail in DOT-113 vessels, single T-75 ISO containers placed on a rail car, and double T-75 ISO containers placed on a rail car. For each of these scenarios, risk is presented to compare for the same leak sizes, weather conditions, example route, and surrounding population of 1,000 people/mile^2. The FN curve for the three scenarios, which depicts the cumulative frequency versus severity (SR), is provided in Figure 4. This plot indicates that 10 fatalities (resulting from a single incident) may occur every 112,600, 3,824,600 and 1,781,140 years for the DOT-113, single T-75 ISO and two T-75 ISOs.

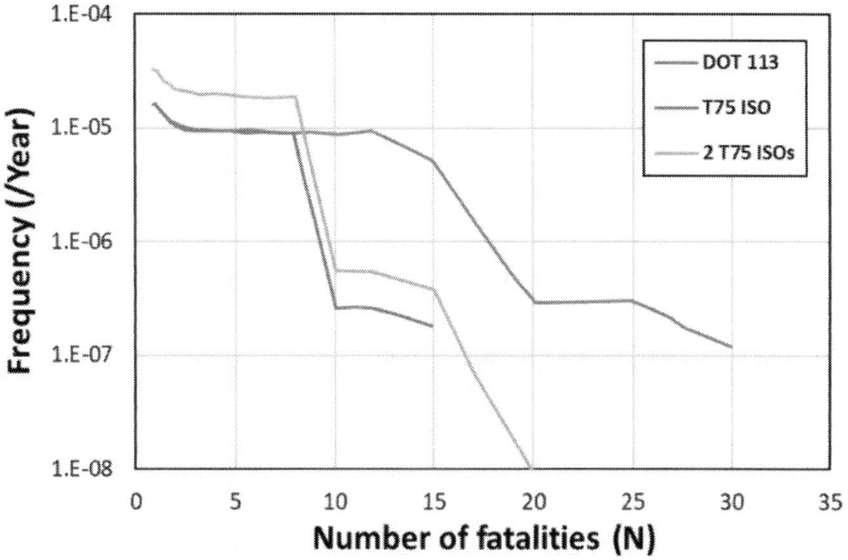

Figure 4. Cumulative SR for Train Scenarios: DOT-113, 1 ISO, and 2 ISO.

The sum of the area under the FN curve is calculated in the QRA model as the "SR Integral". The SR Integral and Maximum individual risk are shown below in Figure 5. Overall, transporting LNG in a single T-75 ISO container on a rail car poses both the lowest societal and individual risk.

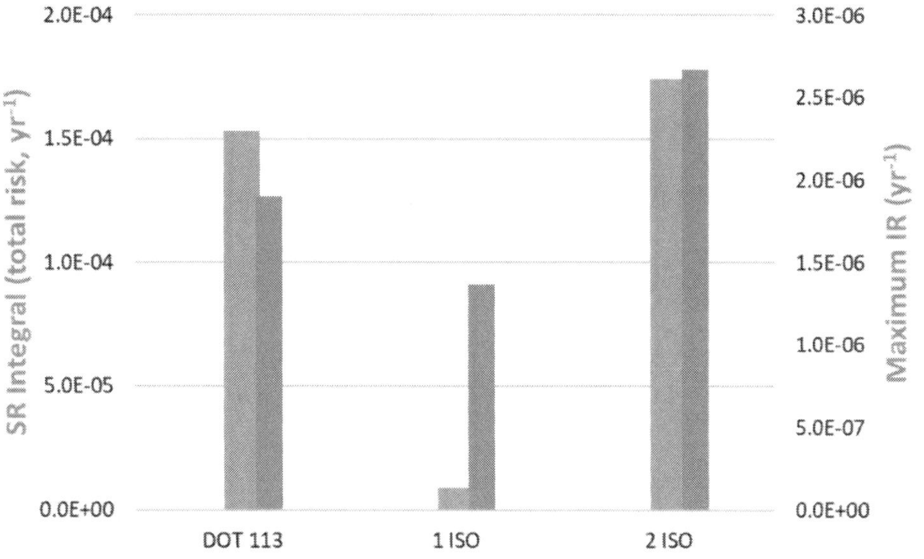

Figure 5. Total SR integral risk and Maximum IR per mile for Train Scenarios: DOT-113, 1 ISO, and 2 ISO.

Although the single ISO container poses the lowest risk, doubling the same ISO containers per rail car is not as intuitive as doubling the risk due to the change in train length. The last method of risk comparison across the different transportation methods are the distances to Zone 3 specified in

NFPA 59A. These distances are shown below in Figure 6, which shows the single and two T-75 ISOs have the lowest distances to Zone 3 with no restrictions specified in NFPA 59A.

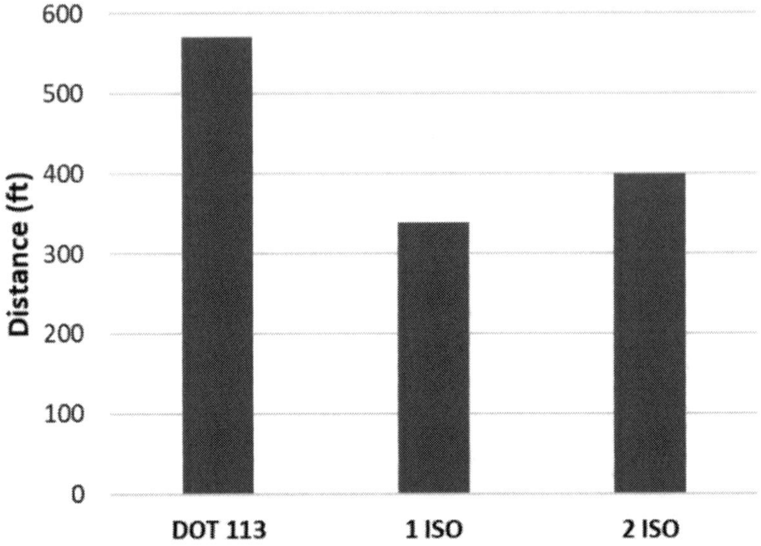

Figure 6. Maximum Distance to NFPA 59A Zone 3 - 3×10^{-7} IR (ft)

Although this QRA kept constant the total volume transported per trip, leak sizes, rail car placement, weather conditions and more, the total volume on each rail car position does vary due to vessel size. While the overall risk is comparable for the different release scenarios, the cost benefits of the different scenarios can be further analyzed if risk is deemed acceptable.

For the proposed mainline route, the risk results varied with population density along the railroad. The underlying accident likelihoods and release scenarios are independent of the route demographics; thus, the calculated shipping risk is directly related to the route length and population along the route.

6 Summary

With the updated CFR, LNG shipment is now allowed in DOT-113 class tank rail cars. These new regulations increase the methods for LNG shipment via rail, with previous shipment occurring in T-75 ISO containers via Special Permit. With this new regulation, an increase in LNG shipment via rail may occur. A QRA was performed to determine the safest method to ship LNG via rail: in a single DOT-113 vessel, a single T-75 ISO container per rail car, or two T-75 ISO containers per rail car. In this analysis, a novel multinomial probability model to approximate the number of tank cars with their release rate for LNG multi-car incidents was utilized. The risk for each of these scenarios was comparable but the single T-75 ISO container per railcar resulted in the lowest societal risk, individual risk, and distance to the NFPA specified zone per mile route per 1,000 people/mile^2. This study can be used to better understand the risk associate with the different LNG shipping methods via rail and identify potential opportunities for risk reduction through understanding the operational limits.

7 References

[1] A. Ibarreta, H. K. Kytömaa and D. R. ". Morrison, "LNG Small Scale and Transportation: Navigating the Risk," *Exponent Oil & Gas Newsletter* , 2015.

[2] "49 CFR, Parts 172, 173, 174, 179 and 180," [Online]. Available: https://www.federalregister.gov/documents/2020/07/24/2020-13604/hazardous-materials-liquefied-natural-gas-by-rail.

[3] "LNG Tank Car SR-603, Bulk Transport Unit, Chart Industries.," [Online]. Available: www.ChartLNG.com.

[4] "Specification of 110 psi T75 UN Portable Tank. Specification Number: CASIS-40-110(S/S)-LNG-SP, Rev. R6 03 May 2018. CIMC Tank.," [Online].

[5] J. M. Morris, D. T. Morrison, and R. J. Hart, "Sensitivity Analysis of Transport Conditions on Liquefied Gas Hazards.," in *2019 Spring Meeting and 15th Global Congress on Process Safety. AIChE,*, 2019.

[6] R. J. Hart, M. Garcia and D. Morrison, "What Is the Safest Way to Move LNG?," in *AICHE Spring Meeting and 14th Global Congress on Process Safety*, 2018.

[7] R. Hart and . D. ". Morrison, "The hazard we know: Comparing transportation risk of LPG and LNG," in *Spring National Meeting and 11th Global Congress on Process Safety*, 2015.

[8] "49 CFR § 213.9 Classes of track: operating speed limits," [Online].

[9] "Federal Railroad Administration Office of Safety Analysis," [Online]. Available: https://safetydata.fra.dot.gov/OfficeofSafety/default.aspx.

[10] "Accident Data as reported by Railroads, Rail Equipment Accidents (6180.54)," [Online].

[11] "FRA Office of Safety Analysis, Report 1.02 – Operational Data Tables.," [Online].

[12] "49 CFR § 174.85 - Position in train of placarded cars, transport vehicles, freight containers, and bulk packagings.," [Online].

[13] "Pipeline and Hazardous Materials," [Online]. Available: https://portal.phmsa.dot.gov/phmsapub/faces/EnrollmentV2.jspx?_afrLoop=812325885745104&_afrWindowMode=0&_afrWindowId=xn1piccu7&_adf.ctrl-state=6pw87wsa1_10.

[14] M. v. V. e. Waterstaat, "Section 3.3.3.3, Railways, page 3.13 in Guideline for Quantitative Risk Assessment, Part Two: Transport (Dutch Purple Book), Publication Series on Dangerous Substances," 2005.

[15] "PHMSA, LNG Plant Requirements.," [Online]. Available: https://www.phmsa.dot.gov/pipeline/liquified-natural-gas/lng-plant-requirements-frequently-asked-questions.

[16] [Online]. Available: https://mathworld.wolfram.com/MultinomialDistribution.html.

[17] "Chapter 4.7, Ignition, in Guideline for Quantitative Risk Assessment (Dutch Purple Book), Publication Series on Dangerous Substances, Ministerie van Verkeer en Waterstaat," 2005. [Online].

[18] "PHAST Risk Technical Documentation, "MPACT Theory," DNV Software, page 103," 2010.

[19] "Chapter 1, Damage Caused by Heat Radiation, in Methods for the Determination of Possible Damage (Dutch Green Book), Publication Series on Dangerous Substances, Ministerie van Verkeer en Waterstaat.," 1992.

[20] in *Chapter 5, Modeling Exposure and Damage, in Guideline for Quantitative Risk Assessment (Dutch Purple Book), Publication Series on Dangerous Substances, Ministerie van Verkeer en Waterstaat*, 2005.

[21] "PHAST Risk Technical Documentation, "MPACT Theory," DNV Software, pages 66-94," 2010.

SPRING21 +17TH GCPS
A Joint AIChE and CCPS Meeting

Evaluation Protocols for Flammable and Toxic Dispersion Models

Filippo Gavelli
Blue Engineering and Consulting
6011 University Blvd, Suite 220, Ellicott City, MD 21043
fgavelli@blueeandc.com

Bryant Hendrickson
Blue Engineering and Consulting
6011 University Blvd, Suite 220, Ellicott City, MD 21043
bhendrickson@blueeandc.com

Prepared for Presentation at
American Institute of Chemical Engineers
2021 Spring Meeting and 17th Global Congress on Process Safety
Virtual
April 18 - 22, 2021

Evaluation Protocols for Flammable and Toxic Dispersion Models

Filippo Gavelli
Blue Engineering and Consulting
6011 University Blvd, Suite 220, Ellicott City, MD 21043
fgavelli@blueeandc.com

Bryant Hendrickson
Blue Engineering and Consulting

Keywords: model evaluation; model validation; dispersion; LNG

Abstract

The siting of LNG facilities requires several types of hazards to be quantified so the risk to the public and public property in areas surrounding each facility can be properly assessed. In order to quantify the hazards from potential accident scenarios, computational tools must be used. Given the multiplicity of tools available, ranging from empirical models to 3D computational fluid dynamics packages, the agency reviewing the project application may not have the methodologies or protocols necessary to determine the suitability of a given model to a certain scenario, its accuracy and any other setup requirements.

The Pipeline and Hazardous Materials Safety Administration (PHMSA) commissioned Blue Engineering and Consulting and the Gas Technology Institute to develop a set of Model Evaluation Protocols (MEPs), that will allow the review of modeling tools for several different hazards, including flammable and toxic dispersion, source terms for dispersion models, vapor cloud explosions, and BLEVEs. The project will provide PHMSA with the means to properly evaluate modeling tools for these hazards, which are required to be evaluated in an LNG facility siting study. It will also provide clarity to LNG plant developers and their safety consultants on which models may be used and which safety factors need to be applied and will increase the public's confidence in the regulatory review process of LNG facilities.

This paper discusses the first two MEPs completed as part of this project to evaluate flammable dispersion and toxic dispersion models. The paper will discuss the evaluation methodology, including the information to be submitted by the model proponents and the review process, and the model validation databases compiled for each protocol.

1 Introduction

The safe siting of LNG facilities requires project developers to evaluate the potential consequences to the public and public property from accidental loss of containment and release of hazardous materials (e.g., flammable and/or toxic). These consequences include: pool, jet or flash fires; vapor cloud explosions; toxic gas exposure; and catastrophic vessel failures. Calculating the distance to which hazardous conditions may extend for each type of scenario requires the use of computational models.

Assessing the quality of models, especially those used to evaluate the consequences of hazardous scenarios on the public, has long been a concern to regulatory bodies. This concern led, over time, to the development of several protocols to evaluate models in a consistent and transparent manner. In the context of LNG facility siting in the United States, a Model Evaluation Protocol (MEP) and Model Validation Database (MVD) were developed in 2007 [1] and updated in 2016 [2] to address the modeling of vapor dispersion from LNG spills (this is commonly known as the "LNG MEP"). However, with the evolution of the LNG industry and the resulting adjustments to regulatory requirements, it has become evident that the scope of model evaluation needs to extend beyond what is covered by the LNG MEP. Therefore, the Pipeline and Hazardous Materials Safety Administration (PHMSA) commissioned Blue Engineering and Consulting and the Gas Technology Institute, via research project #854 titled: "Develop an Evaluation Protocol for Non-LNG Release Hazards – Modeling", to develop a set of MEPs, that will allow the review of modeling tools for several different hazards, including flammable and toxic dispersion, source terms for dispersion models, vapor cloud explosions, and BLEVEs.

This paper provides a summary of the model evaluation protocol for flammable and toxic dispersion models. The complete documents ([3], [4]) can be obtained from PHMSA's project website at: https://primis.phmsa.dot.gov/matrix/PrjHome.rdm?prj=854.

1.1 Scope of the Model Evaluation

The scope of the flammable and toxic dispersion MEPs is to evaluate models that calculate the extent of flammable or toxic dispersion hazards from accidental releases. The objective of evaluation is to allow the regulatory body (i.e., PHMSA) to determine whether a given model should be approved or rejected for use in dispersion modeling within the context of LNG facility siting studies. Additionally, the evaluation should allow the regulatory body to specify any restrictions on the use of an approved model, as well as any "validation factors" to be applied to the modeling results. Comparing models against one another and establishing rankings of preferred models are not objectives of the MEP.

Each model evaluation is specific to a single model and to a single version of that model. As a new version is released, a new model evaluation must be conducted; however, a goal of these protocols is to minimize the time required to evaluate minor model updates.

Separate MEPs and MVDs were developed for flammable and toxic dispersion models. In fact:

1. The concentrations of interest are quite different: flammable concentrations are typically on the order of 1-5% by volume for the lower flammability limit, whereas the thresholds

for toxic hazards are typically in the range of 10-1,000 ppm (0.001-0.1%). A model could be performing significantly better over one range than the other, therefore, separate evaluations can provide clearer information on the behavior of the model.

2. As discussed in more detail later, due to the residence time of the hazardous condition, flammable hazards are better represented by short time average modeling, whereas toxic hazards are better captured with long time averages. Separating the two evaluations allows this distinction to be preserved.

1.2 Types of Release Scenarios

The characteristics of the release can have a significant effect on the extent of the dispersion hazards. There are numerous parameters affecting the release, including: pressure, temperature and phase of the fluid prior to the release; hole size, shape and elevation; etc.

For the purpose of the dispersion MEPs, releases are grouped into three categories based on how the release mechanism influences the cloud dispersion:

1) Momentum-driven releases ("jets") are those in which the material is introduced into open air with sufficient net momentum in one direction to drive the initial dispersion of the hazardous cloud. The most common momentum-driven release scenarios encountered in LNG facility siting are:
   a) Gas jets: releases of pressurized gaseous material through a hole on the order of a few inches diameter. In most practical cases for LNG facility siting, the gas pressure is high enough to result in choked flow conditions and a sonic discharge.
   b) Liquid jets: releases of pressurized liquid material through a hole on the order of a few inches diameter. As the liquid exits the hole, it tends to atomize into a spray of droplets which then progressively evaporate as the jet expands and entrains ambient air. Part of the spray may settle onto the ground before all droplets have evaporated: the deposited liquid fraction is called "rainout" and effectively creates a new source of vapor. In most practical cases, the pressurized liquid is at temperatures higher than the normal boiling point; therefore, upon discharge to atmosphere, a fraction of the release immediately flashes to vapor while the rest atomizes into a droplet spray as just described: this type of release is commonly known as a "flashing jet" or "flashing and jetting" scenario.
2) Low-momentum releases are those in which the material is introduced into open air with negligible net momentum in any direction. Therefore, the cloud dispersion is driven by the wind, gravity and obstacles. The most frequent examples of low-momentum releases encountered in LNG facility siting are:
   a) Spills: liquid releases from low pressure storage (e.g., LNG storage tanks) or from large breaches, which result in negligible flashing and atomization of the stream. The liquid spill typically falls to the ground (although spills on water are also considered in some cases) and spreads following the local topography.
   b) Catastrophic releases: volumes of liquid or gas released instantaneously from the sudden disappearance of a storage vessel's walls. The material contained in the vessel is free to expand in all directions, with zero net initial momentum.
3) Tracer releases may be momentum-driven or low-momentum releases, whose rates are small enough to be rapidly diluted to low concentrations (e.g., less than 1,000 ppm). Examples include small release rates of toxic gas into a highly ventilated area or releases of a mixture

with only a small toxic fraction. Tracer releases are not often considered in LNG facility siting studies.

1.3 Dispersion Areas

Once a cloud is formed by the release of a hazardous material, it begins to disperse driven by its own momentum (if any) and the wind, and it interacts with the terrain and any obstacles or obstructions it encounters. Every facility layout and topography has a different effect on the cloud; however, for the purpose of combining general effects, the following general categories are defined:

1) Unobstructed terrain: flat terrain without any obstacles or obstructions. While this is clearly a theoretical case, it represents a very frequent approximation in dispersion modeling. In fact, even the simplest models can handle dispersion over flat terrain (in many cases, "flat" is the only terrain a model can simulate); additionally, modeling an unobstructed dispersion case often results in longer (i.e., conservative) hazard distances than an obstructed one.
2) Obstructed terrain: flat terrain with one or more obstacles located downwind of the release (e.g., walls or barriers), for the specific purpose of obstructing or redirecting the cloud.
3) Complex geometry: any combination of a realistic (i.e., non-flat) terrain, obstructions due to facility components (e.g., piping, equipment, structures), and obstacles placed to control cloud dispersion (e.g., vapor barriers). Complex geometry is the most realistic type of dispersion area but also the most challenging for models to simulate accurately.

1.4 Phenomena Affecting Dispersion

Given the wide range of dispersion scenarios that can be encountered in an LNG facility siting study, dispersion models need to account for several phenomena. What follows is a brief overview; interested readers can find more details in the published literature, for example in the Center for Chemical Process Safety (CCPS) book "Guidelines for the use of vapor cloud dispersion models" [5].

1.4.1 Material Properties

The thermodynamic properties of the released material have a significant impact on the behavior of the hazardous cloud; therefore, they should be built into the model or allowed to be specified by the user. Given the potentially large pressure and temperature ranges as well as the possibility of phase change, variable properties are preferred over constant properties.

Mixtures of components are frequently encountered in siting studies, therefore, models should be capable of tracking the dispersion of vapor mixtures, either by performing multicomponent dispersion calculations or by defining a pseudo-component with the properties of the mixture.

1.4.2 Source Term

The ability to calculate the source term (i.e., release rate, temperature, phase, etc.) for a dispersion scenario is not a requirement for a dispersion model; in fact, several models opt for the user to calculate the necessary source term parameters and enter them as inputs or boundary conditions in the dispersion model. Nonetheless, a dispersion model should be capable of accounting for the properties and physical characteristics of the source term, such as:

- Temperature;
- Density;
- Phase;
- Composition (i.e., pure fuel or fuel/air mixture);
- Vapor generation rate;
- Source area, elevation and direction;
- Source momentum.

1.4.3 Ambient Conditions

The behavior of a flammable cloud depends heavily on the ambient conditions into which it is released. Dispersion models should therefore be able to account for the effects of ambient temperature, wind speed and atmospheric turbulence (including their variation with elevation) on the evolution of the cloud. Relative humidity can also affect gas dispersion, both due to the change in air density and to the potential for water vapor condensation ("fog") and re-evaporation.

The ability to model the effects of the atmospheric boundary layer (e.g., through surface roughness and Pasquill-Gifford stability classes) on the dispersion of dense clouds is critical for LNG facility siting. Wind direction is also important, particularly for models simulating dispersion in complex geometry and for scenarios in which the wind direction is not aligned with the momentum of the release.

1.4.4 Cloud Dispersion

Several different mechanisms come into play as a hazardous cloud is formed and begins to disperse, including:

- Gravity-driven spreading, including terrain effects;
- Advection by the wind;
- Mixing due to turbulence at the cloud/air interface;
- Turbulence suppression due to dense cloud stratification;
- Cloud/obstacle interaction (e.g., vapor holdup by barriers);
- Mixing due to obstacle interaction;
- Heat transfer from the ground;
- Water vapor condensation (fog) and evaporation.

Dispersion models should therefore be capable of accounting for as many of these effects as possible, through either numerical solution of physical equations or specific submodels.

1.4.5 Averaging Times

Dispersion models solve one or more equations that represent an approximation of the physical phenomena. Part of the approximation includes the time interval over which the stochastic fluctuations of the physical phenomena are averaged. Since flammable dispersion hazards (i.e., ignition of the flammable cloud) can occur instantaneously, short time averaging is generally more appropriate for flammable dispersion models as it can capture short-lived fluctuations that may extend and be ignited farther from the source. On the other hand, toxic dispersion hazards are typically associated with prolonged exposure (particularly at the concentrations of interest in siting studies), therefore long time averaging is generally more appropriate for toxic dispersion models.

2 The Model Evaluation Protocols

The scope of the flammable and toxic dispersion MEPs is to define an objective and transparent methodology to determine the ability of a model to accurately quantify dispersion hazards in the context of LNG facility siting. The MEPs are guidance documents which outline the tasks required by the model petitioner and the reviewer. Specifically, they provide guidance to the petitioner regarding:

- Information to be submitted (e.g., user guide, technical references, etc.);
- Scenarios to be simulated for validation (including any sensitivity analysis); and
- Feedback that should be expected once the evaluation is completed (e.g., whether the model is accepted or not, restrictions on the applicability of the model, or validation factors to be applied to the modeling results).

The protocol also provides guidance to the evaluator, including:

- Requirements that an evaluator must meet to be qualified;
- Tasks to be performed during the evaluation; and
- Reporting requirements.

The dispersion MEPs follow the general model evaluation methodology described by Gavelli et al. [6] and includes the following tasks:

1) Scope of the evaluation
2) Evaluator qualifications
3) Model description
4) Scientific assessment
5) User-oriented assessment
6) Verification
7) Validation, which includes:
    a) Model Validation Database, with:
        i) Datasets for model validation
        ii) Key variables and physical comparison parameters
        iii) Statistical performance measures
        iv) Acceptability criteria
    b) Sensitivity analysis
    c) Qualitative performance assessment
    d) Quantitative performance assessment
8) Model Evaluation Report, which includes:
    a) Summary of the evaluation (i.e., steps 4 through 7 above)
    b) Best practice guidance

c) Recommendations for improvement

For each MEP, the model validation (step 7 above) is included in a separate document. This allows the validation data sets, statistical performance measures and acceptability criteria to be updated as more information becomes available (e.g., new experimental data sets or results from model evaluations) without requiring a complete revision of the model evaluation protocol.

3 The Model Validation Databases

Model validation is the process of comparing the predictions of a model against experimental data relevant to the types of scenarios that the model is intended to predict. Since "validation" in the true sense of the word cannot be accomplished by comparing a model against a finite number of scenarios [7], the purpose of a model validation database is to identify a set of scenarios to be simulated with a computational model, in such a matter that qualitative and/or quantitative comparisons may be made between model predictions and actual observations.

The following criteria were applied to datasets reviewed for inclusion in the flammable and/or toxic dispersion MVDs ([1], [7], [8]):

- The test configuration should be representative of realistic scenarios. For example, the release of a pressurized stream from a small orifice (e.g., 2-inch diameter) can be representative of many scenarios typically included in facility siting studies. The scale of the experiment (e.g., release flowrate, dimensions of the test area, etc.) should also be comparable to actual scenarios.
- The tests should be described in sufficient detail for modelers to set up their simulations in an accurate and consistent manner, with as few assumptions as possible.
- Meteorological data (i.e., ambient temperature, humidity, wind speed and atmospheric stability, etc.) should be available from sensors in/near the area of interest. Location and height of sensors should be provided. The resolution should be sufficient to define the atmospheric boundary layer profiles (e.g., ground roughness, atmospheric stability), or those values should be provided in the test series report.
- The test series should include a wide range of meteorological regimes.
- Sensor data should include measurement of the key variables: in the case of atmospheric dispersion experiments, gas concentration measurements must be available; temperature measurements may also be useful.
- Measurements should be of a sufficient quantity to be statistically representative.
- Concentrations should be available at more than one distance downwind, with sufficient lateral resolution to document the spatial structure of the cloud. Vertical resolution should also be included, to define the cloud stratification.
- Uncertainty of all measured and derived quantities should be provided, together with a description of the method used to define such uncertainty.
- If the raw data is not provided, the time-averaging applied to the data must be specified.
- The data must be publicly available.

In order to optimize the database, the test series were selected to span a broad range of conditions and to strike a balance between the number of trials included for each type of scenario. For that purpose, the test series were grouped according to several parameters as follows:

- Material – Flammable Dispersion:
  - LNG
  - Flammable (non-LNG)
  - Non-flammable
- Material – Toxic Dispersion:
  - Toxic
  - Non-toxic
- Release Type:
  - Liquid Spill or low-momentum gas source
  - Jet (flashing liquid or gas)
  - Tracer – For toxic dispersion only
- Dispersion Area:
  - Unobstructed (e.g., "rural")
  - Obstructed (e.g., barriers)
  - Complex Geometry (e.g., "urban" setting, or topography effects)

Numerous test series were reviewed for each MVD. In order to select a representative yet manageable set of data, each series was "rated" on a 5-point scale according to several categories, following an approach first introduced by Skjold et al. [9]. The different ratings were then combined into an average score, which informed the selection of data sets for the validation database.

The following categories were used to rate experimental datasets for the flammable and toxic dispersion MVDs:

- Relevance to actual scenarios: this category evaluates whether the experimental set up replicates dispersion scenarios that are likely to be modeled in an actual hazard analysis. For example, a pressurized liquid release from a small orifice would be rated highly, whereas a gas release from a tall stack would receive a lower score.
- Material: this category evaluates the relevance of the fluid released during the experiments to those typically evaluated in an actual hazard analysis. For example, spills of LNG or pressurized propane releases would be rated highly, whereas the release of SF_6 or other chemicals not used in LNG facilities would be rated low.
- Scale: this category compares the scale of the experimental set up to typical hazard analysis scenarios. "Scale" should be intended as both the physical dimensions of the test area (i.e., the distance from the source to the measurement stations) and the strength of the source term (i.e., release flowrate). Previous MEPs argued that experimental data from all scales should be considered in the validation of a model, as poor performance at small scales could be indicative of a potential problem in the model. However, the purpose of this project is to evaluate models for a very specific type of application, therefore, if a deficiency only manifests itself at small scales it is not considered relevant to the purpose

of the evaluation. For this reason, "full" scale trials were rated higher than reduced scale or wind tunnel tests.
- Repeatability: this category addresses the inherent uncertainty associated with atmospheric dispersion. For example, if a field scale experiment is run just once, the measurements will be affected by stochastic conditions specific to that run, which cannot easily be quantified; if the same experiment were to be repeated multiple times, an ensemble average of the data could be calculated as well as a measure of the stochastic uncertainty. Typically, field experiments are rarely repeated due to the variability of ambient conditions; smaller scale experiments (e.g., wind tunnel or laboratory scale) can be repeated more easily and are subject to more controlled ambient conditions.
- Data quality/availability: this category evaluates whether the data publicly available is adequate to evaluate a model's predictions. Data quality considers the type of measurement performed, the instruments used and the sampling rate. Data availability considers the number and placement of sensors, and whether time traces are available or just peak values.

For example, the Coyote test series [10], [11], which consisted of LNG spills onto a 58 m diameter water pond with the resulting vapor clouds allowed to disperse over mostly flat terrain, was rated as follows:

- Relevance = 5. LNG spills are considered highly relevant since they are required in every LNG facility siting study.
- Material = 5. LNG releases (spills or others) are considered highly relevant, since LNG is currently the only fuel whose flammable vapor dispersion area represents an exclusion zone.
- Scale = 5. The spill rate and dispersion area for the Coyote trials scale well with typical facility siting scenarios.
- Repeatability = 2. Repeatability is inherently difficult in field trials, due to the variability of ambient conditions.
- Data quality/availability = 3. The overall quality and availability of data is considered acceptable for the purpose of the MVD.
- Overall rating = 4.0. The Coyote test series is considered suitable for inclusion in the flammable dispersion MVD.

A similar review and rating process was performed for each test series reviewed for inclusion in the flammable and toxic dispersion MVDs. The details can be found in the respective MVD reports ([3], [4]).

3.1 Flammable Dispersion MVD

Over 20 experimental test series were reviewed to define the database for flammable dispersion model validation. Table 3.1 lists the test series that were included in the database and Table 3.2 shows them grouped by type of release (rows) and type of dispersion area (columns). Test series involving LNG releases are in bold text; non–LNG flammable releases are in plain text; non-flammable releases are shown in underlined text. The number in parentheses following each test series name represents the number of trials included in the MVD. A summary of each test series reviewed and a detailed description of each test series included in the MVD can be found in the flammable MVD report on PHMSA's project website.

Table 3.1. Experimental test series included in the Flammable Dispersion MVD.

Series	No. of Trials	Fluid	Phase	Release Type	Dispersion Area
BA-Hamburg	9	Sulfur Hexafluoride	Gas	Low-momentum	Obstructed (4) Unobstructed (5)
Burro	3	LNG	Liquid	Spill	Unobstructed
CHRC	3	Carbon Dioxide	Gas	Low-momentum	Obstructed (2) Unobstructed (1)
Coyote	3	LNG	Liquid	Spill	Unobstructed
Desert Tortoise	4	Ammonia	Liquid	Flashing jet	Unobstructed
Falcon	3	LNG	Liquid	Spill	Obstructed
Jack Rabbit II	6	Chlorine	Liquid	Flashing jet	Unobstructed (2) Complex Geometry (4)
Kit Fox	8	Carbon Dioxide	Gas	Low-momentum	Complex Geometry
Lathen (EEC)	7	Propane	Liquid	Flashing jet	Obstructed
Maplin Sands	3	LNG	Liquid	Spill	Unobstructed

Table 3.2. Flammable Dispersion MVD test series, grouped by Release Type and Dispersion Area.

	Unobstructed	Obstructed	Complex Geometry
Low-Momentum	**Burro (3)** **Coyote (3)** **Maplin Sands (3)** BA-Hamburg (5) CHRC (1)	**Falcon (3)** BA-Hamburg (4) CHRC (2)	Kit Fox (8)
Jets	Desert Tortoise (4) Jack Rabbit II (2)	Lathen (7)	Jack Rabbit II (4)

As shown in Table 3.2, the flammable dispersion MVD includes a total of 49 trials, of which 32 (65%) are spills or low-momentum releases and 17 (35%) are jet releases. Cloud dispersion occurred over flat terrain in 21 trials (43%), whereas 16 trials (33%) included obstructions (e.g., barriers) and 12 trials (24%) included an urban style "complex" geometry. LNG was released in 12 trials (24%) which were all included in the previous LNG spills MEP; no jet release experiments involved LNG. Ten trials (20%) involved propane releases; the remaining 27 trials (56%) used non-flammable materials. The following observations can be drawn from Table 3.1:

- A much larger number of dispersion experiments were conducted using non-flammable fluids than flammable ones (LNG, propane, etc.). Several of these test series were included in the flammable dispersion MVD in order to obtain a sufficiently large database.
- Numerous test series evaluated dispersion over unobstructed areas, which is reasonable because such configuration allows a simpler evaluation of models; it is also understandable because most of these test series were conducted in the 1980s, when modeling capabilities were limited. Given the number of CFD models currently available, additional data on dispersion in obstructed or complex geometries would be useful to provide better validation for more realistic scenarios.
- A data gap exists for jet releases and complex geometries, which typically represent the largest number of flammable dispersion scenarios in a siting study.

3.2 Toxic Dispersion MVD

Over 20 experimental test series were reviewed to define the database for toxic dispersion model validation. Table 3.3 lists the test series that were included in the database and Table 3.4 shows them grouped by type of release (rows) and type of dispersion area (columns). Test series involving toxic streams are in bold text; non-toxic streams are shown in plain text. The number in parentheses following each test series name represents the number of trials included in the MVD. A summary of each test series reviewed and a detailed description of each test series included in the MVD can be found in the toxic MVD report on PHMSA's project website.

Table 3.3. Experimental test series included in the Toxic Dispersion MVD.

Series	No. of Trials	Fluid	Phase	Release Type	Dispersion Area
BA-Hamburg	3	Sulfur Hexafluoride	Gas	Low-momentum	Obstructed (2) Unobstructed (1)
FLADIS	3	Ammonia	Liquid	Flashing jet	Unobstructed
Goldfish	3	Hydrogen Fluoride	Liquid	Flashing jet	Unobstructed
Jack Rabbit II	6	Chlorine	Liquid	Flashing jet	Unobstructed (2) Complex Geometry (4)
Kit Fox	8	Carbon Dioxide	Gas	Low-momentum	Complex Geometry
Lathen (EEC)	4	Propane	Liquid	Flashing jet	Obstructed
MUST	6	Propylene	Gas	Jet	Complex Geometry
Prairie Grass	5	Sulfur Dioxide	Gas	Tracer	Unobstructed
Thorney Island	2	Freon-12/Nitrogen	Gas	Low-momentum	Unobstructed

Table 3.4. Toxic Dispersion MVD test series, grouped by Release Type and Dispersion Area.

	Unobstructed	Obstructed	Complex Geometry
Low-Momentum	**BA-Hamburg (1)** Thorney Island (2)	**BA-Hamburg (2)**	Kit Fox (8)
Jets	**FLADIS (3)** **Goldfish (3)** **Jack Rabbit II (2)**	**Lathen (4)**	**Jack Rabbit II (4)**
Tracer	Prairie Grass (5)		MUST (6)

As shown in Table 3.4, the toxic dispersion MVD includes a total of 40 trials, of which 13 (33%) are spills or low-momentum releases, 16 (40%) are jet releases and 11 (28%) are tracer gas releases. Cloud dispersion occurred over flat terrain in 16 trials (40%), whereas 6 trials (15%) included obstructions (e.g., barriers) and 18 trials (45%) included an urban style "complex" geometry. A comparison of Table 3.1 and Table 3.2 shows a large overlap: in fact, 5 test series (BA-Hamburg, CHRC, Jack Rabbit II, Kit Fox, and Lathen) and 20 trials overall are included in both databases. This was done to reduce, where possible, the computational burden for models submitted for review for both flammable and toxic dispersion.

4 Model Acceptance Criteria

In addition to specifying data sets for model validation and providing the information necessary to set up the simulations, the MVDs include the criteria for comparison of the model predictions with the observations.

4.1 Variables and Physical Comparison Parameters

The quantity most frequently measured during dispersion experiments is the gas concentration in air, which also represents the variable of concern in siting scenarios. Other quantities, such as oxygen concentration or gas temperature, are sometimes measured in addition, or in alternative, to gas concentration; however, while measurements based on these quantities may allow to quantify the gas concentration in the lower flammability limit range (approximately 1-5%), they are likely to result in large uncertainty at the concentrations typical of toxic hazard thresholds (10-1,000 ppm).

The selection of physical comparison parameters (PCPs) is critical to ensure that model performance is properly evaluated. Based on earlier model evaluation protocols for atmospheric dispersion models, the PCPs for flammable dispersion were selected as [2]:

1) Maximum point-wise concentration;
2) Maximum arc-wise concentration;
3) Cloud width;
4) Predicted distance to a measured maximum arc-wise concentration;
5) Predicted distance to the LFL concentration;

6) Predicted concentration at the measured distance to the LFL.

The toxic dispersion MVD utilized the same PCPs, with the obvious exception of those associated with the LFL concentration. The cloud width was also excluded from the toxic dispersion PCPs: the main reason for the omission is that the concept of cloud width loses meaning when cloud dispersion is affected by obstacles or complex geometries, which together represent nearly half of the trials in this database and significantly limit the applicability of this PCP to toxic dispersion scenarios. Additionally, even when the cloud width can be calculated, it represents a less discriminating test than the peak concentration [7].

4.2 Statistical Performance Measures and Acceptability Criteria

The quantitative evaluation of a model performance against experimental data is based on a set of statistical performance measures (SPMs) that compare predicted and observed PCPs. Each MVD includes a validation spreadsheet, which calculates the SPMs for each trial as well as for groups of trials with similar characteristics (e.g., all flashing jets dispersing over flat terrain). Duijm et al. [12] suggested that a suitable set of SPMs should:

- Provide an indication of the model's bias (i.e., whether on average it under- or over-predicts the measurements);
- Provide an indication of the level of scatter (i.e., deviation from the average);
- Give equal weight to all measurements/predictions, regardless of their absolute values.

Chang and Hanna [13] further suggested that multiple performance measures should be applied, as each measure has its advantages and disadvantages. The SPMs should also be consistent with previous work in order to build on prior experience and increase confidence on the specified acceptability criteria (or revise their values). The SPMs selected to evaluate flammable and toxic dispersion models are listed in Table 3.3.

Table 3.3. Statistical Performance Measures for Flammable Dispersion Models.

SPM	Definition
Mean Relative Bias	$MRB = \langle \frac{c_m - c_p}{\frac{1}{2}(c_m + c_p)} \rangle$
Mean Relative Square Error	$MRSE = \langle \frac{(c_m - c_p)^2}{\frac{1}{4}(c_m + c_p)^2} \rangle$
Fraction of Predictions Within a Factor of 2	$0.5 \leq \left(\frac{c_p}{c_m}\right) \leq 2.0$
Geometric Mean Bias	$GMB = e^{\langle \ln\left(\frac{c_m}{c_p}\right) \rangle}$
Geometric Variance	$GV = e^{\langle \left(\ln\left(\frac{c_m}{c_p}\right)\right)^2 \rangle}$
Concentration Safety Factor	$CSF = \langle \frac{c_p}{c_m} \rangle$
Concentration Safety Factor to the LFL (Flammable dispersion only)	$CSF_{LFL} = \langle \frac{c_p}{LFL} \rangle$
Distance Safety Factor	$DSF = \langle \frac{x_p}{x_m} \rangle$
Distance Safety Factor to the LFL (Flammable dispersion only)	$DSF_{LFL} = \langle \frac{x_{p,LFL}}{x_{m,LFL}} \rangle$

Acceptability criteria for atmospheric dispersion over unobstructed terrain were first defined by Chang and Hanna [13] and later adopted by Ivings et al. [1]. Only a few models have since been tested against these criteria; therefore, the same acceptability ranges are specified in this MVD. Hanna and Chang [14] later defined a separate set of acceptability criteria for dispersion through an urban environment, recognizing the increased challenge in modeling such scenarios due to the interaction of the cloud with multiple obstacles and obstructions. No models have since been tested against these criteria; therefore, the same acceptability ranges are specified in this MVD.

There is currently no information available to define acceptable values for the more recent SPMs (distance and concentration safety factors). Ivings et al. [2] deemed model predictions to be acceptable if within a factor of two of their respective measurements; therefore, they specified the acceptability range for the distance and concentration safety factors to be the same as for the "fraction of predictions within a factor of 2". For consistency with the LNG MEP, the same acceptability range is specified in this MVD.

The acceptability criteria for the SPMs used in this model evaluation protocol are listed in Table 3.4. It is important to note that all the acceptability criteria previously defined apply to SPMs based on maximum arc-wise concentrations. Note that "simple" geometry includes unobstructed dispersion as well as dispersion obstructed by a single barrier, whereas "complex" geometry represents dispersion through urban or industrial environments.

Table 3.4. Acceptability Criteria for Flammable Dispersion Model Validation.

Statistical Performance Measure	Arc-Wise, Simple Geometry	Arc-Wise, Complex Geometry	Point-Wise, Simple Geometry	Point-Wise, Complex Geometry
Mean Relative Bias	$\|MRB\| < 0.4$	$\|MRB\| < 0.67$	$\|MRB\| < 0.4$	$\|MRB\| < 0.67$
Mean Relative Square Error	$MRSE < 2.3$	$MRSE < 6.0$	$MRSE < 2.3$	$MRSE < 6.0$
Fraction of Predictions Within a Factor of 2	$0.5 \leq FAC2$	$0.3 \leq FAC2$	$0.5 \leq FAC2$	$0.3 \leq FAC2$
Geometric Mean Bias	$0.67 < MG < 1.5$	$0.5 < MG < 2.0$	$0.67 < MG < 1.5$	$0.5 < MG < 2.0$
Geometric Variance	$VG < 3.3$	$VG < 7.5$	$VG < 3.3$	$VG < 7.5$
Concentration Safety Factor	$0.5 < CSF < 2$	$0.5 < CSF < 2$	$0.5 < CSF < 2$	$0.5 < CSF < 2$
Concentration Safety Factor to the LFL	$0.5 < CSF_{LFL} < 2$	$0.5 < CSF_{LFL} < 2$	$0.5 < CSF_{LFL} < 2$	$0.5 < CSF_{LFL} < 2$
Distance Safety Factor	$0.5 < DSF < 2$	$0.5 < DSF < 2$	$0.5 < DSF < 2$	$0.5 < DSF < 2$
Distance Safety Factor to the LFL	$0.5 < DSF_{LFL} < 2$	$0.5 < DSF_{LFL} < 2$	$0.5 < DSF_{LFL} < 2$	$0.5 < DSF_{LFL} < 2$

Finally, it is important to note that the quantitative assessment is just one of several tasks in the model evaluation. Therefore, it is not necessary for a model to meet all SPM acceptability criteria in order to be approved for use. However, a model falling outside the acceptability range for one or more SPMs should warrant a closer review in order to understand the cause of such outcome and possibly identify areas for improvement of the model.

5 Conclusions

This paper described the scope of the Model Evaluation Protocols for flammable and toxic dispersion, which are the first two in a set of several protocols aimed at providing PHMSA (as well as other regulatory bodies that may choose to follow the same approach) with the tools to evaluate models for the different hazards associated with the siting of LNG facilities.

The paper provided a summary of the main section in each MEP as well as a brief description of the test series included in each model validation database. The readers were directed to a public website which hosts the full MEP and MVD reports for further information.

6 References

[1] M. J. Ivings, S. F. Jagger, C. J. Lea, and D. M. Webber, "Evaluating vapor dispersion models for safety analysis of LNG facilities," Health and Safety Laboratory, MSU/2007/04, 2007.

[2] M. J. Ivings, S. E. Gant, S. F. Jagger, C. J. Lea, J. R. Stewart, and D. M. Webber, "Evaluating vapor dispersion models for safety analysis of LNG facilities, 2nd Edition," Fire Protection Research Foundation, MSU/2016/27, Sep. 2016.

[3] F. Gavelli, B. Hendrickson, C. Marsegan, J. Piekarz, R. Kooy, and B. Kolade, "Model Validation Database for Flammable Releases.pdf," Pipeline and Hazardous Materials Safety Administration, 03903-RP-002, Aug. 2020.

[4] F. Gavelli, B. Hendrickson, C. Marsegan, J. Piekarz, R. Kooy, and B. Kolade, "Model Validation Database for Toxic Releases," Pipeline and Hazardous Materials Safety Administration, 03903-RP-004, Dec. 2020.

[5] S. R. Hanna, P. J. Drivas, and J. C. Chang, *Guidelines for use of vapor cloud dispersion models*, 2nd ed. New York: Center for Chemical Process Safety of the American Institute of Chemical Engineers, 1996.

[6] F. Gavelli, B. Hendrickson, J. Kolade, and R. Kooy, "Towards a comprehensive model evaluation protocol for LNG hazard analyses," College Station, TX, Oct. 2020, p. 14.

[7] S. R. Hanna, D. G. Strimaitis, and J. C. Chang, "Hazard response modeling uncertainty (a quantitative method). Vol II - Evaluation of commonly used hazard dispersion models," Sigma Research Corp., AD-A270 095, Mar. 1993.

[8] D. Karaca, "Model Evaluation Protocol," European Cooperation in Science and Technology, COST ES1006, Apr. 2015.

[9] T. Skjold *et al.*, "A Matter of Life and Death: Validating, Qualifying and Documenting Models for Simulating Flow-Related Accident Scenarios in the Process Industry," *Chem. Eng. Trans.*, vol. 31, 2013.

[10] H. C. Goldwire Jr *et al.*, "Coyote series data report LLNL/NWC 1981 LNG spill tests dispersion, vapor burn, and rapid-phase-transition. Volume 1," 1983. [Online]. Available: https://www.osti.gov/biblio/5381934.

[11] D. L. Ermak, R. Chapman, H. C. Goldwire, F. J. Gouveia, and H. C. Rodean, "Heavy Gas Dispersion Test Summary Report," Lawrence Livermore National Laboratory, ESL-TR-88-22, Jun. 1989.

[12] N. J. Duijm, S. Ott, and M. Nielsen, "An evaluation of validation procedures and test parameters for dense gas modeling," *J. Loss Prev. Process Ind.*, vol. 9, no. 5, pp. 323–338, 1996.

[13] J. C. Chang and S. R. Hanna, "Air quality model performance evaluation," *Meteorol. Atmospheric Phys.*, vol. 87, no. 1–3, Sep. 2004, doi: 10.1007/s00703-003-0070-7.

[14] S. Hanna and J. Chang, "Acceptance criteria for urban dispersion model evaluation," *Meteorol. Atmospheric Phys.*, vol. 116, no. 3–4, pp. 133–146, May 2012, doi: 10.1007/s00703-011-0177-1.

Lifting the Fog Off Hydrocarbon Mist Explosions

Stephanie El - Zahlanieh[1]
LRGP, French National Centre for Scientific Research, University of Lorraine
Nancy, France
stephanie.el-zahlanieh@univ-lorraine.fr

Shyarinya Sivabalan[1], Tribouilloy Benoit[2], David Brunello[1], Alexis Vignes[2] and Olivier Dufaud[1]
[1] LRGP, French National Centre for Scientific Research, University of Lorraine
Nancy, France
[2] INERIS, Accidental Risks Division, Parc Technologique ALATA
Verneuil-en-Halatte, France

Prepared for Presentation at
American Institute of Chemical Engineers
2020 Spring Meeting and 17th Global Congress on Process Safety
Virtual
April 18 - 22, 2021

Lifting the Fog Off Hydrocarbon Mist Explosions

Stephanie El - Zahlanieh[1]
LRGP, French National Centre for Scientific Research, University of Lorraine
Nancy, France
stephanie.el-zahlanieh@univ-lorraine.fr

Shyarinya Sivabalan[1], Tribouilloy Benoit[2], David Brunello[1], Alexis Vignes[2] and Olivier Dufaud[1]
[1] LRGP, French National Centre for Scientific Research, University of Lorraine
Nancy, France
[2] INERIS, Accidental Risks Division, Parc Technologique ALATA
Verneuil-en-Halatte, France

Keywords: Fuels, Hazard Identification and Risk Analysis, Process Safety, Explosion, Mist

Abstract

Explosion hazards do not only include gases, vapors and dusts, but also mists. Indeed, previous incident surveys identified dozens of accidents related to hydrocarbon mists generation and ignition. The ongoing occurrence of mist explosions demonstrates the need to assess the relevance of the current approaches for the evaluation of the mist ATEX risks and to determine reliable and standardized safety parameters for hydrocarbon mists.

For this study, different fluids with a high industrial interest were selected (e.g. ethanol, kerosene, and diesel) to be tested in a new apparatus based on the standardized 20L explosion sphere. A new fluid injection system was deployed using siphon/gravity-fed spray nozzles comprising a Venturi junction and proposing a wide variety of dispersion performances. This system was controlled using a specifically dedicated program, which ensures the versatility of the apparatus and its adaptability to the different liquids tested. It allows a fine control of the gas carrier flow, the liquid flow and both the injection and ignition times, which makes it possible to change the dilution rate for a desired droplet size distribution. The droplet size distribution (DSD) was determined for the various fluids dispersed into the 20L sphere. Several tests were performed to determine the minimum ignition energy (MIE), the lower explosivity limit (LEL), the maximum explosion pressure (P_{max}) and the maximum rate of pressure rise (dP/dt_{max}) of hydrocarbon mists.

For instance, for a DSD centered at 55µm, a maximum pressure of 9.2 bar and maximum rate of pressure rise up to 600 bar.s^{-1} were found for ethanol. As for kerosene, for a DSD centered at 80µm, 8.8 bar and 570 bar.s^{-1} were found respectively. Experimental results obtained between the LEL and the stoichiometric concentration were compared with numerical data obtained with Chemical Equilibrium with Applications and showed that they were coherent with the theoretical calculations and that tests were at least performed up to the stoichiometric concentration. A sensitivity analysis was conducted to stress the influence of parameters such as the chemical nature of the fuel, the DSD, the ignition energy or the mist temperature (gas/mist ratio). For instance, tests performed on kerosene showed that as the DSD increases, the mist cloud becomes harder to ignite and hence the LEL increases. These first results already allow to propose a new protocol to determine safety parameters for hydrocarbon mists and provide tools to support the evaluation of mist ATEX risks.

1 Introduction

In the chemical and petrochemical industries, various processes are susceptible to lead to the formation of flammable hydrocarbon aerosols under certain conditions [1]. These unintended generations of hydrocarbon mists may ignite and give rise to explosions. Dozens of such explosions have been reported throughout the years despite the efforts taken to mitigate such incidents. In fact, in 1955, Eichhorn shed light on such matter in one of his publications in the Petroleum Refiner entitled "Careful! Mist can explode" [2]. The author has since introduced the possibility of mists igniting at temperatures well below their flash points. A few years later, in 1995, Eckhoff published a literature survey, in which the author reviewed studies concerning the generation, ignition, combustion, and explosion of flammable mists [3]. Interest in such a crucial subject has, since then, grown throughout the years. For instance, Santon [4] published an incident survey reporting 37 mist incidents, among which 9 explosions lead to 29 fatalities. Ten years later, Lees et al. [1] notably showed that 10% of reported releases on offshore oil and gas installations in the United Kingdom involved sprays or mists. More recently, a comprehensive review was performed by Yuan et al. [5] and led to the proposal of a systematic strategy to investigate aerosol explosion, which is in full accordance with the work presented here.

What defines mists? According to the Globally Harmonized System of classification and labelling of chemicals (GHS) [6], mists are defined as liquid droplets, generally of sizes ranging from 1 to about 100 µm, of a certain substance or mixture suspended in a gas - usually air. The alternative term "smoke" is then used for aerosols with diameters lower than 1 µm [7], while "sprays" consist of dispersed droplets of diameters generally greater than 50 µm. Other definitions are also encountered. For instance, Ballal and Lefebvre [8] consider the term "mist" to be suitable up to 100 µm [9]. As proposed by Eckhoff [10], the terms "spray" and "mist" are used arbitrarily in this paper as they are both relevant in this case.

The ongoing occurrence of mist explosions demonstrates the need to assess the relevance of the current approaches for the evaluation of the mist ATEX risks. Indeed, the importance of hazardous area classification (HAC), as well as the lack of tools which correlate the dispersion of mists with their flammability and explosion severity, is abundantly expressed in literature [5,9]. For instance, European ATEX regulations require assessing the risk of formation and ignition of explosive atmosphere (ATEX) associated with the production of a flammable mists. Nevertheless, due to the lack of tools, risk analysis and area classification prove difficult. This lack of knowledge is clearly highlighted in some guides such as the EI15 [11] in which it is stated that "there is little knowledge on the formation of flammable mists and the appropriate extents of associated hazardous areas".

In order to mitigate the occurrence of such explosions and to assess the flammability and explosion severity of hydrocarbon mists, the factors and criteria of liquid handling, as well as the fluids' safety parameters should be identified. If such definitions are achieved, the classification of hazardous areas (HAC) would be a step closer for mists, and the improvement of current ATEX standards and regulatory provisions concerning mists would then be achievable. In other words, "lifting the fog off hydrocarbon mist explosions" may finally be possible.

In this study, different fluids with a high industrial interest were selected to be tested in a new apparatus based on the standardized 20 L explosion sphere. The mist generated was characterized, notably by its droplet size distribution, and explosion tests were performed to determine the minimum ignition energy (MIE), the lower explosivity limit (LEL), the maximum explosion pressure (P_{max}) and the maximum rate of pressure rise (dP/dt_{max}) of hydrocarbon mists. The main goals of this article are i) to demonstrate through some examples that it is possible to characterize the flammability and explosion severity of mists using modified standard set-ups such as the 20 L sphere, ii) to show the influence of some operating conditions on the safety parameters of hydrocarbon mists.

2 Materials and Methods

2.1 Fuel Selection

A large variety of fuels has been involved in mist explosion incidents, ranging from vegetable oils to hydraulic oils and crude oils [4]. During an oil mists Joint Industry Project (JIP) led by the Health and Safety Executive (HSE) [12], a list of fluids was proposed comprising fuels of industrial interest and specific physicochemical properties. In addition, based on the properties that were identified to be the most affecting on oil mist hazards, a fluid classification system was developed by the Health and Safety Executive. This system divided the chosen fluids into four release classes based on their flashpoint and their ease of atomization, a parameter represented by the Ohnesorge number. For this paper, three fuels were chosen to be studied under certain conditions: ethanol, Jet A1 kerosene and diesel fuel. The choice of each fluid can be justified by both industrial and scientific considerations.

Ethanol

The industrial value of ethanol is continuously growing throughout the years with its emergent use as an engine fuel or fuel additive for automobiles. Such demand for ethanol or ethanol-fuel blends is and will keep leading to an increase in its production and transport, hence increasing the requirements to manage fire or explosion risks. Moreover, flammability and explosion severity studies performed on ethanol vapors as well as mists, such as Mitu and Brandes [13], Thimothée [14], and Pinheiro et al. [15], can be considered abundantly sufficient to help calibrate the testing methods chosen for this paper.

Jet A1 kerosene

In 2009, Santon showed that 7 incidents, over 29 detailed in his review, were related to kerosene mists. They were mostly related to transportation activities, from cargo accident to aviation kerosene explosion [4]. One incident that highlights the dangers of flammable aviation kerosene and air mixtures is the Trans World Airlines (TWA) flight 800. In fact, this tragic incident led to the loss of 230 lives in an airplane explosion resulting from the ignition of the flammable fuel and air mixture in the center wing fuel tank [16]. Bowen and Shirvill [17] also pointed out the risks behind the potential atomization of kerosene.

Diesel fuel

In their incident review, Lees et al. [1] mentioned that data from the UK Hydrocarbon Release Database (HCRD) showed that 20% of mist reported incidents took place from diesel sources. In fact, the authors stated that out of 48 mist/spray release flash fires reported between 2000 and 2005, 11 involved diesel releases.

Table 1 shows the main physicochemical properties of the selected fuels. It should be highlighted that Ohnesorge number (Oh), used for the HSE classification, represents the ratio between the viscous forces to surface tension and inertial forces, i.e. at a given Reynolds number, increasing Oh will improve liquid atomization.

Table 1. Physicochemical properties of the three selected fuels

	Ethanol 96%*	**Aviation fuel Jet A1****	**Diesel*****
Flashpoint (°C)	13-17	≥ 38	> 52
Density ($kg.m^{-3}$)	790	750 – 840	750 – 850
Viscosity (cSt)	1.2	8	1 – 2.5
Surface Tension ($kg.s^{-2}$)	0.022	0.026	0.027
HSE Release Class	'Unclassified'	Class I (Oh ratio ≥ 2, Flashpoint < 125°C)	Class I (Oh ratio ≥ 2, Flashpoint < 125°C)
Flammability limits (%)	3.3 – 19	0.6 – 6	0.5 – 8
Auto-ignition temperature (°C)	363	210	232 – 260

** MSDS from ThermoFisher Scientific ** MSDS from Neste *** MSDS from HOLLYFRONTIER*

2.2 Mist Generation System

The accidental formation of an explosive flammable mist cloud can be caused by several phenomena. For instance, it can be due to the evaporation of fuels in heated areas, their subsequent circulation and condensation in colder areas [3]. In addition, leaks or ruptures due the damages or corrosion in vessels and pipelines can lead to high pressure releases and hydrocarbon mist generation. To better represent a spray or mist caused by a leak or rupture, various generation methodologies could be adopted. The wide use of sprays in industrial applications lays out several options to choose from. Nevertheless, the experimental procedure proposed to characterise the ignition sensitivity and explosion severity of mists aims to be standardized, so that the results can be compared and that generic safety measures can be proposed. Therefore, the selected generation system should be easy to find and implement but should also be able to mimic "idealized" conditions of a mist release.

In this work, a siphon/gravity-fed set-up comprising a Venturi junction with two inlets (an air inlet and a liquid inlet) was used to generate mist. This generation system's main component is a spray nozzle set composed of a fluid and an air cap (Spraying Systems) through which the liquid/air jet passes and fragments. Mist spray generation was investigated as a function of three relevant parameters stated by Kooij et al. [18]: nozzle type, spraying pressure and fluid properties. Three nozzle sets, presented in Figure 1, of different orifice diameters were chosen to represent three ranges of droplet size distribution. In addition to the nozzle set, the air injection pressure, which shows important influence on the droplet size, was varied at least between 3 and 5 bar. These pressures were chosen according the maximum pressure tolerance of the three nozzle sets.

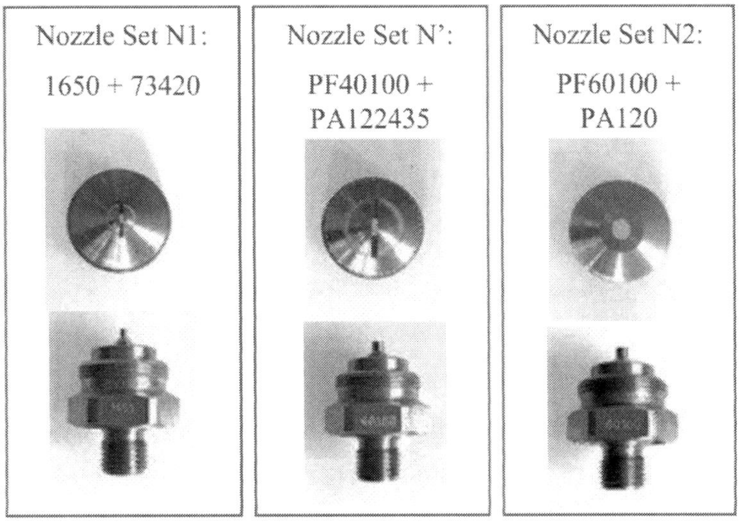

Figure 1. The three spray nozzle sets: (a) N1 (b) N' (c) N2

2.3 Mist Characterization

Gant et al. [9] pointed out that characterizing a mist properly before its ignition is of great importance since the safety parameters of mists are highly affected by their droplet size distribution, concentration and turbulence. These three parameters are studied by using an in-situ diffraction laser sensor as well as by implementing Particle Image Velocimetry; in order not to overload this manuscript, PIV experiments will not be developed here.

Droplet Size Distribution (DSD)

The time evolution of droplet size distributions was determined by in-situ laser diffraction (Helos/KR-Vario by Sympatec GmbH) and will be later supplemented by APS spectroscopy (aerodynamic diameter measurement - TSI) to characterize the submicron droplets. The Helos laser sensor used for such measurement is designed to analyse the droplet size of using 3 high-resolution measuring ranges (R1, R3 and R5) from 0.5 μm to 875 μm. The apparatus measures the droplet size distribution (DSD) directly through the transparent windows (borosilicate glass) of a modified 20L open sphere similar to the standard explosion vessel [19]. R3 lens was mainly used during this study as it covers a range of droplet diameters from 0.5/0.9 μm to 175 μm. The acquisition frequency was set at 2 distributions per millisecond. The measurements given by the sensor are notably the volume diameter d_{10}, d_{50}, d_{90} and the $D_{3,2}$ (Sauter Mean Diameter, SMD). In order to have an approximation of the DSD near the kernel spark produced by an ignition source described in more details in Section 2.4, the height of the sensor was adjusted to a height corresponding to the location of the ignition source.

2.4 Ignition Sources

Potential ignition sources in accidental fires or explosions can vary from electrical discharges or sparks to hot surfaces or to malfunctioning electrical circuits. In previous studies on explosion limits and deflagrations pressures, ignition sources such as exploding wire ignitors, spark ignition, pyrotechnical ignitors, and hot spot ignitors have been used [8, 20, 21].

For this study, wide ranges of energies, starting from 1 mJ, were required. Therefore, both spark ignition as well as chemical pyrotechnical ignitors were chosen. In tests performed with spark ignition, two stainless steel electrodes were placed in the center of the 20 L explosion sphere with a separating distance of 6 ± 0.1 mm. The electrodes were insulated from the sphere's wall using Teflon plugs. Both electrodes were connected to a KSEP 320 high voltage unit, which was also bypassed by a custom control system. The maximum power generated by such system is 225 W, i.e. 225 $J.s^{-1}$. Hence, to deliver an energy of 100 J, a permanent spark would be generated during 444.44 milliseconds.

To avoid generating a spark for a long duration and leading to the sedimentation of droplets, permanent spark ignition was used up to 100 J. For higher energies, chemical pyrotechnical ignitors (Sobbe GmbH) of energies ranging from 100 J to 5 kJ were used. These ignitors were actuated electrically by a low-voltage electrical signal sent by the KSEP 310 unit. Both ignition sources were compared at 100 J to ensure that they both deliver approximate results.

2.5 Mist Flammability and Explosion Severity

The Modified 20 L Ignition and explosion Severity Test device (MIST)

The standard 20 L explosion sphere used for dust explosion tests has been modified with the installation of the mist generation system at the bottom of the sphere, the removal of the dust container, as shown in the figure below, and the installation of two electronic valves to control the inlet flowrates as well as the liquid/air ratio. Before injecting the fuel/air mixture, the sphere was partially vacuumed to a calculated pressure, so that, when the mist was fully injected, an atmospheric pressure would be attained. The two ignition sources mentioned in Section 2.4 were used for this test. The control of the system as well as the acquisition of the data were performed using a new control system and a LabView program prepared by the LRGP. This system controls the KSEP 310 and 332 units (Cesana AG), the inlet electronic valves (gas and liquid) and allows a safe operation of the test equipment and an optimum evaluation of the explosion results with an acquisition frequency of 5000 measurements per second.

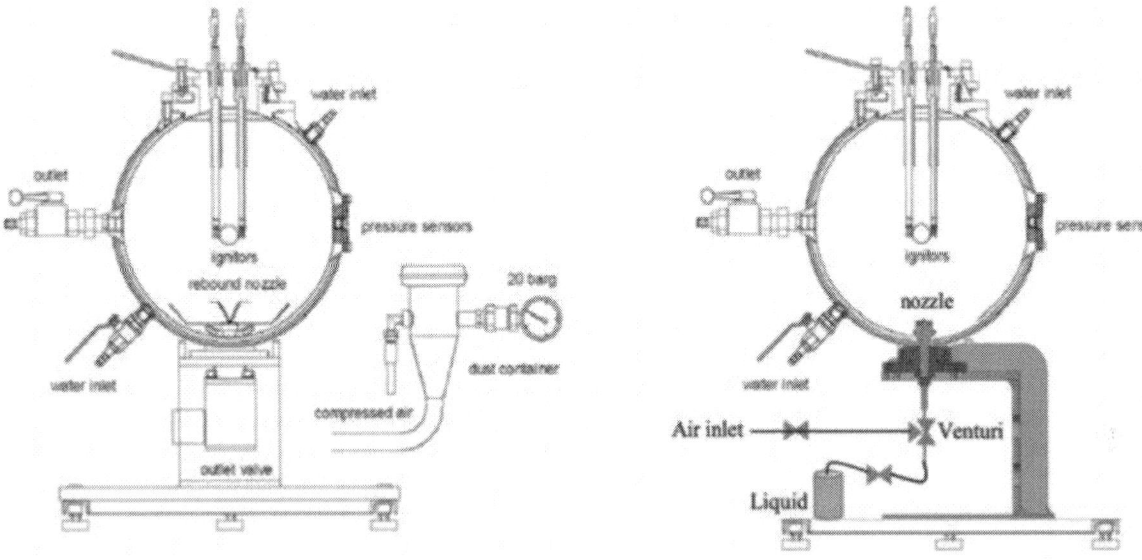

20L sphere (gas, dust) modified 20L sphere (MIST)

Figure 2. Modification of the standardized 20L explosion sphere
Left: standard 20 L sphere – Right: modified 20L sphere

3 Results and Discussion

3.1 Droplet Size Distribution (DSD)

Ethanol mists were generated in the open 20L sphere using the three nozzle sets presented in Section 2.2. DSD measurements were performed during a 4 second generation period and up to 1 second after the valves' closure, with intervals of 50ms (2 DSD per ms and an average value each 50 ms). As it is shown in Figure 3, three ranges of droplet diameters can be obtained by changing the nozzle set. Nevertheless, it should be noted that with nozzle sets N' and N2, a peak corresponding to smaller diameters (between 6 and 18μm) was always observed, demonstrating the persistence of 'primary droplets' for nozzle N' and of a fragmentation phenomenon for N2. It should be noticed that, in the case of mist generation through the N' nozzle set, hybrid mixtures (vapor – liquid phase) can be created during the ignition step due to the presence of fine droplets being easily vaporized along with larger droplets.

Since fluids' properties are one of the many relevant parameters affecting the droplet size distribution of a mist [18], a sensitivity study has been undertaken to study the influence of preheating or pressurizing the liquid on the DSD. A metallic reservoir that holds up to 10 bar of internal pressure and 100 °C was designed to store the liquid hydrocarbons. Granulometry tests were also performed on kerosene JetA1 and diesel fuel mists and demonstrated similar DSDs as these shown in Figure 3.

Being able to independently control the DSD and the flow (i.e. the fuel equivalence ratio) is an essential point of any study on the risks associated with mists. This will allow to control and study the effect of the fuel equivalent ratio, the droplet size distribution and the chemical nature of the fuel, and to determine their impact on the ignition sensitivity and the explosion severity of hydrocarbon mists.

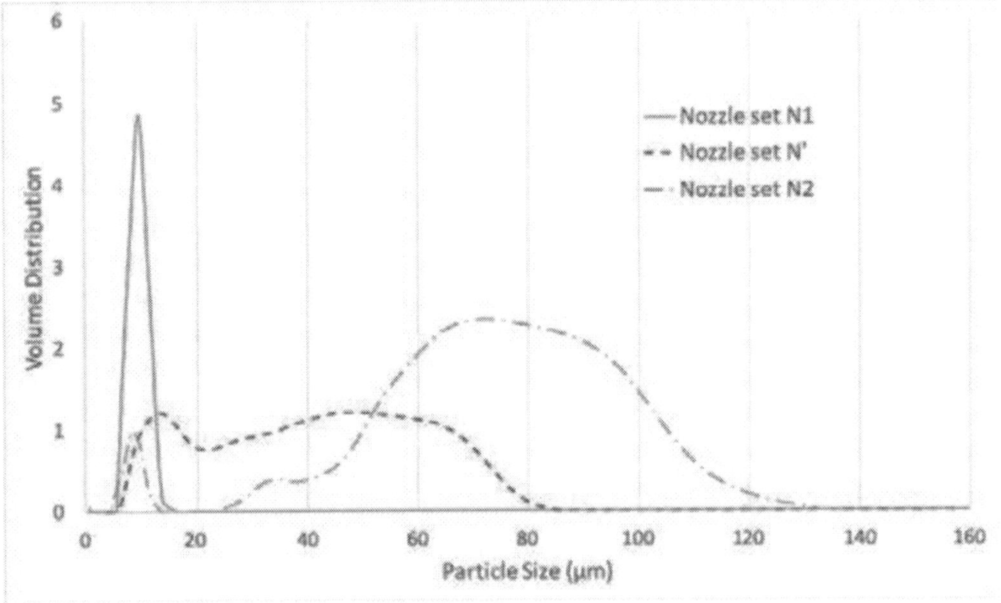

Figure 3. Droplet size volume/mass distribution at t = 2 s of ethanol mist generated at P=3bar using nozzle sets N1, N' and N2

3.2 Minimum Ignition Energy

The minimum ignition energy is the smallest amount of spark energy needed to ignite the most easily ignitable mixture of a flammable substance in air. Such parameter is one of the most important parameters for the assessment of explosion risks and hazardous areas. Krishna [22] stated that a mist becomes more easily ignitable as its mean droplet diameter decreases. This statement can be justified by the fact that the liquid must be partially vaporized before it ignites, and with small diameter droplets, the high surface-area-to-volume ratio leads to a faster evaporation process [9]. Therefore, the nozzle set N1, which generates the smallest mist droplets (Figure 3), was chosen for this test.

For instance, Jet A1 mists were dispersed into the modified 20 L explosion sphere during a fixed injection time, corresponding to the stoichiometric concentration. As the MIE was lower than 100 J, the ignition source used for this part was the spark ignitors. Results showed that the MIE of JetA1 mists of a DSD around 8 – 10 µm lies between 1 and 3 J. On the contrary, previous studies, such as Shepherd et al. [23], showed that energies between 5 mJ and 1 J are sufficient to ignite a mixture of Jet A vapor fuel in air. This large dissimilarity in values shows how mists can behave differently from vapors and how the presence of droplets can affect the ignition sensitivity. It should be noted that the electrode shape, spark gap, circuit capacitance and the spark duration are parameters that influence the values of the minimum ignition energy [9, 24].

3.3 Lower Explosive Limit

The lowest concentration of a mixture at which a propagation of a flame is supported away from the ignition source is defined as the Lower Explosive Limit (LEL). For d_{50} ranging from 8 to 10 μm, the LEL of the three selected fluids was determined as shown in Table 2. Burgoyne [25] stated that the lower explosive limit of hydrocarbon mists, specifically tetralin, is around 50 $g.m^{-3}$. The difference between the results obtained and those found by Burgoyne [25] can be explained by the different sources used to ignite the mists (a pilot flame in the case of Burgoyne). Moreover, the experimental apparatus used to measure the LEL might influence the values [9]. It should also be noted that a single average diameter is often given in order to describe the whole DSD, whereas it can greatly vary as a function of the turbulence and time due to coalescence phenomenon. In 1995, Eckhoff concluded that the LEL of a spray will range between 100 $g.m^{-3}$ and 500 $g.m^{-3}$ regardless of the fuel volatility [3]. Dufaud et al. [26] also found a LEL in the order of 250 $g.m^{-3}$ for lube oil mists. Table 2 shows that lower values of LEL can be obtained.

Table 2. Results for the lower explosive limit (LEL) for ethanol, Jet A1 and diesel

	Ethanol 96%	**Aviation fuel Jet A1**	**Diesel**
LEL ($g.m^{-3}$)	77 ± 5	94 ± 5	73 ± 5

As stated by Gant [9], the LEL of a mist is linked to the size of the droplets. In fact, Burgoyne [27] studied the effect of the droplet size on the lower limit of flammability and showed that in upward flame propagation tests, the LEL values tend to decrease with increasing drop diameter; however, with a downward flame propagation such indications become harder to pinpoint with increasing drop diameters due to the presence of droplets falling vertically downwards. Tests on kerosene JetA1 mists generated with the three nozzle sets and ignited with 100 J chemical ignitors showed that as the DSD increases, it becomes harder to ignite the mist cloud and hence the LEL increases (Table 3), a result well coherent with that found by Zabetakis [28].

Table 3. Effect of the DSD on the LEL of kerosene JetA1 mists

Nozzle set	**Mist DSD range (μm)**	**D_{10} (μm)**	**D_{50} (μm)**	**SMD (μm)**	**LEL ($g.m^{-3}$)**
N1	8-10	7	9.2	9	94 ± 5
N'	40-60	10	15	14.5	127 ± 5
N2	80-100	8	63	17	166 ± 5

3.4 *Explosion Severity*

3.4.1 Validation Tests

As presented in Section 2.5, the 20 L sphere was modified to allow the generation of mists and chemical ignitors of 100 J were used as ignition source. In accordance with the European standard PR EN15967:2020 "Determination of the maximum explosion pressure and maximum rate of pressure rise of gases and vapors", a MATLAB program was prepared to determine P_{max} as well as the maximum rate of pressure rise dP/dt_{max} using both two-point and five-point derivatives.

Preliminary tests were performed on ethanol to calibrate the test method used. They show that the maximum explosion pressure reaches 7.5 barg, whereas the maximum rate of pressure rise reaches 567 $bar.s^{-1}$ (Figure 4) for an ethanol mist concentration of 128 $g.m^{-3}$ (6.3$\%_{v/v}$)of a DSD of about 8 - 10 μm generated by the nozzle set N1 with an injection pressure of 3 bar and at ambient temperature. These values are in good agreement with Mitu and Brandes [13] who found an absolute maximum explosion pressure of about 8 bar for an explosion of vapor ethanol/air mixture with a fuel/air ratio $\varphi = 1.15$.

Chemical Equilibrium with Applications (CEA) was also used as a comparison tool between experimental results and theoretical calculations of liquid or gas phase combustion. Numerical data on the combustion of ethanol at different fuel/air ratios were obtained using CEA (Figure 5). Such results are consistent as the explosions performed in the MIST sphere are not under adiabatic conditions; therefore, it is normal to obtain an experimental P_{max} well below the theoretical adiabatic overpressure. Figure 5 also shows that tests were at least performed up to the stoichiometric concentration. The use of the CEA program permits to obtain theoretical equilibrium compositions of combustion products which can help in modeling mist explosions more finely.

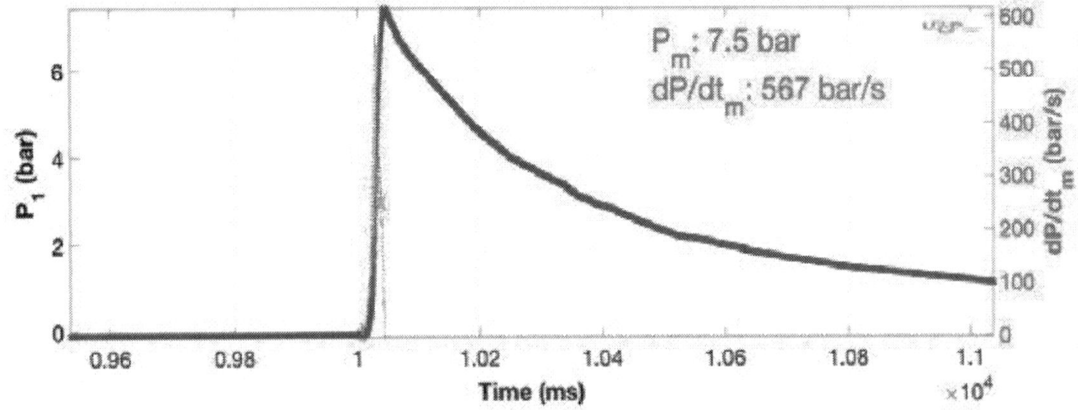

	dP/dt 2p	dP/dt 5p
1	570.00	567.50
2	560.00	560.83
3	560.00	555.00
4	550.00	550.83
5	550.00	547.50
6	550.00	545.83
7	540.00	545.00
8	540.00	543.33
9	540.00	543.33
10	540.00	539.17
11	540.00	539.17
12	540.00	536.87

Figure 4. Time evolution of the explosion pressure of ethanol mist in the 20L explosion sphere
(generation at ambient temperature using nozzle set N1)

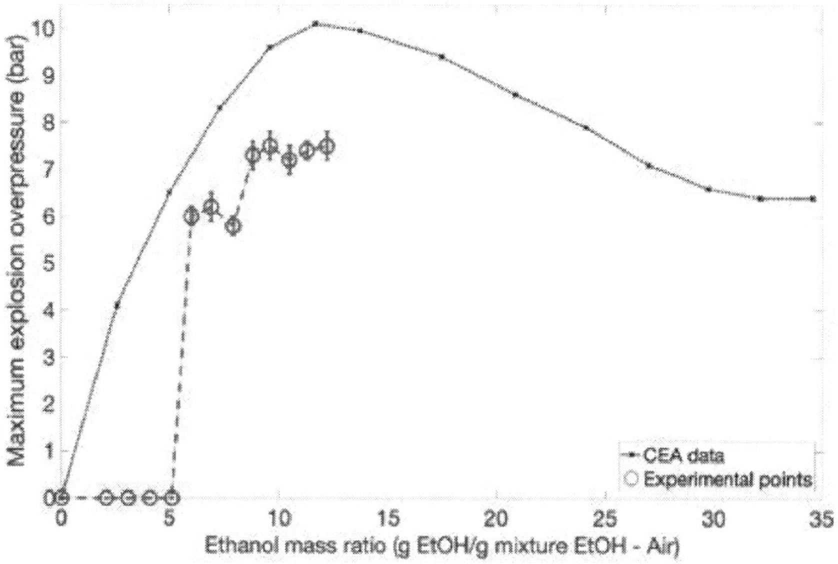

Figure 5. Comparison between theoretical CEA overpressure values and experimental results

3.4.2 Effect of the Chemical Nature of the Fluid

The three fluids presented in Section 2.1 were tested under the same injection conditions (Nozzle set N1, ambient temperature, P_{inj} = 3 bar) and using the same 100 J chemical ignitors. Figures 6 and 7 present the variation of both the maximum explosion overpressure and the maximum rate of pressure rise for diesel fuel, kerosene Jet A1, and ethanol. It should be noted that the lines are only arbitrary point-to-point connections and that a fitting will be adopted later on. As it can be seen, in the figures below, the three mists behave very differently once ignited. For ethanol, a maximum value of P_{max} of 7.5 barg and a maximum rate of pressure rise of 567 bar.s^{-1} were found at a mist concentration of 128 g.m^{-3}. However, in the case of kerosene JetA1, 6.4 barg for P_{max} and 356 bar.s^{-1} for dP/dt_{max} were found at a mist concentration of 159 g.m^{-3}, which is consistent with tests performed by Zheng et al. [20] on high flash point jet fuels (RP-5 and RP-3) resulting in values of about 6.5 bar as Pmax. As for diesel fuel mists, even though its values of rate of pressure rise are not much elevated, it can be seen that diesel can indeed ignite and produce an explosion at a temperature well below its flashpoint.

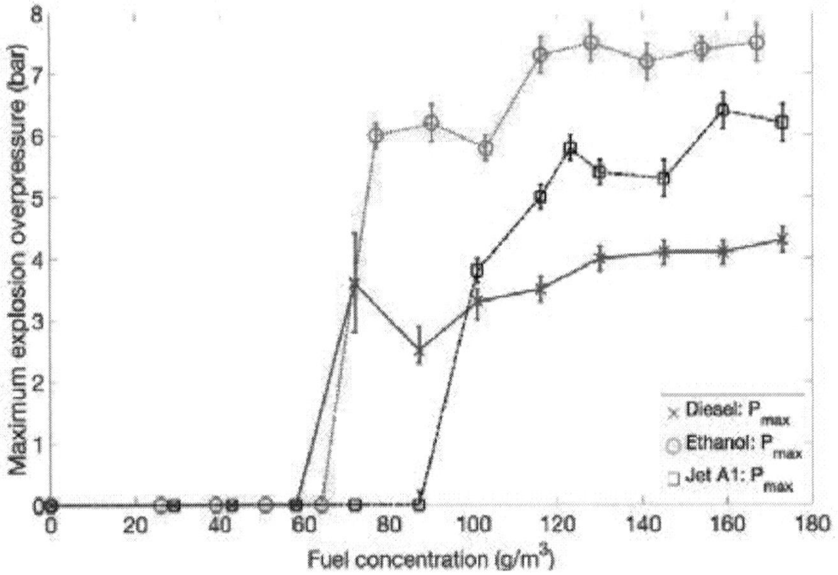

Figure 6. Variation of the maximum overpressure as a function of fuel concentrations in the 20L sphere
(generation at ambient temperature using nozzle set N1)

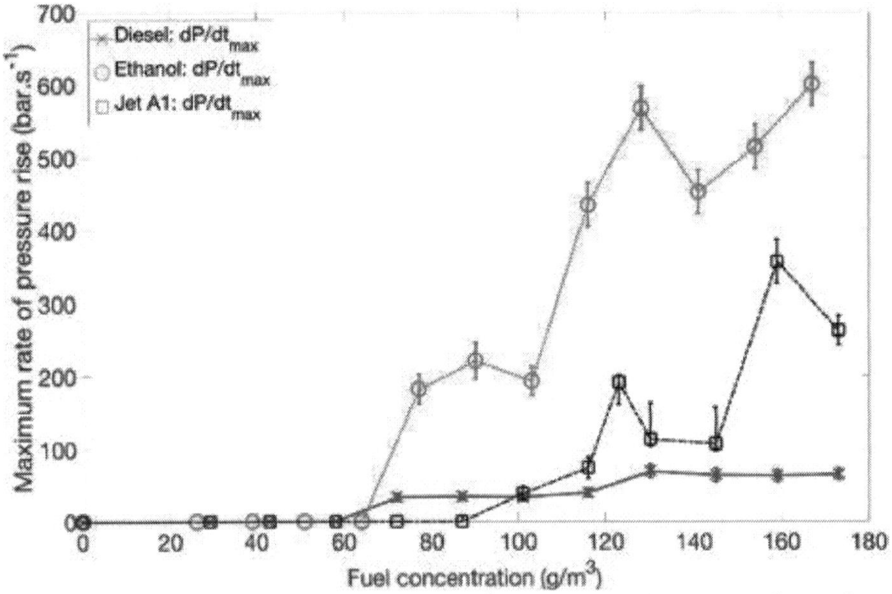

Figure 7. Variation of the maximum rate of pressure rise as a function of fuel concentrations in the 20L sphere
(generation at ambient temperature using nozzle set N1)

3.4.3 Effect of the Ambient Temperature

The MIST sphere is equipped with a water jacket to control its temperature. To study the influence of the ambient temperature on the explosion severity of mists, the temperature of the sphere was notably increased to 60°C for kerosene JetA1, which corresponds to a temperature greater than its flashpoint. Tests were performed at various concentrations with spark ignition source at 100 J. Results showed how ambient conditions can affect the rate of explosion pressure rise from values of 168 $bar.s^{-1}$ at T = 20°C to about 540 $bar.s^{-1}$ at T = 60°C at a mist concentration of 87 $g.m^{-3}$. This change can be explained by an increase of the vapor concentration around the droplet before ignition, reaching the lower explosive limit. The ignition step is then governed by a gas combustion regime and not limited by the droplet evaporation as it is the case at 20°C. It can be seen in Figures 8 and 9 that this effect is more pronounced on the maximum rate of pressure rise than on the maximum explosion pressure, highlighting the impact of the temperature on the mist combustion kinetics. It should also be underlined that the effect of the temperature is less perceptible at high mist concentrations. In fact, the vapor pressure of kerosene increases from 4 to 18 mbar when the temperature rises from 20 to 60°C [29]. This implies that, even at 60°C, kerosene cannot be fully vaporized for concentrations greater than 125 $g.m^{-3}$. Moreover, apart from the very first moments of ignition, the flame temperature will rapidly rise over the auto-ignition temperature, the impact of the initial temperature being then negligible. As a consequence, for mist concentrations greater than 125 $g.m^{-3}$, it seems consistent that the difference between vapor and mist explosive behavior becomes smaller.

In addition to the augmentation of the sphere's temperature, pyrotechnical ignitors of high energies (1, 2, 5 and 10 kJ), which permit to increase the temperature locally, were tested. The results were consistent with that obtained by increasing the sphere's temperature.

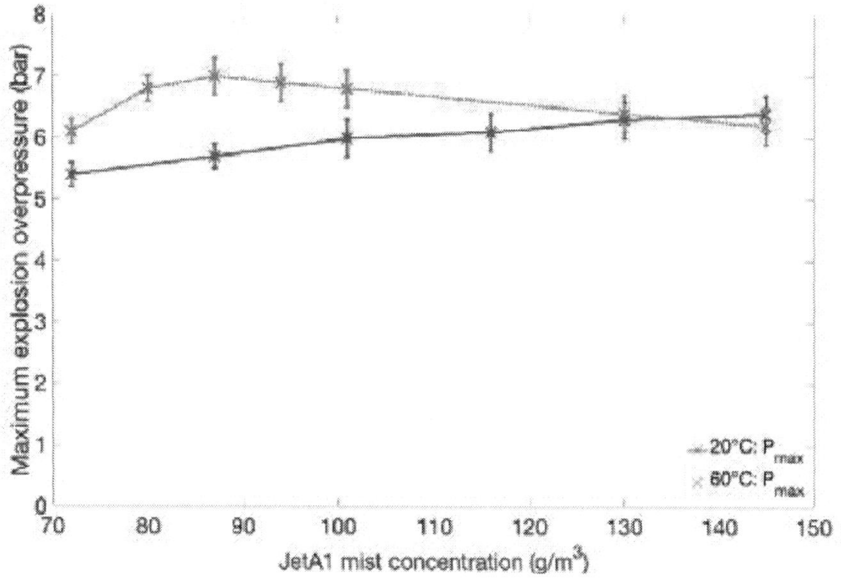

Figure 8. Variation of the maximum overpressure as a function of Jet A1 mist concentrations in the 20L sphere at two different temperatures
(generation using nozzle set N1)

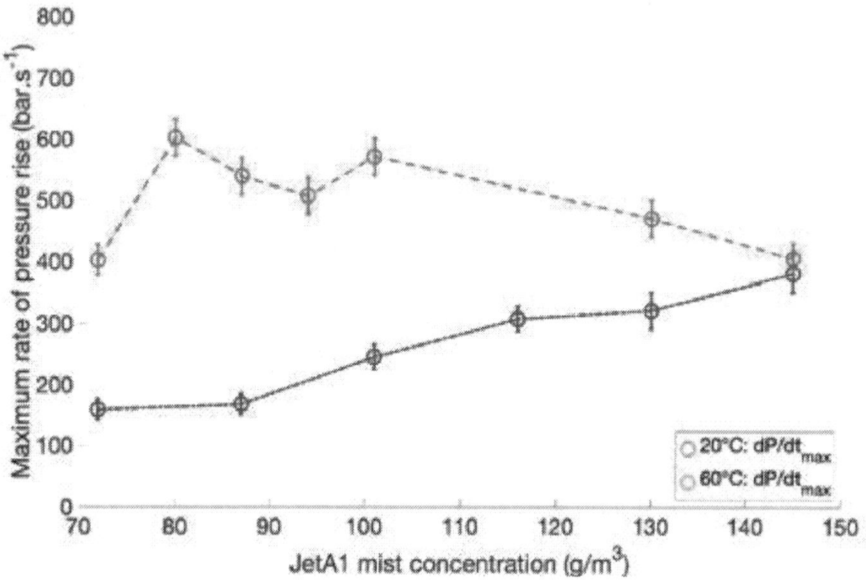

Figure 9. Variation of the rate of pressure rise as a function of Jet A1 mist concentrations in the 20L sphere at two different temperatures
(generation using nozzle set N1)

3.4.4 Effect of the Droplet Size Distribution

The droplet size distribution is one important parameter to be taken into account while studying the flammability and the explosion severity of a mist. In fact, in low-momentum sprays or quiescent mists, large droplets are affected by gravity and hence fall faster than smaller droplets, which tend to decrease the average mist mass concentration [9]. Such effect can be seen in Table 4 where both the P_{max} and dP/dt_{max} decreased with increasing droplet size distributions at a mist concentration of about 165 g.m^{-3}. Obviously, this is not the only phenomenon to consider and the droplet surface can also play a significant role in the evaporation phase (d^2 combustion law) as well as in the heat transfer, especially for radiative transfers.

Table 4. Explosion severity of JetA1 using the three nozzle sets for mist generation

	Nozzle set N1	**Nozzle set N'**	**Nozzle set N2**
P_{max} (barg)	6.2	3.5	0.8
dP/dt_{max} (bar.s^{-1})	263	46	28

4 Conclusion

This study aims to demonstrate that it is possible to characterize the ignition sensitivity and explosion severity of hydrocarbon mists with a single set-up. These tests should be performed on simple or/and standard equipment which can be found in industries or laboratories, in order to be able to compare the results and propose adequate solutions for explosion risk management. The Modified 20 L Ignition and explosion Severity Test device, developed in this study, can be a suitable apparatus to assess mist explosivity.

Experiments carried out on ethanol, kerosene JetA1 and diesel have allowed to validate the procedures and setups and show the influence of some operating conditions, such as the ambient temperature, the fuel equivalence ratio and the droplet size distribution, on the safety parameters of hydrocarbon mists. For instance, MIE values for JetA1 mists of 8 - 10 μm were found to lie between 1 and 3 J, a value much greater than that corresponding to JetA1 vapors (50mJ). The LEL of the three hydrocarbon mists showed to be 77 $g.m^{-3}$ (3.8$\%_{v/v}$), 94 $g.m^{-3}$ (1.2$\%_{v/v}$) and 73 $g.m^{-3}$ (0.86$\%_{v/v}$) for ethanol, JetA1 and diesel mists respectively. Moreover, it was notably demonstrated that the LEL of JetA1 mist increases with an increasing DSD. The rate of pressure rise is also greatly influenced by the combustion kinetics and its rate-limiting step (evaporation, fuel oxidation…). The effect of the sphere temperature was also studied, showing an increase in both P_{max} and dP/dt_{max} of JetA1 mist explosions when the temperature of the sphere was increased from 20°C to 60°C and thus highlighting the influence of the fuel phase (gas, gas-aerosol, liquid aerosol) before ignition. It was also shown that, over the tested diameter range, increasing DSD leads to a diminution of the explosion severity.

Additional tests are currently performed at various initial turbulence levels and on a larger range of DSD with several conventional fuels or hydrocarbons (e.g. iso-octane, biodiesel, light fuel oil, lube oil…). This experimental approach will be coupled with both CFD simulation, to better understand the hydrodynamics within the explosion vessel, and combustion modeling, to highlight the influence of the rate-limiting step of mist explosion on the safety parameters.

Acknowledgements

This work was supported financially by the French Ministry for the Ecological Transition and the French Ministry of Higher Education, Scientific Research and Innovation. The authors would also like to thank the Health and Safety Executive (UK) for the fruitful exchanges on the topic. The contents of the work, including any opinions and/or conclusions expressed, are those of the authors alone and do not necessarily reflect French Ministries policies.

5 References

[1] P. Lees, S. Gant, R. Bettis, A. Vignes, J.-M. Lacome and O. Dufaud, Review of recent incidents involving flammable mists, (2019) 23.

[2] J. Eichhorn, Careful! Mist can explode, Petroleum Refiner. 34(11) (1955) 194–196.

[3] R.K. Eckhoff, Generation, ignition, combustion and explosion of sprays and mists of flammable liquids in air: a literature survey, 1995.

[4] R.C. Santon, Mist fires and explosions - an incident survey, (2009).

[5] S. Yuan, C. Ji, H. Han, Y. Sun and C.V. Mashuga, A review of aerosol flammability and explosion related incidents, standards, studies, and risk analysis, Process Safety and Environmental Protection. 146 (2021) 499–514. https://doi.org/10.1016/j.psep.2020.11.032.

[6] ST/SG/AC.10/30/Rev.8, Globally harmonized system of classification and labelling of chemicals (GHS)., 8th revised version, UNITED NATIONS, New York and Geneva, 2019.

[7] R.P. Fraser and P. Eisenklam, Liquid Atomization and the Frop Size of Sprays, Institution of Chemical Engineers, 1956.

[8] D.R. Ballal and A.H. Lefebvre, A general model of spark ignition for gaseous and liquid fuel-air mixtures, Symposium (International) on Combustion. 18 (1981) 1737–1746. https://doi.org/10.1016/S0082-0784(81)80178-6.

[9] S. Gant, Generation of flammable mists from high flashpoint fluids: literature review, Health and Safety Executive, Research Report RR980. (2013). https://www.hse.gov.uk/research/rrpdf/rr980.pdf.

[10] R.K. Eckhoff, Chapter Five - Explosions of Clouds of Combustible Liquid Droplets in Air, in: R.K. Eckhoff (Ed.), Explosion Hazards in the Process Industries (Second Edition), Gulf Professional Publishing, 2016: pp. 209–232. https://doi.org/10.1016/B978-0-12-803273-2.00005-0.

[11] Energy Institute, Model code of safe practice: Area classification code for installations handling flammabe fluids, (2015). https://publishing.energyinst.org/topics/asset-integrity/ei-model-code-of-safe-practice-part-15-area-classification-for-installations-handling-flammable-fluids.

[12] G. Burrell and S. Gant, Liquid classification for flammable mists, Health and Safety Executive, Research Report RR1108 (2017).

[13] M. Mitu and E. Brandes, Influence of pressure, temperature and vessel volume on explosion characteristics of ethanol/air mixtures in closed spherical vessels, Fuel. 203 (2017) 460–468. https://doi.org/10.1016/j.fuel.2017.04.124.

[14] R. Thimothée, Experimental characterization of a laminar flame propagation in a two-phase medium (aerosol) in microgravity conditions, Theses, Université d'Orléans, 2017. https://hal.archives-ouvertes.fr/tel-01994264 (accessed February 28, 2020).

[15] A.P. Pinheiro, J.M. Vedovoto, A. da Silveira Neto and B. G.M. van Wachem, On ethanol droplet evaporation in the presence of background fuel vapor, (2019) 8.

[16] T. Al Mughanam, Operational Flammability Limits of Commercial Fuels – A Review, MATEC Web Conf. 249 (2018) 03013. https://doi.org/10.1051/matecconf/201824903013.

[17] P.J. Bowen and L.C. Shirvill, Combustion hazards posed by the pressurized atomization of high-flashpoint liquids, Journal of Loss Prevention in the Process Industries. 7 (1994) 233–241. https://doi.org/10.1016/0950-4230(94)80071-5.

[18] S. Kooij, R. Sijs, M.M. Denn, E. Villermaux and D. Bonn, What Determines the Drop Size in Sprays?, Phys. Rev. X. 8 (2018) 031019. https://doi.org/10.1103/PhysRevX.8.031019.
[19] A. Santandrea, M. Gavard, S. Pacault, A. Vignes, L. Perrin and O. Dufaud, 'Knock on nanocellulose': Approaching the laminar burning velocity of powder-air flames, Process Safety and Environmental Protection. 134 (2020) 247–259. https://doi.org/10.1016/j.psep.2019.12.018.
[20] L. Zheng, L. Changbo, A. Gaojun, X. Chunhua, Z. Youjie, R. Lianling, W. Xudong, X. Lifeng, W. Gang, W. Feng and S. Limin, Comparative Study on Combustion and Explosion Characteristics of High Flash Point Jet Fuel, Procedia Engineering. 84 (2014) 377–383. https://doi.org/10.1016/j.proeng.2014.10.447.
[21] S. Yuan, C. Ji, A. Monhollen, J.S.-I. Kwon and C. Mashuga, Experimental and thermodynamic study of aerosol explosions in a 36 L apparatus, Fuel. 245 (2019) 467–477. https://doi.org/10.1016/j.fuel.2019.02.078.
[22] K. Krishna, Measurement and Prediction of Aerosol Formation for the Safe Utilization of Industrial Fluids, 2003. http://psc.tamu.edu/files/library/center-publications/theses-and-dissertations/kk_phd_chen.pdf (accessed September 24, 2020).
[23] J.E. Shepherd, J.C. Krok and J.J. Lee, Jet A Explosion Experiments: Laboratory Testing, (1997) 74.
[24] C.J.R. Coronado, J.A. Carvalho, J.C. Andrade, E.V. Cortez, F.S. Carvalho, J.C. Santos and A.Z. Mendiburu, Flammability limits: A review with emphasis on ethanol for aeronautical applications and description of the experimental procedure, Journal of Hazardous Materials. 241–242 (2012) 32–54. https://doi.org/10.1016/j.jhazmat.2012.09.035.
[25] J.H. Burgoyne, Mist and spray explosions, Chemical Engineering Progress. 53 (1957) 121–124.
[26] O. Dufaud, A. Charvet, G. Mougel, M. Molière, S. Luthun, D. Brunello, L. Perrin, S. Delimoges and M. Couchot, Generation, Characterization and Ignition of Lube Oil Mists, in: ASME Turbo Expo 2015, ASME, Montreal, Canada, 2015: p. Paper No. GT2015-43524, p. V04BT04A039; 8 pages. https://doi.org/10.1115/GT2015-43524.
[27] J.H. Burgoyne, The flammability of mists and sprays, Proc. 2nd Symp. on Chemical Process Hazards. (1963) 1–5.
[28] M.G. Zabetakis, Flammability characteristics of combustible gases and vapors, Bureau of Mines, Pittsburgh, PA (United States), 1964. https://doi.org/10.2172/7328370.
[29] J.E. Shepherd, C.D. Nuyt and J.J. Lee, Flash Point and Chemical Composition of Aviation Kerosene (Jet A), (1999) 38.

619219 Large Volume Liquid Hydrogen Releases: Key Results and Outcome of Modelling Exercises

Abstract:

In line with energy transition commitments made by a growing number of governments worldwide, the decarbonising of ocean-going shipping and transportation remains a significant challenge. The difficulties in storing electrical energy from renewables in significant capacity makes it of little use against the requirements of running a ship at sea. Liquid hydrogen (LH_2) forms one potential solution [1] where the required energy density / volume can be met for ships of all sizes. Over and above it's use as a fuel in the space industry, the use of LH_2 in the maritime sector poses challenges in the scale of operations, number of usage points and proximity to personnel. In order to enable pilot projects and to identify and quantify safety related issues which may need to be resolved, it is necessary to conduct research. This research needs to produce data for the validation and development of predictive models for the hazards involved with LH_2 at scales representative to the processes needed for the application. Some data at smaller scales exists [2 and 3] but the effects of scaling to the larger scales investigated here are not known.

The Norwegian Public Roads Administration (NPRA) is an administrative agency of the Norwegian government, responsible for planning, construction, operation and maintenance of Norway's road networks at the national and regional levels. The agency also grants subsidies for Norway's extensive system of ferry operations. In 2018, NPRA signed a contract for the operation of a hydrogen-electric ferry to serve one of the routes along the coast of Norway [4]. One concept for such zero-emission ferry transportation involves liquid hydrogen (LH_2) as a concentrated form of hydrogen storage.

DNV GL Spadeadam Research and Testing was commissioned by the Norwegian Defence Research Establishment (Forsvarets forskningsinstitutt, FFI) and the NPRA to conduct a series of experiments investigating the behaviour of LH_2 when large releases occur to the atmosphere and into a closed ventilated enclosure [5].

The project included two experimental campaigns to produce data related to outdoor leakage and also leakage into ventilated rooms. The present paper details the experimental arrangement, programme and results pertaining the outdoor leakage studies only. The quality of the data is discussed when compared to expected behaviour for outflow, pooling, dispersion and fire. The expected behaviour is illustrated by predictions from DNV GL's GasVLE, FROST and Phast software packages. Discussion of the performance of the models is given.

A total of 7 large releases (up to 50 $kg.min^{-1}$) from a liquid hydrogen storage tanker were performed with variations in source tanker pressure, flow rate, orientation, ignition and atmospheric conditions. The data has been shown to be credible and suitable for model evaluations. Given the number of variables and relatively small number of experiments, it was not possible to entirely quantify all of the phenomena associated with the releases but this paper summarises observations of the following:

- **Outflow / Flashing**
  - Assessments of the mass flow versus pipeline and outlet conditions are made including assessments of the level of flashing (liquid mass fraction) within the pipe
- **Pooling / Rainout**
  - Measurements of ground surface and sub-surface temperature around the releases were made and interpreted for the presence of liquid hydrogen or condensed products of air
  - Comparison with pooling models in FROST is made

- **Dispersion**
    - Determination of the extents of the flammable limits near to ground level from each release is made from the measurements and comparisons with model predictions.
- **Thermal Properties**
    - Thermal radiation measurements recorded around each ignited release are interpreted against simple models and existing correlations and guidance for hydrocarbons. Discussion of the flux within the flame is included.

References:

[1] Decarbonising Shipping: Setting Shell's Course. https://www.shell.com/energy-and-innovation/the-energy-future/decarbonising-shipping.html [Accessed 20-11-2020]

[2] Releases of unignited liquid hydrogen, HSE Research Report RR986, M Royle and D Willoughby, 2014.

[3] Ignited releases of liquid hydrogen, HSE Research Report RR987, J Hall, 2014

[4] NPRA Press Release on Ferry Contract Award: https://www.vegvesen.no/om+statens+vegvesen/presse/Pressemeldingsarkiv/Vegdirektoratet/hydrogenferje-avtale-gir-nullutslipp-i-ferjesektoren [Accessed 20-11-2020]

[5] NPRA Announcement of Results: https://www.vegvesen.no/fag/trafikk/ferje/utviklingskontrakt-hydrogen/testing [Accessed 20-11-2020]

HIRA in Mining Companies Is Necessary and a Big Challenge

Presenter/Principal Author
Américo Diniz Carvalho Neto
RSE CONSULTING
Brazil
americodcn@rsem.com.br

Other Author(s)
Afonso Costa
RSE CONSULTING
Brazil
Afonso.costa@rsem.com.br

Prepared for Presentation at
American Institute of Chemical Engineers
2021 Spring Meeting and 17th Global Congress on Process Safety
Virtual
April 18 - 22, 2021

HIRA in Mining Companies Is Necessary and a Big Challenge

Presenter/Principal Author
Américo Diniz Carvalho Neto
RSE CONSULTING
Brazil
americodcn@rsem.com.br

Other Author(s)
Afonso Costa
RSE CONSULTING
Brazil
Afonso.costa@rsem.com.br

Keywords: HIRA Project,

Abstract

The implementation of a risk-based process safety management system is a turning point for any company that intends to have a better understanding of the risks associated with their process. However, full implementation of all elements takes time, and in the meantime, catastrophic scenarios should be managed accordingly, without wasting time.

To fulfill both criteria, having long and short run implementation strategies, HIRA (Hazard Identification and Risk Analysis) process was developed for mining companies worldwide. Based on the Critical Control Management (CCM), HIRA process is focused on "quick wins" of process safety, management gaps that must be eliminated shortly, to keep high consequences scenarios from happening. Those scenarios are known as MUE (Material Unwanted Event).

HIRA for mining companies is based on the methodology developed by ICMM (International Council of Mining and Metals), which is comprised of 9 steps, 6 of them are only to identify scenarios and their causes, consequences and mainly the critical controls associated. The methodology intends not only to identify critical controls, but also to check their status (implemented or not implemented), identify items that can make them more reliable and to assign responsibility for the management of the control, throughout its lifetime, to guarantee they will perform their function when demanded.

To accomplish such tasks, HIRA process widely uses Bow Tie Analysis, a very intuitive and visual risk management tool, which allows even people who do not have a lot of experience in the subject to easily understand the graphic representation of the MUE, such as cause, consequence, top event and mainly the controls and their status. Bow tie sessions are followed by field verification of each critical control identified, to check coherence between records (data), integrity and reliability, of the controls.

As a result, both existing and possible controls (known as Future State) are identified, according to the MUE analyzed and, comparing information from field verification, a qualitative risk assessment is carried out for three different categories – Inherent, residual and forecast risk.

By the end of the HIRA process, many types of controls are identified, such as procedures, safety valves, inspection plans, guard rails, alarms, interlocks…and so on. They now will be managed differently, as they are classified as critical controls – items that are crucial do prevent the MUE form happening or mitigate its consequence.

1 Main section 1 heading (Introduction)

Many companies have improved their performance on safety-related issues using a wide range of rules, procedures, and standards; however, Material Unwanted Events, such as fatal accidents, rare catastrophic events and circumstances of over-exposure that are harmful to health, are still subject to occur. For this reason, the ICMM (International Council for Mining and Metals) developed an approach called Critical Controls Management (CCM), aiming at the continuous improvement of management over rare events, but potentially catastrophic, emphasizing critical controls. The methodology to improve the management control of these types of events, called Material Unwanted Events (MUEs) follow below:

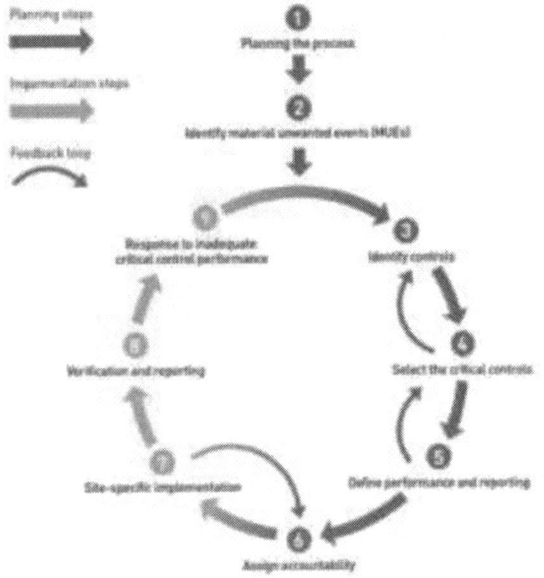

ICMM Critical Controls Management Process (CCM)

CCM's approach is based on some STEPS as follow:

Step 1: Develop a project plan that contemplates the organizational context, objectives, team responsibility and the individuals and business areas affected in the implementation.

Key actions for STEP 1

• Develop a plan that describes the scope of the project.

This includes:
– organizational context
– project objectives

– responsibilities
– business sections involved.

• Develop methods to:

– identify potential hazards and unwanted events
– assess risk
– review MUEs
– select critical controls
– assess objectives and performance of critical controls
– investigate critical control underperformance
– measure impact of the project
– identify ownership and accountability.

Step 2: Identify the MUEs to be managed and summarize the main information for each MUE. In order to identify, verify if they present material risk, evaluate the opportunities to eliminate it by improving the designs and describing the MUE completely (hazard, materialization mechanism and possible consequences).

Key actions for STEP 2

• Understand major hazards and identify potential MUEs.
• Apply selection criteria to MUEs with a focus on the consequences.
• Identify design opportunities to address the hazard, reducing the potential consequences and eliminating the MUE from the CCM process.
• Describe the identified MUE, including the relevant hazard, mechanism of release and nature of the consequences.

Step 3: Identify the controls, developing Bow Ties diagrams and evaluating them.

Key actions:

• Identify the controls.
• Prepare a bowtie diagram.
• Assess the adequacy of the bowtie and the controls.

Step 4: Select the critical controls for each identified MUE, summarizing its main information.

Key actions For STEP 4

• When identifying critical controls, apply the critical control definition and guidance in this section.
• Consider the performance requirements of the potential critical controls and how they could be verified.

• The final set of critical controls for an MUE should represent the critical few that, when managed using CCM, can effectively manage the MUE risk.

Step 5: Define the objectives and Critical Controls performance requirements, how they will be checked and the activities that will be affected by them.

Key actions for STEP 5

• Define objectives and performance requirements for each critical control.
• Identify current activities that affect the critical control's performance.
• Describe activities to verify performance and reporting requirements.
• Identify what would trigger immediate action to stop or change the operation and/or impose the performance of the critical control.

Step 6: Develop a list of those responsible for the MUE, for the Critical Control and for its verification, including the preparation of the final information report.

Key actions for STEP 6

• Assign owners for MUEs, critical controls and verification activities.
• Describe reporting plan for the health of critical controls.
• Assign owner for review of reports.

Step 7: Define a verification plan and reporting plan for each MUE, implementing it according to the specific needs.

Key actions for STEP 7

• Critical control information must be specific to a site or asset.
• Adjust the critical control definition, performance information and verification requirements as necessary to suit the local context.
• Site-specific planning for implementation may involve an iterative process.
• Site-specific planning should include establishing a foundation for CCM that includes leadership, communication and appropriate development of knowledge and understanding related to the critical controls.

Step 8: Implement verification activities and report the process, defining the status of each critical control.

Key actions for STEP 8

• Undertake verification activities for critical controls as described in MUE/critical control verification and reporting plan (developed in Step 5).
• Report a summary of verification activity results to the critical control owner.

• Report critical control verification status to the MUE owner.
• Reports should highlight priority information succinctly using traffic light system.
• Action initiated if critical control performance drops below the defined triggers (established in Step 5).

Step 9: Performance investigation if critical controls are performing below expectations.

Key actions for STEP 9

• Take action when critical control performance is inadequate (below the defined trigger threshold).
• Investigate the causes of unacceptable critical control performance.
• Information and data from the investigation should be used to continuously improve the CCM.

The companies risk map, in most cases, does not fully reflect the main risks to which the operation and the external population are exposed. This is due to the lack of a unified methodology to identify, quantify and measure risks. In view of this reality, one company can initiate the HIRA Project (Hazard Identification and Risk Assessment), to conduct a quick technical scan of risks in its operations to identify relevant and unmitigated situations.
HIRA was designed to quickly identify and verify the main risks of the company units operations. This is a global project to be implemented and will be at the center of the operational risk agenda in the coming months. The HIRA Project will focus on the very critical risks of severity in the company operations and its subsidiaries and its results, will be reviewed frequently by the Executive Committee for Operational Risks, the Executive Board and the Board of Directors. His results will guide the main actions to reduce the organization's risk profile.

GOALS

The current study aims to identify and analyze operational safety risks with a very high-risk magnitude or very critical severity of a company site. In addition, the identification, definition of performance criteria and the establishment of criteria to guarantee the performance of associated critical controls are part of the scope of this work. From this, it will be possible to help the company to produce a complete and accurate inventory of the relevant unwanted events of the organization (MUEs) and, later, based on the identified MUEs, identify critical controls, associated performance requirements and means of verification.

For this purpose, this study will include:

1. A consolidated list of operational safety risks assessed with a very high risk magnitude or very critical severity (MUEs) for the unit;

2. Bow Ties based on the identified MUEs and a corresponding list of critical controls identified for each (following the suggested ICMM decision tree and based on the group discussion);

3. One-Pager for each critical risk control.

4. List of action plans to be developed based on the prioritization of risks.

5. Updated MUEs risk matrix of the unit.

METHODOLOGY

The Bow Tie Analysis (BTA) can be used to identify and assess hazards with the potential to cause major accidents involving company site. This study can be conducted through presential or remote meetings through the TEAMS videoconferencing platform, by multidisciplinary teams composed of corporate company members and company site members and can be supported by external consultants.
Initially, happens a Workshop session, jointly between the consultants of RSE Consulting and company employees present at the meeting. The site MUEs were defined taking into account that the chosen MUEs would be among the most at risk in the facilities analyzed. In these sessions, MUEs were selected and evaluated from a previous list of more than 170 different MUEs, from different units of the company, including units in different countries which the HIRA project will be developed. Other new scenarios were also discussed and included as discussed with the local technical team. After determining the likely MUEs applicable to the site, these were then categorized as follow:

• Category 1: Scenarios in which, all the information presented during the Workshop, was sufficient for validation as a MUE. These are considered priority scenarios;

• Category 2: Scenarios in which, despite the information presented during the Workshop, these were not sufficient to validate the MUE in remote mode, thus requiring field visits and more information for validation. Also, in this category are included scenarios that can be considered as developments of scenarios classified as Category 1;

• Category 3: Scenarios in which, despite the information presented during the Workshop, these were not enough for validation as a scenario that meets the HIRA criteria, to be considered MUE. An on-site visit would be necessary to complement information and analysis. Thus, the team present at the MUE Workshop session with different specialists concluded in a list some (%)

MUEs classified as Category 1, that is, priority to the units of the complex. These meetings can be remotely, and the list was
subsequently validated by company leaders. It should be noted that the identification of MUEs needs to consider the historical events (Accidents and Near Miss) of the site, company and sector, in addition to predictable events, given the operations and activities at individual locations. Therefore, the identification of MUEs needs to include personnel with adequate experience and a review of relevant data.
Then, meetings were held remotely to analyze the MUE and generate project deliverables for the Complex. The analyzes were composed by members of the company corporate members, site regional members and external RSE consultants, where the consequences, threats and the respective existing controls for each MUE of the site were identified. During these meetings Bow Ties from other company units were analyzed with similar processes as a reference. Details about the Bow Tie technique are presented of this paper.

To identify the critical controls and controls of the MUEs, the proposal decision tree was followed:

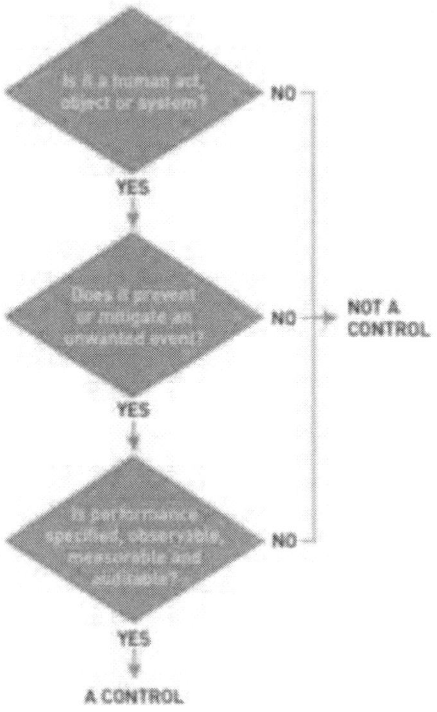

Parameters for defining a control

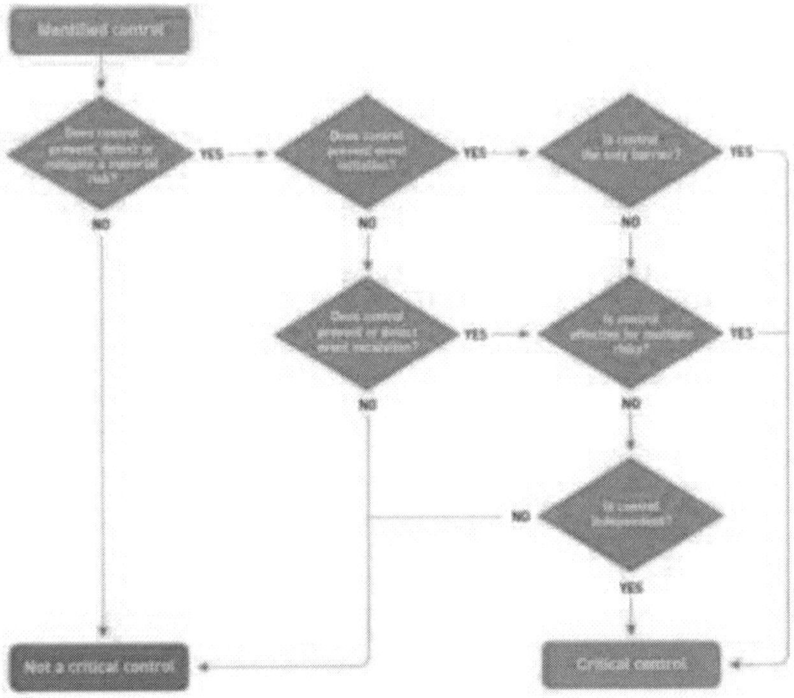

Parameters for defining a Critical Control

Once the critical controls were identified, a "One Pager" was generated for each one, that is, a quick briefing of just one page for each critical control (except "Future State" critical controls) containing the following information:

- Objectives of related critical control
- Performance requirements to meet objectives
- Activities that can enable or strengthen control
- Activities that can be verified to attest to the performance of a critical control
- Performance for critical control
- Trigger to stop the unit, revision or investigate, depending on the performance of the control
- Critical Controls attributes

The current levels of effectiveness (status) of performance of these critical controls were classified initially without field verification, based on the opinion of company specialists present at the meetings, to generate previous actions for the company, according to the methodology established for carrying out remote HIRA, during the COVID-19 pandemic. However, these classifications must be rectified / ratified in another phase of HIRA, when the field visit is done.
Remote risk classification meetings were also held, categorized as Inherent, Residual and Forecast, classifying them as follows:

- Inherent risk: premise = the existing critical controls to prevent each cause or mitigate each consequence are null (there are no controls);

• Residual Risk: premise = the existing critical controls to prevent each cause or mitigate each consequence are fully implemented or not implemented, depending on the classification of the status of each critical control determined through remote meetings by the Vale specialists team

• Forecast Risk: premise = critical preventive controls, mitigating controls, future controls (Future State) and all actions and recommendations captured during the meetings are fully implemented and functional, that is, they are compl ete and reliable.

It is important to note that risk assessments were carried out, remotely, without the field verification by the HIRA teams to validate, based on the One Pagers, the critical controls. Therefore, critical controls were evaluated in future scheduling, considered the next phase to be carried out at HIRA on site.

For the risk analysis of each MUE, the risk acceptability criteria of the company must be adopted. Below follows a brief explanation of Bowtie methodology selected in this Project.

BOW TIE TECHNIQUE

Bow Tie is a qualitative risk assessment method used to assess how high-risk scenarios develop. The essence of Bow Tie consists of plausible risk scenarios surrounding a certain hazard, and ways in which the organization prevents these scenarios from happening. The method takes its name from the shape of the diagram you create, which looks like a male bow tie.

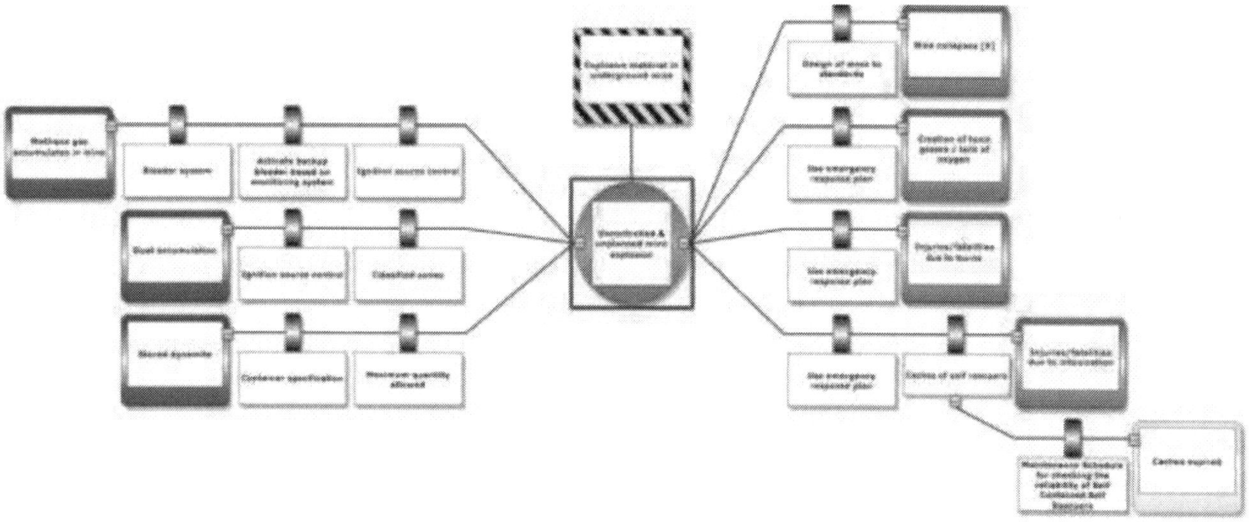

The Bow Tie methodology has several objectives:

• Provide a framework for systematically analyzing a hazard.

• Helps to decide whether the current level of control is sufficient (or, for those who are familiar with the concept, whether the risks are as low as reasonably practical or ALARP).

• Helps to identify where and how to invest in resources that could have the greatest impact.

• Increases communication and risk awareness.

General step by step construction of the Bow Tie diagram:

Hazards Identification The purpose of hazard identification is to find, recognize and describe the risks that can help or prevent an organization from achieving its goals. Hazards can be operations / activities (operating rotating machines, driving a car),

I. substances (chemicals, hot fluids, etc.) or situations (a suspended load) that we deal with in the normal processes of our business. As long as these hazards are under control, they will not cause harm, but introduce the potential for harm.

II. Definition of the Top Event / MUE When control over a hazard is lost, it is usually possible to identify the exact moment when a normal situation changes to an abnormal situation, that moment is called Top Event. The Top Event is not yet a catastrophe, but the company is now exposed to the potential for harm from hazard.

III. Identification of Consequences The consequences are classified as the possible negative results directly linked to the top event, which can affect people, the environment, the facilities and the company's reputation.

IV. Threats Identification Factors that can lead directly to the top event are called threats. These are capable of causing the Top Event independently and are not associated with barrier failures.

V. Identification of Preventive Controls Risk management is about controlling risks. This is done by implementing barriers to prevent certain events from forming. A barrier (also called control) can be any measure taken that acts against some unwanted force or intention, in order to maintain a desired state. Preventive barriers, identified on the left side of the Bow Tie, can prevent the top event from happening or can eliminate the threat.

VI. Identification of Recovery Controls Controls or recovery barriers identified on the right side of the Bow Tie can prevent the top event from resulting in unwanted consequences or mitigate the consequences.

VII. Define Escalation Factors Once the controls have been identified, it is recommended, according to the Bow Tie method, to identify specific conditions or actions that reduce the effectiveness of the control. These are called escalation factors.

VIII. Identification of Escalation Factors Barriers Once the escalation factors have been identified, the last step is to look at what barriers we have to manage these factors. These barriers have the same concept as all the barriers previously discussed, but, now they do not prevent / mitigate a top event or consequence, they will prevent a barrier from failing. The BowTieXP software, version 9.2, developed by CGE Risk, was used to construct the Bow Tie diagrams. BowTieXP is the most widely used risk assessment software based on the Bow Tie methodology that allows you to easily create risk assessment diagrams.

ANALYSIS RESULTS

Table attached shows the Material Unwanted Events (MUEs) that were hipotetical identified and jointly defined by RSE Consulting and the company team. As defined by the company, for an event to be considered as MUE it must have the potential to cause one or more fatalities or the potential to cause a major environmental impact that goes beyond the limits of the company.

In addition to these available scenarios (Category 1), the MUEs Workshop, conducted with a multidisciplinary team, generated potential scenarios for MUEs whose information presented was not sufficient to validate them in the remote mode, requiring additional information through field visits for validation (Category 2 and 3).

In total, 221 existing controls were defined to prevent and / or mitigate MUEs and 138 of these were classified as critical controls for these scenarios. These numbers consider the proposed future controls and the repetition of the same control between the scenarios.

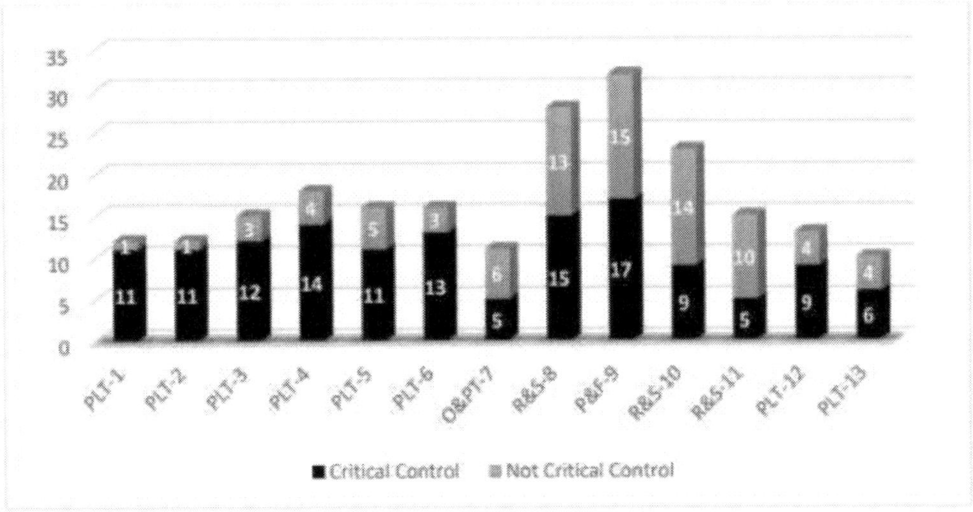

Controls and Critical Controls

According to the criteria adopted for the status of critical controls, the following were identified:

✓ 98 Critical controls classified as "Implemented"

✓ 09 Critical Controls classified as "Not Implemented"

✓ 31 Critical Controls classified as "Future State"

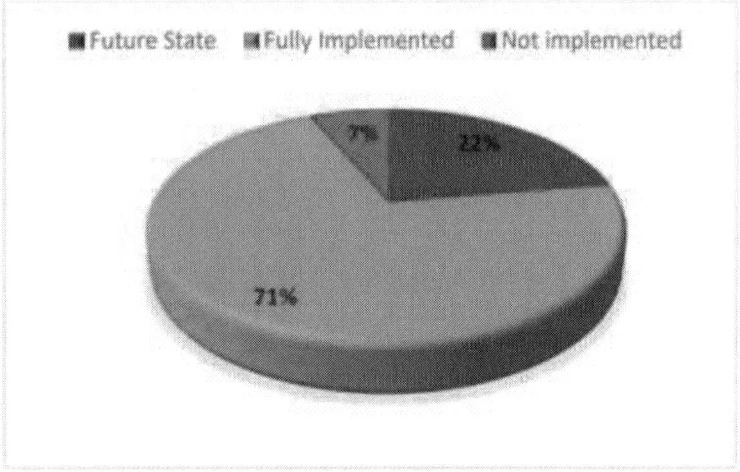

Percentage distribution of the status of Critical Controls

Regarding the type of critical controls, the following were identified:

✓ 104 Critical Controls classified as "Preventive";

✓ 34 Critical Controls classified as "Mitigatory".

This distribution demonstrates the greater effort of the professionals who participated in the assessments and identification of risks / controls in the Search for preventive solutions instead of mitigating unwanted scenarios, which satisfies the initial premise of HIRA.

Additionally, considering a larger sample of 19 sites, where 252 MUEs were analyzed, one can have a better understanding of the results.

• Critical controls - for the 252 MUEs evaluated in this sample, 4283 controls were identified, of which 2487 were considered critical controls. In this number are included the controls considered Future State, that is, the proposals identified by the group to reduce the risk of the assessed scenario. This means that 58% of the controls were considered crucial in preventing the occurrence or mitigating the consequences of the MUE, according to the evaluation team. This is always a point of attention when evaluating controls because it is not uncommon for members to tend to classify a large percentage of controls as critical. This is because companies prioritize resources to ensure the reliability of their critical controls, and the Operation team can use this criterion as a factor for prioritizing resources, going in the opposite direction to the methodology, which is a big mistake. It is up to the consultant to ensure that the flow of classification of critical controls is properly followed.

A rule used, which has served for the present project, is to consider as critical controls something between 45% to 55% of the controls. If the numbers indicate something greater than this, a ratification with the evaluation team is recommended.

For the same project, 1115 controls considered critical were classified as administrative controls, that is, they are basically controls that are highly dependent on human action to fulfill their required function. Examples of these controls are procedures and inspections. The latter is one of the main points of attention because it presents many opportunities for improvement as inspections are common not only in mining but in any company. However, many operational inspections carried out do not have the required reliability, that is, they are not able to identify deviations in integrity that can lead to the MUE.

These inspections are characterized by having a high frequency (sometimes daily) of execution and low level of details. On the other hand, there is the integrity inspection, which has a low frequency and a high level of details. The latter is usually performed by certified inspectors, has an associated procedure and inspection report issued at the end of each inspection. That is, depending on the type of inspection carried out, it is possible that this process is very effective in identifying structural damage or just a simple routine daily verification process, without the required level of detail.

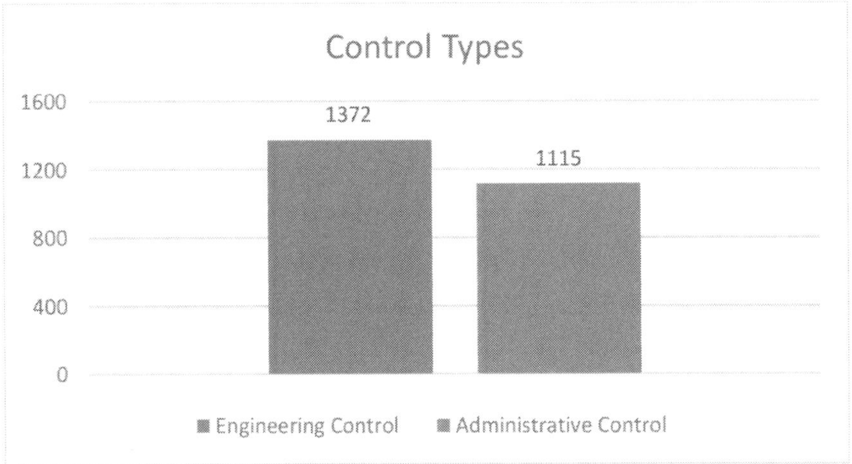

Graph 2 – Types of Critical Controls

It is highly recommended that this information about the items that can enable or strengthen critical control, in the case of operational inspection, are clearly described in the One Pager.

Preventive or mitigating controls - Considering the type of control as preventive or mitigating, again for the project considered, it was found that 1874 (75%) were preventive and 613 (25%) were mitigating. The mitigating controls are basically the emergency action plan and fire fighting system. Less frequently, dykes also appear as important controls for the MUE of loss of fuel containment.

It is important to say that, although they represent only 25% of the identified critical controls, mitigating controls are responsible for a higher percentage of recommendations. One of the most recurrent is the failure to consider the MUE in emergency pre-plans and drills.

Conclusion:

Throughout HIRA´s remote work of the company site, risk scenarios considered more critical were evaluated, with the potential for multiple fatalities or significant environmental impact or level 1 process safety event. Although the result of critical controls considered as implemented was in higher percentage, deviations were identified in the management of some pillars of Process Safety, such as Mechanical Integrity, and Risk Management. For a better understanding of the deviations listed, the list of actions is done.

A larger number of "automatic" or engineering critical controls type were identified in relation to those with human interface, such as operational procedures, inspections, maintenance plans, among others. For some of these controls, actions were proposed to ensure greater reliability on demand, activities such as systematic inspections, maintenance, calibration plan, looping control tests, fail-safe mode, etc.

About the critical controls identified with human actions, it is necessary the following requirements to the reduction of the failures: to reach deadlines of procedure's review, elaborate training plan/program with knowledge retention (evidence with records), recycling and periodic evaluation of the team performance, with actions for "under the established standard", as well as investigation records of near miss incidents.

About company site, it is important to highlight:

1. Treat the critical controls as critical items into the SAP or the specific management software.

2. Ensure proper inspection and maintenance for the automatic controls (safe interlock systems);

If the field verification of critical controls is performed in the company site made it possible to diagnose opportunities for improvement, strengthening and implementation of Process Safety Risk Management. Considering the concepts of hazard identification, risk analysis, change management, implementation of Process Safety Management System will be essential for the company site to manage its risks associated with the MUEs, to keep them at an acceptable level.

2 References:

[1] RSE Consulting HIRA methodology process, Americo Diniz Carvalho Neto, 2017.

[2] RSE Consulting Barriers Audit methodology, Americo Diniz Carvalho Neto, 2007.

[3] Health and safety critical control management Good practice guide, ICMM.

[4] CCPS. *Guidelines for Bowtie in Risk Management.* Center for Chemical Process Safety, American Institute of Chemical Engineers, New York, NY, 2018.

[5] CCPS. *Guidelines for Hazard Evaluation Procedures.* Center for Chemical Process Safety, American Institute of Chemical Engineers, New York, NY, 2008.

616590 Title: Silencing a Noisy Agitator Seal By Managing Stress

Abstract:

Title: Silencing a Noisy Agitator Seal by Managing Stress

Abstract: Did you know the sound that comes when you run your finger on the rim of a wine glass is referred to as "partial boundary layer lubrication"? Did you know this same phenomenon also occurs commonly on buffer and media tank agitators in bio-processing facilities? However, on agitator seals the sound can amplify to over 90 decibels, which exceeds OSHA's personal exposure limits.

Using an FEA model, John Crane found, that the noise from partial boundary layer lubrication is a result of vibration caused by the application of an even or consistent stress to the seal face, which is consistent with field trials since the seals rarely squealed at startup. Instead, the seals squeal after some run time when the right amount of stress is applied. This stress could come from the pressure of the vapor space in the tank or vessel, which also explains why the noise level and pitch can come and go depending on the pressure in the tank, agitator speed, as well as other scenarios.

This led to the discovery that the squealing can be reduced, or even eliminated, by managing the stress imposed on the agitator seal faces. Subsequently, John Crane developed a solution that results in an active seal face, meaning the degrees of flatness change with the stresses applied during operation. In other words, the rim of the wine glass changes as you run your finger around it preventing the noise condition from ever taking place.

Take Home Message from Poster: The noise from a media or buffer tank agitator seal can be reduced, making a quieter and safer workspace.

619172 Use of Data Analytics to Obtain Cost Reductions and Performance Improvements in a Regulatory Application

Abstract:

All operating nuclear power plants (NPPs) in the United States are required to monitor and assess the performance of plant structures, systems and components (SSCs) as specified in the Maintenance Rule (10CFR50.65). This rule requires that all licensees identify the occurrence of any SSC failure that results in a loss of safety function, often referred to as a Maintenance Rule Functional Failure (MRFF). All MRFFs are subsequently evaluated through the plant corrective action program (CAP) as required by 10CFR72.172. To ensure these requirements are met, all NPPs document events that occur which are evaluated to identify impact on the plant. A typical dual unit generating facility generates as many as 15,000 of these incident reports (IRs) per year which each must be evaluated by experienced plant staff. Of these IRs, only 0.1-0.2% typically are MRFFs. NPPs have found the monitoring of plant SSCs for compliance with the maintenance rule to be expensive and resource intensive.

An alternative approach which has the potential to reduce these costs is to automate the review of IRs through the use of data analytics and machine learning (DA/ML). Advances over the past several decades in computational capabilities and computer programming techniques / algorithms now permit application of DA/ML techniques to analyze the large historical data sets that contain information on the performance of SSCs over years of plant operation. Unfortunately, these techniques have not been widely adopted within the nuclear industry due to its highly regulated nature. A consequence of this structure is that any DA/ML models used in applications subject to regulatory oversight must provide a high degree of confidence in the predictions due to potential regulatory impacts associated with an incorrect determination.

The MRFF Analyzer is a DA/ML application which significantly improves the efficiency of the maintenance rule screening process. The model uses a combination of natural language processing (NLP), Bayesian statistics, and machine learning to identify the likelihood that the incident described by an IR represents a MRFF. The software was trained, tested, and validated on four years of historic incident reports (2015-2019) generated across Exelon Generation's fleet of operating NPPs. This data set consisted of approximately 650,000 IRs. Approximately 420,000 IRs from 2015 through the summer of 2018 were used to train the model, and 80,000 IRs from summer-winter 2018 were used to test the model. Model performance in the field was validated during 2019 with each IR generated by the sites analyzed by the software with review performed by on-site subject matter experts as specified in the plant program implementing procedures. Once the performance of the MRFF Analyzer software was verified as adequate to meet the requirements of the Maintenance Rule, the parallel use of review by human experts was discontinued.

This paper presents the development, training, and deployment of the MRFF Analyzer software across Exelon Generation's fleet of operating NPPs. The algorithms in use in the software are discussed in detail. The challenges dealing with highly skewed data are presented and the approaches used to address these challenges discussed. The results of a detailed sensitivity assessment are presented and used to estimate the expected statistical performance of the software. The overall performance of MRFF Analyzer since its deployment at Exelon Generation is discussed and estimates of the cost savings are provided.

617819 Search: A Symptom-Based Expert for Advanced Response to Chemical Hazards

Abstract:

Chemical exposure accidents pose a risk of enormous personal injury and property damage if initial response is not appropriate. In Korea, from 2014 to October 2019, 1,000 chemical accidents occurred, and 20 people died and 444 were injured. In addition, the number of chemical accidents shows a trend of increasing from 233 in 2014 to 277 in 2018. Existing chemical substance detection technology is focused on the development of response technology for leaks at known sites, and research on the detection at dynamically changing environment and knowledge service for hazardous chemicals considering symptoms is insufficient. Therefore, there is a need for research on a real-time hazardous chemical detection and diagnosis system and knowledge service that protects workers or exposed people from unexpected chemical accidents in the field, by detecting exposed chemicals quickly and helping to cope with dangerous situations early.

In this study, we develop an AI-based system for chemical identification from actively collected exposure symptoms, for real-time response to hazardous material incidents. In order to analyze the symptoms of chemical exposure accidents or contact with substances, information related to chemical substances (symptoms, toxicity, structural information, etc.) is collected and extracted from WISER, PubChem, MSDS, etc. to systemize the related information. The knowledge base of 1,000 core substances is built by designing ontology for symptom knowledge of accident-involving chemical substances. In case of WISER developed by NIH, chemical substances are estimated by simply matching symptom words manually inputted, but the proposed system actively collects the symptoms and predicts the candidate chemical substances (with probabilities) through the knowledge base inference considering the exposure situation. The knowledge base inference uses a knowledge graph as knowledge base that is easy to search, and uses a reasoning method such as the dynamic RDFS++ reasoner, which is easy to infer dynamic data provided by AllegroGraph, and OWL2RL materializer, which is easy to infer static data. In the version 1 of SEARCH, the dynamic RDFS++ reasoner is mainly used to proceed with inference.

Using Machine Learning (k-NN, Decision Tree, SVM, etc.) and Deep Learning, it also predicts toxicity from chemical substance information (structure, etc.) or symptoms from toxicity [1,2] for building expanded symptom KB. The accuracy of the chemical symptoms estimated from the structure and toxicity of the chemical substance is derived from the model with good performance among the previously used models, developing a system with an accuracy of 80% or higher that can estimate 43 symptoms for 1,000 chemicals. Through the combination of two network models that predict toxicity from the structure of chemical substances or predict symptoms from toxicity, it is possible to provide symptom knowledge to be used for symptom detection even for new chemical substances.

SEARCH provides a wide range of information on hazardous substances, including substance identification support, physicochemical characteristics, human health information, and containment and suppression advice. The prototype system is serviced with AI speakers as the main interface and demonstrated at the Gyeonggido Disaster and Safety Headquarters, and plans to combine the established knowledge base and chemical substance-symptom estimation model. The voice recognition interface using AI speakers enables effectively acquire information by saving time and effort rather than directly inputting and matching information manually.

617983 Application of Machine Learning Algorithm for the Analysis of OSHA Accident Database to Identify Trends in Performance Indicators of Process Safety Incidents

Abstract:

Many workplace injuries and fatalities occur in facilities dealing with hazardous chemicals due to unexpected leaks, exposure to dangerous process conditions and other loss of containment events. The U.S. Occupational Safety and Health Agency (OSHA)'s Process Safety Management (PSM) program has been in effect for almost three decades in the United States, requiring companies to establish comprehensive management programs to deal with such hazardous chemicals with the target to assure safety and health in the workplace. Analysis of the trend of process safety incidents in these industries can be a good performance indicator for the effectiveness of the PSM program.

For a long time, OSHA has been maintaining and publishing accident data collected from various industries since the 70's and currently has recorded information of more than 130,000 incidents involving fatalities and injuries arising from various causes. Analysis of such a large database to identify process safety incidents manually is extremely difficult and time consuming. Current study focuses on the development of a machine learning algorithm to screen the database and identify process safety incidents through use of the keywords and incident descriptions associated with each accident provided by OSHA.

In order to validate the machine learning process, the large OSHA database is filtered to analyze incidents from 1984 to current time and only chemical manufacturing and petroleum industries are considered through use of industry classification codes. More than 2,200 incidents are screened out in this manner and are analyzed to determine those that were process safety related incidents. The analysis was done both manually and through use of the developed machine learning algorithm to validate the capability of the latter. The keywords are classified into various categories to aid the algorithm in the learning process. This is the beginning step to the application of artificial intelligence to analyze big databases such as the OSHA database to determine common patterns behind injury and fatal incidents and provide industry with performance indicators that will provide an understanding of where the current laggings are and enable proactive measure to be taken to prevent future incidents.

617448 Game Based Learning – Making the Message Memorable

Abstract:

A perennial frustration for organizations working within high hazard industries is that despite going to great time, cost, and trouble to communicate important process safety messages across the entire workforce, incidents continue to happen. Making these messages "stick" is essential. This presentation will outline a novel approach to developing process safety knowledge and competence through Game Based Learning (GBL).

Based on research that suggests people remember 90% of what they 'do', GBL takes your message out of PowerPoint and broadcasts it interactively to the trainee. A GBL session typically lasts 30-60 minutes, is hands-on and, by using every-day analogies for technical subjects and grounding them in the real world, it is accessible to all personnel across an organization. Indeed, GBL for major accident awareness has been run for school children as part of Science, Technology, Engineering and Mathematics (STEM) teaching for years. GBL comes into its own in high hazard industries where, generally speaking, improvement is about small margins. This presentation will include some examples of games developed by Risktec, and cover a wide range of common process safety themes, to help prompt the audience's ability to replicate similar ideas in their own workplace.

SPRING21
+17TH GCPS
A Joint AIChE and CCPS Meeting

Process Safety Cards – A Good Deal Safer

David Hatch
Process Safety Integrity
Derby, UK
david.hatch@psintegrity.com

Prepared for Presentation at
American Institute of Chemical Engineers
2021 Spring Meeting and 17th Global Congress on Process Safety
Virtual
April 18 - 22, 2021

Process Safety Cards – A Good Deal Safer

David Hatch
Process Safety Integrity
Derby, UK
david.hatch@psintegrity.com

Keywords: Process Safety Cards, Bowties, Gamification, Microlearning, Workforce Engagement.

Abstract

To improve the awareness of less technical personnel to potential Loss of Containment (LoC) scenarios and increase their vigilance to such events & their potential causes so they can be promptly addressed; memorable images are shown on familiar playing cards representing typical LoC events.

1 Introduction

Process Safety Cards have a simple goal – to improve the awareness of a broad range of personnel to potential major accident scenarios and increase their vigilance to such events & their potential causes so they can be promptly addressed. These memorable images represent typical LoC events to deliver knowledge literally into the hands of those who might contribute to and/or be affected by Process Safety Incidents.

With an ageing workforce and increased outsourcing of maintenance and operations activities, it is vital that time and knowledge poor stakeholders (who typically fall outside corporate process safety training strategies) are engaged and equipped to ensure they play their part in sustaining the integrity of assets.

2 Challenge

Process Safety Incidents, Major Accidents or Loss of Primary Containment Events continue to occur in the 21st Century despite better technology and bigger data. For example, on 28th January 2021, a liquid nitrogen release at the Foundation Food facility in Gainesville, US resulted in 6 fatalities and multiple injuries.

A review (by the author) of the 100+ CSB Investigations [1] indicates 200+ fatalities occurred of which approximately 40% were Contractors. This does **not** suggest or imply that the Contractors were at fault, however it does highlight that 3rd parties are often involved in major accidents.

Although Process Safety initiatives, typically referencing Process Safety Management frameworks including OSHA [2], CCPS [3] or the Energy Institute [4], are now commonplace, these may exclude external resources including outsourced Operations & Maintenance personnel who are therefore not as equipped as their internal counterparts to recognize and address potential Process Safety incidents.

We must also acknowledge that wisdom is being lost with the ageing workforce and that assets are getting older whilst those responsible for them are getting younger.

Are we possible focusing on the wrong people and/or protection measures?

How do we put knowledge into the right hands?

3 Workforce Engagement

In order to gain and sustain the attention and participation of the workforce, which may range from novices who need to be informed to veterans who need to be reminded, it is becoming more common to use Gamification methods. These employ game design elements e.g., point scoring, competition with others, rules of play to improve user engagement, learning and knowledge retention.

However, many games require considerable time to setup and play or often cannot be played in the workplace due to the cost and complexity of deploying electronic devices e.g., laptops or tablets to host the games safely (flammable atmosphere ignition potential).

Process Safety Cards offer an alternative approach which offers the benefits of gamification without the time and cost burdens and is based on Microlearning e.g., small learning units absorbed in small time units.

Players are encouraged and helped to learn and apply simple principles:

- REMEMBER Events,
- RECOGNISE Causes (Threats) of Events,
- RESPECT Controls (Barriers) which prevent or mitigate Events.
- REPORT Concerns about the presence or performance of the Controls/Barriers,

Thereafter there is an obligation on the duty-holder or owner/operator to:

- RESPOND to Feedback (concerns, questions, or suggestions).
- RECOVER Protection to its original state.

Using this simple & memorable 'R' model aims to embed these principles into the minds & memories of frontline personnel. It is critical that the RESPOND & RECOVER steps are performed otherwise there can be no faith that contributions are respected and that improvements are being made.

4 Deployment

Process Safety Cards represent 52 different LoC events that occur in the Process industries which are presented as familiar playing cards split into 4 separate suits.

- DEGRADATION Failure within design conditions or "under-strength"
- DISCHARGE Open route to atmosphere
- DAMAGE External impact
- DEVIATION Failure beyond design conditions or "over-stress"

These events are drawn from industry publications including the HSE [5] and CCPS [6] and whilst they are not claimed to be an exhaustive list, they can be attributed to familiar major accidents across a variety of industries.

Cards make it easier to communicate in all directions, top-bottom, side-to-side and inside-outside the organization by providing a common visual language. This gets and keeps the workforce onboard with the Process Safety or Asset Integrity mission.

Sample cards are shown in Figure 1 below:

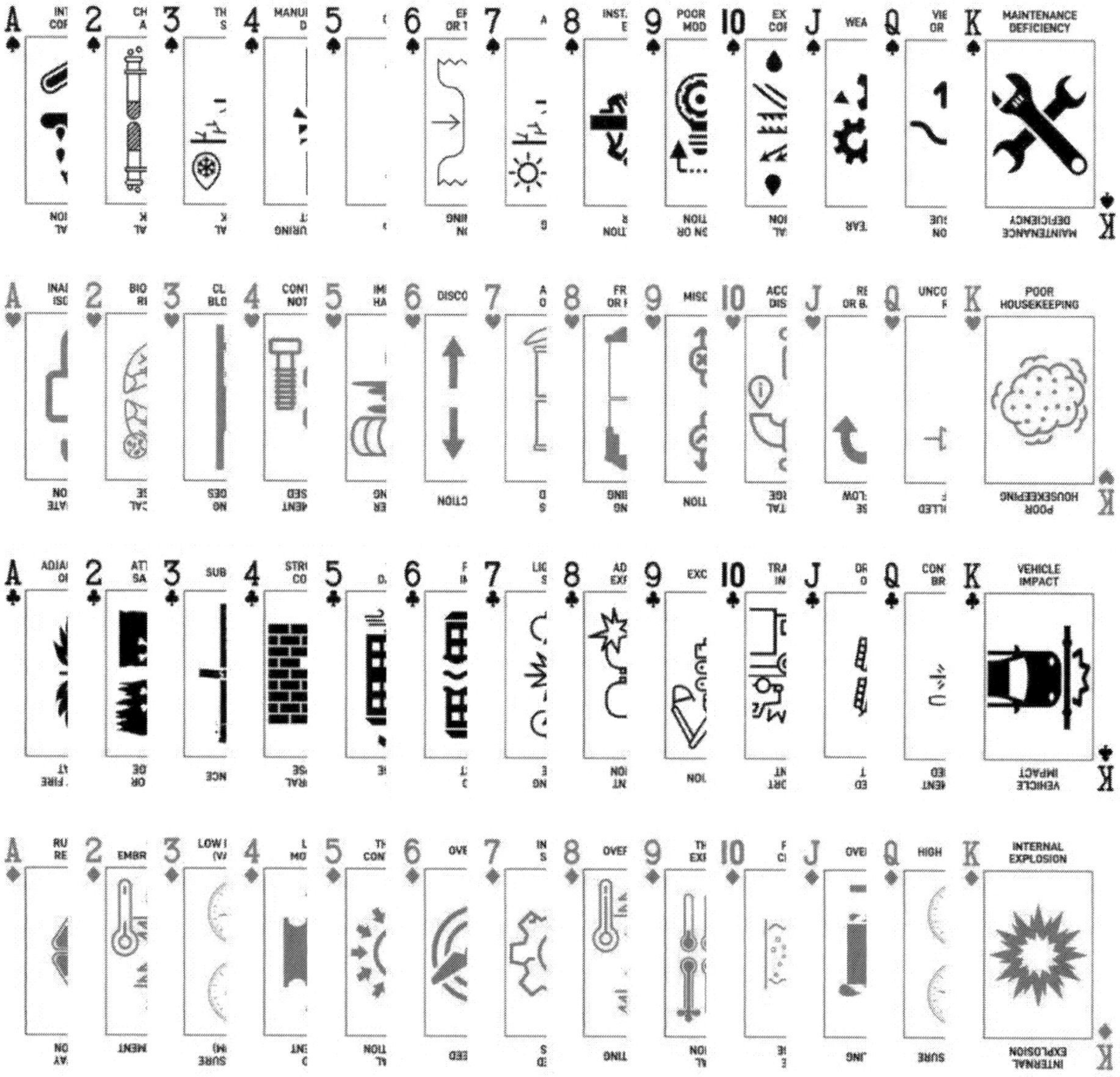

Figure 1 Sample Playing Cards

One of the core concepts of this approach is to "Know your Enemy" [7] where hazards are not the enemy (these are part of doing business) – it is Loss of Containment that is everyone's opponent.

There are several different opportunities to deploy the cards in operational arenas, because they are easy to carry & exchange:

- Simple passive learning (by osmosis) via familiar, informal card games.
- Toolbox talks, e.g., explain your experience of the card you've been dealt.
- 'Pop' or quick quiz e.g., name the event on the card (without title)
- Hazard spotting or surveys e.g., punch hole in cards and tag/tie them onto the plant at vulnerable locations

- Hazard reminder e.g., attach relevant cards to Permit to Work.
- Inductions e.g., visitors, suppliers or short-term contractors during outage, turnaround etc.

The cards are not limited to frontline personnel as they can also be used to engage engineering or management personnel.

- Safety moments e.g., discuss your role related to the card you've been dealt.
- PHA/HAZOP validation e.g., ensure all deviations (cards) have been discussed.
- Performance dashboards e.g., incidents categorized according to LoC type (or card)
- Hazard potential e.g., sum (add) the values of each card i.e., Ace = 11, King/Queen/Jack = 10 & other cards numbered
- Campaigns e.g., each week showcase specific events or cards.
- Relate cards to client or industry incidents e.g., CSB investigations [1].

Just as a pack of cards can create a variety of different games e.g., poker, blackjack, bridge etc. Process Safety Cards utilize the same playing cards elements (or 'pips') which offers an organic approach to gamification that can grow with peer or community feedback i.e., facilities, organizations or industries can develop their own games based on their technology and culture.

5 Scenario Summaries

Playing cards increase awareness of the potential LoC events, however in order to address these, an appropriate level of insight is necessary. Visualizing the LoC scenarios in bowtie format provides a clear, concise summary which is more easily understood and retained than textual or numeric data.

A one-page representation of each type of LoC event i.e., playing card, provides typical:

- Cause (Threat)
- Controls (Barrier)
- Effectiveness or Vulnerabilities (by barrier type)
- Degradation factors (by barrier type)

These are not intended to replace existing analyses and should not be considered sufficient to properly monitor & manage barriers, however they do provide memorable context with a particular focus on the Degradation Factors which create or enlarge the holes in the Swiss Cheese model and are controlled via asset and/or safety management systems e.g., competence, information, maintenance change etc.

A sample is provided in Figure 2 below:

INTERNAL CORROSION

Typical **CAUSES** (Threats)

◇ Contamination
- Process e.g. carbon dioxide, water, acids & other chemicals.
- Microbiologically Influenced Corrosion (MIC)

◇ Stress corrosion cracking
- Chlorides, sulphides, alkali & nitrates

◇ Chemical embrittlement
- Hydrogen assisted/induced or Liquid metal (mercury)

Typical **CONTROLS** (Barriers) & **VULNERABILITIES** (Degradation Factors)

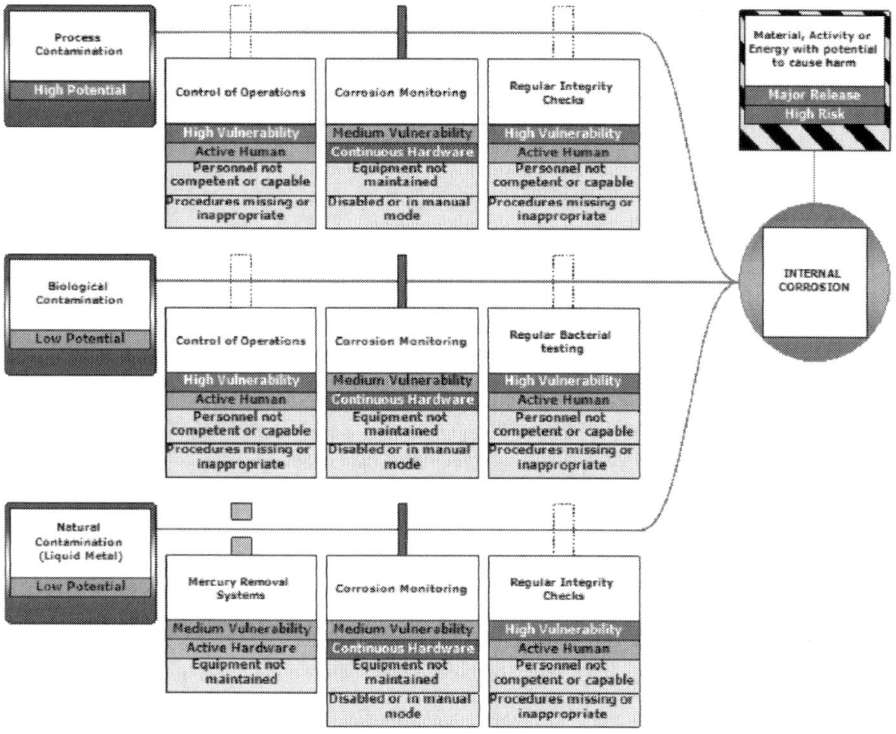

Version 2.1 Process Safety Cards © Process Safety Integrity 2020 www.ProcessSafetyCards.com 1

Figure 2 Sample Scenario Summary

Scenario summaries can be posted inside or outside populated or high traffic areas such as control rooms, canteens, workshops, or meeting rooms and rotated on a campaign basis e.g., a different LoC per week. They can also be published online e.g., intranet, SharePoint, or newsletter.

6 Online Collaboration

Although Process Safety Cards are designed to be physically handled, the current COVID-19 pandemic prevents or discourages such personal interaction, therefore digital solutions can be used to use or play the cards via collaborative platforms e.g., virtual whiteboards to connect remote or isolated team members.

7 Conclusions

Although based on gamification principles, the Process Safety Cards are not a game in themselves i.e., there are no rigid rules to follow. They simply provide a knowledge exchange toolkit to facilitate communications (in both directions) between management and frontline personnel to share learning using universal images and local languages.

These are not a new strategy (although they are a novel idea), they simply help deliver or reinforce established Process Safety (Asset Integrity) initiatives which are designed to 'Protect the Protection' i.e., look after the systems (hardware & humans) that look after you.

The cards increase awareness, the bowties enhance understanding, knowing your enemy improves vigilance which drives protection and asset integrity.

8 References

[1] U.S. Chemical Safety Board (CSB) Investigations (https://www.csb.gov/investigations/).

[2] U.S. Department of Labor, Occupational Safety & Health Administration (OSHA) Process Safety Management of Highly Hazardous Chemicals (29 CFR 1910.119) (https://www.osha.gov/process-safety-management).

[3] American Institute of Chemical Engineers (AIChE), Center for Chemical Process Safety (CCPS) Risk Based Process Safety (RBPS) Framework (https://www.aiche.org/ccps/resources/publications/summaries/summary-guidelines-risk-based-process-safety).

[4] Energy Institute High Level Framework for Process Safety Management (https://publishing.energyinst.org/topics/process-safety/leadership/high-level-framework-for-process-safety-management).

[5] UK Health & Safety Laboratory "Loss of Containment Incident Analysis" HSL/2003/07.

[6] American Institute of Chemical Engineers (AIChE), Center for Chemical Process Safety (CCPS) "Guidelines for Chemical Process Quantitative Risk Analysis".

[7] Sun Tze "The Art of War".

618386 Are Your Safety System Barriers Still Work for You?

Abstract:

Process Hazard Analysis is the tool to assess the systems in terms of the process safety scenarios and to determine the ALARP level of the risks that may arise. All the barriers in the Process Hazard Analysis is assumed to be working in the normal operations. However, how you manage to sustain the ALARP level? Do you know any deficiencies and by-pass during the operation? How do they effect the ALARP level? Are you still in the safe side?

All the barriers like Emergency Shut Down Systems, Pressure Safety Valves, Fire and Gas Detection Systems work for your system and you need to follow up in terms of any deficiencies or by-passes made on these. This process starts with reliability of these systems and continues to follow up during the operation. Emergency Safety Systems by-pass are undesired activities but during the operation there would be a need to do this. Here the important thing is how you manage the system with these by-pass and how you follow up these by-pass procedures. Industry and automation systems are improving and there are lots of ways to simultaneously cover all these changes and do the risk analysis to operate the systems safely during these by-pass times. So many companies would be using different ways of controls, in our refinery we have improved a system to follow up all the emergency safety systems with a hierarchy of control and approval automatically including the "Plan B" risk assessments for safe operation of this time periods.The important point here is the type of by-pass (conditional or non-conditional), duration of by-pass (approval hierarchy), what else do you have to operate safely, what other parameters should be followed up by the operator during the by-pass and does he know it?

These automated systems should be covered and controlled during the manual operation or by-pass time intervals in order to eliminate the human errors and not to have any kind of process related incidents. We do need to remember important Process Safety Incidents related with the by-pass of emergency shut down systems like Chernobyl. This solution and automated system would be a barrier not to have such incidents.

Effective response to a power loss event is critical to the safety and mechanical integrity of a manufacturing plant. Power failures can lead to accidents, injuries, loss of primary containment (LOPC), government recordable events, mechanical failure and loss of production. Due to the unpredictability of a power event, there is no single solution or response plan that can mitigate all of the potential issues that a power event creates.

A power event can cause alarm showers/floods which prevent the operator from seeing and reacting to high consequence alarms that could prevent a serious incident if responded to in a timely manner. Based on this, we performed investigation into the last several power failure events at our sites with a focus on the distributed control system (DCS) and its impact on response to power events. This presentation covers the findings from this investigation which include when power events occur, alarm counts/metrics during the events, the severity of each different power event, and findings during root cause investigations (RCI).

We then discuss the different tools that have been implemented into the DCS specifically to aid in operator response to a power event. Some of the tools that are covered in this presentation are DCS graphics "overview" screens for power failure events which provide one-click capability to layer 3/4 process graphics, alarm hiding/suppression concepts, bulk alarm acknowledgement/unlatching techniques, alarm silencing, and automated actions in logic/programming to build more resiliency against an event.

Development of Mobile App for Behavioral/Near Miss/PSE Reporting: Executive Summary

Ecolab, a global petro/chemical manufacturer wanted to improve their safety data reporting procedures, and create a consistent process for collecting/analyzing data.
The company also had the goal of incorporating more digital technology into their work practices and process production areas.

After a needs analysis, leadership recommended:

1. Development of a consistent Safety reporting procedure for all global plant sites to follow
2. Incorporate digital technology by use of a mobile app with shared database entry
3. Standardized analysis of safety data to quickly address trending safety events

Development of a consistent Safety reporting procedure

Global plant sites weren't required to use a consistent tool or process for safety data reporting. The project team analyzed numerous safety data reporting practices at sites, to look for consistencies and best practices. These were used to develop a consistent reporting procedure.
The procedure included types of events for reporting, as well as roles and responsibilities of employees at the plant sites. Required data would include Behavioral observations, Process Safety Events (PSE) and Near Miss Events. The overall goal of the procedure was to eliminate unsafe conditions and behaviors (leading indicators) before they escalated to more severe events (lagging indicators).

Incorporate digital technology by use of a mobile app with shared database entry

With the company goal of digital transformation, the project team looked for a solution that would eliminate paper-based practices at the sites. A mobile app solution was recommended, and pre-made options were explored. Nothing was found to meet the needs of the plant sites, so an internally developed app was recommended.
The project team used Microsoft Power Apps as a low-cost solution to develop an app in-house. The app included data entry fields for plant site, type of event, location event occurred, and many other important data points. The app was translated into several languages.
Safety events could be reported either using a mobile device, or on computer terminals located throughout the plant site. All reported safety events were downloaded from the app into a shared database for easy data analysis, and easy access by many different plant sites.

Standardized analysis of safety data to quickly address trending safety events

Part of the reporting procedure included roles and responsibilities for analyzing safety data reports. Safety data was analyzed to identify trends. Specific recommendations/actions were made to eliminate the unsafe conditions/behaviors found in the safety data trends.
PowerBI was used to quickly identify safety data trends during the analysis process. Trends were communicated through the PowerBI dashboard by using graphs and other easily interpreted visual elements.

Project Pilot

The pilot consisted of a 90 day "Safety Challenge" where a specific group of plant sites reported as many near miss and PSE observations as they could identify. Trends in the data were tracked through system generated reports, and displayed using a PowerBI dashboard. Plant site safety leaders immediately saw the benefits of a standardized safety reporting procedure, and ease of use with the mobile app.

The pilot has expanded to include additional plant sites in various countries. This has created a new set of challenges for deploying the Safety reporting procedure, expanding use of the app, translations, and identifying best practices for mobile devices. The pilot is set to complete in early 2021

617938 Using Technology to Improve Safety

Abstract:

As the world grows and changes, our business processes evolve, and we must change to meet our new needs. Recent technologies like smart phones, tablets, computers, smart televisions, smart homes, and applications that can manage every aspect of daily life are now widely available in any employee's personal life. These devices and programs reduce the time needed to do routine tasks, providing new opportunities to focus on things the individual may feel are more important or fun. In some cases, these processes learn from previous transactions and tasks and require less of the individual's input as it continues to operate.

However, in a person's professional life, it is common to have a significant portion of one's day composed of filling out paperwork and performing hands-on tasks. Although work instructions, procedures, and policies for everyday tasks are now available electronically, a field employee needs access to a computer to view, access and print documents. This limits the productivity and effectiveness of employees since it forces them to access all information at a single access point or having them carry around paperwork that can easily be damaged by work conditions and the environment. One limitation is unexpected equipment failure. In such cases, it is rare that employee would have all technical information readily available to help troubleshoot in the field immediately, except for minimal visual HazCom labels and warnings. This decreases overall process efficiency by inadvertently creating restraints – specifically, having to go back and collect all required paperwork- on the employee which can be overcome with everyday technology.

Understanding how so much information is instantly available outside of work is important to replicate that level of access and ease in the workplace. At Rain Carbon Inc., our mission is to be "a progressive industrial company committed to collaborative science-driven innovations that result in safe, reliable and efficient operations to ensure outstanding results for our customers, partners, employees and investors." With this as our guide, a new system was implemented to target several areas: safety, reliability, information availability and efficiency.

At our Candiac advanced materials site in Quebec, Canada, integrative software was implemented in 2017 as a pathway to achieving the company's mission. This software provides instant access to process information through phones, tablets and computers, including work instructions, policies, procedures, ability to complete round check sheets and work permits, action assignment and tracking, picture and video capability, report unsafe conditions and incidents, equipment malfunctions, and the management of workforce skills. The capabilities provided by the software have proven to be most effective when used on mobile devices as it allows for information to be accessed directly from any location within the plant helping us improve in all the below points:

- Enhancing site management by providing a live view of notices, work requests and unsafe conditions to immediately communicate and respond to the situation at hand;
- Increasing employee involvement in safety due to higher visibility and lower response times to address unsafe acts/conditions;
- Reducing redundancies of having to go back and forth between paper and electronic work; and
- Decreasing unplanned downtime by providing all relevant technical information instantaneously enabling employees to address deficiencies without delay.

Rain Carbon has pushed forward with innovation and technology by implementing QR codes for all equipment linked to the integrative software solution. This adds a significant amount of efficiency to work processes on the shop floor. When unplanned work is to be performed, the employee can take any mobile device with the software installed on it and scan the QR code on the respective equipment. This will take the employee to a screen with all necessary information regarding the respective equipment, including the work permit and list of materials needed to perform the work in question. A process that would, in most cases, take a matter of hours is reduced to a few minutes and a couple of taps.

Although reliability and efficiency have increased since the system was implemented at the Candiac facility, the most significant improvement has been observed in safety. Prior to implementing this software, safety walks were completed in an electronic or printed version of a Word document, safety incident notifications would be accomplished through phone calls and emails, and near misses would be recorded in a separate system. The centralization of reporting and the enhanced information availability highly encouraged employees to report all safety items with the backing of a more engaged management group, thus enhancing employee safety. Reporting increased with the introduction of the program by 772% on the first year and by 2,150% on the second year. Similarly, during that timeframe, near misses decreased by 40% and first aid cases by 33% compared with the current year.

Results:

From 2017 to 2019, there was a clear increase in reporting after the implementation of the software. Through management support and commitment, we can see a clear decline from 2019 to 2020 YTD of overall identified items, increasing the quality of work and safety across the site.

Implemented in 2017, the program has significantly improved the safety performance of Rain Carbon's Candiac site by engaging management and ensuring all safety items are properly communicated and addressed with as little downtime as possible.

617740 Facility Siting Revalidation - a New Tool

Abstract:

Facility siting studies require periodic review at specified intervals during the life of a facility. As stated in 29 CFR Part 1910.119 (e) (6), revalidation is required every five years based on the previous full study completion date. While revalidation is a regulatory requirement, little is often known about the process of undertaking the initial study or review and, in some cases, this process can become challenging.

A new and simple revalidation tool is available that helps operators determine if their facility siting study needs a complete redo, a retrofit, an update, or a simple revalidation. This paper introduces and explores this simple revalidation tool, providing examples of how it can be used in a plant environment.

617756 Quantifying the Conditional Dependencies between Layers of Protections in Event Tree Analysis

Abstract:

An approach to include conditional probabilities when calculating probabilities of disaster in quantitative risk analysis is presented.[1] Since not all layers of protection are independent of each other, using conditional probabilities leads to more accurate calculations of probabilities of disaster in a chemical process. Based on event tree analysis, probabilities of disaster are calculated assuming that events are statistically independent (independence assumption), and separately considering that events affect each other (conditional probabilities), in three cases with increasing complexity and number of layers of protection. First, a simple alarm – operator case is studied, then two cases involving a chemical reactor and operator with increasing number of layers of protection are analyzed. For all cases, we considered multiple scenarios to show that results do not depend on the input probabilities of particular events. For the latter two cases, the input probabilities of each event were randomly sampled with Monte Carlo simulations. Probabilities of disaster calculated with the assumption of statistical independence were compared to probabilities of disaster calculated with conditional probabilities. In most scenarios of the three cases, findings show that the probabilities of disaster calculated using the independence assumption are lower than those with conditional probabilities, which can lead to undercounting the layers of protection that are needed. Process engineers need to be aware that assuming that events are statistically independent can lead to the possibility of undercounting of layers of protection needed for a chemical process.

Reference: Mott TC, Kivistik PM, Panorska AK, Cantu DC. Comparing conditional probabilities and statistical independence in layers of protection analysis. Proc Safety Prog. 2020;e12215. https://doi.org/10.1002/prs.12215

617799 Comparison of Methodologies for Dropped Objects Analysis in Shallow Water Drilling Rigs in the Gulf of Mexico.

Abstract:

The reform of the energy sector undertaken in 2013 by the Government of Mexico to empower foreign investment in the hydrocarbon sector led to the arrival of private companies with new projects and technology to the Gulf of Mexico. The projects managed in this region move continuously from the exploration stage to the production stage. During this stage transition, Offshore Projects require development of safety evaluations for the design of the platforms that were not common in the Mexican market; among these evaluations there is "Dropped Objects Analysis".

The accidental dropping of objects is considered among the main risks on offshore production platforms since its occurrence can affect personnel (injuries and / or deaths) and damage the facilities (causing spills and / or fires). Thus, as part of engineering design development, there is a need to assess the potential risk associated with impact from oscillating objects and loads that could fall during routine and non-routine lifting activities.

The analysis focuses on the possible damage to the area of operation of a given platform, structures, process equipment and subsea pipelines, evaluating the frequency of dropping combined with the type of damage that would result from the impact. However, for Dropped Object Analysis there are different methods that involve complex calculations, including solving integral equations, which for practical purposes can result in a large time investment.

This work addresses the use of a numerical method that allows a complex calculation routine to be carried out towards a solution with simple arithmetic, thus facilitating the evaluation of the frequency and impact of dropped objects on the platform and reducing the solving time.

As an application case, the Dropped Objects Analysis is presented in two offshore platforms for the exploitation and transportation of hydrocarbons installed in southeast Mexico by a private operator. During the study, the frequency and consequence of dropped objects were evaluated to issue protection recommendations in the design of both platforms and were also compared with the use of traditional calculation methods.

617993 CFD-Based Ventilation Analysis - Basis for Conducting Consequence and/or Risk-Based Assessments

Abstract:

The combination of powerful capabilities in the development of Computer-Aided Design (CAD) drawings; i.e., 3D models; and the performance of formidable computer processors confirm the execution of robust Computational Fluid Dynamics - CFD-based simulations in manageable short-duration periods of time; i.e., minutes. Using CFD tools allow considering the fluid-geometry interaction, key factor for identifying specific features of the site section being analyzed. This information is not possible to be foreseen when using screening tools based on experimental data correlated and/or phenomenological methods.

The present paper illustrates a case study that provides examples of how a CFD-based ventilation analysis can provide insight information of areas that require special attention when a consequence or risk-based assessment is to be performed. The identification of regions with air recirculation, stagnant flow areas with low wind speeds, congestion volumes, etc. verify leak locations and orientations to be evaluated when performing advanced dispersion, fire, and explosion analysis.

SPRING21
+17TH GCPS
A Joint AIChE and CCPS Meeting

SYSTEMATIC APPROACH TO PROCESS SAFETY MANAGEMENT ON ROAD LPG TRANSPORTATION

Presenter/Principal Author
Américo Diniz Carvalho Neto
RSE CONSULTING
Brazil
americodcn@rsem.com.br

Other Author
Adriano Nonato
RSE CONSULTING
Brazil
ext.adriano.nonato@rsem.com.br

Prepared for Presentation at
American Institute of Chemical Engineers
2021 Spring Meeting and 17th Global Congress on Process Safety
Virtual
April 18 - 22, 2021

SYSTEMATIC APPROACH TO PROCESS SAFETY MANAGEMENT ON ROAD LPG TRANSPORTATION

Presenter/Principal Author
Américo Diniz Carvalho Neto
RSE CONSULTING
Brazil
americodcn@rsem.com.br

Other Author(s)
Adriano Nonato
RSE CONSULTING

Keywords: LPG transportation Riks; PSM in LPG distribution, LPG Transport

Abstract

Process Safety Management is systematically adopted in sectors of Brazilian industry, such as oil and gas and petrochemicals, with clear operating safety gains, based on performance indicators. In common, they all use fixed structures and are siting centered to carry out their operations. Also important in the energy chain in Brazil, bulk LPG transportation supplies industrial chain, and business and residences as well. With all this importance requirements for robust process safety management are still missing in LPG distribution sector.

This work aims to implement a safety management system within a company currently responsible for the largest share of the bulk LPG market, considering the particularities of this type of operation, totally dependent on road transport for LPG distribution. To achieve this the first step was identifying the main activities that represent the bulk system's operation, involving bobtail trucks, as follows: supply through road routes; unloading, which is the supply to the customer and the operation of the bobtail truck itself. The strategy of implementing the safety management system considered sequential application of the following steps: 1) Reassessment of the bobtail truck project, aligning it with the latest technological solutions; 2) Conducting a hazard identification study for supply and unloading activities, proposing a customized hazard identification model for supply; 3) barriers auditing to evaluate unloading components reliability; 4) Auditing operating activities based on CCPS RBPS in order to identify existing gaps. The fifth and final step is the definition of the Process Safety Management approach aligned with road transportation best practices worldwide and in compliance with legal and corporate requirements.

The studies already developed demonstrate that a careful analysis of the main normative reference identified gaps in the bobtail project, with emphasis on passive safety systems, absent in the original project. In the supply risk assessment stage (routes assessment), a specific model for route risk analysis was developed using a customized Excel spreadsheet for risk analysis through predefined menus. This model reduced the assessment time required and the level of subjectivity, allowing the company to develop hundreds of analyzes with the use of video conferencing tools, essential in this case, due to its nationwide scope.

This is a work in progress and the next steps, when completed, will make the company achieve a degree of excellence in Process Safety Management through monitoring for the implementation of all RBPS elements that have been identified as absent or deficient. Many other industrial sectors that carry out transport of dangerous products can benefit from this methodology for implementing RBPS, through the use of the described models, adapting them to the specific modal.

Introduction

It is recognized that chemical transportation risk management is more than just a service; but rather is a wide-ranging system of activities that involves numerous stakeholders with varying backgrounds and interests. To effectively manage transportation risk the entire supply chain, spanning the delivery and handling of raw materials through the offloading and unpacking of finished products, needs to be included and evaluated. In addition to safety and business continuity, the evolving threat of security concerns are now included as part of a comprehensive transportation risk management program. Given the complexity and global reach of many transportation networks, as well as the tradeoffs between safety and security, the management of transportation risk can challenge even the best business plans and programs.

Considering all these aspects and risks involved in LPG transportation, RSE Consulting developed a PSM project with one of the biggest gas distribution companies in Brazil, o order to develop a diagnostic, and create a PSM system to be implemented in the company.

1 Bulk LPG transportation risk assessment

This stage aimed to evaluate a pilot project of adequacy of LPG transport vehicle. The use of common risk analysis methodologies and verification of compliance with international benchmarks were applied.

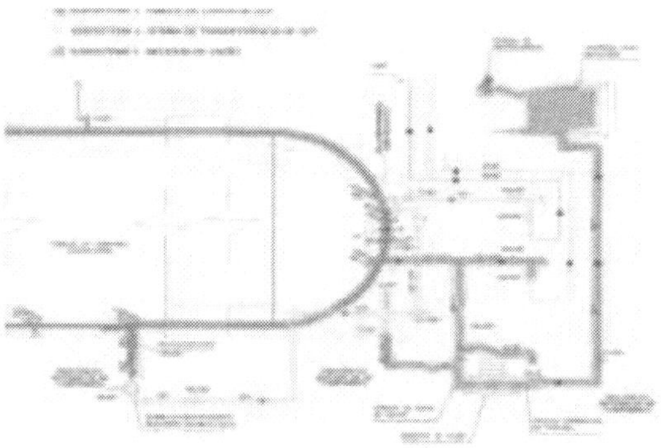

A HAZOP of the system was carried out to identify operational weaknesses

A revised project was obtained considering improvements in management and safety elements, some highlighted below:

Include the identified critical components into the inspection and maintenance plan (Bobtail). Ensure that inspection and maintenance plans have effective tests and tasks for identified and potential damage mechanisms.

Conduct training with operators on the new operating demands (Bobtail), reinforcing the concept of critical components, defining the points that should be evaluated during inspection routines, in order to ensure that the trailers transporting LPG meet the appropriate conservation conditions in relation to the Preventive Barrier of Periodic Inspection of Trailers.

Ensure that truck fleet will have a supply pump interlock system when the minimum level of 5 % of the volume of the LPG tank is reached, to avoid cavitation of the pump.

Every operational unit across the country should comply with managerial based actions concerning standardized execution of maintenance and inspection of critical components.

Include as managerial based actions the adoption of indicators for assets of integrity and reliability as stated on API 754

Define a maximum and minimum speed range to be adopted during the hose rewinding activity on the reel in order to ensure operation compatible with operator safety.

Ensure that a relief valve is installed on the swivel joint. This valve has the function of relieving pressure due to LPG expansion that can occur between two shutoff valves. Ensure that its construction and pressure class meet NFPA 58 requirements

Evaluate the feasibility of deploying a portable remote device for activate emergency activation system.

Modify emergency activation system so that the entire system is shut down or otherwise enforced to fail safe state (pump, pneumatic valves and engine and truck electrical system).

Reevaluate the removal of LPG liquid supply line check valve, which, as designed, will be replaced by a pneumatic-driven shutoff valve.

2. EVALUATION OF TRANSPORTATION ROUTE

The company under assessment supplies LPG for the vast majority of brazilian states

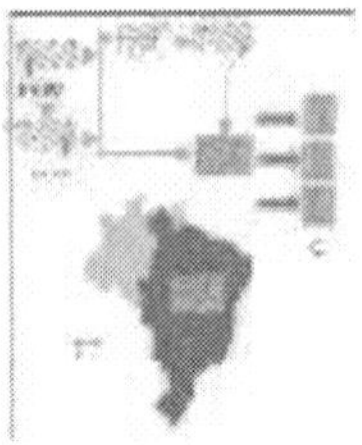

Process safety management on this company is still on its infancy ant still being circumscribed to the staff on Brazilian southwest headquarter. The staff engaged on conducting the evaluation had to overcome the lack of any previous knowledgeabout Risk Analysis of the operating team further hampered by mobility limitations imposed by the pandemic.

A method was developed for route risk analysis as recommended by the CCPS.

All relevant attributes that impacted the risk of the routes were identified and assigned weights for them.

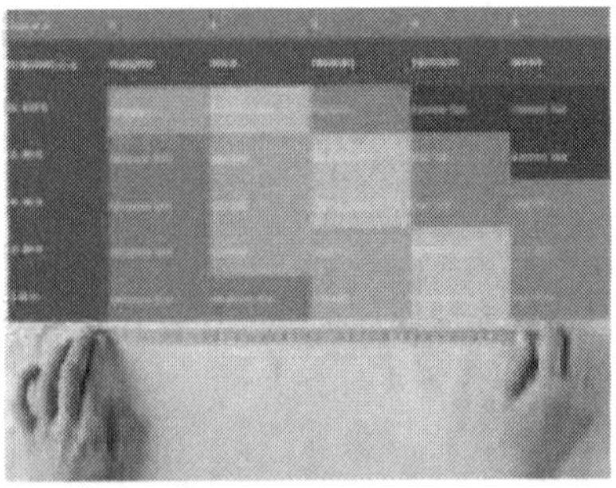

The weights have been calibrated to ensure consistency with the normative risk matrix.

An Excel spreadsheet was developed to assess risk, deploying the afore mentioned risk attributes, allowing teams to engage with maximum objectivity and time savings. Another advantage of the spreadsheet was its smooth learning curve, enabling the use for different teams of the company and possibility of future improvements to evaluate management elements

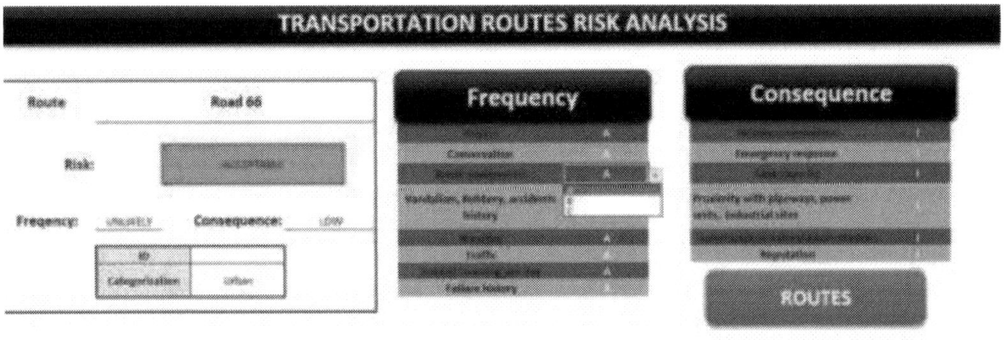

A total of 22 routes were analyzed. It is expected that teams can perform a much higher number of analyses with a minimum of tutoring, given the smooth learning curve for using the tool. Improvement points "appear" from the analysis of the result graphs, allowing recommendations to be properly addressed, as follow:

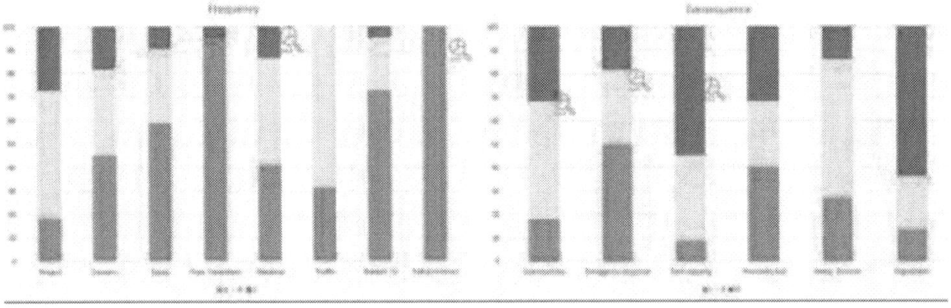

3. CUSTOMER RISK ANALYSIS

The 3rd stage of the project (in progress) will be the application of a bow tie diagram to the customer facilities.

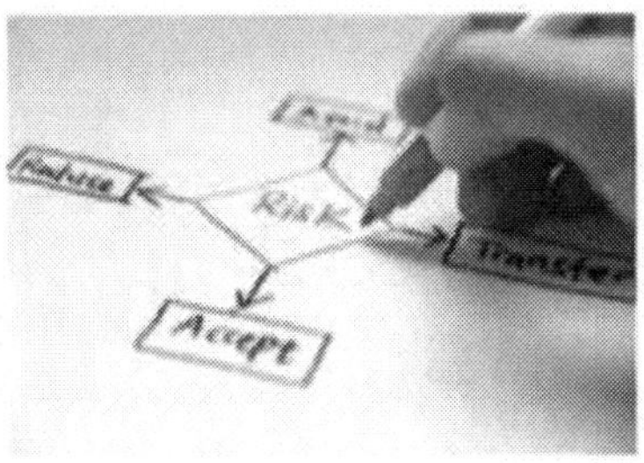

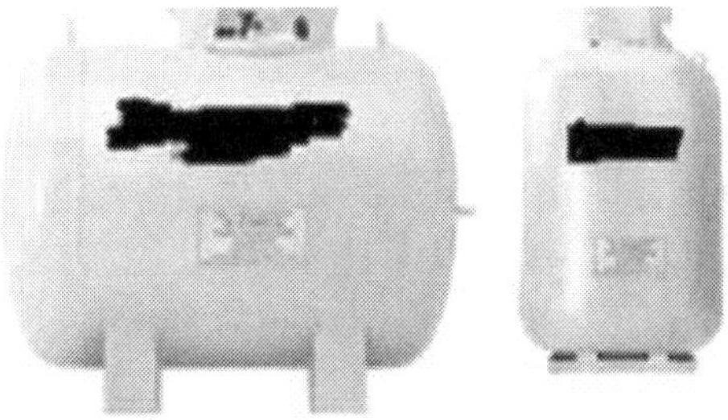

The bow tie will facilitate the process of identifying barriers for subsequent analysis.

4. BARRIER AUDIT

Step to be performed as follows below:

- Definition of critical risk scenarios for the preparation of protocol d andaudit;
- Preparation of audit protocols of critical scenarios and preventive and mitigating barriers;
- Audit of barriers / critical controls of the scenarios defined in the 1st step
- The systems to be analyzed are those of customer facilities. CCPS Bow Tie Approach was the chosen method.

5. IMPLEMENTATION OF PSM

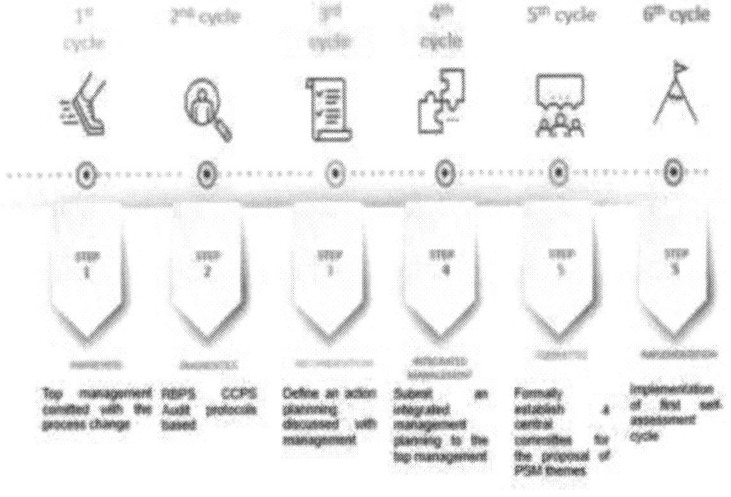

The PSM system implementation strategy is structured in the following steps:

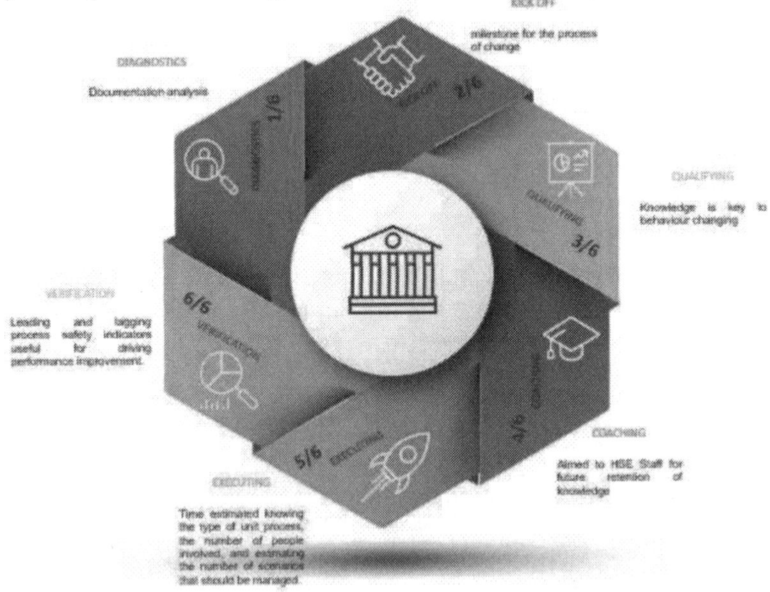

Conclusion

The area of transport of dangerous products, specifically the transport of LPG in large cities, presents many positive contributions to society, however this logistical process has many risk scenarios with high severity in all stages, from loading to the process bases, distribution of the LPG via tank trucks, and refueling at customers' different facilities. This Project allows the organization to not only make a technical diagnosis of the risks of the entire process, but also to set up a Risk Management System that, through the implementation of the requirements of the Management System and procedures, will be allowed to achieve a high performance with the elimination of leakages. gas with the potential for catastrophic accidents.

2 References (examples given to show format).

[1] J.A. Klein and B.K. Vaughen. “Implement an Operational Discipline Program to Improve Process Plant Safety,” June 2011.

[2] CCPS. *Guidelines for Chemical Transportation, Safety, Security and Risk Management.* Center for Chemical Process Safety, American Institute of Chemical Engineers, New York, NY, 1995.

[3] RSE. RSE Consulting procedure for Risk transportation qualitative and quantitative risk analysis, 2019. Developed by Professor Americo Diniz Carvalho Neto.

618466 The Evolution of Emergency & Crisis Management in Petronas

Abstract:

Managing emergency and crisis effectively is a paramount aspect in Process Safety Management, especially in the complexity of oil and gas business. Optimum readiness is aimed to efficiently address potential impacts arising from incidents; immaterial of size; to contain the extent of exposure as quickly as possible, with the primary intent of saving lives and protecting the environment. Often we pay too much attention on the left hand-side of the bowtie (control barriers) that we often overlook the right hand-side of the bowtie (recovery barriers). The objective of this paper is to share the evolution of emergency and crisis management underwent by PETRONAS, with the ultimate aim to ensure our interests (people, environment and asset), and reputation are protected.

PETRONAS operates globally in more than 50 countries, hence effective establishment and implementation of emergency and crisis plans are vital. Historically, emergency and crisis management and preparedness has been a reactive science. We had our own share of incidents, and multiple lessons learnt had been consolidated to further improve our systems and barriers. The integration of emergency and crisis management as part of the HSE Management System (HSE MS) and Mandatory Control Framework (MCF) built a strong foundation and governance, and ensures continual improvement, taking into account all the internal and external lessons learnt. Conducting and monitoring the number of emergency drills and exercises, specifically based on the Major Accident Hazard (MAH), is imperative to ensure effective response in case of real emergency and crisis. Learning from real cases raised the vital need of coordinated response and putting the rights interventions and controls in place to minimize the impact of incident. Effective emergency and crisis management is crucial to support process safety implementation throughout the asset life-cycle covering design, operation and maintenance stage.

Throughout the years, we are also continuously enhancing our capability in terms of resources and people. In a nutshell, this paper will talk about the evolution of emergency and crisis management in PETRONAS, where it started with system establishment and capability building, whilst moving towards digitalization to better manage emergency and crisis. This paper will share some real incident cases and how we learn from these incidents to further intensify our emergency and crisis management.

617018 Pressure Relief System Force Balance Assessment – a Review of Its Effectiveness

Abstract:

Design documentation requirements for spring-loaded pressure relief systems are very detailed, based on the definition in API Standard 521, 6th edition. The Installation Documentation section of API Standard 521 recommends provision of inlet and outlet piping pressure drop calculations as part of a complete analysis.

Inlet pressure drop is an important factor, since it is considered that excessive inlet pressure drop will affect pressure relief valve performance, and potentially result in chatter which may damage the valve and piping. In situations where inlet pressure drop exceeds 3% of set pressure for spring loaded pressure relief valves, previous editions of API Standard 520 Part 2 have allowed for an "engineering analysis" to be conducted for these systems - however the content of the engineering analysis has never been well defined.

The latest version of API Standard 520 Part II provides a more detailed description of an engineering analysis and allows a Force Balance Assessment to be conducted for spring loaded pressure relief valves exceeding 3% inlet pressure drop. This method can predict valve stability during relief flow, and identify whether chatter, flutter or stable flow is likely to occur. Valve characteristics such as valve opening and closing times are critical input parameters for this analysis. If stable operation can be demonstrated, then expensive mitigation efforts, such as modifications to piping or the valve itself, can be avoided.

This paper presents an overview of the force balance assessment in a step-by-step manner. Additionally, a sample of spring-loaded pressure relief systems which exceed the 3% inlet pressure drop criteria is reviewed. The aim of this review is to determine the effectiveness of conducting a force balance assessment, and to identify which systems show stable operation, and which systems were identified as likely to chatter.

618390 Chemicals Management inside a Refinery - Does Everyone Knows the Hazards of the Chemicals on Site?

Abstract:

Chemicals are widely used in Oil&Gas industries. Each of them has different phsyical and chemical properties and these differences make it hard to store and use them in a proper way.

In the refineries, those widely used chemicals require a good storage and transportation standards. In order to create a safe environment for those areas, Tüpraş has created its own "Chemical Interaction Matrix". In this matrix, various aspects of chemicals are handled and each Unit has its own matrix.

As a starting point, each of the chemicals are checked as per their stability and reactivity properties from their safety data sheets. The chemicals that do not have detailed information are reacted to each other in the laboratory in order to see if there is a dangerous reaction.

Chemicals that may cause a dangerous reaction to people and environment are stored seperately and the properties of each are shown in the matrixes. These matrixes are hanged on the unit boards so that everyone can see and check if the storage is provided correctly.

It is important to keep these matrixes up-to-date. The matrix is updated when a new chemical arrives in the unit and everybody is informed about update. The update and information is achieved wit the actions that we tracked via our MoC (Management of Change) system.

617653 Operational Risks Involved on Integrating Flare Gas to Fuel System

Abstract:

Flaring is a safe industrial process to dispose waste or excess gas from a plant. But, it has limitations in energy conservation and environmental loss. Flare gas integration to cogeneration plant benefits from utilizing high calorific value waste gases as supplemental fuel to cogeneration plant or gas turbines. Earlier studies on flare gas integration have shown environmental incentive and economic viability. However, the impact of flare gas quality on the operation of cogeneration plants are yet to be identified.

This study delves into the operational hazards and risks arising from the variability of flare gas stream. A novel framework is proposed which can be applied to new or integrated processes which has inadequate information on its operational performance and associated hazards. The study identifies the technical hazards through design data collection, process simulation and sensitivity analysis. Followed by a systematic approach to evaluate the critical operational event occurrences and their impacts through scenario development and quantitative risk assessment.

An important outcome from this study is the identification of critical flare-fuel gas parameters affecting the cogeneration operation through process simulation. High hydrogen content and rich fuel-air ratio in the boiler can affect the boiler operating conditions. Increase in the hydrogen content in fuel system can increase the boiler gas temperature, combustion mixture and flame stability inside the firebox. Risk analysis through Bayesian Network showed a significant risk escalation. With a presumed frequency of 12 hrs/year of high and medium hydrogen gas flaring, there is a substantial rise in the occurrence of boiler gas temperature exceeding design limit and rich fuel mixture formation in the firebox. These top events has major influence on the operational time as well as loss control events - flame impingement and tube rupture.

618388 Process Hazard Analysis Methodologies and Refinery Applications

Abstract:

Process Safety Management System starts with managing the process related risks and from the first step of establishing a new process unit that is needed to evaluate the risks and define the necessary measurements. Process Hazard Analysis methodologies are the systematic tools which help to evaluate the risks in a process. For different needs there are several methodologies like DOW FETI, HAZOP, What if, HAZID, Fault Tree and Event Tree Analysis, Scenario Modelling, Facility Siting etc. Some of these methodologies are semi quantitative some are quantitative methodologies. In order to define and prevent major industrial accidents in our refinery we use DOW FETI, HAZOP, FTA and ETA and Scenario modelling. Moreover for some kind of projects we may also apply What if and HAZID analysis instead of HAZOP depending on the complexity. Facility Siting is another tool for risk related studies which is used to evaluate the risks for buildings in order to protect people inside the buildings.

The workflow for a full unit PHA study starts with DOW FETI calculation which is used for risk ranking of al static equipment inside the unit. DOW FETI calculation is done in accordance with the equipment temperature, pressure, hold-up amounts, components and also reaction types if there is any. Then risk assessment process continues with HAZOP studies which is for scenario identification. In HAZOP, first step is chopping the unit into small parts to be evaluated which is called nodes. After that all nodes are evaluated using P&IDs and PFDs by keywords like temperature, pressure, flow, level etc. HAZOP helps to understand what kind of scenarios are possible inside the unit and these critical scenarios are taken through FTA and ETA studies. HAZOP is a semi quantitative risk assessment method where FTA and ETA are quantitative methods. FTA is applied for critical scenarios in order to find out the probability of top event then ETA is then applied in order to find out the probability of scenarios like fire, explosion, toxic dispersion etc. caused by this top event. FTA and ETA analysis all use the statistical PFD and failure rate values of all initiating events and barriers defined in HAZOP for the related scenario.

Scenario Modelling is the following step which is used for understanding the effect of the probable scenario. The most probable scenarios coming from ETA is modelled to see the effect zones. These effect zones are used to plan Emergency Response Studies for all these potential major accidents.

Facility Siting is another check point in a unit PHA study all potential scenario modelling results are used to evaluate the risks for the buildings and also if there are any in the effect zones of fire, explosion or toxic release then there should be actions defined.

619040 Identification of the Process Hazards By Way of Technical Integration:

Abstract:

Undisclosed threats can lead to **death** and **devastation** in the process industry. Right and constructive hazard recognition supports security procedures, employment health and risk control techniques. It thereby is essential to the health and security of all company employees and to effective business practices.

Without understanding of process hazards, risk cannot be controlled and risk assessment is compulsory for incident prevention. Managers want their workers to be able to recognize not only apparent but evolving threats effectively. Risk assessment needs various **strategies**, **skilled** and testing teams to identify risks and create obstacles to mitigation in their operating environment. Different approaches can enhance the detection of hazards to the risk assessment process.

This paper explains the approach used in the implementation of combined **traditional** and **non-conventional**, **qualitative** and **quantitative** approaches, to identify process hazards, risk assessment and risk management. An insight on the non-conventional techniques *(Temporary repairs reduction & effective permit to work system)* will enable conference audience from all sectors of process industry, in developing integrated process hazard identification and evaluation

617089 Keep the Lion inside the Cage

Abstract:

Keep the Lion inside the Cage

All Chemicals, that we utilize to produce our final Products, are considered as Lions. We have to maintain their healthiness in good condition & under control inside the Plant (Cage) elements (Pipe, Equipment, Vessel, Tank). Process Safety Engineer's Career is an adventure where all safeguards, Barriers & mitigations shall be designed, constructed & operated safely in order to achieve the Process Safety ultimate Goal: "Prevent Loss of Containment". When the chemical leaks, the Lion is free to have high Consequences. With integration for all Reliability, Process Safety & Asset Integrity, we can keep the Lion inside the Cage & eliminate all Probability to have any Consequence whatever its Risk Level.

In order to achieve our ultimate goal, several aspects shall be managed & steward-shipped by the own Organization. There are many Process Safety pillars starting from "hazards identification" till removing the Risk or reducing it to insignificant level; including The Process Safety Competency Development for all Organization levels, Process Safety Knowledge Enhancement, System Support, Technical Guidelines, Risk Management, Process Safety Performance Monitoring and end with the Emergency Preparedness & Response Capability.

The beauty of the Process Safety Concept is the alignment with the Emotional Intelligent that we feel during the dealing with Chemicals (Lions). They shall be always contained & maintained inside their containments. In the same manner of having a specific Process Safety Competency Development Plan for All members of Organization to maintain the Chemicals inside the containment, it shall be available for all members in order to have the Lions inside the Cage. Our ultimate goal is to complete the Journey in safe manner & to maintain the Healthy condition ... in order to keep "Chemicals" inside the pipe & to keep the Lions inside the cage.

617835 Beware of Blind Risk -What You Do Not Know Can Hurt You

Abstract:

This paper discusses the concept of blind risk, how it arises, its dire consequences, and how to best avoid it.

Blind risk is most often due to incomplete, inconsistent, or out-of-date asset and process safety information. Most often, these information quality issues are not obvious or visible, and as such they are a blind risk. They lie dormant, like a beartrap in the undergrowth of a forest until someone inadvertently steps on it. They become gaps in our "layers for protection" -the systems and processes we trust to stop incidents before they happen. But blind risk is also a choice, and it is a measure of a company's tolerance for risk.

Safety, engineering, capital, and operating decisions are only as good as the information they are based on. Most companies may not ***objectively*** know how good (read reliable and trustworthy) the quality of their underlying asset and process safety information is. It does not have to be this way.

There are two things you can do today that will have a significant impact now and into the future:

- Use standardized information quality metrics and tools to measure, monitor and report the true state of asset and process safety information across multiple repositories, applications, and formats. Let all stakeholders including executive management know the good, the bad and the ugly. Use this as a baseline to measure your improvement against.
- Monitor and quantify the cost of poor asset and process safety information. Use this information to justify funding for information quality improvement initiatives.

619136 Relief Systems Work Is a Money Pit: Change My Mind

Abstract:

As a relief system engineer, I revalidate and correct pressure relief system designs daily. Many times, the owner/operator views the pressure relief systems efforts as a 'bottomless money pit' that never returns value and is seemingly never complete. This presentation will dispel this notion and explain how to ensure your pressure relief systems efforts turn into cost-avoidance measures and high-value money-makers.

This three-part presentation will summarize the business purposes of pressure relief system efforts, describe how good relief systems design bases are converted into cost-avoidance measures, and finally share how targeted relief systems efforts can turn into net-positive returns for your facility.

Example content includes safety and regulatory obligations, keeping pace with industry, insurance impact, risk reduction, reputation, safe operating limit changes, and debottlenecking.

ISO 9001-2015 and Risk Mindedness in the Management of Heat Exchangers in Ammonia Synthesis Process

Festus Addo-Yobo, Adrian Lutchman, Sandra Maharaj, Devon Indar
The University of Trinidad and Tobago, Point Lisas Campus, Esperanza Street, Brechin Castle,Trinidad and Tobago.

Introduction

The ammonia synthesis process is one of the most important in the manufacturing Industry because ammonia is a feedstock for diverse products including fertilizer. This process is an interconnected system of production units which include: generation of hydrogen, separation of air to produce nitrogen, ammonia synthesis using catalysts, ammonia condensation and recycling. In the twenty first century efforts were concentrated on the optimizing of the processes either independently or as groups of processes providing inputs to various stages of the processing plant (CEP 2015). As integrated design has been largely successful in optimizing the production plant that is delivered to the producer, attention is now focussed on running the plant so as to optimize its performance. Plant safety, operational efficiency, effectiveness of operation, utilization of resources and productivity are desired outcomes. To achieve these ends, Process Control Engineers must necessarily have conversations with custodians of Quality Management Systems (QMS) in order to optimize the plant performance at a higher management level.

Bayer Technology Services offer Online Optimization & Operational Excellence solutions . Others utilize the TQM and Six Sigma quality assurance models. Nowadays, most plants are run by Distributed Control Systems (DCS) and have modelling software such HYSYS for carrying out such tasks. Thus the opportunity to optimize plant performance online has become feasible. BASF Services have teamed up with Haldor Topsoe to offer Ammonia producers new optimization opportunities online (Process Worldwide 2020).

Ammonia plants have optimized energy recovery through design. Waste heat in the process stream exiting the front end, exiting the secondary reformer, is recovered via heat exchange to produce high pressure steam. Similarly the heat produced by the highly exothermic conversion reaction in the backend is also recovered to produce high pressure steam. The heat exchangers used in these sections (shown in red boxes in the adjacent figure) operate under high temperatures (>400 ºC). They thus are susceptible to failure mechanisms not usually encountered in most petrochemical operations.

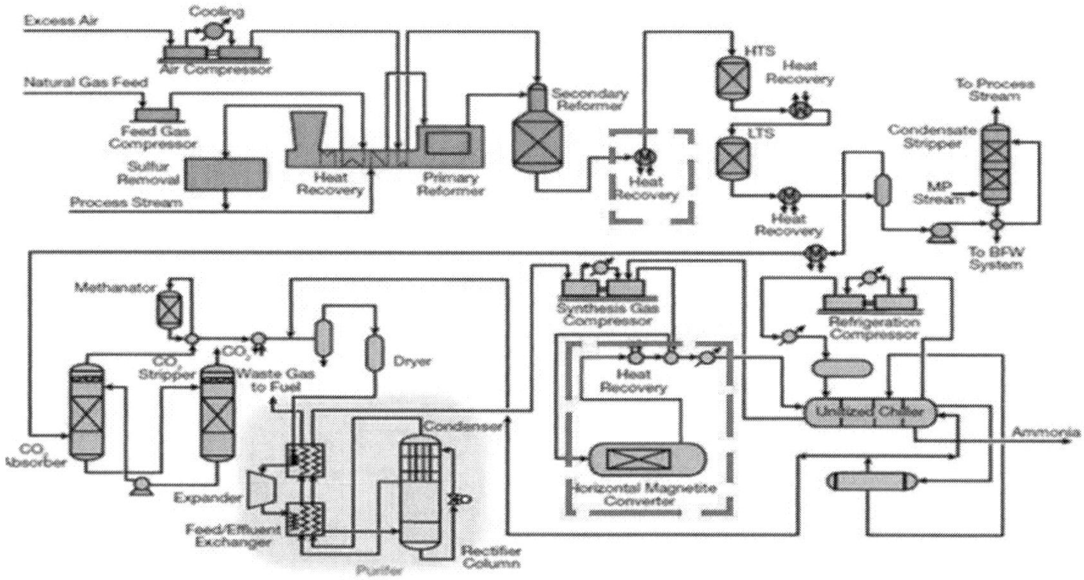

The Quality Management System

An important development in the standard to which quality management systems for production processes must conform is the revision of the ISO 9001 Standard to the 2015 version which makes it a requirement to address risk in all aspects of the QMS and provide anticipatory remedial actions. The standard places the responsibility on top management. Addressing risk is identified in the Standard as having the potential to reduce cost and also providing opportunities. One important way to operationalize risk-mindedness as advocated in the standard is to model the plant performance or the performance of relevant components and investigate scenarios to identify the risks. Heat exchanger tube thinning In the nuclear power generation industry is well researched and reported (Gao et al – 2017, Jonas – 2004, Snachez-Carrera – 1984). Remedial action in the nuclear industry includes scheduled inspection and replacement of heat exchanger tubes prior to their rupture according to protocols using Computer Programmes e,g,CHECWORKS^{TM} (Smith & Horowitz – 2014). In the Ammonia Industry, non-destructive testing can identify tubes that have thinned significantly, which are then blocked off as a preventive measure to avoid water leaking from them to the process stream.

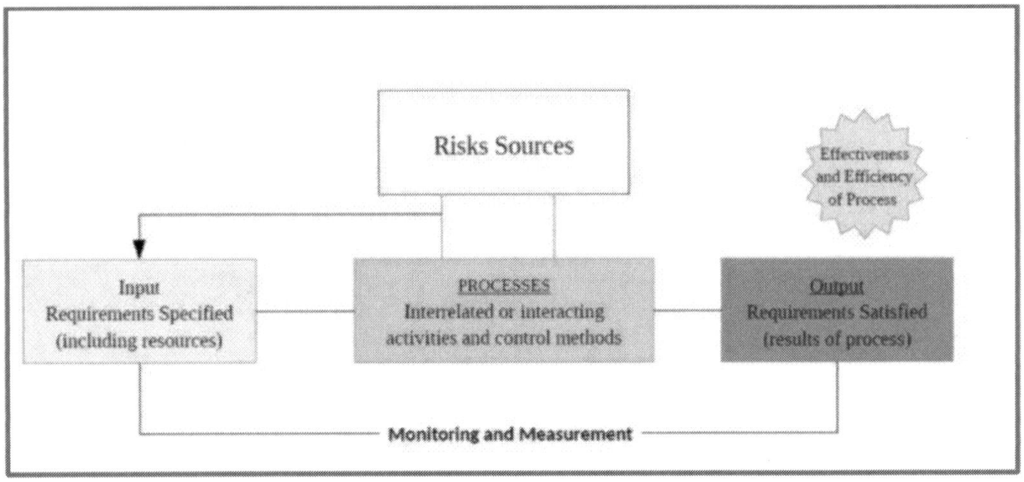

Juxtaposing the great effort and successes in ammonia process optimization to the experience of plant operators, demonstrates the existence of a gap. Even when plants are controlled at desired set points, heat exchanger tubes can and do fail leading to days and weeks of unplanned downtime. Given the high throughput of the plant (600 tonnes/day), this represents significant reduction in plant productivity. Such failures are risks that can be expected, though the events cannot be predicted with accuracy. Heat exchanger networks form an important part of the ammonia manufacturing process. Therefore, these risks should be managed by modelling the operation of the networks. In this work we use HYSYS to model the performance of the heat exchanger network used in typical Ammonia Plant (Chemical Engineering Progress 2015) before and after heat exchanger tubes are blocked off. We investigate numerically, the influence using Boiler Feed Water (BFW) to manage four of the interacting process variables which influence the risk of tube failure: process gas temperature, the velocity of the steam-water mixture in the tubes, the average temperature and the quality of the steam-water mixture in the boiler tubes.

Identification of Risk Factors

The mechanism of platelet erosion has been identified as leading to the thinning of ductile stainless such as the high carbon steel used in the nuclear industry. Abrasive particulates scrape the surface and take off the protective oxide which is swept away by the current of fluid flowing in the tubes, leading to further corrosion. This is known as flow assisted corrosion. Levy (1986) shows that above about 250ºC, temperature speeds up tube thinning process exponentially. Other factors which contribute to Flow Accelerated Corrosion (FAC) are the velocity of the particle, their hardness, the shape and the impinging angle. The geometry of the pipeworks is very important because bends and fittings serve as targets for high velocity dislodged protective oxide.

The HT exchanger tubes which carry BFW for cooling process gases are made of Stainless steel 304L whose mass fraction of chromium are as high as 18-19%. Chromium forms an oxide layer up to 300nm which protects the steel. Should this layer be removed, chromium would react with dissolved oxygen in the feed water to form Chromium Oxide or an even more protective layer comprising a solid solution of Fe, Ni, Cr oxide. Even though the BFW is purified by ion-exchanged to remove chlorides, the possibility of chloride contamination by downstream cooling seawater cannot be ruled out. Chromium can be leached from the protective layer by presence of alkali chloride, which could react with oxygen to release chlorine which further attack chromium oxide and leach out the chromium at temperatures above 500ºC. At lower temperatures, it is postulated that the chloride could react with the oxygen and chromium oxide according the eqn below.

Cr_2O_3 (s) +4KCl(s/l)+$2H_2O$ +1/2 $O_2 \leftrightarrow 2K_2CrO_3$ (s/l) + 4HCl.

The chromate formed would either remain in solid phase in the absence of liquid water or dissolve in the hot BFW water and be leached away. In either case, no contiguous layer of protective chromium oxide layer would remain and the iron underneath would rust. This mechanism would lead to thinning. Thus, the risk factors are: the contamination of the BFW with chloride, high temperatures and the oxygen levels in the BFW. The steam weight would also constitute a risk factor. The presence of nitrates is a risk factor as it known to corrode SS304L. These two mechanisms would apply to all boiler tubes in all the heat exchangers and thus all the tubes would be equally exposed to this risk.

In an effort to increase heat transfer coefficients in heat exchangers in the ammonia process, "static" mixers are often inserted in boiler tubes. The increase in the heat transfer coefficients is accompanied by the disruption of the laminar boundary sub-layer which also leads a higher mass transfer coefficient, thus increasing the potential for leaching chromium and iron ions from the metal. The additional risk factor associated with presence of static mixers in tubes is the higher velocities in the pipes they cause by occupying space.

The tube inserts pose additional risks introduced at the design stage. They are pendula bodies which will undergo complex vibrations and translational motions during plant operation. The amplitude of the vibrations will depend on the flexural rigidity of the metal. Their motion pose two additional risks of tube thinning: they can impose forces normal at the tube wall which could disrupt the formation of replacement protective oxide layer (F. Addo-Yobo Unpublished work); they could also make contact with the tube wall, peel off existing protective layer and set off vibrations in the walls which can continually weaken the replacement oxide layer and increase its porosity to migrating metal ions thus accelerating erosion and corrosion thinning of the steel underneath.

Thus the operational risk factors are: steam Weight, temperature in boiler tubes, velocity in the boiler tube. Design risk factors are: presence and motion of inserts for promoting turbulence, vibrations; and Utility risk factors are: contamination of process water with chlorides and nitrates after the ion-exchange stage.

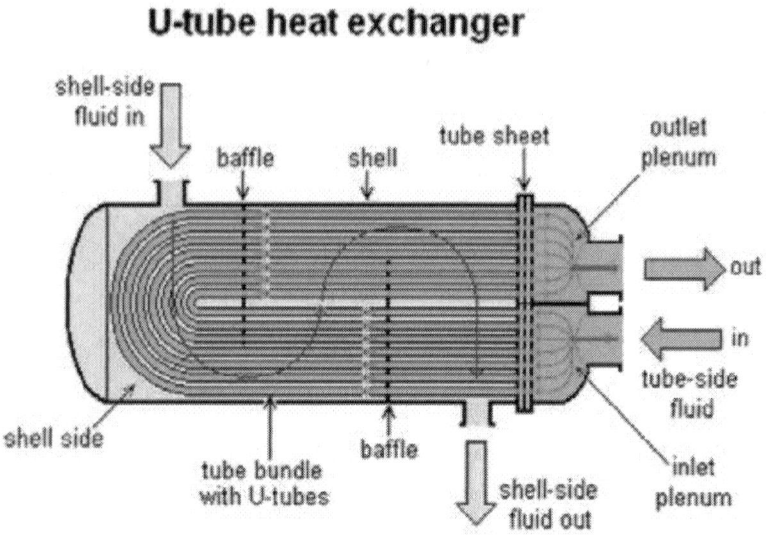

The Models in Hysys

he heat exchangers enclosed in the dotted red boundaries in the Process Flow Diagram above are modelled as a network. E2A is immediately downstream of the ammonia converter, E2B is the second in that stream. E111 is the exchanger in the front end stage of the process. The flow of Boiler Feed Water for cooling, in the boiler tubes in the heat exchanger E2A, E2B and E11, commence as single phase then becomes two phase with the generation of steam as heat is transferred from the process gases on the shell side. The dispersed two phase model in HYSYS was adopted to account for this.Whereas standard HYSYS models for heat exchanger could not accommodate finned and unfinned tubes in one exchanger, the rigorous EDR models were modified successfully to include this capability without changing the hydraulic and heat transfer characteristics. We were also able to incorporate variations in the process gas specific heat with composition without much computational expense.

From heat balance, the weight fraction of steam in the tubes (v_f) relates the process gas flowrates ($\dot{m}_G$) and temperature drop (T^G), BFW flow rates ($\dot{m}_G$) and changes in enthalpies (H^W). The steam weight was computed as shown in the adjacent equation.

$$v_f = \frac{\frac{\dot{m}_G}{\dot{m}_w}\left[C_{pc2}(T_1^G - T_2^G) + C_{pc1}\left(T_{1,in}^G - T_{2,out}^G\right)\right] + H_1^W - H_2^W}{H_2^v - H_2^W}$$

Thus confirming that the flow rate of the BFW is a suitable manipulated variable for controlling gas outlet temperatures. As stated above, the velocity of BFW in the tubes is important and its value can pose a risk. Its value varies with the weight of steam. The dispersed two phase model for steam-water mixes in HYSYS was used for modelling the flow and heat transfer to BFW in the tubes of the heat-exchanger. The shell side heat transfer coefficient in HYSYS are predicted by the well-known Dittus-Boelter equation.

The models built in HYSYS accurately predicted the output pressures and temperatures of the process gas entering and leaving the three heat exchangers, under standard conditions, within 3% percent of the expected values, as shown in the Table1. The simulations were carried out under standard operating conditions at 100% plant rate. The table also shows the risk values of these deviations from the TEMA values, which will be discussed later.

Results and Discussion

Often process Engineers find the need to increase the throughput of a Plant beyond the design values in order to improve productivity. However, this may carry additional risks of shortening the lifespan of the heat exchangers. In order to protect the heat exchangers and avoid overloading them, the product stream can be spilt and a percentage diverted around exchangers E2A and E2B. Table 2 compares output variables when model plant is operated with higher throughputs than the design levels. Columns 4,5,6 of Table 2 below show that the tubeside temperatures can be maintained within 0.3% of the design values with the bypass arrangement. However, steam weight increases beyond the prescribed TEMA values of 30%, thereby increasing potential operational risk. The weight fraction of steam increases beyond prescribed TEMA values with additional feed rate of 7% to 17%. Columns 5 & 6 of the above table show that at around 17% excess feed rate, increasing the cooling water rate by around 8% reduces the steam weight fraction to values close to the prescribed limit, thereby reducing operational risk and at the same time raising production rate.

The studies show that when 5% of tubes in E2A are blocked to mimic the response of industry to tube thinning beyond a certain level, to eliminate the risk spilling water from burst tubes wetting the product gases, the steam weight remains at reasonable levels close to the TEMA limit. However, the velocity in the tubes increase to values close to the prescribed TEMA limit because of the reduced cross sectional area available to the flow of BFW. The risk values of plugging tubes in exchanger E2A on exchangers E2B and E11 are negligible. Tables 3 summarizes the predictions of output variables caused by the plugging 10% of the tubes in E2A on E2A. The simulations show a reduction in the area available for heat transfer, in exchanger E2A by about 3% accompanied by a reduction in the heat duty of ca 4.3%. The reduced cross-sectional area available for flow on the tubeside results in 11% increase in the mean velocity through the tubeside and a 7% increase in the maximum velocity in the tube. The latter is close to the prescribed TEMA limit thereby increasing the operational risk posed to the plant by the blocking the tubes. The effect of this increase in velocity could be a shortening on the service life of E2A.

Table 1: Comparison of the predicted plant output variables with the design historic values.

	Mass Flow Rate	% Deviation of predicted Temperature from Standard TEMA values	Risk Value	% Deviation of predicted Pressure from Standard TEMA values	Risk Value
EE11 Gas Inlet	Standard	0.00	None	-1.85%	None
EE11 Gas Outlet	Standard	-0.21%	None		None
EE11 BFW Inlet	BFW rate	-0.39%	None		None
E2A BFW Inlet	BFW rate	0.18%	None		None
E2A Gas Inlet	Standard	0.00	None	1.31	None
E2A Gas Outlet	Standard	-0.05%	None	2.62	None
To ** (E2A BFW Outlet)**	BFW rate	-0.29%	None	-1.33	None
BFW	BFW rate	2.53			None
E2B Gas Outlet	Standard	1.57%	Negligible	2.80	None

Table 2: Effect of Increasing the Plant Throughput

KEY PROCESS PARAMETERS	Design Specification or Historic data	100% PLANT RATE	7% increase in feed process gas	17.4% increase in feed process gas	16.7% increase in feed process gas

Process Gas Rate (kg/s)	Standard	No deviation	+7%	+17.4%	+16.7%
Deviation in Water circulation rate (Wt/Wt)	Standard	No deviation	-17.2%	-21.9%	-15.7%
Deviation in steam weight $\frac{Predicted - Tema\ limit}{TEMA\ limit}$	-23.3%	-23.3%	+33.73%	+46.8%	+13.4%
Risk Status E2A due to steam weight	None	None	Highest	Highest	Highest
Deviation in tube-side outlet temperature of E2A	None	-0.29%	-0.14%	-0.143%	-0.15%
Risk value of tube temperature E2A	None	None	None	None	None
		Mean/Max			
E2A TUBESIDE % Velocity deviation	No Data	-61%/-20%	-64%/-3.7%	-63%/+3.6%	-63%/-3.7%
Risk Value E2A %Vel deviation	N/A	No Risk	/Localized Medium Risk	/High risk	/Medium risk
E2B % dev in TUBESIDE VELOCITY	-51%	-59%/-57%	-66%/-64%	-66%/-64%	-63%/-61%
Risk value	No risk	No risk	No risk	No risk	No risk
E11 TUBESIDE VELOCITY (m/s)	1.92	-60/-58	-66/-64	-66/-63	-64%/-61%
Risk value in E11 tube side velocity		No risk	No risk	No risk	No risk

Table 3: The Impact of blocking off 5% (10%) of the tubes Exchanger E2A of Performance of E2A

Heat Exchanger Characteristics	E2A with no plugged tubes at	E2A with plugged tubes	Percent Deviation

	theoretical conditions	at conditions from DCS	
Effective Surface Area available for Heat Transfer (m^2)	*	*	-3.07
Heat Duty (kW)	*	*	-4.35
Dev from Standard in Mean Temperature of Outlet of tubeside E2A	-0.15%	-0.16%	
Risk value	No Risk	No risk	
Corrected Mean Temperature Difference (°C)	*	*	7.34
Deviation in Heat Transfer Coefficient	*	*	-8.1%
Mean Shellside Velocity (m/s)	-56.2%	-56.1%	0.28
Maximum Shellside Velocity (m/s)	-53.2%	-53.2%	0.00
Risk value of deviation in maximum shell side velocity	No risk	No risk	
E2A TUBESIDE % mean Velocity deviation $\frac{mean\ velocity - Tema\ vel\ limit}{TEMA\ velocity\ limit}$	-63%	-59%	
Risk Value E2A %Vel deviation	None	None	
Dev maximum Tubeside Velocity (%) $\frac{Predicted\ max\ velocity - Tema\ vel\ limit}{TEMA\ velocity\ limit}$	-6.7%	-0.23%	
Risk value of Predicted maximum velocity	Medium risk	Medium risk	

Discussion, Conclusion & Recommendation for Future work

The studies have demonstrated the usefulness of building a rational model of the operation of the heat exchangers in the Ammonia process based on the diagrams presented in CEP (2015). They confirm the remedial actions taken by the Industry to improve productivity and maintain safe operation are sound. However, they point to the unfortunate fact that the quest to increase the heat transfer coefficients in the U-tube heat exchangers by inserting turbulent promoters are fraught with design risks that might not be possible to eliminate.

The simulations coupled with the present understanding of the mechanism of corro-erosion of high chromium stainless steel products (Folkeson et. al. 2011) suggest that once tube thinning is allowed to commence, caused by a risk factor introduced at the design stage, e.g. the inclusion of unfastened tube inserts, which increase heat transfer coefficients but which disturb the protective oxide layer, the exchangers may have to be eventually de-commissioned because of the difficulty of laying down a contiguous protective layer as shown in the schematic shown above right.

In order to firm up on the mechanisms that trigger tube thinning by corro-erosion of stainless steel boiler tubes in the Ammonia Process, further research into the risks posed by tube inserts and the careful weighing up of the cost of mitigating this factor as required by the Standard need to be carried out.

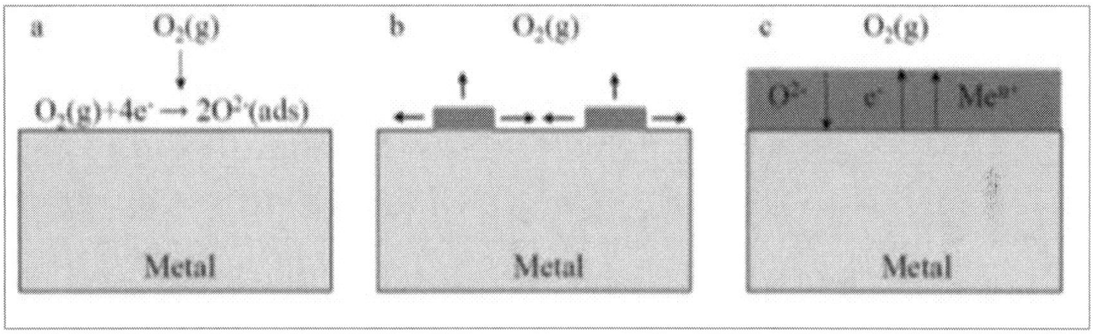

Schematic of the initial formation of an oxide scale by adsorption of oxygen, oxide nucleation and growth to form a continuous oxide scale (stages a–c).

References

1. Venkat Pattabathula & Jim Richardson 2016, Introduction to Ammonia Production, Chemical Engineering Progress pp 69-75
2. Process Worldwide: 12:08:2020 Editor Ahlam Rais. Haldor Topsoe, BASF to Offer Ammonia Producer new Optimisation possibilities. ID 46800904 https://www.process-worldwide.com/haldor-topsoe-basf-to-offer-ammonia-producers-new-optimisation-possibilities-a-955495/
3. The Role of CHECWORKS in an Effective FAC Programme, David Smith and Jeffrey Horrowitz , ICONE 22, 22nd International Conference on Nuclear Engineering, 2014 Prague
4. Gao, Weimin, Li, Yungang and Kong, Lingxue 2017, Numerical investigation of erosion of tube sheet and tubes of a shell and tube heat exchanger, *Computers and chemical engineering*, vol. 96, pp. 115-127, doi: 10.1016/j.compchemeng.2016.10.005.Boiler tube failure in Heat Exchangers for Power generation in the Nuclear Industry

5. L.E Sanchez-Carrera, "Mechanism of Corrosion-Erosion in Steam Extraction lines in Power Station", PhD Thesis Dept of Mechanical Engineering MIT 1984

6. Jonas, O. (2004). FAC and Cavitation: Identification, assessment, monitoring, prevention. *Combined Cycle Journal.*

7. Suez Water Technologies and Solutions. (2020). *Handbook of Industrial Water Treatment Chapter 11- Preboiler & Industrial Boiler Corrosion Control.* Retrieved from https://www.suezwatertechnologies.com/handbook/chapter-11-preboiler-and-boiler-corrosion-control

8. Md. Moniruzzaman Bhuyan1, Ujjwal K. Deb, M. Shahriar, Simul Acherjee 2017, Simulation of Heat Transfer in a Tubular Pipe Using Different Twisted Tape Inserts, Open Journal of Fluid Dynamics, 2017, 7, 397-409, http://www.scirp.org/journal/ojfd

9. Folkeson, N., Jonsson, T. Halvarsson, M. Johansson, L.-G., Svensson J.-E., The influence of small amounts of KCl(s) on the high temperature corrosion of a Fe-2.25Cr-1Mo steel at 400 and 500°C. *Materials and Corrosion*, (2011), 62(7), pp. 606–610

10. F, Addo-Yobo 2021, Motion of Pendula inserts in heat exchanger tubes in the Ammonia Process and their impact on exchanger risk, Unpublished work.

617202 Optimal Placement of Gas Detectors Using Dispersion Modeling

Abstract:

Historically, gas detector placement was performed by experienced designers using rules of thumb and judgment based on fundamental principles of gas cloud behavior. Unfortunately, designs using these approaches have been inconsistent and, in many cases, poorly performing. Historical data from the UK HSE shows that certain fixed gas detection systems routinely were only able to detect less than 70% of "major" gas releases. In the past decade, advances in computer modeling and 3D graphics has enabled the implementation of chemical process quantitative risk analysis (CPQRA) in the service of gas detection system design. The general process for using CPQRA to develop and verify a gas detection performance metric called coverage, which is essentially the fraction of the all release scenarios that are detected by a gas detector array, has been documented in the International Society for Automation (ISA) technical report ISA TR84.00.07 – Guidance on the Evaluation of Fire, Combustible Gas, and Toxic Gas System Effectiveness. This tutorial provides an overview level explanation of the workflow for the process of gas detection mapping, including the risk-based selection of performance targets and quantitative verification of the achievement of the coverage performance targets using dispersion modeling results and leak rate estimations that are similar to the proven approaches used in CPQRA. The tutorial goes through the entire process of gas detector placement for a sample facility where liquefied petroleum gas products are stored, beginning with risk assessment and going all the way through a completed gas detector layout.

Presenter Information	Contact Information
Presenter Name: Bejoy K. Bharatiya	**Contact Name**: Bejoy Bharatiya Telephone No.: +61 413 098 008
Presenter Biography: Bejoy Bharatiya was Process Safety Advisor at DNV GL Offshore and Industrial Services (B) Sdn. Bhd. He has over 30 years of professional experience in the upstream oil & gas, refining, petrochemical, and mineral processing industry. Prior to joining DNV, he held various positions, Process Safety Group Lead at Saudi Aramco Shell Refinery Company (SASEF), Senior Process Safety Engineer at SANTOS (South Australian Oil Search) Pty Ltd. Australia, Senior Operations Engineer with Saudi Aramco Gas Operations, and the Principal Process Engineer of Kellogg, Brown & Root (KBR), Australia.	Fax No.: E-mail Address (Required): **bejoy2014@gmail.com** **pinkubejoy@iinet.net.au** Your Company/Department:

Abstract (max of 1500 words)

HSE Philosophies in Capital Projects of Petroleum Refinery and Petrochemical Complex - A Practical Approach

Bejoy K. Bharatiya

The petroleum industry has long since recognized that certain risks are involved in the major hazard industries. It becomes a major challenge to minimize risk by comprehensive HSE framework in the project execution of a major process facility in the hydrocarbon industry. Over the years, the oil and gas industry has learned lessons from a number of process safety events and various hazards such as loss of control and loss of containment top events that could result in: fires, vapour cloud explosions (VCE), toxic and asphyxiant exposure, which occurred in various parts of the world and are not limited to the country of origin. The design and operation of process facility for the refinery and petrochemical complex have been the subject of major hazard to the oil industry.

It has been the continuous effort of the industry to update the design and operating philosophies to improve the inherent safety of the processing units. The present paper is concerned with defining the major safety considerations of international oil companies, and Industry Standards in designing and operating the refinery and petrochemical process facilities. HSE philosophies are required to guide and drive the design intent of HSE critical systems. The main safety philosophies include: ESD and blowdown, fire and gas detection, toxic gas (H_2S) detection, active and passive fire protection (including guidelines for firefighting systems and fire protection), layout HSE philosophy, process safety and containment etc.

This paper provides the basic safety design guidelines for a major project but does not attempt to cover in detail every aspect of safety design. Reference should be made, therefore, to specific discipline philosophies and specifications as appropriate.

The primary environmental philosophies include venting and flaring of hydrocarbons and H_2S/SO_2, CO_2 production, solid waste and effluent wastewater disposal, energy efficiency etc. In the project planning and execution, these philosophies must be effectively embedded in the design through the HSE premises document.

For the major project execution, the HSE philosophies are written in a way to instruct the design functional specifications to be compatible and aligned with the existing facility, but are updated to take into account of new technology, new knowledge and other industry requirements.

HSE philosophy shall be considered throughout all the project development phases, engineering, construction, start-up, and operation via dedicated and well-structured efforts. The safety philosophy, and resulting protection of persons and assets, is combining all provisions undertaking during all project phases.

HSE Philosophy for the detailed engineering phase is prepared with the objective to elaborate the Project design philosophy with respect to identification and elimination of hazards and to promote a design that is inherently safe. Objectives of this HSE philosophy are:

- Provide a safe working environment for personnel
- Minimize the potential for hazardous occurrences
- Avoid exposure of personnel to potential hazard
- Contain and minimize the effects of hazards
- Provide adequate safe access and escape away from hazards
- Protect the environment
- Provide a design that is fit for purpose for field life with the required integrity for safety and environmental protection

HSE Design Philosophies

During the detailed engineering, and subsequent construction and commissioning, particular care will be taken by achieving:

- Good engineering practice
- Application of an all-encompassing QA system
- Development and implementation of a safety management system
- Design reviews at 30%, 60% and 90%, layout review, HAZID, Design Safety Review (DSR), accessibility reviews and other reviews.
- Qualitative and quantitative risk assessments and IPF review
- Design of installation for extreme weather conditions
- Execution of an extensive pre-start up review of the plant before introduction of feed (Hydrocarbon, gas) covering both hardware and, control systems and procedures.
- Conformity check with local regulations, project standards and other applicable industry standards and directives.

Hazard assessment – Hazard assessment is based upon the hazard of the materials present in the processing facility. Some of the highest-ranking typical hazards included in the risk assessment (to be taken from facility Hazard Register) are:

Hazard	Top Event	Consequence
All Liquid Hydrocarbons (HC) including condensed flare gas	Loss of Containment	Fire
Toxics / H2S / N2 /BTX (Aromatics-Benzene, Toluene, Xylene)	Loss of Control or Containment	Personal Exposure
All HC, including H2 enriched HC	Loss of Containment	Generation of vapour cloud leading to VCE or fire
Heavier than air HC	Loss of Containment	Vapour cloud and/or Fire resulting from heaving than air HC

The design for the active/passive fire protection systems, fire detection systems, gas detection systems and general HSE requirements shall comply with the HSE Philosophy, project standards and referenced codes/standards, specifications, and regulations.

The fire safety of a plant is the degree of reduction in vulnerability to fire and/or explosion incidents in terms of their probability of occurrence and associated potential damage. This degree is maximized by the integrated application of fire prevention, fire protection and fire-fighting measures.

The philosophy is to achieve the desired degree of fire safety primarily by fire prevention measures, which will also result in an optimum availability of the facilities to produce. Where despite the preventive measures potential fire hazards remain, fire protection measures are taken. The basic objective of fire protection is to limit or prevent the escalation of a fire to avoid risk to life and to minimize material damage.

The Project shall be divided into separate fire risk areas of possible fire involvement. This separation may include major pipe racks, roadways, or vacant space. For estimating firewater demand, only one major fire occurring in one fire risk area at any time is to be considered. The fire water system shall be developed to meet the highest firewater demand of the worst-case firewater scenario. Extension of the existing firewater system on the Refinery to cover the new installations will be considered. Harmonization of fire safety measures shall be maximized.

In addition to fire protection measures, the plant fire-fighting organization is available to control and/or extinguish fires that occur despite the fire prevention and fire protection. All shift operators to be trained in industrial firefighting form the first intervention team (FIT) and additional fire fighters will be available from the Fire Station Crew. The combination of FIT and the existing Fire Station Crew will form an effective team, available within minutes after an incident is detected. If required, extra support can be provided through the external fire brigade.

There shall always be more emphasis on fire prevention, fire protection and fixed and semi-fixed firefighting systems than on the capacity of any fire-fighting organization.

The existing fire safety facilities in terms of mobile fire-fighting equipment and fire brigade organization of the refinery complex will be used to a maximum extent.

Furthermore, as the refinery hydrotreating and Sulphur Recovery Units process/handle large quantities of H_2S/SO_2, special emphasis shall be made during the design phase for the installation of suitable system for H_2S service such as high mechanical integrity (e.g. materials, pipe types, double pumps seals etc.), safeguarding, H_2S detection and automatic isolation/shutdown system. Design of this system shall be as per project standards and actual HSE philosophy for H_2S detection system.

Safety relief venting shall be provided for equipment, which is subject to internal pressure build-up from an external heat source. Pressure relieving devices shall be designed in accordance with project standards and/or industry standards.

Plant layout and spacing shall be developed based on project standards and industry practices, e.g. API RP/STDs, General Electric GAP Services (GE GAP) 2.5.2 "Oil and Chemical Plant Layout Spacing. Safety distances will be optimized to reduce the area utilized once more detailed design is progressed and risk studies have been completed (e.g. Physical effect modeling).

This paper discusses industry experience and summarizes the salient design considerations and risk assessment to improve safety integrity in the execution of capital projects for refinery and petrochemical complex. An effort has been made to highlight the HSE methodologies adopted by major oil companies using project standards, and industry best practices and standards like API, ASME, NFPA, IEC, ISO etc. Emphasis has been made to increase the awareness among the discipline design engineers and process safety engineers to develop safer designs of the refinery and petrochemical facilities.

619065 Addressing Aging Processes and Equipment

Abstract:

Number of the fertilizers plants across the globe have been installed and are in operation since many decades. The operation with the old equipment and conventional process techniques is being managed coupled with budget constraints. Hence the old aged plant equipment have high risk of failure under the shed of outdated processes, and may cause severe damage to the plant assets as well as to human. Consequently, one has to bear more and repeated maintenance cost and regular replacements of equipment.

Fatima Fert Plant (Previously Daud Hercules) also commissioned in 1969 and is in operation to date. Dealing with the challenge of old aged equipment and process techniques, very strict and robust reliability plans have been established and implemented to inspect the equipment with the advanced ultrasonic techniques and to make sure that the asset is in operation under controlled parameters. Based upon the finding of inspections, 10 year rolling plan have been developed to replace the equipment that are prone to failure.

The paper will be providing the details of the challenges and risks faced due to old age equipment and the way forward to mitigate and control that risk with optimized budget utilization.

619392 Why Is There a Regularity of Process Incidents Occurring, or Is It All Just a Re-Occurrence of the Past?

Abstract:

We attend conferences to hear the experiences of others, learn lessons from past events and take that experience and learning back to our place of work. Yet a simple analysis of suggest there is a plateau in the progress to reducing the frequency of process safety incidents.

So perhaps the question to address is, "Why are lessons not being learnt"?

This paper will highlight various data demonstrating the "plateauing" of performance; highlight particular recurring indicators that need to focused on and look at how learning may be distributed to those that need to increase their knowledge of PSM. There will be a particular emphasis of the use of case studies in this approach.

The paper will particularly interest those still on the learning journey in PSM and those questioning alternative approaches.

618397 How You Manage the Actions Coming through Process Safety Management Systems?

Abstract:

For improving process safety in the plant, we are putting great afford managing task of operators. Alarm management is important for control operator's tanks. Quantity of the alarms should not be more then control operator can manage.

However, there is gap in engineer's tasks management. Workload of an engineer is an important parameter in complex plants. This workload can be rapidly increase with HSE related tasks. As a result of Management of Change, Incident investigations, ... most of the tasks are given to operations engineers, projects engineers, maintenance engineers.

If the workload is too high it is easy for the engineer to miss the important actions. With HSE Action Portal, we aim to give management a view of task load for each department. The department responsible can focus on the overall tasks of the department

HSE Action Portal make possible to reach actions and tasks coming through PSM system elements such as Management of Change, Pre-Start up safety review, incident/accident investigations and process hazard analysis in the same platform. Users are able to analyse their action/tasks in terms of number and completion percent. In this way, users are triggered to follow closely their actions and complete them in a correct way.

It is an useful tool also for top management in order to collect data and create statistics periodically. It can be seen actions/task situations for each refinery and department.

617997 Guidelines for Risk Ranking Hazardous Scenarios When Conducting PHAs

Abstract:

Most of the companies in the process industry use corporate guidelines, or dedicated criteria and principles for conducting Process Hazard Analysis (PHAs). Examples of factors addressed are the following: nodes selection, process deviations to be brainstormed, level of detail when writing causes/consequences, safeguards categorization, risk matrix to be used, and how to risk rank the identified scenarios.

This paper collects different approaches from leading PHAs at several companies. An analysis comparing these approaches is provided and strengths and weaknesses are identified. Finally, based on safeguards and consequences categorization, guidelines are proposed for consistently risk rank hazardous scenarios.

619107 Practical Guide to Operationalize & Visualize Cumulative Process Safety Risks Perspective

Abstract:

The paper describes how Process Safety Management can be seamlessly integrated with Operational Risk Management. It demonstrates how operational processes like Permit to Work, MOC, Job Hazard Analysis, Incident Management and Process Safety Management can be brought together to support day-to-day decision making based on the actual cumulative risk status of the facility.

Key thinking is developed to answer questions such as *How do we make Process Safety Risk as visible as HSE* and *How can we prevent previously identified hazardous sequences of events from reoccurring?* , thus increasing the entire system safety.

This paper presents practical approaches in overcoming the challenges in collecting data from disparate sources and learnings on realising Dynamic Operational Risk Management (ORM) from various client implementations. In addition, it touches upon topics such as *How to analyze the data* and *Use of machine learning to predict future risk* .

The paper presents a risk calculation methodology that calculates barrier status and visualizes cumulative risk:

in a standardized barrier model, based on IOGP 544 standard, in Swiss-Cheese view

geographically by site and area/location using site graphics

per critical scenario in a BowTie view

It connects and visualizes process safety risk by automatically collecting data for Tier-1,2 and 3 safety indicators that are covered by various processes spread across the organization.

Furthermore, it shows how the cumulative risk status can be used in operational processes, potential conflict detection and planning of activities to manage and minimize the actual risk.

Keywords: Barrier Management, Process Safety Management, BowTie Methodology, Swiss-Cheese Model, Operational Risk Management, Project Implementation Methodologies, permit-to-work, MOC, JHA/RA, Machine Learning

ENGINEERING & OPERATIONS

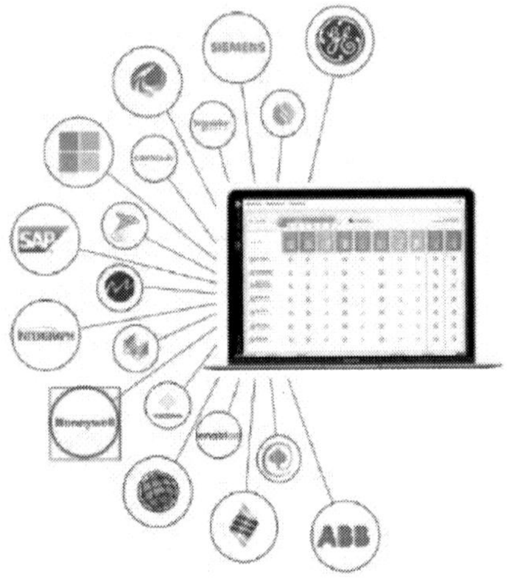

Wolters Kluwer Enablon Proprietary and Confidential

BARRIER MANAGEMENT INTEGRATION DATA POINTS

Integrity Data: CMMS, Audits & Assurance, Competence Management, Inspections, Engineering data.

Operational Intervention: Process Control System / Historian, Shift Logs, Operator Rounds, Permits, Temporary defeats.

Barrier Challenges & Failures: Incident Management, Management of Change, Process upsets, PSM events.

BARRIER MANAGEMENT – PROCESS SAFETY MANAGEMENT

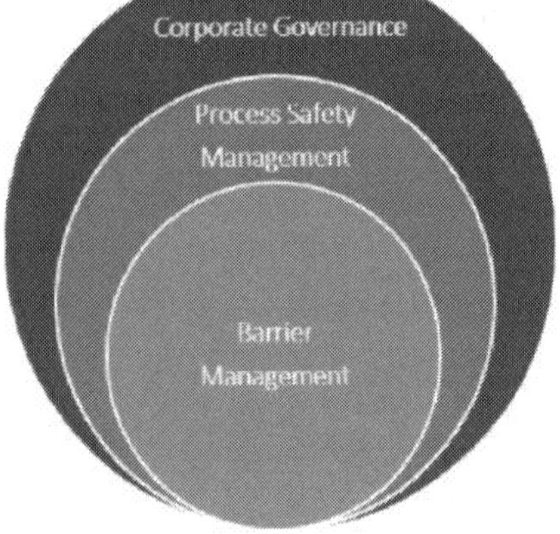

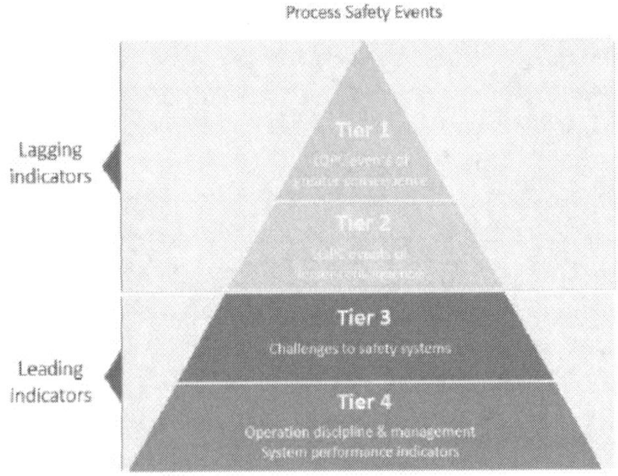

The focus of Barrier Management

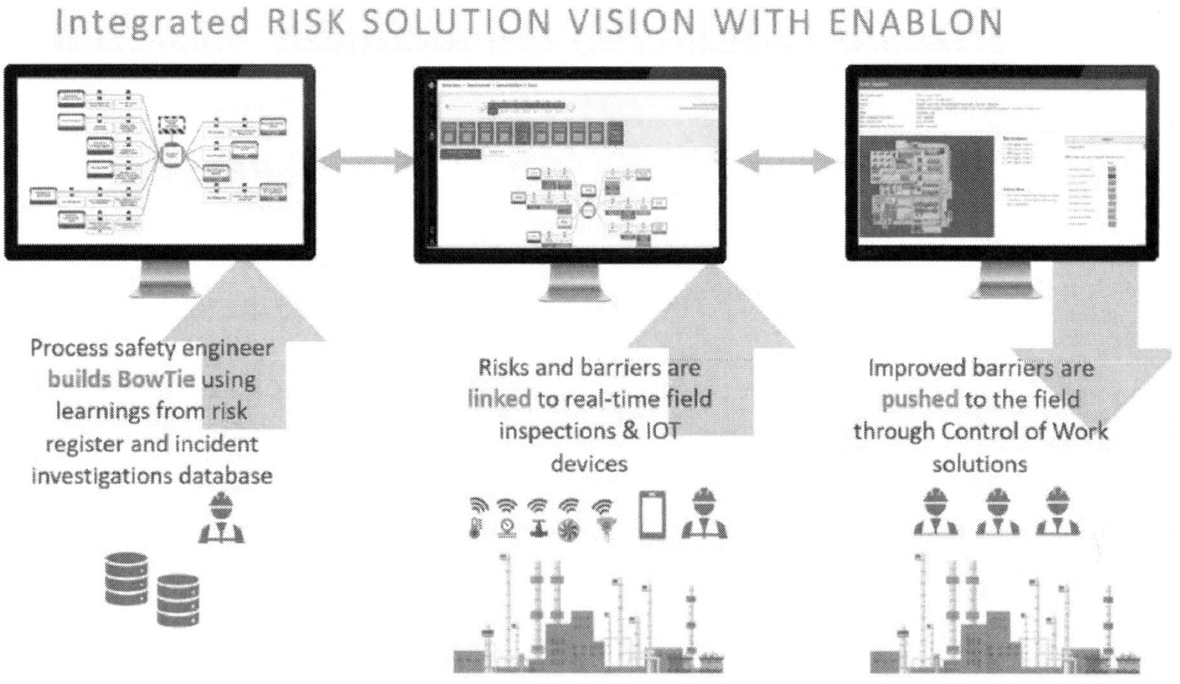
Integrated RISK SOLUTION VISION WITH ENABLON
Process safety engineer builds BowTie using learnings from risk register and incident investigations database
Risks and barriers are linked to real-time field inspections & IOT devices
Improved barriers are pushed to the field through Control of Work solutions

619126 Controlling Human Performance Between Both Unplanned and Planned Tasks within Abnormal Operation Mode

Abstract:

There is confusion in terminology used in the chemical-related industry for the class of procedures commonly referred to as Abnormal Mode of Operation and Abnormal situation management. This paper provides a clear definition of each mode of operation and gives examples of how the human performance is controlled for each.

1. Normal – either a continuous mode or a normal batch mode of operation.
2. Planned Non-routine (Non-Normal)– Startup, shutdown, are online maintenance are the main non-routine or non-normal modes of planned operation. But planned Temporary procedures, with time limits, are also part of these.
3. Abnormal– These include those activities covered by generalize procedures or guides (planned in a general sense) the activities that don't have written, step-by-step procedures (but noting that many companies have guides for most of these abnormal situations). A further breakdown is possible:
    1. Response to upsets using a Trouble-shooting Guides(TSG) for handling deviations from the operating window (these or normally triggered on an alarm); these may ultimately lead to a shutdown, safe park, or emergency shutdown (all of which are also proceduralized), if the deviation cannot be corrected in time.
    2. Response to failures using a Temporary procedure - such as how to run in bypass mode if the flow controller fails and you want to keep running in manual mode.
    3. Response to Unanticipated Events - The FIRST occurrence of this is handled by what is generally referred to as Emergency-MOC, which is really saying we will make a change immediately, and do the risk review of the change later, and then learn from this one case and proceduralize into a Temp Procedure or TSG for the next time it comes up.
4. Emergency Operations- a diminished or reduced operating plan; normally a Temporary procedure (similar to 3.b. above)
5. Emergency Shutdown - a TSG that fails to resolve the issue in time will go to this; or for some events such as a sudden loss of containment, we go straight to these
6. Emergency Response- normally in conjunction with 4 or 5, but focused on the protection of people, assets, environment, given the release or other imminent harm is in play.

This paper will provide a framework to ensure no mode of operation is overlooked and it will help sites understand what is needed to control risk during each mode of operation. Most of the time in the paper and presentation will be focused on classification 3, Abnormal Mode of operation.

619218 Proactive Lopc Indicator Critical Analysis through Power BI Dashboard

Abstract:

This work applies Power BI dashboard in order to proceed a critical analysis of proactive loss of primary containment indicator so that possible process problems can be detected before greater events occur by prognoses and decision-making to direct resource management in a more assertive way.

A process safety event is an unplanned or uncontrolled release of any material including non-toxic and non-flammable materials from a process, or an undesired event or condition that under slightly different circumstances could have resulted in a release of material (API 754). The pyramid model can represent the predictive relationship between lower and higher events.

It is important to guarantee the good register of pyramid base events (Tier 3) to achieve enough information, identify the fail installation's weakness, avoid higher process safety events, reduce losses and improve the facility performance. Nevertheless, the only register is not enough to achieve these improvements. Its's necessary to quantify and analyze the data to guide the decision-making. A critical analysis involves the following parameters: leaked product, fail causes, equipment that failed, engineer discipline, LOPC location, consequences.

The first step to make the critical analysis is having a reliable database. This step involves training the operational team in leading and lagging process safety indicators and process safety management. The communication must be direct and clear. Furthermore, it is necessary to provide a satisfactory registration platform with the necessary fields to carry out the subsequent critical analysis.

After that, it is possible to extract all the information to proceed the proactive LOPC indicator critical analysis. At this point Business Intelligence (BI) presents itself as an adequate tool. Power BI, from Microsoft, is one of the most prominent BI tools. It is a collection of software services, applications and connectors that work together to consolidate, make information from different data sources coherent, visually engaging and interactive (from an Excel spreadsheet to a cloud-based data collection).

Through the LOPC critical analysis using Power BI dashboard it is possible to evaluate the Tier 3 incidents, focused on developing a learning approach. Therefore, it is possible to prevent and monitor failures trends and effectiveness of implemented actions. The critical analysis permits to direct the resource management by identifying the equipment and systems with recurrent failure events. It is worth mentioning the effective contribution for Process Safety Culture, since the leadership, operation and related areas could see the cause-effect relationships between the minor events and the daily operations.

619342 A Case Study on Ensuring "ALARP" Demonstration to Justify Continued Operation with Respect to "Mah" By Integrating Process Safety Techniques for Existing Process Facility

Abstract:

A lot have been written on process safety. However, practical applications of all these theories & studies has its own challenges. At the end, after overcoming the challenges as well as developing, implementing & monitoring strategies, a risk to existing facilities can be reduced to ALARP/ acceptable level.

This paper will throw light & discuss about requirements to integrate different process safety strategies, challenges during implementation & their solutions to demonstrate that existing oil & gas facility is safe for operation, through a practical case study of one complete process train. The methodology includes;

- Identifications of process trains
- Data collection
- Data Review
- Review past changes & their implementation
- Update records & P&IDs
- Carry out/ Update Integrated PHA with process simulation
- Development of Failure Scenario
- Risk Calculation considering existing safeguard (preventive and mitigating measures)
- Risk submission and ALARP demonstration
- Compare risk
- Identification of safety critical elements (SCEs) & Safety Critical Tasks (SCTs)
- Development of Performance Standard of SCEs
- Developing management system of PLAN-DO-CHECK-ACT
- Integrating the same in Audit/Inspection
- Develop training through simulations

Challenges faced by team for existing facility to achieve optimum process safety & ensure ALARP level of risk, and their solutions will be discussed.

Conclusion

This exercise provides evidence that wherever practical, the design and operations of the existing oil & gas facility has attained a level of risk that Justifies Continued Operation with respect to major fire and explosion events. This demonstration of Justification for Continued Operation is based on the number of safety measures inbuilt at design stage, Quantitative Risk Assessment and various other process safety studies revalidated during operational phase of the oil & gas process facility as well as after implementing changes through MoC. It must be noted that a separate process of action tracking is in place to prioritize and monitor implementation of recommendations provided to reduce the risk further to As Low As Reasonably Practicable (ALARP). A systematic approach of Plan-Do-Check-Act is developed to ensure that, safety critical tasks are carried out in timely manner by competent personnel, to ensure safety critical elements are in place and functional in line with performance standard.

616668 Towards Management Systems Perfection – Successes with Implementing a Layer of Verification Analysis to Continually Treat Process Safety Risks

Abstract:

Trust, but verify, by continually checking everything! This paper will demonstrate the importance of verification as a critical step in continually improving the treatment of process safety risks, including implementing inherently safer designs and operating practices, and controlling residual risks.

A well-established design concept for management systems is the continual improvement lifecycle involving the steps: Plan-Do-Check-Act (PDCA). Similar to standards for managing quality, environmental, occupational health and safety, and security, models for effective process safety management (PSM) systems also utilize the PDCA cycle. At a higher level, the PSM elements of Mechanical Integrity, Incident Investigation and Compliance Audits address verification in the "Check" step; while every management process of the PSM System should incorporate a "Check" step with verification processes and procedures to monitor and assure compliance and performance. This paper will present a taxonomy of activities for meticulous verification against policies, objectives, plans and other requirements for a vibrant PSM system, showing the importance of examining, analyzing, assessing, confirming, inspecting, testing, and reviewing planned activities, desired conditions and achieved results.

Attendees will benefit from practical takeaways of how each management system element can be developed with "verification designed in" and a verification culture created to treat process safety risks at three opportune times: before the incident (prevention), during the incident (emergency management) and after the incident (incident investigation, root cause analysis and lessons learned).

A comprehensive case study with practical examples will be presented to demonstrate a proven technique of Layers of Verification Analysis (LOVA) as a best practice integrated into the conduct of a process hazard analysis (PHA). The case study will show the audience "how to" enhance the PHA facilitation wearing an "auditor's hat" to raise questions and verify the effectiveness of existing safeguards, all of which are managed by the PSM System. Borrowing from the basic principles of management systems auditing, LOVA asks these questions for each hazard scenario: are systems, processes and procedures associated with causes (of human errors and equipment malfunctions), with consequences (of physical phenomena and resulting impacts), and with safeguards (for both engineering and administrative controls) *(1) in place?, (2) adequately designed? and (3) effectively implemented?* LOVA has also been used for supplemental analyses within the PHA to address human factors, previous incidents, facility siting and changes (e.g. MOCs). Recommendations from the LOVA in PHAs can be used as lessons learned for improving the implementation of the PSM system on a daily basis, without having to wait on the next periodic audit or PHA revalidation. The LOVA technique has been shown to be equally suitable for any stage of the process lifecycle, including inherently safer design evaluations.

The paper will also present three incident case histories to show how missing, inadequately designed or ineffectively implemented verification processes and activities have contributed to process safety incidents.

This paper will draw upon the authors' extensive experiences in designing, developing, implementing and auditing PSM systems, as well as conducting PHAs, building siting evaluations, and incident investigations.

617112 Study and Experimental Validation of CO2 Blowdown at Cryogenic Conditions

Abstract:

CO_2 solidification is one of the critical issues being identified in cryogenic technology process, particularly during the blowdown process. This poses critical challenge in developing the technology for separation of CO_2 from natural gas as there are limited studies and data to support the design and simulation of the process. Depressurization of high CO_2 fluid at cryogenic temperature would lead to possibility of CO2 solid formation, hence potential blockage of process equipment and venting line.

The new cryogenic technologies are currently being tested for the proof of concept. Hence, a pilot plant, which is a scaled down version of the technology was developed. One of the major challenges faced during the pilot plant testing is the emergency depressurization philosophy as the process involves CO_2 solids handling which is uncommon to the industry standard. Lack of experimental data of CO_2 blowdown at these conditions presented a technical gap which needs to be addressed to properly predict the solidification and mitigations, hence avoiding over-design of the plant and chances of accidents and unscheduled downtimes can be minimized.

Therefore, this paper will focus on the experimental study which have been done on CO_2 blowdown scenarios at a pilot plant. The pressure and temperature profile provide real insight of the solidification of CO2 during several blowdown scenarios. Finally the paper would validate a simulation data conducted using the experimental data as one of the potential prediction tools for CO2 blowdown process.

617588 The Supervisors Next Big Challenge: How to Ensure Operational Discipline on the Work Fronts

Abstract:

The Supervisors Next Big Challenge:

How to Ensure Operational Discipline on the Work Fronts

There is a problem that the petrochemical and oil and gas industries have not solved effectively: how to transform work fronts in environments based on operational discipline.

All the commitment and effort that an organization mobilizes in its risk management, finally, is updated in what happens between supervisors and operators on the work fronts. That is why it is so necessary to work on the alignment of the workforce of operating companies and contractors.

In the last decade, Senior Management became more aware of its role: defining what were tolerable risks and assuring resources. Process Safety, Integrity and Environment ceased to be discursive themes, to effectively have a genuine place at the small table where strategic decisions are made. The biggest challenge has been getting it to operate according to those definitions.

Supervision is carried out on the work front, the true executive arm of the organizations in the field. And for this organizational segment to ensure a disciplined operation, they need to be able to manage the culture.

A common mistake in the risk management approach has been to think of the lack of operational discipline of supervision as a problem of isolated individuals, and not of the context in which these people work, that is, of the context that the activities provide. organizations, from the context produced by culture.

When analyzing the supervision segment, one should not think of undisciplined people, but cultures that tolerate them, or that reinforce risky behaviors and shortcuts to the established norm. If supervision does not ensure operational discipline on the work fronts, it is because the organization does not provide the conditions for this to occur. People are the reflection of the culture in which they operate. Individuals do not impose a personal style of managing risks in isolation, but adapt to group agreements, displaying safe or unsafe behaviors depending on what is collectively tolerated. Field supervisors only take risks that the culture allows them to take.

The change of mindset in the supervision segment should not be built on the basis of shortcuts or "crash plans", but rather in avoiding in a progressive and lasting way the repetition of incidents, through a transformative process: generating a culture of discipline, learning and mutual trust in the work fronts, based on group agreements between operators and contractor companies, managed by a trained supervision to make decisions based on risks, combat the normalization of deviance and keep in mind the sense of vulnerability.

The key is to generate the conditions for effective learning. In addressing group problems, and not individual ones. The challenge is to create the conditions so that operational discipline is not a declaration of principles but a reality in the field. For this they do not need new knowledge, but to deepen what they already know, in the right context.

The pandemic, paradoxically, provided an opportunity to improve the quality of supervision development, personalizing it. Although online self-training has been around for a long time, it is not the most appropriate tool to generate changes in operations. For that, culture needs to intervene.

The next decade seems to keep a central challenge: transforming supervisory teams into custodians of culture and guarantors of risk management, through work fronts based on operational discipline.

Presenter Information	Contact Information
Presenter Name: Bejoy K. Bharatiya	**Contact Name**: Bejoy Bharatiya Telephone No.: +61 413 098 008
Presenter Biography: Bejoy Bharatiya was Process Safety Advisor at DNV GL Offshore and Industrial Services (B) Sdn. Bhd. He has over 30 years of professional experience in the upstream oil & gas, refining, petrochemical, and mineral processing industry. Prior to joining DNV, he held various positions, Process Safety Group Lead at Saudi Aramco Shell Refinery Company (SASEF), Senior Process Safety Engineer at SANTOS (South Australian Oil Search) Pty Ltd. Australia, Senior Operations Engineer with Saudi Aramco Gas Operations, and the Principal Process Engineer of Kellogg, Brown & Root (KBR), Australia.	Fax No.: E-mail Address (Required): bejoyb@hotmail.com bejoy2014@gmail.com Your Company/Department:

Abstract (max of 1500 words)

Project Assurance Review and Audit Framework for Capital Project of a Petroleum Refinery - A Process Safety Approach

Bejoy K. Bharatiya

The petroleum industry has long since recognized that certain risks are involved in the major hazard industries. It becomes a major challenge to minimize risk by **comprehensive project assurance/audit framework** in the project execution of a major process facility in the hydrocarbon industry. Over the years, the oil and gas industry has learned lessons from a number of process safety events and various hazards such as loss of control and loss of containment top events that could result in: fires, vapour cloud explosions, toxic and asphyxiant exposure, which occurred in various parts of the world and are not limited to the country of origin. The design and operation of process facility for the refinery and petrochemical complex have been the subject of major hazard to the oil industry.

It has been the continuous effort of the industry to update the design and operating philosophies to improve the inherent safety of the processing units. The present paper is concerned with defining the major safety considerations of international oil companies, and Industry Standards in designing and operating the refinery and petrochemical process facilities. Inherently safe design and structured project assurance and audit program are mandatory to guide and drive the design intent of HSE critical systems. The main safety philosophies include: ESD and blowdown, fire and gas detection, toxic gas (H_2S) detection, active and passive fire protection (including guidelines for firefighting systems and fire protection), layout HSE philosophy, process safety and containment etc.

This paper provides the **assurance/audit framework** for a major project but does not attempt to cover in detail every aspect of safety design. Reference should be made, therefore, to specific discipline philosophies and specifications as appropriate. (Refer to **Figure 1**).

Project audit or assurance program shall be applied throughout all the project development phases, engineering, construction, start-up, and operation via dedicated and well-structured efforts. The safety philosophy, and resulting protection of persons and assets, is combining all provisions undertaking during all project phases. (Refer to gated process, 1-6 **Steps** explained below).

HSE Design Philosophies

During the detailed engineering, and subsequent construction and commissioning, particular care will be taken by achieving:

- Good engineering practice
- Application of an all-encompassing QA system
- Development and implementation of a safety management system
- Design reviews at 30%, 60% and 90%, layout review, HAZID, Design Safety Review (DSR), accessibility reviews and other reviews.
- Qualitative and quantitative risk assessments and IPF review
- Design of installation for extreme weather conditions
- Execution of an extensive pre-start up review of the plant before introduction of feed (Hydrocarbon, gas) covering both hardware and, control systems and procedures.

- Conformity check with local regulations, project standards and other applicable industry standards and directives.

Gated Process During Project Development and Project Execution Phases

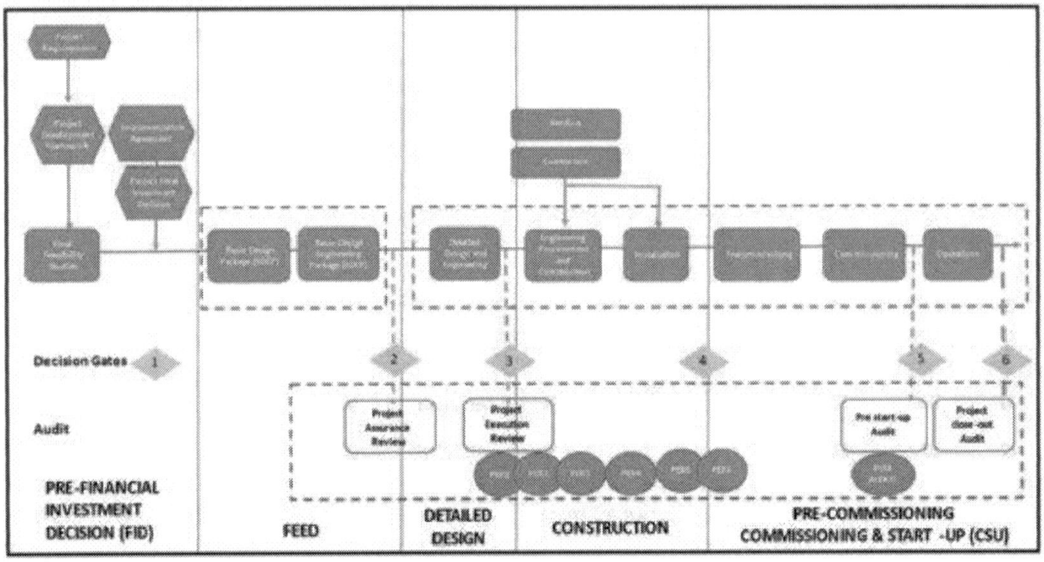

Figure-1 Gated Process for Project Assurance/Audit Framework for Project Execution

Project Assurance Review/Audit is required at all stages of project execution.

1. **Pre-Financial Investment Decision (FID) stage:**

1) Management: review critical contractor's capability, business license, & project experiences, PEP, schedule.
2) QA/QC: review critical contractor's ISO 9000 certificate, QA/QC plan, standards/policies/procedures (**SPP**), sample check of work records/reports etc. (see Table-1 International Codes and Standard)
3) HSE: review critical contractor's ISO 18000 certificate, project HSE plan, SPP, sample check of work records/reports etc.
4) Technical: review Feasibility Study, Environmental Impact Assessment (EIA) reports., compare critical design area engineering standards against international standards as well as best industry practices, etc.

Deliverable: review comments or audit findings

2. **Front-end Engineering Design (FEED) Stage:**

1) Management: review critical contractor's capability, business license, & project experiences, PEP, schedule
2) QA/QC: review critical contractor's ISO 9000 certificate, QA/QC plan, SPP, sample check of work records/reports etc.
3) HSE: review critical contractor's ISO 18000 certificate, HSE plan, SPP, sample check of work records/reports, including **PHAs/HAZOP/QRA** etc.
4) Technical: review FEED engineering design, & risk assessment e.g. HAZID, HAZOP, QRA, etc.

Deliverable: audit findings on Statement of Review for FEED, Statement of Review for FEED Safety Study.

3. **Detail Design Stage:**

1) Management: review critical contractor's capability, business license, & project experiences, PEP, schedule

2) QA/QC: review critical contractor's ISO 9000 certificate, QA/QC plan, procedures, standards/policies/procedures (**SPP**), sample check of work records/reports etc.
3) HSE: review critical contractor's ISO 45001 certificate, HSE plan, standards/policies/procedures (**SPP**), sample check of work records/reports etc.
4) Technical: review Detail Design engineering design, design variation, & risk assessment e.g. closeout of HAZOP recommendations, etc.

Deliverable: review comments or audit report on Statement of Review for Detail Design, Statement of Review for Detail Design Safety Studies.

4. Procurement Stage:

1) Management: review critical contractor's capability, business license, & project experiences.
2) Technical: review specifications of safety critical equipment and QA/QC process.
3) Engineering requirement e.g. specification, database, drawings etc.
    - Manufacturing documents e.g. drawings, inspection & test plan, witness Factory Acceptance Test (FAT), review FAT report & Manufacturing Databook/Spare parts, etc.
    - Inspection & Test Organization documents e.g. Inspection Report, 3rd party certificate etc.

Deliverable: review comments or audit findings.

5. Construction Stage:

1) Management: review critical contractor's capability, business license, & project experiences, PEP, schedule, etc.
2) QA/QC: review critical contractor's ISO 9000 certificate, QA/QC plan, standards/policies/procedures (**SPP**), sample check of work records/reports etc.
3) HSE: review critical contractor's ISO 45001 certificate, HSE management system, plan, procedures, standards/policies/procedures (**SPP**), sample check of work records/reports etc., construction environmental monitoring, waste treatment plan, emergency response plan
4) Technical: compare critical design area project engineering standards against international standards, review construction methodologies, check list, punch list, site construction/safety/environmental/QA/QC activities, design variation, mechanical completion, handover report etc. (**see Mechanical Completion Certificate Checklists** for conducting site audits

Deliverable: review comments or audit report, Statement of Review for Construction

6. Pre-Commissioning & Commissioning Stage:

1) Management: review pre-commissioning & commissioning plan schedule, integration plan etc.
2) Technical: review pre-commissioning & commissioning plan, isolation list, punch list, Site Acceptance Test (SAT), red-line markup drawings, design variation, witness pre-commissioning & commissioning activities, Prestart-up Safety Review (PSSR), ready for startup inspection etc.

Table – 1 International Codes and Standards

IEC 79	Electrical Apparatus for Explosive Gas Atmospheres
API RP 520	Recommended Practice for the Sizing, Selection and Installation of Pressure Relieving Devices (Part 1 and 2)
API RP 521	Recommended Practice for Pressure Relieving and Depressurizing Systems
NFPA 10	Portable Fire Extinguishers
NFPA 11	Low Expansion Foam
NFPA 13	Installation of Sprinkler Systems
NFPA 15	Water spray Fixed systems for Fire Protection
NFPA 16	Deluge Foam Water Sprinkler and Foam Water spray systems
NFPA 17	Dry Chemical Extinguishing systems
NFPA 24	Installation of private fire service mains and their appurtenances
NFPA 30	Flammable and Combustible Liquids Code
NFPA 72	National Fire Alarm Code
NFPA 101	Life Safety Code
NFPA 2001	Clean Agent Extinguishing System
NFPA	Fire Protection Guide to Hazardous Materials

IP Model Code of Safe Practice Part 3	Refinery Safety Code
IP Model Code of Safe Practice Part 19	Fire Protection at Petroleum Refineries and Bulk Storage Installations
BS 5839	BS 5839 Part 1988 for Fire Detection and Alarm Systems for buildings
IEC 61511	Functional Safety, Safety Instrumented Systems for the Process Industry
ISO 14692	Petroleum and Natural Gas Industries – Glass Reinforced Plastics (GRP) piping

This paper discusses industry experience and summarizes the salient design considerations and risk assessment to improve safety integrity in the execution of capital projects for refinery and petrochemical complex. An effort has been made to highlight the ***project assurance/audit framework*** *with HSE methodologies adopted by major oil companies using project standards, and industry best practices and standards like API, ASME, NFPA, IEC, ISO etc. Emphasis has been made to increase the awareness among the discipline design engineers and process safety engineers to develop safer designs of the refinery and petrochemical facilities*

619336 Case Studies – Beware of Common Errors during Risk Assessment of Powder Handling

Abstract:

Case studies – Beware of Common errors during Risk Assessment of Powder handling

"Single size shoe does not fit all," sounds so much in synchrony when we want to assess risk related to powder handling. Powders, unlike liquids have varied challenges right from it's transportation, receipt storage and handling.

Dust handling can have primarily 2 major hazards:

1. Dust explosion
2. Dust thermal decomposition

Few of the important parameters to understand for dust explosion are:

1. Fuel characteristics of the powder:
2. MIE: Minimum energy that can ignite dust cloud in the air.
3. MIT: lowest temp of a hot surface that will cause dust cloud (not dust layer) to ignite.

- Particle size: Smaller size poses higher risk

1. Kst: Max Rate of pressure increase. lower the kst number means the lesser severity 0 bar m/s means no risk while >300 bar m/s means high risk and bigger explosion severity.
2. Ignition source:
3. Hot Surfaces
4. Mechanical sparks/friction

- Electrical motors/contacts

1. Static electricity

Few important parameters to understand and control thermal decomposition in powders is:

1. Onset temperature of decomposition
2. Ageing of powders in vessels that the in process powders are handled such as Dryer, pipelines etc
3. Bulk storage stability of powders

Powders have challenge in terms of dissipation of heat, sampling of representative samples and uniform temperature indication and thus control.

The case studies will focus more on common errors during defining of controls for powder handling and will focus on how preventive controls can be built in to avoid major incidents.

Sr.No	Case Study Title	Controls in brief
1	Drying of Powders with detonation property	Selection of right type of Dryer to ensure metal tramp is not an issue Inert atmosphere
2	Aging of powders and thus lowering of thermal decomposition onset	Thermal studies post commissioning to ascertain periodic cleaning of process vessels such as dryers to avoid decomposition at lower temperatures
3	Handling of shock sensitive powders	Handling the powders in situ in inert atmosphere Ensuring that all process lines such as filter to dryer are flushed every batch Higher authority control for permit to work on such systems Avoid use of agitators for such processes.
4	Handling low MIE powder formulations including drumming out	Static charge generation is a concern. Some preventive controls to be considered are: Ensuring that drop height of powder is less than 3 mtr. Use of earthing continuity stem such as earthrite system. Ensure personnel use antistatic body protection and shoes. Use of conductive FIBC or drums Good housekeeping

All the above studies are based on Author's 10 years of experience in process hazard analysis for bulk chemical manufacturing and is expected to benefit the audience at large.

YES You Can: Case Study of Implementing a Center for Chemical Process Safety Risk Based Process Safety (CCPS RBPS)

Nestor Humberto Sposito
PPR Consultora
Bahia Blanca, Buenos Aires, Argentina
pprconsultora@gmail.com
Juanita Corredor
Process Safety Leader Brinsa
Cajicá, Colombia
juanita.corredor@brinsa.com.co
John Murphy
Process Safety Services
Punta Gorda, FL
hamjfm@embarqmail.com
John Diaz
Quality Director, Brinsa SA,
Cajicá, Colombia
john.diaz@brinsa.com.co

Prepared for Presentation at
American Institute of Chemical Engineers
2020 Spring Meeting and 17th Global Congress on Process Safety
Virtual
April 18 - 22, 2021

YES You Can: Case Study of Implementing a Center for Chemical Process Safety Risk Based Process Safety (CCPS RBPS)

Nestor Humberto Sposito
PPR Consultora
Bahia Blanca, Buenos Aires, Argentina
pprconsultora@gmail.com
Juanita Corredor
Process Safety Leader Brinsa
Cajicá, Colombia
juanita.corredor@brinsa.com.co
John Murphy
Process Safety Services
Punta Gorda, FL
hamjfm@embarqmail.com
John Diaz
Quality Director, Brinsa SA,
Cajicá, Colombia
john.diaz@brinsa.com.co

Keywords: Process Safety, RBPS,

Abstract

This paper discusses the challenges and opportunities that arise in the implementation of a risk-based process safety management system in a hazardous materials facility in the chlorine-alkali industry, starting from a relatively low level of compliance with standards, rules of good art, and a general lack of knowledge of the basic principles of process safety.

The motivation to begin the implementation of risked-based process safety (RBPS) came from an external audit requested by the company in November 2017 to evaluate its compliance with several pamphlets from the Chlorine Institute. At that time, the organization did not have a process safety sector and safety efforts were oriented towards personal safety. The final report included a long list of recommendations, each of them associated with good practice from the CI pamphlets. These recommendations were also associated with one of the 20 elements of the CCPS RBPS.

The authors discuss how the implementation plan was developed, the awareness and training of the top and middle levels of the organization's leadership to support the project, the mechanism for selecting priorities, and the process of preparing the schedule including scope, responsibility, and compliance dates.

The paper discusses examples of improvement in facilities, equipment, systems, and procedures in the operational, maintenance, engineering, etc. sectors, the incorporation of systematic and rigorous hazard identification and risk analysis techniques, and the benefits of implementation and monitoring the recommendations suggested in the studies. The strong leadership commitment and the process monitoring methodology by the company's senior management drives the implementation and gives visibility to the management system within the entire organization.

The authors also discuss the economic results of the implementation and quantify the impact of the improvements.

The pandemic unleashed in the world, altered the course of implementation of the process safety management system. The paper discusses the alternative and exceptional measures adopted to mitigate this impact and the results of those measures.

While it is true that the progress toward total implementation of RBPS is very important, it also important to understand what is missing to achieve the full implementation of the 20 elements of RBPS, the appropriate level of maturity necessary in the culture of the organization to achieve continuous improvement and what is the plan to address the challenges to reach each stage of implementation.

The authors encourage all organizations that have not yet implemented a process safety management system to understand that it can be done, even in less than ideal conditions, with very old facilities and a poor or non-existent process safety culture, if there is a strong commitment from the top leadership of the organization.

1 Introduction

A process safety management system that is robust and reliable is necessary to ensure the safe operation of a facility involving highly dangerous products and operating conditions, such as the ones found in a chlor-alkali chemical plant.

Below, there will be a brief introduction of the installation's, which was the subject of this study, most important characteristics.

Brinsa is a Colombian company, leading in salt chemistry, with 25 years of experience. They extract and refine the salt, and from this raw material, they develop the chemistry for the chlor-alkali which supplies the whole industry. The company has manufacturing facilities in Colombia, Costa Rica, and the Dominican Republic, as well as a commercial presence in Latin America.

The chemical plant has been increasing the historical monthly production record of chlorine-soda, through electrolytic cell technology, in order to achieve the separation of the sodium chloride molecule. This is cutting-edge technology that improves process efficiency as well as environmental control.

There are inherent risks involved in the chlor alkali processing because of the hazardous substances involved in the process. Some are classified as corrosive, toxic, and flammable.

The chemical plant has an installed capacity of 100 ton/day of liquid chlorine, 225 ton/day of 50% soda, 160 ton/day of hydrochloric acid, 180 ton/day of 15% sodium hypochlorite (batch production), and 260 ton/day of 5.25% sodium hypochlorite (continuous production). Additionally, there is a storage capacity for 1540 tons of sodium hydroxide, 503 tons of hydrochloric acid, 421 tons of 15% sodium hypochlorite, 302 tons of 38% sodium chloride, 327 ton of lineal sulphonic acid, and finally, and 141 ton of chlorine as a finished product.

The plant has 4 boilers that supply steam at 400 PSIG and 400 °C. Part of the steam that is produced is used in the plant's turbo generators to produce up to 8.5 MW, achieving energy self-sufficiency. There is also a sulfonation plant, which produces raw material for the manufacturing of the cleaning products, and a blowing plant that fabricates packages and lids.

The company has a total of 1400 workers, between fixed and temporary workers and 400 permanent contractors.

2 External Audit. Trigger for the Process.

In November 2017, the company's leadership, decided to conduct third party audit of the chlorine production plant, including all of its stages (membrane, compression, storage, packaging, shipping, and distribution) to access the level of the compliance to the Chlorine Institute's Pamphlets applicable to chloro alkali production.

A protocol was developed (a checklist), based on the guidelines from the Chlorine Institute and the industry's experience. The list was used as an evaluation tool during the 5-day visit to the chlor-alkali process plant.

The evaluation activities in the field included:

- A tour of the facilities
- Review of the following programs with the plant's personnel:
  - Incident reports
  - Emergency preparation and response plan
  - Standard Operating Procedures
  - Plan for personnel training and development assurance
- Review of the pertinent Chlorine Institute checklists, with the plant's personnel.
- Presentation and discussion of the CCPS Risk-Based Process Safety program to the plant's personnel and leadership.

2.1 Findings and Priority Recommendations.

Below, are itemized the most recommendation that was given, as mandatory, by the auditors:

- Develop a written program for the investigation and report of process safety incidents. (1)
- Develop a written program for the involvement of concerned parties, including information on the process dangers, the preparation and emergency plan, self-protection measures, the alarm system, medical treatment, etc.
- Develop a plan for the performance of a periodical risk identification and analysis for the high-risk process areas.
- Develop a written plan on the reliability and mechanical integrity of the plant's assets. Define the critical equipment and an inspection and maintenance plan.
- Develop and implement a Management Of Change program (MOC)
- Develop plans of electrical areas classification in accordance with international standards
- Develop a solid culture of process safety. Establish process safety as a core value
- Provide solid leadership from all levels, starting with the board of directors (lead by example)
- Improve the operative procedures management system
- Establish a Process Safety role

The audit results were presented and shared on the last day with the facility leadership team. The team showed itself very willing to implement all of the recommendations, especially those cataloged as mandatory, and to launch a project to implement a process safety management system that didn't exist at that time in the organization.

3 Awareness & Planning

The company decided to hire the services of a process safety specialist consultancy to advise and monitor the implementation of the Risk-Based Safety System

The concept of process safety was new for all of the organization levels. Initially, a broad training and awareness plan was launched on process safety topics using different courses and workshops encompassing the different organization levels.

The plan was designed to bring them the basic process safety knowledge and to achieve the commitment from all the people to the future program implementation.

Workshops were designed with different scopes and approaches in the function of the target audience, starting with the management and leadership levels.

The first workshop was targeted at leaders from different sectors of the organization. During the 3 days training, after 24 training hours was completed, all participants were summoned to demonstrate their commitment to the system by answering the following questions:

> From your role, what is your commitment to Brinsa's Process Safety implementation?
> What strategy will you use for the implementation?

Indicate which are the 5 activities you will carry out immediately to improve the Process Safety development in your area or sector

All of the resulting material was collected in a report which was subsequently raised to general management and was a significant contribution to the further planning and implementation

Additionally, at this first stage, a 4-hour workshop was developed for the top management of the company, where top management members took part, including the company's CEO. This was key for having the commitment of the highest authorities in the organization on the management system implementation.

Table 1 shows the training that was carried out covering a total of more than 2000 men-hr of training in safety process topics by the end of 2020. During 2021 it is intended to train operators and personnel in sectors not associated with operations such as contracts, human resources, and purchases through specially designed courses for these groups.

Table 1. PS Training Course

Course Name	Top Management	Directors	Senior Management	Supervisors	Operators	Total Men-hours
RBPS Introduction			168	240		408
PSM Audit		24	168	192		384
PSM Audit Refresh for Auditors		8	56	48		112
RBPS Basic - Virtual			40	32		72
Process Safety Advanced		24	336	288		648
Inspiring Leadership	12	12				24
MOC Sensitization			46	40	180	266
MOC Implementation			46	40	180	266

Total	12	68	860	880	360	2180

At the end of the first training, an 8-hour workshop was held to determine the planning for the implementation of the 4 pillars and the 20 elements of the Process Safety management system. The managers, chiefs, and leaders who participated were divided into multidisciplinary working teams.

The consultant developed a tool that contemplated an assessment for each element the following

- Element's impact on the organization's risk level
- Element's level of development at the organization
- Resources availability

Rating Guide
Very High - The implementation will immediately have a very strong impact on the prevention and mitigation of the risks of the installation (Weighted as 4)
High - The implementation will immediately have a considerable impact on the prevention and mitigation of the risks of the installation. (Weighted as 3 points)
Medium - The implementation will immediately have a moderate impact on the prevention and mitigation of the risks of the installation. (Weighted as 2 points)
Low - The implementation will immediately have a low impact on the prevention and mitigation of the risks of the installation. (Weighted as 1 point).

Impacto del Elemento en el Nivel de Riesgo de la Organización				
Process Safety Commitment	Very	Hig	Mediu	Lo
Proces Safety Culture	X			
Standar Compliance	X			
Proces Safety Compdetencies			X	
Workforce Participation		X		
Disclosure to Stakeholders	X			
Current score count by ranked category	3	1	1	0
Percentage of current scores by ranked category	60%	20%	20%	0%
Average Score	3.40			
Understanding Hazards and Risks	Very	Hig	Mediu	Lo
Knowledge Management Process	X			
Hazard Identification and Risk Analysis	X			
Current score count by ranked category	2	0	0	0
Percentage of current scores by ranked category	100%	0%	0%	0%
Average Score	4.00			
Risk Management	Very	Hig	Mediu	Lo
Operative Procedures	X			
Safe Work Practices	X			
Asset Integrity and Reliability	X			
Contractor Management		X		
Guarantee Training and Performance	X			
Management of Change	X			
Operational Availability	X			
Conduct in Operations			X	
Emergency Management	X			
Current score count by ranked category	7	1	1	0
Percentage of current scores by ranked category	78%	11%	11%	0%
Average Score	3.67			
Learn from Experience	Very	Hig	Mediu	Lo
Incident Investigation	X			
Measurements and Metrics		X		
Audit		X		
Management Review and Continuous Improvement	X			
Current score count by ranked category	2	2	0	0
Percentage of current scores by ranked category	50%	50%	0%	0%
Average Score	3.50			

Figure 1. Element prioritization for PSM

The teams had to classify each elements:

- High
- Medium
- Low

Then the tool performed a weighted calculation to determining the implementation priority of the each element of the management system.This was done to achieve the greatest possible impact in reducing the risk of the installation, to assure short-term implementation progress results and to maintain a very strong commitment of the organization. The workshop results and recommendations were presented to the company's management team and finally, after a few requested changes, it was approved. The plan contemplated 3 stages with priorities 1, 2, and 3 for the implementation of the different elements. The elements Process Safety Culture, Management Review, and Continuous Improvement (Stage 0) they would start in stage 1 and continue to develop during the entire implementation process. It was estimated that all the stages would be completed in 3 years. Subsequently, based on the experience, the maturation of the program, and the impact of the pandemic, it was updated extending to 5 years.

4 Organizational Strategy for Implementation.

A Company has to strive to make sure that Process Safety activities are completed in such a way that it minimizes the risk for employees, the public, and the environment, whichever the risk level. The activities that have high-risk levels could require higher attention levels from Management and higher resources assignment. The RBPS methodology intentionally guides the companies to direct efforts to the highest-risk levels sectors.

In compliance with the RBPS strategy, the organization decides to start with the Chemicals Plant, due to the high risk of the process and the results obtained from the diagnosis assessment. Likewise, a specific task force was created, in accordance with the auditor's recommendations and the industry best practices.The focus of the task force would be to develop the implementation methodologies for each of the elements of the RBPS system, follow up on the work plan, ensure interdisciplinary participation from the different sectors in the organization and ensure the system sustainability. The "Process Safety" team was created under the Quality Management integrated by a Chief, 3 professionals, and 1 analyst.

In parallel, the maintenance area team designed its preventive maintenance model and strategy. The Assets Integrity and Reliability task force, which was assigned to the Engineering and Maintenance Management, composed of 1 chief and 5 engineers. It had specific focus on preventive and predictive maintenance. The Technical Documentation center attendants were included in this work team,.They are responsible for centralizing and managing all the organization's projects and processes information.

With the emphasis on the implementation of the elements, a work plan was established with different activities, responsible and execution dates to strengthen the implementation process and to follow an effective realistic schedule.

The Process Safety Area, in alignment with the other company areas, will follow up on the progress and closure of the assigned activities. In Figure 2 a general schedule is presented of each of the elements with its implementation general plan. It should be noted that the Process Safety Culture, Management Review and Continuous Improvement elements are to be addressed the entire 5 years of implementation. Other elements started in 2018 will continue their work strategy with assistance of the consulting firm.

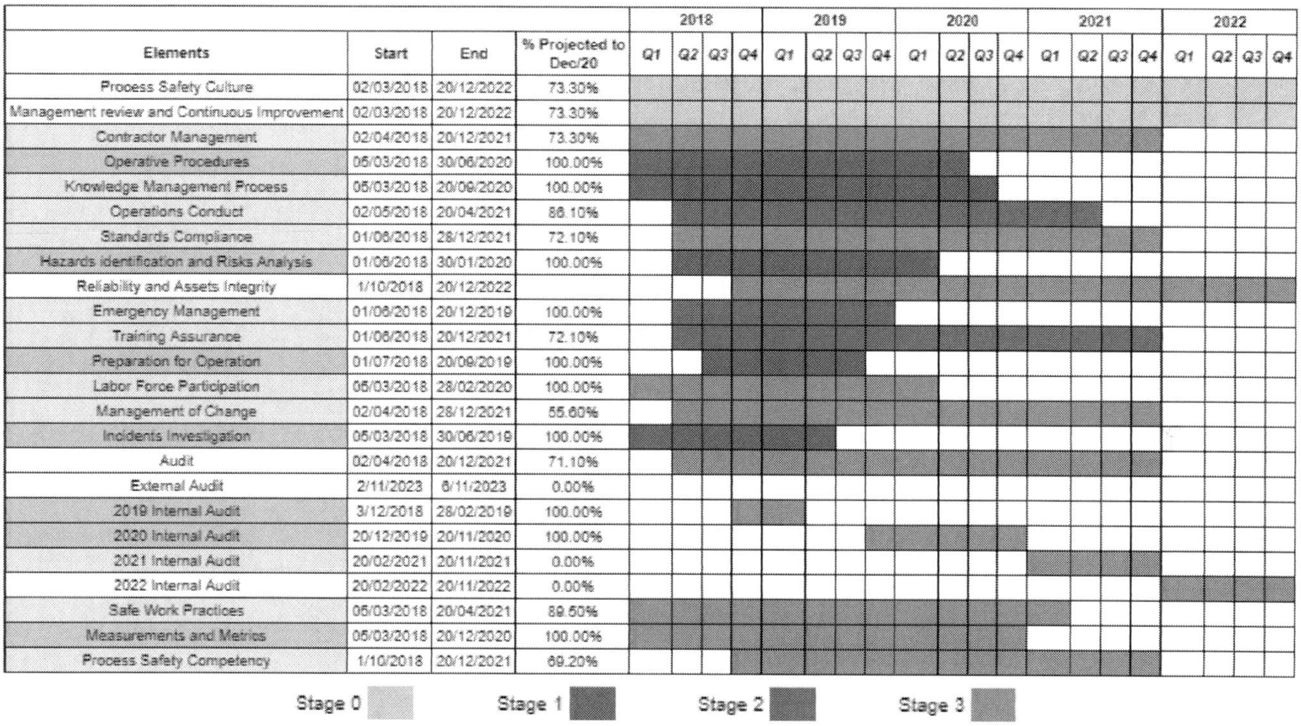

Elements	Start	End	% Projected to Dec/20
Process Safety Culture	02/03/2018	20/12/2022	73.30%
Management review and Continuous Improvement	02/03/2018	20/12/2022	73.30%
Contractor Management	02/04/2018	20/12/2021	73.30%
Operative Procedures	06/03/2018	30/06/2020	100.00%
Knowledge Management Process	05/03/2018	20/09/2020	100.00%
Operations Conduct	02/05/2018	20/04/2021	86.10%
Standards Compliance	01/06/2018	28/12/2021	72.10%
Hazards identification and Risks Analysis	01/06/2018	30/01/2020	100.00%
Reliability and Assets Integrity	1/10/2018	20/12/2022	
Emergency Management	01/06/2018	20/12/2019	100.00%
Training Assurance	01/06/2018	20/12/2021	72.10%
Preparation for Operation	01/07/2018	20/09/2019	100.00%
Labor Force Participation	05/03/2018	28/02/2020	100.00%
Management of Change	02/04/2018	28/12/2021	55.60%
Incidents Investigation	05/03/2018	30/06/2019	100.00%
Audit	02/04/2018	20/12/2021	71.10%
External Audit	2/11/2023	6/11/2023	0.00%
2019 Internal Audit	3/12/2018	28/02/2019	100.00%
2020 Internal Audit	20/12/2019	20/11/2020	100.00%
2021 Internal Audit	20/02/2021	20/11/2021	0.00%
2022 Internal Audit	20/02/2022	20/11/2022	0.00%
Safe Work Practices	05/03/2018	20/04/2021	89.50%
Measurements and Metrics	05/03/2018	20/12/2020	100.00%
Process Safety Competency	1/10/2018	20/12/2021	69.20%

Figure 2: Management System Work Plan Implementation schedule

Next, the results of some elements of the system are presented, and the advances obtained during their development and implementation.

4.1 Hazard Identification and Risk Analysis.

The second pillar of the RBPS system "Understanding Hazards and Risks" is supported by two elements:

- Knowledge management
- Hazard identification and Risk Analysis

This pillar was identified at the diagnostic stage as High Priority(See Figure 2). Thus both elements start to be developed simultaneously, given the interdependence that exists between them. The organization decided to adopt HAZOP as an evaluation tool, complementing the company's risk matrix. HAZOP studies were developed for 10 chemical plant processes, with the participation of the entire interdisciplinary team. This made it possible to identify the current situation of the plant and assess the hazards according to the organization's risk matrix. The study was divided into two stages. Onstage implemented in 2018 included the Chlorine Circuit, Electrolysis, Hydrogen, Hydrochloric

Acid, and Sulfonation processes resulted im 162 recommendations categorized as very high and high risks The second stage started in 2019 included the CEDI (Chemical Distribution Center by its Spanish acronym meaning), Sodium Hypochlorite, Caustic Soda, Calcium Chloride and Demercurization processes resulted 176 very high and high risks recommendations.

In general, there's a high risk at the analyzed plants. 72% of the identified scenarios are at extreme or high risk. The recommendations generated by the HAZOP team when implemented significantly reduce the risk. They mainly focused on strengthening the designs with engineering measures to prevent and mitigate incidents.

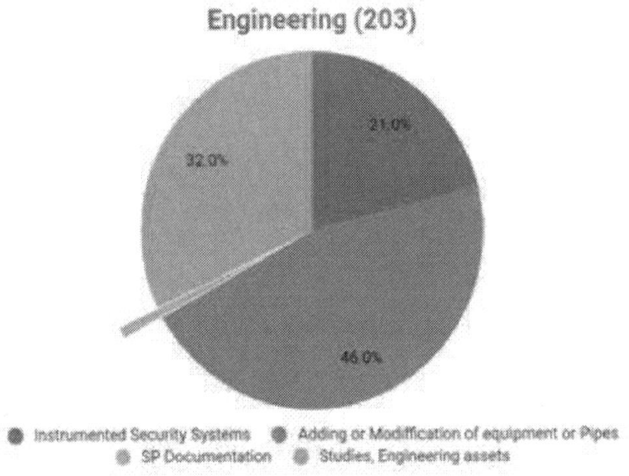

Figure 3: Engineering Recommendations

About 60% of recommendations are engineering improvements, which include among others, instrumented safety systems installation (21%), modification or adding new equipment or piping (46%), engineering studies (32%).

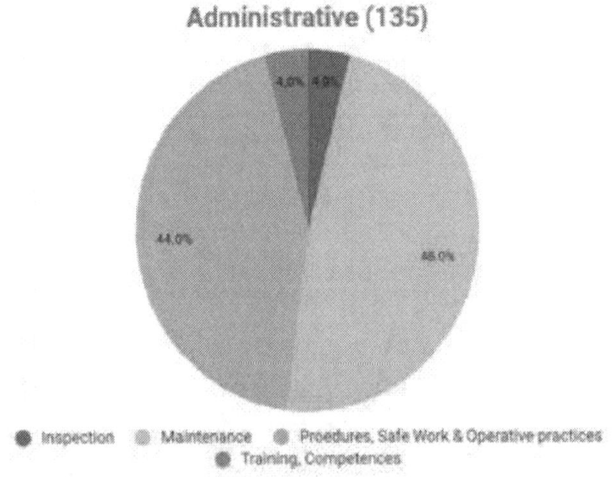

Figure 4: Administrative Recommendations

40% of remaining recommendations were administrative barriers improvements such as inspections, maintenance, training and qualification, procedures, operational and safe-work practices, and in general, process safety standards (Figure 4).

A 3-year work plan was defined for closing gaps, with budgetary assignments and defining specific task forces for gap closing.

4.2 Operating Procedures and Performance Assurance.

The RBPS element that ensures adequate development, timely maintenance, and the consistent use of operative procedures (procedures) is one of the 9 elements at the RBPS pillars: Risk Management.

Good operative procedures, among other things, describe the process, the hazards, tools, protective equipment, control systems and also provides troubleshooting instructions when a non-expected behavior emerges. They also address abnormal situations such as emergency shutdowns and temporary operations.

Operative procedures at Brinsa didn't comply with many of the requirements outlined above. Therefore, a standard is generated for drafting operative procedures according to the CCPS recommendations. This way, the participation of the operators in drafting procedures, created synergy within the personnel expertise and the information related to the operation manuals, the review by the supervision, and the coordination and approval by the Plant management. This procedure ensures that the input of different management levels is guaranteed and the documentation is appropriate. The drafting of all operative procedures required about 2 years. The final result was much more user friendly procedures, with adequate language for easy comprehension by users. Pictures were also included and unambiguous instructions of the steps to follow during the operation.

By the end of 2019, after the information and procedures updated and approved for the processes of Compressors, liquefaction, drying and Chlorine storage, de-chlorination and brine treatment, sodium hypochlorite production, calcium chloride and hydrochloric acid, a personnel retraining program was implemented. It included the technical skills assessment of the operators. it included 3 parts:

- Performance list: It is an evaluation of the knowledge of operational practices in the field.
- Theoretical evaluation of the process: The questions are asked taking into account the operating procedures.
- Cause-Effect evaluation: It measures the ability of the operator to understand the main risks that can be produced by the process and his role to minimize, delete or correct the risk.

After completing the first evaluation, feedback on weaknesses was carried out and subsequently, training was given to each operator. Finally, a second evaluation was carried out, emphasizing the non-conformities. When the training was completed all of the operators had a rating above 90%.

After updating the operative procedures, the first phase of generating the Pre-startup Safety Review (PSSR) lists was started, with the same creation, review and approval flow. In this first phase, the lists only contemplate start-ups after programmed stoppage or non-programmed stoppages without changes or modifications of the process were included. The documentation was developed during 2020. A second phase is planned for 2021that will ensure that the RFSR lists will be also mandatory for new projects.

4.3 Incident Investigation.

Developing, sustaining, and enhancing the organization's incident investigation competency is one of four elements in the RBPS pillar of ***learning from experience.*** (1)

In accordance with RBPS methodology, the company implemented a process for reporting, tracking and investigating incidents that include: 1. a formal process for investigating incidents, including staff provision, conducting, documenting and tracking process safety incidents & 2. Incident and incident investigation data trends to identify recurring incidents. This process also manages the resolution and documentation of recommendations generated by the investigations.

From the implementation of the process, a culture of reporting process safety incidents, including near misses, gradually began to emerge. Reports include, but are not limited to, chlorine leaks, spills (or drips) of liquid and corrosive substances, safety valve activation, fire outbreaks, among others. These events are investigated by interdisciplinary teams (mechanical, electrical maintenance, reliability, operation, process safety) through different investigation methodologies, with the use of Tap Root® being preferred in the organization.

The investigations involve the socialization of the lessons to be learned from the events that occurred at all levels in the operation, through daily operation meetings (DOM), and the lessons are left at different points of the plant for a free consultation. In Figure 5, the lesson-to-learn model currently being socialized

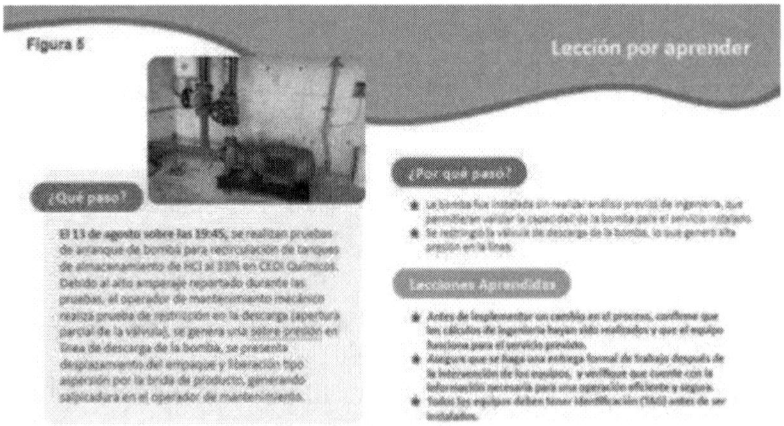

Figure 5: Lesson-to-learn model being socialized

With the implementation of this element, different levels of the organization have been reached, a culture of both reporting and interdisciplinary understanding of the events is being generated.

In Figure 6, the increase in process safety incidents reported by the Chemical Plant is evidenced. The conclusions and recommendations of these investigations led to substantial improvements in the processes, increasing safety and operational reliability.

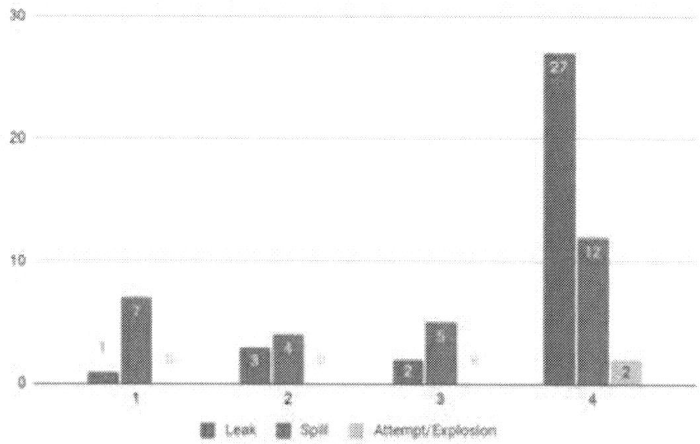

Figure 6: Process Safety reported incidents

During the first 3 years of implementation, activities of all the main elements that have allowed a systematic advance in the implementation have been developed. This has also made it possible to unite efforts from different fronts, create a comprehensive system and receive genuine support from the senior management team.

4.4Management Review and Continuous improvement

Since 2018, monthly monitoring is carried out by the management team and operations leadership to review the progress in the implementation of the process safety management system, opportunities for improvement, delays, budget execution and redefinition of strategies that leverage the system implementation. In these evaluation meetings, the gap closures identified in risk analysis and incident investigation are also reviewed. The results and conclusions of these evaluations are presented and analyzed every 4 months at the company's shareholders' meeting.

5 Investment

The RBPS implementation requires allocating human, material and financial resources to implement a robust, consistent and sustainable system over time.

The leadership of the organization, understanding its responsibility, approved for the first 4 years of implementation an approximate budget of 10,000,000 USD. This budget was assigned, among other things, to Training and performance assurance, incorporation of resources, developing the procedures and policies required by the RBPS and implementing improvements in the plant arising from the risk studies, the requirements of the management system and the conclusions of the investigation of process safety incidents.

Some improvements and changes involve large projects, and others are "minor" improvements that seek to reduce the human factor as a source of risk.

Below we will describe some of the main improvements made to the installation:

Around 30 level transmitters were installed in different process tanks, to eliminate the level glasses with which the level visualization was conducted (Figure 7).

BEFORE	AFTER
15% Sodium Hypochlorite Preparation tanks 688 / 689. Level visualization through a sight glass, without an alarm system to alert the overflow of the tanks.	Installation of level transmitters with alarms, giving the operator greater control over the operation and identification of substandard conditions, both due to overflow and lack of power.

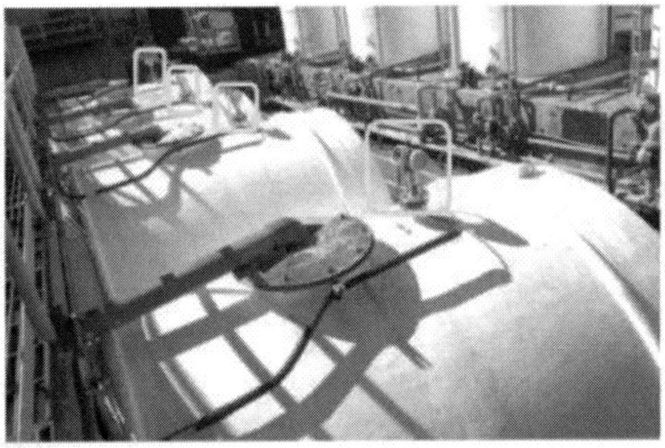

Figure 7: Level Transmitter, 15% Sodium Hypochlorite Preparation Tanks

A project was created for the installation of new liquid chlorine storage tanks, meeting the minimum requirements defined by the Chlorine Institute pamphlets, the EUROCHLOR technical definitions, and the available ASTM and ASME construction standards, in replacement of existing ones that did not comply with the RAGAGEP standards mentioned above. The storage system will start operations in the second half of 2021 (Figure 8).

Figure 8: Liquid Chlorine Storage Tanks

The implementation of the safety management system allowed the organization to know in greater depth the hazards associated with the handling of hydrogen and for this reason it was decided to carry out a specific HAZOP study for the system with the advice of a hydrogen expert. The study incorporated the risk rating using the company's matrix and based on this, approximately 40 recommendations for improvement were obtained. Most have already been implemented, especially those associated with the elimination of equipment and systems that were not operating after the change of technology from mercury cells to membrane cells.

BEFORE

Equipment in the out-of-service area of the hydrogen filtration system, excessive obstacles in the area.

AFTER

Dismantling of equipment, cleaning the area, facilitating access to operational areas. Image was taken on May 30/20

Figure 9: Hydrogen Filtration System dismantling

In the HAZOP study carried out on the hydrogen system, it was identified that in the hydrochloric acid receiving tanks there was an imminent risk due to the presence of an explosive atmosphere (accumulation of hydrogen gas). This recommendation was verified with the technology provider, and an engineering study was started to implement the improvement. Tank blanketing, or padding with N2 was installed.

BEFORE	AFTER
Mersen Recommendation (HCl Furnace Supplier): Ensure nitrogen purge in the process tanks with minimum flow control to avoid an explosion due to hydrogen accumulation. In previous years one of the tanks had an explosion with the expulsion of part of the tank. Rec. 315. Consider adding nitrogen as a blanketing gas to these Receiver Tanks.	Connections and adjustments are made in the receiver system and gas scrubbing to maintain a permanent purge to both receiver tanks of HCl guaranteeing the displacement of remaining gases and avoiding any accumulation of hydrogen inside.

Figure 10. Blanketing gas to the hydrochloric acid receiving tanks.

Several processes in the facility still have a highly manual operation. During the HAZOP study, several recommendations were made associated with the automation of some processes. The figure shows the implementation of the automatic valve to guarantee the start-up of the backup fan of the sodium hypochlorite fan, to avoid a plant shutdown and a chlorine leak due to the pressure increase in the chlorine collector.

BEFORE	AFTER
Rec. 20. Implementation of the automatic start of the spare fan in case of fan failure in service. Risk: Loss of vacuum in the absorption tower and	Automatic valve installation and control loop programming was performed.

loss of receipt of Cl2 with stoppage of the chemical plant due to increased pressure in the Cl2 receipt lines and subsequent risk of Cl2 leakage into the atmosphere with risk to people of the area.

Figure 11. Automatic startup of the sodium hypochlorite backup fan.

Consideration of the "human factor" is a key aspect of a process safety management system. This discipline takes into account, in the work environment, the interaction between people, facilities and equipment, and management systems. (2) Based on these concepts, the organization decided to incorporate them into many aspects of the management system, for example, to improve the consistency of all technical documentation, including operating procedures and PSSR checklists, which includes updating of P&IDs, datasheets, among others, a program was launched to improve the identification of all equipment, instruments and pipes in the plant. (See Figure 10, Figure 11, Figure 12).

BEFORE

Unidentified valves can lead to their misalignment and therefore a mixture of chemicals that can give off dangerous gases Cl2, H2S, generating risk to people in the area.

AFTER

Valves are tagged, minimizing operational errors due to confusion.

Figure 12. Valves tagging

BEFORE

Tanks 769 and 779 unidentified.

AFTER

The identification of the auxiliary storage tanks of sodium hypochlorite at 15% is carried out.

Figure 13. Tanks tagging

BEFORE

Unidentified pipe, an event occurred due to mixing of raw materials due to misalignment of valves.

AFTER

The identification of the pipes is carried out to minimize errors in the direction of fluids.

Figure 14. Pipes tagging

6 AUDIT

The only way to determine that a PSM program is working properly is to thoroughly examine both its design and implementation on a regular basis. Therefore, a PSM audit represents a way to measure the effectiveness of the PSM program and also allows a comprehensive comparison with predetermined PSM metrics. If the PSM program is not measured carefully, it cannot be controlled or improved. (3)

A successful PSM program audit begins with a commitment by the Company's senior and facility management to conduct periodic audits of these programs. This is essential to allocate appropriate resources to conduct the audits, ensure that findings and recommendations are actively and carefully addressed promptly, and ensure that on-site activities allow a successful development of the audit.

(Successful PSM program auditing according to CCPS begins with a commitment by the senior company and site management to perform periodic audits of these programs, to allocate the appropriate resources to perform the audits, to ensure that the findings and recommendations are actively and carefully addressed in a timely manner, and to ensure that the activities on-site during the audit are arranged to support the audit to the maximum extent possible.)

These principles were strongly adopted by the leadership of the organization and based on this, the process safety audits program was established to accompany the implementation of the system in order to control its effectiveness and make adjustments and corrections promptly. The plan first contemplated an internal training process and the development of audit protocols for the 20 elements of the RBPS. All this material was based on the CCPS audit book

Then it was decided to start with internal audits for partial elements of the management system. So far, 2 internal audits have been carried out, one in person in 2019 and one in 2020. Despite the challenges posed by the global situation due to the pandemic, the company decides to continue with the internal audit plan in 2020, counting with remote assistance from the consulting firm and blended participation by the audit team and the audited team, seeking to guarantee the biosafety protocols defined by Brinsa.

The audited elements in 2020 are in an advanced development process with partial implementation; a greater understanding of the spirit and scope of the audit and a more agile development were observed.

Ultimately, and beyond the qualification and findings, there is evidence of growing knowledge and commitment of most sectors of the organization with the process safety subject.

Procedures and tools have been incorporated that are strengthening the entire process safety management system. The Asset Integrity and Reliability element shows a significant improvement in its rating (2.44 in 2019 vs. 3.19 in 2020) derived from the important incorporation of procedures and tools for planning and monitoring the reliability and maintenance tasks of the equipment to be throughout their lifecycle.

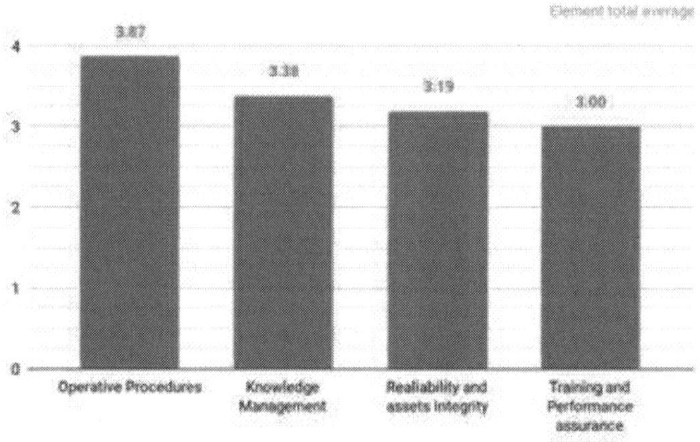

Figure 15: Qualification of audited elements 2020

The "operating procedures" element has notably improved in the second audit compared to the first one. It is worth highlighting the existence of a procedure that establishes the scope, format, responsibilities, etc. for the development of the operating procedures, and promoting the active participation of the operators in carrying out / reviewing them.

7 CONCLUSIONS

The most important conclusions of this case study are:

- An RBPS can be applied with very good results, in an organization with many years of operation and a culture focused only on personal safety.

- Awareness and training in the concepts of process safety before the implementation of the management system, at all levels of the organization, especially leadership, are essential to achieve understanding and awareness of its importance.

- The support and commitment of the organization's leadership is key to implement a process safety management system in organizations that are unaware of the issue and their safety concern is exclusively personal safety. Leadership must provide the human and material resources and make the necessary investments for a correct implementation as happened in this case.

- Implementation poses a huge challenge when starting from scratch. However, it can have very good results in the short and medium-term if it is properly planned and with the active participation of all levels of the organization in such planning.

- The proper selection of the elements of the management system on where to start the implementation is essential to try to quickly reduce the risks of the installation and at the same time show significant progress that maintains the interest and commitment of all levels of the organization, especially the leadership.

- The implementation requires a specific sector of Process Safety to lead it, monitor progress and solve limitations and challenges during the process.

- The implementation of the management system introduced benefits in other areas of the organization such as production by achieving increased reliability and operational continuity through the creation of a previously non-existent asset integrity group.

8 Bibliography

1. **Center for Chemical Process Safety, American Institute of Chemical Engineers.** *CCPS. Guidelines for Risk-Based Process Safety.* New York: Wiley, 2007.

2. **Center for Chemical Process Safety, American Institute of Chemical Engineers. CCPS. .** *Human Factor Methods for Improving Performance in the Process Industries.* New York: Wiley, 2007.

3. **Center for Chemical Process Safety, American Institute of Chemical Engineers. .** *Guidelines for Auditing Process Safety Management System.* New York: Wiley, 2011.

SPRING21
+17TH GCPS
A Joint AIChE and CCPS Meeting

BS Team: Finding Competence Gaps

Salvador Ávila
UFBA
avilasalva@gmail.com

Luiz Fernando Lopes de Souza
COPENOR
lflsouza@globo.com

Gunther Josua Costa
UNIVASF
guntherjosua@gmail.com

Lucas Menezes Pereira
UFBA
lucasmpereira13@hotmail.com

Prepared for Presentation at
American Institute of Chemical Engineers
2021 Spring Meeting and 17th Global Congress on Process Safety Virtual
April 18 - 22, 2021

BS Team: Finding Competence Gaps

Salvador Ávila
UFBA
avilasalva@gmail.com

Luiz Fernando Lopes de Souza
COPENOR
lflsouza@globo.com

Gunther Josua Costa
UNIVASF
guntherjosua@gmail.com

Lucas Menezes Pereira
UFBA
lucasmpereira13@hotmail.com

Keywords: BS, Competences gaps, Decision models

Abstract

The academy form competences to work in the routine, without worries, without unknown hazards. The rules and objectives in routine activities are clear and require operational discipline to control the cost of production and product quality. The current educational model does not prepare society and groups for an emergency or even for decisions in a crisis, where resources are scarce mainly because the necessary knowledge about danger and its mechanism of action is lacking.

What really happens in the chain event scenario when the danger is unknown? What are the typical models of action of the emergency leader and his team? What are the required actions in crisis management required for a BS type event considering the scarce resources, the high level of stress and the direction of the danger energy? Which models are indicated for the situation of unknown danger and its intense energy flow with high impacts during these events? What is the appropriate mind map of those who lead and execute the actions considering the low level of visibility in the events and the dynamic of geopolitics in the BS scenario where, depending on the preparation for emergency, in case of intense demand for attention and action, possible modifications may occur in the map mental for decisions?

This work intends to review the discussions already performed about new concepts and tools to be used in the crisis prevention area. Between this works we will understand the effect of the stress level on emergency decisions valuating human error and observing team response using LODA tool. The Human elements designed in the organizational culture, can stop the high danger energy that can flow from some industries. The principal aspects to be constructed and monitored are technology (risk and complexity); management (leadership and stress); and behavior (cooperation, commitment, competence, and communication).

By the other hand, human factors are analyzed to avoid situations which hazard energy carriers during top events, as the effect of cultural aspects to chain reaction until the occurrence of disaster. Decision models for the emergency brigade (leader and team) indicate the motor and cognitive gaps that cause lack of control during contingencies. Finally, the analysis of simulation types, by rotating observation of field simulation allow find the deficiencies of the emergency team in a real situation in the future.
The methodology divided in following steps are dynamic: first, rescue the discussions about stress and the BS events and, in second step, recognize the intense hazard energy flow through failure events, antecedents, elements and human factors. In third step, we try to locate the gaps and environmental conflicts influencing losses from accidents. Finally, we assess the current event of covid-19 pandemics to suggest correct emergency preparedness actions in the preparation of leaders and their teams regarding cognitive and motor characteristics in situations of high demand. When comparing the characteristics already described by TALEB about BS with the event COVID-19, we intend to carry out a case study to indicate the possible causes this global impact crisis, or even, understand the technological disaster causes where uncontrolled events from chemical, nuclear, and oil industry begins chain reactions until the disaster. Unfortunately, a Crisis situation can be caused from several intricated dimensions, making difficult the contingency and mitigation actions.

1 Introduction

The academy form competences to work in the routine, without worries, without unknown hazards. The rules and objectives in routine activities are clear and require operational discipline to control the cost of production and product quality. The current educational model does not prepare society and groups for an emergency or even for decisions in a crisis, where resources are scarce mainly because the necessary knowledge about the hazard and its mechanism of action is lacking.
The knowledge learned to graduate the engineer is limited to logic and objective concepts. The students do not discuss different concepts, as social culture, legislation, workstation behavior at the same time, mixing objective aspects with subjective ones. We need to understand Cultural traits, geopolitical aspects, intrinsic variables from worker during routine to discuss about response models for emergencies, necessary to avoid disasters.

Unfortunately, we only look for new competences after the disaster happen! Why?

The metrics for Process Safety management are discussed in methods indicated in the CCPS books [1]. Some important issues such as: management and investigation of major incidents [2]; technological changes that cause accidents [3]; lack preventive actions for organizational changes [4] are exemplified and barriers are suggested, so as not to affect the safety of processes with big accidents.
These issues and solutions seek to avoid the accident through management actions, technological systems, and the attempt to control human factors [5] based on the lessons learned by various industries and associations when investigating the events that occurred. This investigation of the past, looking back to the facts, has a limit when external, dynamic environments threaten normality. The crisis scenarios can be very dynamic, not visible and of parallel occurrence, at the same time as those caused by climate change, by social conflicts, by conflicts arising from

degraded geopolitics, and by the effects of multiculturalism in the world of work. and in society - indicating low tolerance behavior.
In 2018, CCPS [6] with the leadership of Eastman, raises the issue of normalization of deviation, indicating repeated situations of acceptance of operational deviations, in the shift routine, are considered normal situations. The lack of research with the indication of bottom-up techniques for the analysis of operational culture and the respective heuristics turns the book into another attempt to guide the occurrence of events looking to the past and without an appropriate view of what is happening in the present. and what will happen in the future, on the factory floor. What are the social emergent that cause deviations to normalize?

What are the social emergent that cause deviations to normalize?

In this context, analyzing the internal and external social threats to avoid environments prone to deviations gains importance. Risk aversion, excessive self-confidence, and fear of guilt for human error are considered cultural events that produce underreporting. This phenomenon changes the shape of Bird's pyramid [7] to the Christmas tree as shown in Figure 1 and causes the accident to occur already with the chain reaction triggered. The assessment and actions resulting from the notification (not performed) could avoid the domino effect or reduce the risk of major events.
The accumulation of deviations without treatment and with high level of underreporting, a cultural event, increases the likelihood of Black Swan events. Failure to recognize the Scenario for hazard identification and risk analysis that implies a lack of insight into the operational context and creates uncontrolled situation with natural, technological, political disasters. The Black Swan Event goes beyond the local scale and reaches a global scale, with unknown dangers, or already known dangers, but due to lack of attention, it has the characteristics of unknown danger because it has not been investigated as to the mechanisms of action. We do not know what the appropriate barriers are. Thus, we classify the Black Swan because of totally unknown or operationally unknown dangers.

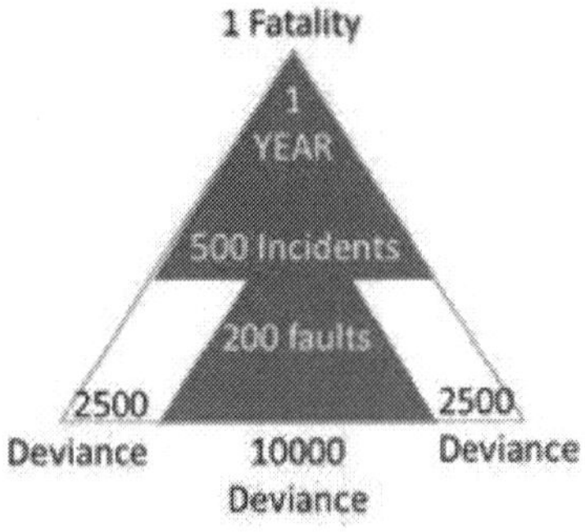

Figure 1. Christmas Accident Tree

2 Literature Review

This work intends to review the discussions already performed about new concepts and tools in the crisis area. Between this works will understand the effect of the stress level on emergency decisions valuating human error and observing team response using LODA tool. The Human elements in the organizational design can stop the high hazard energy flow of some industries. The principal aspects to be constructed and monitored are technology (risk and complexity);

management (leadership and stress); and behavior (cooperation, commitment, competence, and communication).

By the other hand, human factors are analyzed to avoid situations in which hazard energy carriers during events, from cultural aspects to chain reaction until disaster. Decision models for the emergency brigade, leader, and team; indicate the motor and cognitive gaps that can result in lack of control during contingencies. Finally, the analysis of simulation types, by rotating observation of field simulation allow find the deficiencies of the emergency team.

Given the great stress present in the geopolitical situation, in which there is a constant struggle for territory, renewable and non-renewable energy sources, in addition to climate changes intensified by anthropic activity, large flows of Hazard Energy (HE) are generated that are commonly not perceived. The unnoticed concentration of this energy gives rise to what is known as latent hazard, and when it evolves to levels that can no longer control or filter its pending of burst and destruction, a megaevent of disaster occurs, which has recently been called the Black Swan.

Due to the lack of a notion of the hazard to which disaster victims are subjected, it is said that the Black Swan is still unknown to the population, and because they are often unprecedented occurrences, they are considered rare and of high impact in multiple respects. For [8], the victims of the twin tower attack on 9/11 were victims of a Black Swan. It was not possible to locate where the hazard energy was concentrated, nor measure it. Gray Swans, according to the same author, are unlikely but impactful events that are outside the normal scenario.

A Black Elephant, according to [9], is a significant threat, risk or issue that is obvious to everyone, but no one wants to address or deal with it, "they get ignored". When it causes high impact, people get surprised and act as if it were an unpredictable black swan.

There are new models of behavior to include high stress and ways to measure this feature through the radar chart. This chart presents the characteristic of presenting multivariable data in the form of a graph.

The managerial, human-social, and technological aspects can create an environment for unsafe behavior, in which they allow the flow of hazard energy across barriers. The human and organizational factors elaborate, by events, the condition for the transfer of the social hazard towards the physical realization of the accident or incident.

In order to prepare people to deal with risk situations and be able to participate in decision-making processes that report to their own lives, it is necessary investment in: 1) Integrated Communication - Training and language competence are really important. Subjective factors such as judgments, culture of guilt and fear of damage to the image cause the worker to omit information that could contribute to the construction of barriers to avoid incidents and accidents; 2) Stakeholders - person or group who have an interest in a particular business, industry or service; 3) Intellectual Capital - Knowledge of preventive and mitigating actions provides people with adequate decision-making; 4) Competence - Understanding the causal nexus of the phenomena involved in the accident, having experience and experience similar to the activities related to the events.

3 Models & Hypotheses

The discussion about the competence of Black Swan Team beginning answering questions after observing real situations of emergency. After we feel some gaps and try to answer after investigation about the reasons of emergency team perform this behavior. After each question we construct or explain the scenario and answer with new tools that will compound the methodology.

3.1. What really happens in the chain event scenario when the danger is unknown?

We have nothing to perform about the existence of a disaster from an unknown danger as in actual case of a pandemic. The disaster is a truth, we must reduce its impact, investigate the various dimensions involved with this unknown danger, or unknown operational danger considering the cultural and environmental effects during the isolation and flexibilization.
The study of the origin of the disaster allows us to understand its mechanism of action and possible consequences. The danger that causes the accident-disaster of technological (industry) or natural (climate, virus) origin can be acute, being a fast event with high impact, or it can be chronic, lasting a long time with, probably, immense impacts. Thus, the event becomes a public accident when it affects the population. We classify the danger and consequent events of COVID19 as natural, chronic, and public. This discussion is part of Strategic Intelligence that define the need resources and Hazard Identification, and Analysis of General Risk to define and test the best barriers.
While the danger is at work in an external region or country (not close to our region) we will observe and learn about its mechanisms, we seek to install two instances: The Integrated Management Committee and the Strategic Intelligence Committee. The first will join the necessary resources and leaders to take decision and the second will look for the best actions, benchmark, learning from other countries and regions.
In Integrated Management, Specialists, Managers and Practitioners who are part of Public-Private Institutions meet to define studies that assist in the decision on actions. Some tests are suggested to find the mechanisms of action of the hazard. In this way, barriers are built for known dangers, for the unknown and for threatening environments. Indicators and parameters and variables are mined to understand the impact of the pandemic on the economy and social relations.
Strategic Intelligence intends to identify the source of the main hazard and parallel hazards to project the Future Crisis Scenario. Scenarios are designed to test barriers, variables, resources, and knowledge. We will value to learn from the past, and from abroad experience, sending representatives to recognize the mechanism of these dangers through visits to government and institutions. In this moment a good general knowledge about pandemic situation, technological disaster and natural, impacts on social conflicts and economy, will help do draw an initial description and after, inviting specialists, the crisis leader tries to construct and perform effective actions (corrective, preventive and mitigation actions).
After confirmation that the Scenarios indicate an active Pandemic, the Contingency Committees begin in specific and critical regions. In some cases, such as natural disasters involving an earthquake, breaking of barriers in the mining industry and other technological accidents (fire or toxic spill) we need to install a Contingency Committees. In cases of acute accidents, groups are sent for mitigation action. Therefore, operations and local intervention teams will be installed.
The Black Swan Team at the Strategic (Strategic Intelligence), Tactical (Integrated Management) and Operational (Command and operations) levels need the classification of the type of event and impact for the work of the teams. One of the cases may be the study of technological accidents such as the breaking of barriers in mining or the leakage of toxic gases near the city (similar to Bhopal). The other case can be related to public events of high capillarity about impact, a virus that contaminates quickly and through unknown routes, the current COVID19, or even a series of earthquakes that promote tidal waves and affect the functioning of cities. Or even, the possibility of multiple events occurring involving the technological and the public dimension.

3.2. What are the typical models of action of the emergency leader and crisis management required for treat a black swan type event considering the resources, the level of stress and the direction of the danger energy?

LODA (PROGRESSIVE STRESS MULTI-LYER MODEL) shows that the mind map change in extremely high stress with wrong decisions.

Ávila [10], knowing about the possibility of high impact of technological accidents near cities and after confirming the risk to the risk affected by the work team and the population through underreporting. Ávila confirms the occurrence of the phenomenon of cultural normalization of deviation and for the analysis of the consequence of progressive stress, he designed an accident involving an LPG fireball near a city in Brazil.

This event constructed from real events and deviations that occurred, although the top event did not occur, was composed of 65 diversions and successive events, sociotechnical failures in the form of the local routine and complemented with those that occurred in New Mexico. The Accident prepared for the Risk Analysis impacted with 40 fatalities (this simulation was structured in the thesis and published as a tool [10].

The analysis model involved 5 levels of stress in the team, where 1st is uncontrolled process with operational deviations and the last was the crisis scenario in the next city. Progressive stress disperses or disorganizes the mind maps that affect the emergency leader's decisions. In this model it is considered the occurrence of a theater with critical roles where there are communication failures or non-acceptance of the roles in the staging. Observing each phase of the accident and its cognitive stress will indicate social phenomena that represent problems in the treatment of accidents. These problems must be addressed before the actual demand for contingency.

EMERGENCY PERFORMANCE MODEL indicated that there are characteristics for the leader and the team that are intuitive, affective and cognitive and that it was not found in emergency team in these accidents of the history.

The studies by Carvalho [11] and Ávila et al [12] indicate that there are specific characteristics that are appropriate for the leader and the emergency team on real demand. In the case of leadership, we can mention perception of real risks (signs and scenarios); competence regarding the danger mechanisms (fire, virus, mechanical impact); energizing the team to act in contingency; a good memory to reinforce the decision; and speed of command since the time of the chain event, until the greatest impact, is very fast, and evacuation measures can save many lives.

In the case of the emergency team, we can mention: the quick cognitive to retrieve documents, a discussion, a phone call and assist the leadership in the decision through opinion and calculations; quick rescue of memory for the analysis of the scenarios, this aid sometimes brings information avoiding the direction of the danger to where there is greater vulnerability, for example, sending of chemical product to empty the burnt tank where there are visitors or maintenance service - the leader does not knows the level of vulnerability of the areas and depends on the team's support for this analysis; emotional balance brings a better quality of the cognitive and motor apparatus; ability to predict future situations through greater knowledge of the event's phenomena; and the ability to act on time, without delay, bringing firmness to field interventions.

Thus, Carvalho [11] and Ávila et al [13] affirm that the preparation of the emergency team mixes past experiences, attitudes, field and memory exercises, team energization, which are often not considered in normal training for teams of emergency. The results of the team's performance must be observed so that there is continuous improvement in performance under real demand.

COGNITIVE GAP ASSESSMENT IN REAL ACCIDENTS AND SIMULATION where we assess the emergency teams based on public and technologic accidents confirming deficiencies indicated by Carvalho, Ávila e Pichon Riviere.

This research work analyzed the national and international accidents regarding the characteristics discussed in the models for emergency action [11] [12]. The emergency team was analyzed in Accidents: (a) Chemicals and Oil of high impact in number of fatalities and environment (piper alpha - oil; Horizon - Oil; Union Carbide Bhopal - isocyanate); (b) national accidents of shipwreck in Mar Grande and explosion with fatalities at the FPSO São Mateus in the Brazilian state of Espirito Santo; and accident designed in the form of a simulation for the analysis of fireball risk by LPG in northeastern Brazil.

Specialists working in Risk Analysis, Contingency and Emergency Leadership met to classify the teams in terms of cognitive gaps and emotional control. This analysis was published at LACPS in Buenos Aires and it was concluded, in discussions with Process Safety Managers, that the classic way of training for emergency brigade teams does not meet the need to control real demand. Why? Although the situation presented in the training simulations for the Emergency Brigade, the petrochemical, refining, and oil industries seek typical situations for fire, fire, leaks and environmental impact, there are always organizational factors that affect human factors, there are always social factors that they cause conflicts and affect the quality of communication at work, and there are always false indications, a high backlog in the industry today. Thus, events always happen in multiple dimensions and the possibility of something unknown happening is real. Thus, the greatest value to be given is cognitive-affective-intuitive training for the emergency brigade. Lately, Process Safety Managers work only in Cognitive and without surprises. The level of stress and the style of the Leadership bring a level of comfort that may not be satisfactory for the occurrences of real demands.

COGNITIVE FAILURE IN DISASTER FIELD SIMULATION where the team is not always integrated and energized in the hot simulation situation, believing only in the obligation to be present. In this case, preparation through the simulation leaves cognitive and compromising gaps for the contingency in the disaster.

At the CIDEM, Mass Disaster Congress, held in Feira de Santana, UFBA was a rotating observer and investigator during a simulated exercise involving 25 public agencies including Armed Forces (Navy, Army and Air Force), Federal Police, Municipal Guard and State Military Police, SAMU, State Fire Department in addition to the BOPE anti-terrorist squad. Various functions of civil defense, social defense, isolation, and medical care came together in this exercise to prevent events of bomb explosion on buses, tipping of benzene trucks, and kidnapping, actions promoted by terrorist groups, all with an impact on the population and in nature, in the road modal.

The observation of these events indicates that the hierarchy of the agencies is above the recognition of their roles during the simulation and the inter-agency communication. The communication made with noise leads the team to different levels of urgency for mind mapping in the different teams. Some equipment was underused indicating lack of training, therefore, there are human and organizational factors that must be put into discussion to improve the integrated actions of the emergency command. The main characteristic is the lack of common priorities in the agencies promoting a disintegrated management state, where the size of the problem is not known for sure, only when faced with it. We note mainly that communication tools and devices are not used properly. Because it is a cold simulation, prescribed events of common knowledge, many communication failures did not appear. And this is a serious problem when analyzing previous events when municipal, federal, state forces and private governance act around disaster mitigation, for example, by preventing ship collisions. We note that the lack of credit of the great emergency teams on the simulated exercise can induce people to repeat mechanical actions bringing nullity to the Strategic Intelligence team.

3.3. Which models are indicated for the situation of unknown danger and its intense energy flow with high impacts during these events?

SYSTEMIC FAILURE OR FINGERTIP FAILURE MODEL, where the failure is multidimensional and involves from emotional balance, and knowledge - quick cognitive, we must analyze the connection between these properties to invest in selection and training.

The systemic failure models that indicate connectivity points of events of different dimensions to cause the top event. The format of the Systemic failure is like the digital of the fault that was published at the European Reliability Congress (ESREL).

EXERGY OF OPERATIONAL FAILURE AND COMPLEX PROCESSES where the indicators of complexity in the construction of BS are analyzed.

The discussion on the Exercise of Operational Failure [14] presenting that cognitive losses in the task are related to energy losses in equipment and processes. The presentation of The Investigation on Complex Processes through the tools that connect different sources that cause uncontrol in the process, interpreted through: Connectivity Diagram [15] [16]; of the Materialization Diagram where the failures that transform communication noises from management into a disconnected mental map in execution, are transferred into disturbance in the physical dimension; and finally the logical diagram of the fault indicating regions where to better install the barriers, in sociotechnical functions. These are models discussed at the Brazilian Congress of Chemical Engineering and applied in the Chemical Process Industry (aliphatic and aromatic amines in addition to liquefied petroleum gas).

NETWORK OF FACTORS AND HUMAN ELEMENTS, BARRIER EFFICIENCY, AND DANGER ENERGY SATURATION where we need to identify the points where danger energy enters saturation to elaborate the best barriers.

In recent times we have discussed ways to identify the regions in which the greatest flow of danger energy occurs, following the Model of Reason's Swiss cheese [17]. This danger flow circulates among the factors that influence human performance on the sequential, parallel and crossflow route. This multidimensionality of the failure follows the model in a network format of human and organizational factors divided into culture, management, routine and accident. This multidimensionality of the failure also discusses the safeguards of the technological, managerial, and behavioral dimension through the human elements that are design barriers to contain the danger energy that flows until reaching the maximum levels that are the accident and the disaster. Human factors are part of the routine of the operation and human elements are design safeguards. As described in Table 1, in the operation there are different factors that cause incidents and situations of loss. The nature of this event has the following dimensions: Cultural; Managerial; Routine control; and the scale of the accident and disaster. Also, in the same table we present the types of human elements in the project: Management; Technology; and Behavior.

Table 1 – Human Factors and Elements

OPERATION - A. CULTURAL DIMENSION OF HUMAN FACTOR: (1) Social Culture - (2) Social Phenomenon - (3) Organizational and Safety Culture. B. MANAGEMENT: (4) Management - (5) People & Groups. C. DIMENSION OF CONTROL IN THE ROUTINE: (6) Operational Control - (7) Failure Control D. ACCIDENT AND DISASTER DIMENSION: (8) Accident Control - (9) Disaster Control.	DESIGN - E. HUMAN MANAGEMENT ELEMENT (1) Stress - (2) Leadership F. TECHNOLOGY (3) Risk - (4) Complexity G. BEHAVIOR (5) Communication - (6) Competence (7) Cooperation - (8) Commitment

4 Methodology

The investigation was carried out following the steps: (1) Recovery from the discussion about stress in the treatment of emergencies, especially in the case of BS type disasters; (2) Recognition of the intense flow of danger energy through antecedent events, human safeguard elements, human factors of industrial operation and consequent events that hinder the return to normality; (3) to locate the environmental gaps and conflicts that influence process losses and the occurrence of accidents, chain reactions and the difficulties of returning to normality, for this we divided the competences in: Strategic Intelligence, Integrated Management, Command and action in Local Operations, and Organizational Resilience. For each competence, possible gaps, deviations, or human errors are discussed, comparing with the expected standard.
Finally, we assess the current event of covid-19 pandemics to suggest correct emergency preparedness actions in the preparation of leaders and their teams regarding cognitive and motor characteristics in situations of high demand.
Here in the methodology, we will indicate the competencies required for the BS team and the mental map of each competency for further discussion on the case of the COVID19 disaster.

4.1. Competencies for BS

As shown in Figure 2, Strategic Intelligence and Organizational Resilience have the purpose, respectively, to identify the danger and to restore the organization to the condition of normal, controlled operation, as it was before the disaster. Therefore, they outline strategies based on investigations of the past, the present, and the future, in addition to comparing local situations with external ones and in other countries. It is important to identify what the danger is and its mechanism of action, map the impacts to project the recovery. The Intelligence group is composed of risk analysts who work with scenarios of different dimensions (health, politics, technology, climate) and who can defragment dispersed information to build Future Scenarios and explain the events, even if the danger is still unknown, through testing and observation.
Another group that will guarantee the application of resources in the right place and that will effectively communicate the current situation involving the population and cities, industry, commerce, and services, is the integrated management group that will develop appropriate indicators for decision making from the disaster impact curve and from the data and information generated by strategic intelligence.
The local operations and command group works to carry out the activities indicated by the integrated and recommended management group, guided by strategic intelligence. The role of the

leader is important to energize the team and the role of the team providing cognitive and motor support in this achievement.

In this way, the Academy, Public Agencies and Risk activities such as the chemical industry and the oil industry need to structure themselves to prepare competencies that know how to act in the investigation of scenarios, in integrated public-private crisis management, in command and in carrying out operations. contingency and the recovery of local and regional activities.

Cold simulations [18] do not bring the expected result to structure contingency teams. Crisis Managers have not been effective in the application of resources, in communication and in the strategies drawn from the decision tools, as presented in the pandemic of COVID19 [19]. The risk management team is not able to design the scenario because it does not identify the danger and its mechanisms and impacts. We have no skills prepared for Organizational Resilience and we lose control of the situation during the emergency period. Thus, we need to develop the skills mentioned in Figure 2 and confirmed from the team's mind maps.

Mind maps are average informal rules accepted by the group after all the list of organizational priorities, group preferences, and regionalities. This map will then show how the team works in each of the four blocks of competence and how the teams should work.

Note that the models do not characterize the relationship of urgency of time, recognition of danger and its mechanisms, also the importance of the discussion on the types of communication (between groups, with the community, with the population or inserted in a political environment) and care in relation to false-incomplete or anticipated information.

The final analysis is a brief discussion of the mind map for BS events in general and for the specific case of the COVID 19 Pandemic.

5 Case Study

The Research Group on Operational and Dynamic Risks (GRODIN) has prepared a Master Program for Disaster Preparedness and Crisis Management and has been acting as a critical observer of Government actions and people's reactions to address COVID19, discussing tools in the Strategic, Managerial and of Operations.

Unfortunately, new strains of this virus are emerging, in a transformation of unknown origin, which shows once again that we are during a Black Swan Event. We must deal with this matter coldly to act in the face of the terror of mass death, and to avoid the wrong decisions. The skills presented in Figure 2 must be developed and available to act in this Global Crisis:

(1) IE Strategic Intelligence (study of the origin of the danger, its mechanisms, and the construction of future scenarios, where spurious data and warnings about changing the danger line must be valued, investigating the past to understand the present, investigating the future for disaster preparedness.

(2) Integrated Management and Crisis Project - GIP (data flow, organization of resources, development of tools for decision appropriate to the mechanisms of action of the danger and its dynamism when the danger is unknown, communication mode and the treatment of the sub- political-economic and technological notification).

(3) Command of local operations COL with the preparation of resources for the here and now based on the organization of the Integrated Crisis Management in the projections of the future created by the IE (people, structures), reviewing the Integrated Management and the IE with new data from the present coming from the COL, having a preparation of the leadership and the team to avoid simulating extreme stress.

(4) Organizational Resilience RO where, with the control of the situation, seek the causal nexus based on practice and science in order not to lose control of the recovery. Always value spurious data and information and warnings, classifying the quality of the information. Motivate IE for further investigations about the danger, its mechanisms of action and possible impacts before it happens.

Taleb [8] stated that in the face of an unknown danger, "everyone" will try to explain without knowing for certain what he is talking about (we need IE), demonstrating dominance that does not exist. Others who are comfortable with being a subject of scientific knowledge, virology, are not concerned with the preparation for situations not lived in the past, so we consider that unknown danger = known local non-operational danger. Thus, with chronic and growing impacts, we will have uncontrolled scenarios due to lack of resources, infrastructure, and lack of competent people for action.

The Black Swan has several dimensions that occur at the same time, Health, Economy, Politics, Logistics, Climate and Geopolitics, regardless of the dimension of origin of the danger. This multidimensionality gets in the way of decisions since the leaders of knowledge, public and private management, do not have the habit of thinking, investigating and deciding considering the systemic failure and its multidimensionality.

The low quality of information, underreporting and / or addictive data, undermines the role of management and science, which creates addictive or incorrect interpretations following mathematical models that are not practical. Observation of facts and events must be valued in times of Crisis (Disaster) to enable the endless elaboration of multidimensional scenarios to dynamize the Strategies.

In times of crisis, it is necessary to work with two teams that feed and feed each other on information. The IE and RO must be calm and when analyzing fatality data, review future actions and investment of restricted resources. The GIP and COL that see in the here and now practical needs to control the mitigation teams towards operational control and human errors caused by extreme stress (EE).

The IE and RO must be calm in the construction of Scenarios with weekly average of 4-day moving averages of fatalities to assess the effectiveness and effectiveness of previous actions, seek to explain or approach the truth about the danger. IE and RO's lack of calm in analyzing the scenario can vitiate interpretations in a single dimension of the disaster leading to incorrect projections of the future. It is necessary to change the position of the gaze to see what is hidden. Thus, we have developed guidelines to be followed:

(1) Installation of Crisis Committees for IE, GIC, COL and RO with a multidisciplinary and dynamic team; (2) Choosing-developing easy-to-use and effective statistical tools to support decisions, therefore, the weekly average of the 4-day moving average tries to avoid underreporting and spurious data that does not indicate a trend; (3) Enhance research on spurious data or information or warnings that may indicate a change in the danger or its mechanism and, consequently, its impact; (4) Prepare the team to maintain the quality of the cognitive and also adapt to the "time available" to guarantee the effectiveness of the actions. Extreme stress increases human error from 10% to 70%, bringing incorrect decisions and actions. We have to use tools to preserve the cognitive in extreme stress !; (5) Build a knowledge network platform, develop practical tools based on grounded theory, information flow, integration, and critical observation; (6) Following the film World War Z, it is important to treat spurious data and observe-validate unconventional local solutions that can be the difference in the application of restricted resources; (7) Following the UTOPIA Series, recognizing the political, geopolitical movements that have low visibility, can indicate simulacra of information, fake news. When knowing the fake news and underreporting, the safeguards will be more assertive, the aim is to test the hypotheses to protect the population; (8) Even if the situation is more optimistic, maintain recourse to pessimistic

scenarios, considering the opinion of people who are reliable but who project a different future from the majority as in the film World War Z.
I forgot the following aspect about the Leader to avoid Accident-Disaster Situations discussed by Collins [20]: (1) Not wanting the power and nor having vanities of signatures; (2) Extreme humility in dealing with unknown hazards; (3) Active listening for events, warnings and deviations; (4) Fierce determination in decisions and actions; (5) Recognize the possibility of own and team mistakes in strategy and action; (6) valuing the organizational and family unit.

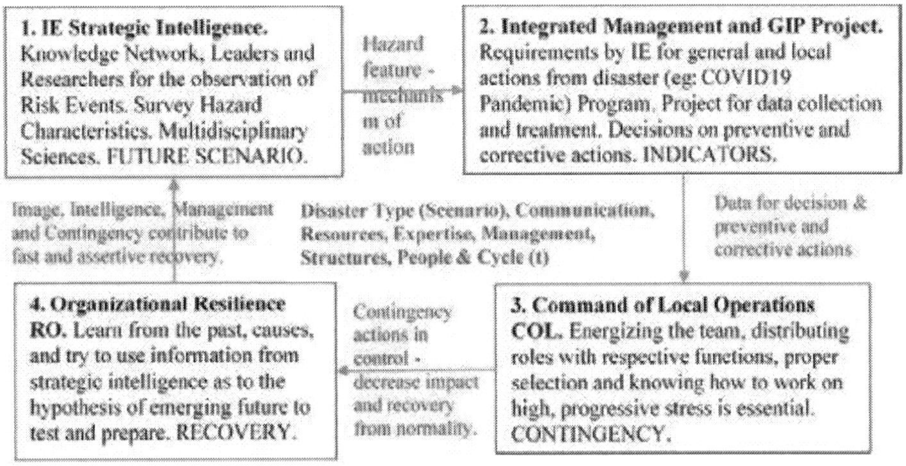

Figure 2. Competencies for the BS Team: Intelligence, Management and Operation.

5.1. Mind Map for BS and COVID19

The characteristics already described by TALEB and Avila using the models from research about the dynamic risk we show the principal competences to attend a BS event. In Table 2, we discuss about the generic pattern and a case of COVID19 BS team.
The informal rules that turn into actions follow mind maps by type of function, Intelligence, Management and Operation. Below are examples of expected patterns and deviations observed in different mind maps.

Table 2. Mind Map for BS

Mind Map	Generic standard	Expected standard (COVID19)	Deviations (GENERIC)
Intelligence	- To study the accidents and incidents;	- Study the disease in depth knowing the form of contagion, consequences, forms of treatment;	- Study superficially without exploring for learning;
	- In the crisis, seek as much information about the products involved, the team exposed, the area affected;	- In the crisis, seek as much information as possible about the disease, exposed teams, affected areas;	- Search for a solution without knowing what we are facing, people involved;
	- Know the risks that are relevant to the business;	- Know the risks that are relevant to the disease;	- Not knowing the extent of the risks and not preparing for it;
Man	- Know the resources available to face the crisis;	- Know in depth all the resources available for coping with the disease;	- Have resources that are not available.

			- Wait to seek resources in crise;
	- Treat accidents and incidents with preventive and corrective improvement actions;	- Predict and seek resources to strengthen the fight against the disease with the creation of scenarios;	- Treat accidents and incidents superficially without seeking a solution;
		- Treat deviations with preventive and corrective improvement actions;	- Treat deviations superficially without seeking a solution;
Operations	- Know and study their role in acting during the crisis;	- Know and study their role in acting during the pandemic crisis;	- Facing the crisis in a way that is not consistent with the procedures causing conflict with the team;
	- Communicate any changes that are made to the procedures by training the others involved;	- Communicate any changes that are made to the procedures by training the others involved;	- Carrying out activities in a manner inconsistent with the procedures, putting the activity at risk;

6 Conclusion

The current educational model prepares society for routine operations only, always with a focus on the main objective of producing with quality and low cost. The safety of the process is studied in a superficial way, focusing on the clearest and expected risks. BS are situations that arise during an event resulting from the sum of risks that may have been studied separately but never together. This moment requires several skills from the teams that work on the front line and coping and mastering the situation can take more time and we know that in an emergency, time and knowledge are fundamental. The intention of this work was to present a review and discussion about new concepts and tools that can be used in the prevention of human errors during crises (LODA) and thus we were able to understand the effects that the level of stress can influence in the decision making during the crisis and compromise coping with the crisis based on the understanding that the mental map changes completely in the crisis.

The presentation of the methodology with the necessary skills to face the BS and the mental map models are fundamental to create the PDCA of the crisis that always starts with strategic intelligence, integrated management, Operations command, and the organizational resilience that will feed the strategic intelligence in a learning cycle and the search for continuous improvement. The more we study this methodology, we see that although the event is unexpected, the knowledge and the feedback of information for the knowledge of the risks, will make that the mastery of the situation is reached more quickly.

The academy needs to better prepare society for unexpected situations. BS events are becoming more common, given the COVID19 pandemic experienced today and methodologies applied at the time of the crisis can save lives.

7 References

[1] T. Overton, M. Broadribb, C.N. Cheung, E. Christie, S. Cowher et al. "Guidelines for Process Safety Metrics", pp. 1-178, 2009.

[2] Center for Chemical Process Safety (CCPS). "Guidelines for Investigating Process Safety Incidents", Third Edition, 2019.

[3] Center for Chemical Process Safety (CCPS). "Guidelines for the Management of Change for Process Safety", 2010.

[4] Center for Chemical Process Safety (CCPS). "Guidelines for Managing Process Safety Risks During Organizational Change", 2013.

[5] Center for Chemical Process Safety (CCPS). "Human Factors Methods for Improving Performance in the Process Industries", December 2006, 248 p.

[6] Center for Chemical Process Safety (CCPS). "Recognizing and Responding to Normalization of Deviance", September 2018, 176 p.

[7] Herbert W. Heinrich, Industrial Accident Prevention: A Scientific Approach, 1931

[8] N. Taleb, "The Black Swan: Second Edition: The Impact of the Highly Improbable". Random House Trade Paperbacks, 2010, 469 p.

[9] P. Ho, "Future Beasts and Where to Find Them". Available at https://www.psd.gov.sg/challenge/ideas/feature/future-beasts-and-where-to-find-them. Accessed on January 22, 2021.

[10] S.F. Ávila. "Dependent Layer of Operation Decision Analyzes (LODA) to calculate human factor, a simulated case with PLG event", 2011.

[11] A.C.F. Carvalho. "Diagnóstico de lacunas nas competências da equipe para enfrentamento de situações fora da rotina". Master's dissertation – Federal University of Bahia, Polytechnic School of Engineering. Salvador, 2016, 103 p.

[12] S.F. Ávila., B. Mrugalska, M. Wyrwicka, M. Souza, J. Ávila, et al. "Cognitive and Organizational Criteria for Workstation Design", 2019.

[13] S.F. Ávila., A.C.F. Carvalho, G.P.J. Portela, C. Costa. "Learning Environment to Take Operational Decision in Emergency Situation". Procedia Manufacturing. vol. 3, pp. 1780-1787, 2015.

[14] S. F. Ávila. "A exergia da falha operacional, uma discussão filosófica com bases práticas". In: Proceedings of 19th Brazilian Congress of Chemical Engineering, 2012, Buzios.

[15] S.F. Ávila. "Análise de falha em processos complexos". In: Proceedings of 19th Brazilian Congress of Chemical Engineering, 2012, Buzios.

[16] C. Perrow. "Normal accidents: living with high-risk technologies". Princeton University Press, Princeton, N.J., 1999.

[17] J. T. Reason. "Human error". Cambridge University Press, 1990.

[18] S. F. Ávila., I. Cerqueira, V. Júnior, J. Nascimento, C. Peres, C. Brunoro, P. Arezes. . Integrated Management in Disaster: Discussion of Competences in a Real Simulation. AHFE 2019.

[19] S. F. Ávila., J.S. Ávila, A. A. Santos, M. C. N. Carmo, L. F. L. Souza. Crisis Decision Tool, fighting against COVID-19. AHFE 2021.

[20] J. Collins. Level 5 Leadership: The Triumph of Humility and Fierce Resolve. Harvard Business. 2005.

619165 Case Study in M+a – Communication and Understanding Hazards and Safety Concepts

Abstract:

A multinational corporation acquired a local consumer products manufacturing facility. Hazard awareness and process safety protocols were lacking at the local company, especially as measured against the parent company's corporate standards. The primary root cause was lack of a strong management of change procedures, as the processes in the facility were originally designed for non-flammable products but over time the facility transitioned to handling flammable liquids and combustible dusts.

Hundreds of unique chemicals/products stored and handled at the facility were analyzed to determine compliance with local fire codes and serve as a design basis for safety program improvements. Specific challenges around flammable liquids and combustible dusts were deemed to be highest risk and were addressed immediately.

This case study presents high level findings and both capital and operational budget related recommendations applicable to companies in similar situations. By applying a combination fire/building code evaluation and hazardous materials / process safety gap analysis, the company was able to prioritize risk reduction actions in alignment with their strategic objectives. This work presents a basis for hazardous material safety in facilities that store and handle a wide range of hazardous materials and aligning risk remediation activities with corporate objectives.

618542 Is Your Plant at Risk from Climate Change?

Abstract:

Climate change is occurring and some of the changes will impact the safety of process plants. This paper summarizes potential adverse impacts on process plants and then outlines how to conduct a Climate Risk Vulnerability Assessment (CRVA) for a process plant. One product of the CRVA is a plan of action to eliminate or reduce adverse climate effects on the process plant.

619167 Is Emergency Response for Facilities Impacted By Climate Change?

Abstract:

Climate change essentially bring about alteration to our natural environment. This can be seen by way of changes in the weather pattern. We see this here in the Gulf region of the US, almost every year during our most infamous hurricane/storm season.

Emergency response plans and preparedness during this season is always critical. It is based on science, best practices and lessons learned. At these times, not only is the shutdown process very critical , however the start up and recovery measures are also critical..

This paper will discuss thoughts and views of the author of changes required to emergency response and preparedness plans that facilities have to account for, due to changes in the weather pattern. The author will also discuss how the risks to people and process during the recovery measures can best be minimized. Irrespective of the conditions and planning, the human lives must be preserved.

619069 Transitioning from One Explosive Fuel to the Next – Safety Considerations in the Emerging Hydrogen Economy

Abstract:

Since the turn of the century, there has been a focus on the development of the hydrogen economy, driven by the desire to reduce greenhouse gas emissions through utilization of alternative energies and cleaner fuels. With both the current and future interest in utilizing hydrogen, it is crucial to evaluate and mitigate the fire and explosion risks posed by hydrogen during storage and transfer. While many other fuels such as natural gas, propane, LNG, and petroleum also present flammability and explosive hazards, hydrogen has several unique characteristics. These include, but are not necessarily limited to, a relatively low ignition energy, low density, and wide flammability limits. These properties present unique challenges for risk management compared to other flammable liquids and gases that will need to be addressed as the hydrogen economy further expands into its emerging market.

In this paper, the properties of hydrogen and how they impact the fire and explosion hazard compared to other fuels will be analyzed. Then, best practices and risk management will be discussed for storage and transferring fuels, with a focus on highlighting differences that are important considerations for hydrogen. Finally, a case study will be presented that reviews an incident involving a hydrogen leak and ensuing jet fire that ignited during the removal of a portable tube trailer. The events that led to the unintended release and ignition of pressurized hydrogen, and the lessons learned from the incident will be discussed. The case study will highlight several aspects related to the safe storage and transfer of hydrogen in commercial applications.

619275 Application of Causal Analysis Based on System Theory (CAST) to the T2 Laboratories Incident

Abstract:

Application of Causal Analysis based on System Theory (CAST) to the T2 Laboratories incident

Joshiba Ariamuthu Venkidasalapathy[a,b,*]

[a]*Artie McFerrin Department of Chemical Engineering, Texas A&M University, College Station, TX 77840, USA*
[b]*Mary Kay O Connor Process Safety Center, Texas A&M University, College Station, TX 77840, USA*

[*]*Presenter email:* joshvenkat03@tamu.edu

Abstract

STAMP (Systems-Theoretic Accident Model and Processes) is an accident causalty model that is based on systems thinking and systems theory. STAMP can be applied for incident investigations through the methodology called causal analysis based on system theory (CAST)[1]. It is a socio-techinial causal analysis method that systematically helps to identify causes that led to the accident. Unlike traditional accident models that rely on linear causation effects such as swiss-cheese model and domino effect models, CAST adopts a control-centric approach that helps account for causes that arise due to nonlinear interactions. Some unique features of CAST include treating safety as an emergent property, and providing the scope to account for software interations and temporal aspects of control actions. However, it is seen that applications of CAST to chemical engineering systems is scarce so far[2-4]. Considering the increasing complexity of chemical processes, accident models based on simple linear causation effects may not be sufficient in identifying the root cause of an incident.

CAST methodology is implemented in five steps: assembling the basic information, modeling the safety control structure, analysing the control structure to identify failed control elements, analysing the systemic factors and control structure flaws and providing recommendations to improve process safety. The second step, which is to model the safety control structure, is the essence of the CAST methodology. The control structure is heirarchical and requires every control block to have well defined functional objectives and control elements to achieve its goals.

In this paper, the objective is to apply CAST to a chemical incident to identify the social and technical causes that led to a chemical indident. The T2 laboratories explosion that occurred in Florida, 2007 is chosen to illustrate the CAST methodology[5]. The explosion led to four fatalities and several injuries. The CAST methodology along with extensive simulation based on MATLAB environment led to identifying both social and technical proximate causes. Most importantly, the results showed that temporal aspects of control actions, such as, how long the action was applied or how early/late it was applied also played an important role in the T2 Laboratories incident.

References

1. Thomas, J. and C. Malmquist, *Learning from Accidents that are a Consequence of Complex Systems.* Presented at the 48th Annual International Society of Air Safety Investigators, 2017.
2. Elliott, J., *CAST analysis of the Buncefield incident.* Procedia Engineering, 2017. **179**: p. 23-33.
3. Leveson, N.G., *CAST Analysis of the Shell Moerdijk Accident.* 2016, MIT. Cambridge: EU Major Accident Hazards Bureau.
4. Muñoz, M.F.T., *The Underestimated Value of Safety in Achieving Organizational Goals: Cast Analysis Off the Macondo Accident.* 2017, Massachusetts Institute of Technology, System Design and Management Program.
5. CSB, *T2 Laboratories Inc. Reactive Chemical Explosion.* 2009.

SPRING21
+17TH GCPS
A Joint AIChE and CCPS Meeting

Process Safety Manuscript Archives Available to the Practitioner

Raymond "Randy" Freeman
S&PP consulting
12303 Lake Shore Ridge
Houston, TX 77041
Rafree@Yahoo.com (for correspondence)

Ronald J. Willey, PE, PhD
Professor and Associate Chair Chemical Engineering
Northeastern University
360 Huntington Ave. 201 CN
Boston, MA 02115
R.Willey@northeastern.edu (for correspondence).

Prepared for Presentation at
American Institute of Chemical Engineers
2020 Spring Meeting and 17th Global Congress on Process Safety
Virtual
April 18 - 22, 2021

Process Safety Manuscript Archives Available to the Practitioner

Raymond "Randy" Freeman
S&PP consulting
12303 Lake Shore Ridge
Houston, TX 77041
Rafree@Yahoo.com (for correspondence)

Ronald J. Willey, PE, PhD
Professor and Associate Chair Chemical Engineering
Northeastern University
360 Huntington Ave. 201 CN
Boston, MA 02115
R.Willey@northeastern.edu (for correspondence).

ABSTRACT

Since 1967, the American Institute of Chemical Engineers (AIChE) has sponsored a series of safety symposia. The goal of these symposia is to reduce the frequency and severity of accidents in the chemical process industries (CPI). The Safety and Health Division of AIChE was established to serve as a central hub for collection and dissemination of this process safety information. Over the years, this dissemination expanded to several process safety symposia tracks at the Global Congress on Process Safety. Three symposia manuscripts have been assembled and archived into a data base accessible on the Internet through the AIChE Academy.

- Loss Prevention Symposium (LPS)
- Process Plant Safety Symposium (PPSS)
- Center for Chemical Process Safety (CCPS) Symposium

The technical papers presented at these symposia cover a wide range of topics such as fire protection, risk reduction, explosion analysis, plant maintenance practices, process safety management and many other topics.

A total of 2,524 papers were scanned and organized into a form which can be used to search for information on process safety topics. A spreadsheet of the contents of the archives was created. This paper describes the organization of the Process Safety Archives and presents an example of how the archives can be used to find information.

1. BACKGROUND

As the English philosopher George Santayana[1] (1905) eloquently stated:

"*Those who cannot remember the past are condemned to repeat it*"

Trevor Kletz[2] (1993) restated the importance of the failure to remember the past in his book entitled: *Lessons from Disaster: How Organizations have no memory and accidents recur.* Only people have memory. Unfortunately, our memories are very poor. Therefore, to preserve knowledge two conditions must be met:

- It must be written down
- It must be findable.

The ability to find information is very important. It does very little good to state that the needle you must rapidly find is located in one of many haystacks in the back of 40 acres of a field. Likewise stating that a document of interest may be in the Library of Congress or the National Archives achieves little in actually finding the document of interest. For information to be useful, it must be timely and rapidly findable.

Previously, the Safety and Health Division sponsored projects to scan and digitize technical papers presented at various meetings. The papers were loaded into a Folio database and made available to the Safety and Health Division membership. Several items are of interest:

- The last update of the Folio database was in 1995.
- The process safety databases (versions 1, 2, 3) use a proprietary FOLIO format and did not include PPSS.
- Better scan technology today than was available 25 years ago.

Why this is important and why is it useful? Many safety research programs were only presented at Loss Prevention Symposia and were never formally published. These include accident investigations, chemical stability tests, hazardous material property test results and other information. Such research reports may only be available in the printed conference proceedings. Rapid access to these reports and papers may improve designs and prevent future process safety incidents.

2. LOSS PREVENTION SYMPOSIUM (LPS)

2.1 History of LPS

The first LPS was organized and held in Houston, TX in February of 1967. A total of 23 technical papers were given on topics such as:

- Building Design Criteria
- Equipment Design
- Reactive Chemicals
- Process Safety Management

The final session of the first LPS was devoted to a review of six process incidents.

Since the first 1967 symposium, the Safety and Health Division has sponsored an annual LPS. More details on the history of the LPS may be found in papers by Davenport[3] (2006) and Freeman[4] (2016).

2.2 Number of LPS papers

In the time period from 1967 to 2005, a total of 1,324 technical papers have been presented and are included in the Process Safety Archives. The papers include many summaries of process accidents, as well as discussions of such topics as fire protection, safety instrumented system analysis, dispersion analysis, and nitrogen inerting to name but a few.

3. CENTER FOR CHEMICAL PROCESS SAFETY (CCPS)

3.1 History of CCPS

In the second half of 1984, several incidents occurred which had international impact:

- In October 1984 a lubricating oil tanker caught fire and exploded immediately after leaving San Francisco. One sailor died and a large oil spill resulted.
- In November 1984 a large fire and explosion occurred in a PEMEX storage and distribution center in Mexico City. Between 500 and 600 people were killed and houses were damaged to a distance of over 5 miles.
- In Early December 1984, water entered a storage tank of methyl isocynate (MIC) at a Union Carbide plant site in the Indian town of Bpohal. The resulting reaction of MIC and water caused in a runaway exothermic reaction. A large cloud of toxic material (MIC and others) was released killing at least 2,000 people and injuring 10s of thousands of people.

As a result of these disasters, 16 U.S. chemical companies approached AIChE with a request to help improve the process safety practices in the chemical industry. This request was the start of CCPS. Details on the early history of CCPS may be found in the paper by Berger[5] (2009).

3.2 Topics and intent of CCPS Symposia

The first CCPS conference was held in 1987. The conference was focused on the following areas:

- Hazard Identification
- Storage and Transfer of Hazardous Chemicals
- Reactive Chemicals
- Post-Release Mitigation
- Human Factors
- Vapor Cloud Modeling

All of these topics are of direct relevance to the Bhopal Incident.

3.3 Number of papers

In the time period of 1987 to 2006, a total of 736 technical papers were presented at the annual CCPS conference. These technical papers are focused on specific problems and included detailed reviews of lessons learned from accidents.

4. PROCESS PLANT SAFETY SYMPOSIUM (PPSS)

4.1 History of PPSS

In 1989, there were two major process incidents in the Houston, TX area.

- A fire/explosion at the Phillips Pasadena plant killed 23 workers
- A fire/explosion at the Arco Channelview plant killed 17 workers

In 1992, the South Texas Section of AIChE organized a technical symposium on process safety. The symposium was held in Houston, TX to minimize the travel to attend the meeting and to encourage engineers and managers from the Houston area chemical and process industries to attend. The first PPSS was a great success with attendance by over 700 people.

4.2 Topics and intent of PPSS Symposia

The focus of the PPSS is on plant operations. PPSS papers are focused on practical application of process safety programs. More detailed technical material is found in LPS and CCPS symposia series. Papers on the management and control of process safety programs are a popular topic. Details on the history of the PPSS may be found in the Congress paper by Edwards[6] (2012). Edwards summarizes the focus of the PPSS as:

"*Practical Application of Process Safety Management at the Plant Level: Ensuring the Protection of People, the Environment, Capital and Business.*"

4.3 Number of papers

In the time period of 1992 to 2005, a total of 515 technical papers were presented at PPSS conferences. These technical papers are focused on specific problems and included detailed reviews of lessons learned from accidents.

5. CONTENTS OF PROCESS SAFETY ARCHIVES

Contents of the process safety archive include the following collections:

- Loss Prevention Symposium manuscripts from 1967 to 1980, originally Published annually as "Loss Prevention Symposium, Vol XX" by AIChE, New York, New

York, where XX went from 1 through 14. 326 titles across 14 volumes are in the archives.
In 1981, Loss Prevention Symposium manuscript numbering changed over to a AIChE session number, alpha character (e.g. 60b), mode of numbering and the AIChE discontinued the annual compilation of manuscripts into a single issue. Fortunately, John Davenport and others collected these papers and when this archive was announced, shared their personal collections with one of the authors where these were scanned in. In 1982, selected papers from the 1981 session forward began to appear in a new AIChE publication, initially called *Plant/Operations Progress*. In 1993 *Plant/Operations Progress* was renamed to *Process Safety Progress*. This collection of LPS manuscripts may also include both the *Plant/Operations Progress* version and the original manuscript delivered at the conference. Approximately 397 manuscripts and publications from 1981 to 1989 have been archived.

In 1990, AIChE moved to posting manuscripts onto digital storage media, and all manuscripts have been extracted from various digital media that existed at the time for the years of 1990 to 2005. Approximately 601 manuscripts exist in this part of the collection. After 2005, AIChE retained manuscripts from various Global of Process Safety on its web site AIChE Academy, and these can be accessed by going to the appropriate year and congress.

- The Center for Chemical Process Safety began congresses in 1987 and all manuscripts from 1987 to 2006 have been found, scanned, and archived in this database. Approximately, 736 manuscripts make up this part of the archive. Manuscripts after 2006 are available via the AIChE Academy.
- The final collection of manuscripts continues with Process Plant Safety Symposium, beginning 1992 and continued to 2005. 515 manuscripts make up this part of the archive.

Manuscripts published after 2005 are available on at www.aiche.org under this web link by year: https://www.aiche.org/conferences/aiche-spring-meeting-and-global-congress-on-process-safety

6. INTERNET ACCESS OF PROCESS SAFETY ARCHIVES

6.1 Overview of Access Protocol

Accessing the Process Safety Archives involves two major steps:

- Determine what you want to look at. This is accomplished by using the Master Index to the Process Safety Archives. The Master Index (Revision 0) is presented as supplementary material to this paper in the form of a Microsoft Excel spreadsheet. Details on how to use the Master Index are presented in section 7 of this paper,
- Once you know which volume of the LPS, PPSS or CCPS that the paper of interest is located, you must purchase the volume. The purchase price for a member of the Safety and Health Division is $0.00 (FREE). Purchase of the volume is via the AIChE transaction portal. The following section of this paper demonstrates the process for obtaining a paper from the Process Safety Archives.

6.2 Detailed Description of Method to Access Process Safety Archives

Important note: this example contains working links that if you are a member of the safety and health division, should take you to the images shown.

The process safety archives can be accessed at AIChE thru this web link: https://www.aiche.org/ccps/resources/safety-health-proceedings.

Once at the web site clicks "sign in" if you are a member of safety and health division (see Figure 1). The Home Page of CCPS will appear.

Re-click this web link: https://www.aiche.org/ccps/resources/safety-health-proceedings. to return to the archives.

This part of the archive is organized sequentially by year from 1976 to 2005. There are 7 subpages. Within each year is a link to the AIChE academy where the papers are actually stored. This approach requires the user to know what year and what conference the paper of interest exists. The easiest way to find out what to download is to use the Master Index as presented in the Supplementary Materials. For example, clicking on page 6 (bottom). Then scrolling, you will find two conferences as shown in Figure 3.

Click "READ MORE" for CCPS 17th conference. You reach this page (Figure 4).

Then you click TECHNICAL PAPERS and Figure 5 will appear.

Click the first title, Figure 6 will appear.

Click Checkout (this link is unique to the user and thus is not included).

This page appears, Figure 7.

Click Process my order. Figure 8 Will appear.

Next click CONTINUE and the following page appears (Figure 9).

You need to re-navigate after paying or receiving permission to access the conference papers. Click TECHNICAL PAPERS, and then click the first title again; this time a pdf file appears in the window (Figure 10).

This page now shows the pdf file of the original paper. Click on the paper abstract and a pdf display of the first page of the paper will appear (Figure 11).

Figure 11 shows the pdf file that can be read in a web reader or downloaded.

Another link, shown back in Figure 9 and highlighted in Figure 12, on the lower right hand side is a zip file download that will contain all papers presented in that conference, as well as the table of contents.

You can also download a zip file with all papers presented in that conference. Simply click on the Downloadable version and put the zip file onto your computer.

7. HOW TO USE THE PROCESS SAFETY MASTER INDEX -EXCEL FILE AN EASY WAY TO FIND A PAPER ON THE WEB SITE

7.1 The Process Safety Master Index

As the project progressed, an Excel File was assembled of year, symposium name, titles, authors, paper numbers. This file, to be available on the process safety web site, brought in two other data bases, the Process *Beacon* collection titles from 2001 to 2019 and the Hyperlinks to all *Plant/Operation Progress* and *Process Safety Progress* Papers published from 1982 to 2020, Issue 2. To the process safety professional, in need to seek past papers on the topic, this spreadsheet can proven to be very helpful. Two examples follow.

7.2 Ammonia Nitrate Example

An ammonia nitrate accident occurs (example Beirut, Lebanon, August 2020). The process safety expert desires to know all titles that have "ammonium nitrate" that are available in the archive:

Following these steps,

- Open the Process Safety Master Index in Microsoft Excel
- Initiate a Find and Replace dialog Box ("Ctrl F") – Figure 13
- Enter the search term "ammonium nitrate", and select the whole Workbook under option button to see all titles: - Figure 14
- Click Find All

The following list appears (in this case two manuscripts related to ammonium nitrate appear as the key word appears twice for each manuscript).

7.3 Ethylene Oxide Example

Let's try one more example, one that will have more results by searching on the combination "ethylene oxide". The following Figure 15 shows 19 titles appearing 7 titles from LPS, 1 title from CCPS (a duplicate appears), and 11 titles from *Process Safety Progress*.

Let's pick paper 3 in the list Ethylene Oxide Explosion at Doe Run. We see in Figure 16 that the paper was presented in the Loss Prevention Symposium in 1968.

Next arrow over to column M and click on the hyperlink provided:

Then, arrow over to column S and click on the hyperlink provided there (See Figure 18):

After clicking the hyperlink in Column S, Figure 19 should appear.

Figure 19. Paper Title Page

The Checkout link exists to purchase this paper. See the previous section for directions on how to purchase. If you have already purchased this volume, the pdf file link should appear below the authors (not shown in this figure).

8. DISPOSITION OF PAPER COPY OF THE ARCHIVES

The paper copies of the Process Safety Archives were scanned by InterActive of Houston, Texas. After scanning, the paper copies of the archives were deposited in the Linda Hall Library of Kansas City, MO for safe keeping. This step is still in progress as of March 2021.

9. ACKNOWLEDGMENTS

The creation of the Process Safety Archives was sponsored by the Safety and Health Division of AIChE. Creation of the Process Safety Archives using historical documents was a major task.

The Process Safety Archives includes over 2,000 technical papers, reports and documents. The Safety and Health Division Committee members were Randy Freeman (S&PP Consulting), Ron Willey (Northeastern University), Tracy Whipple (BP), Philip Myers (Advantage Risk Solutions), and Dan Crowl (Professor Emeritus, Michigan Technological University). Special thanks go to Jing Chen and Gennady Shafirovich from the New York AIChE staff. Scanning, initial file organization, and meta- data creation was completed by Maricela Manteca and her team at InterActive of Houston, TX.

The Proof Checking, Quality Assurance Team was organized by Ron Willey at Northeastern University. This team was instrumental in finding scanning issues, and cross checking the scanned material against the original conference programs to find missing papers. In several cases the QA/QC team identified papers that were missing and were retrieved by contacting the original paper author. This team included Zachary Kylar, Isabella Cardona Barber, Saule Bussing, Michael Nguy, Naa Monoh Odateifio, Joshua Garvey, and Eric Weinstein, as well as others.

9.1. Special Acknowledge for John Davenport

During the period of 1981 through 1989, the papers presented at the Loss Prevention Symposia were not collected by AIChE and published in a book as had been done in the earlier years. The years 1981 through 1989 are called the "Lost Years" by the officers of the Safety and Health Division. During this period, John Davenport of IRI collected the papers from each of the "Lost Years" as well as any associated published technical papers. Thanks to John Davenport, the collection of the "Lost Years" Loss Prevention Symposia papers have been digitized and are included in the Process Safety Archives. In many cases the collection of material by John Davenport contains both the technical paper presented at the LPS symposium and the subsequent paper as published in a journal by AIChE. Thank you John for a job well done!

10. REFERENCES

1. George Santayana, *The Life of Reason: The Phases of Human Progress, Volume 1 – The Life of Reason*, Dover Publications, New York, 1905. (Available on line at the Gutenburg Project, Ebook number 15000.)
2. Kletz, T., *Lessons from Disaster: How Organizations have no and memory and accidents recur*, Institute of Chemical Engineers/Gulf Publishing Co, Houston, TX, 1993.
3. Davenport, J., History of Loss prevention symposia: Forty years, 1967 -2006, *Process Safety Progress,* Volume 25, 2006, pp 268-273.
4. Freeman, R. "History of the Loss prevention symposium: The first 50 years – There are no secrets in safety," *Process Safety Progress*, Volume 35(1), pp 32-38, 2016.
5. Berger, S. *CCPS – 20 Years and more, Process Safety Progress,* Volume 25(1), 2006, pp 83-85.
6. Edwards, et al., *History of Process Plant Safety Symposium, Process Safety Progress*, Volume 31(3), 2012, pp 200–202.

SUPPLEMENTARY MATERIALS

1. Master Index – Excel Spreadsheet

    Contents – LPS, CCPS, PPSS, Beacon List, Index to PSP papers

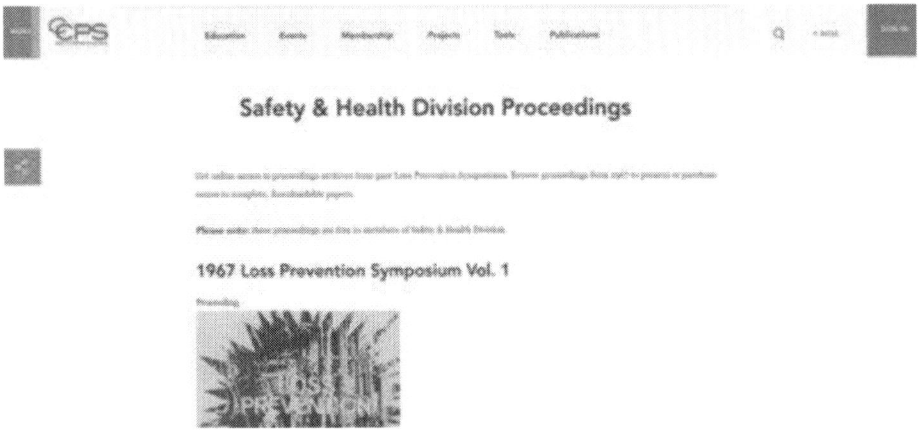

Figure 1. Initial entry page to the Safety and Health Proceedings.

Figure 2. CCPS home page.

Figure 3. Example index page that take one to the AIChE Academy Web Page.

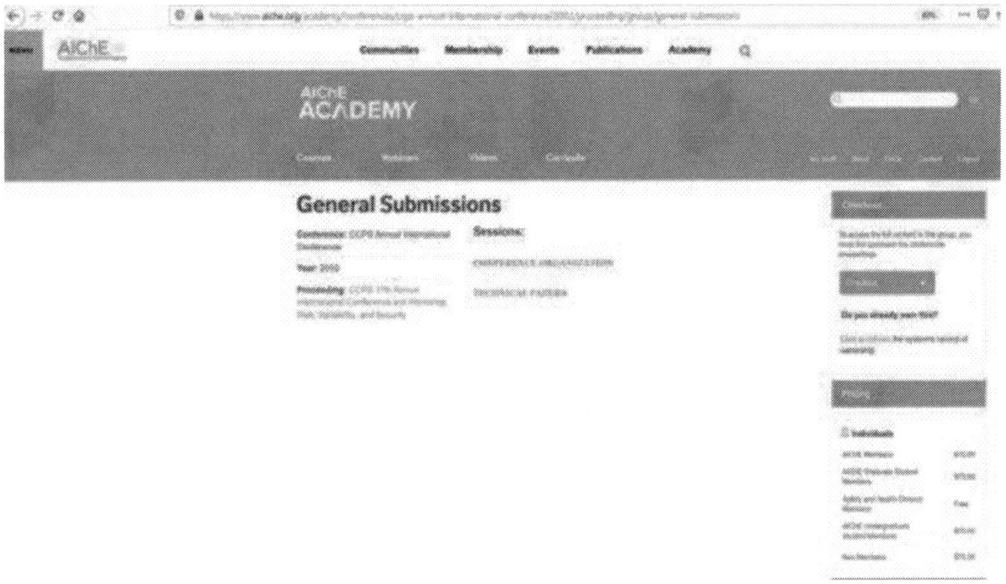

Figure 4. How to reach technical papers.

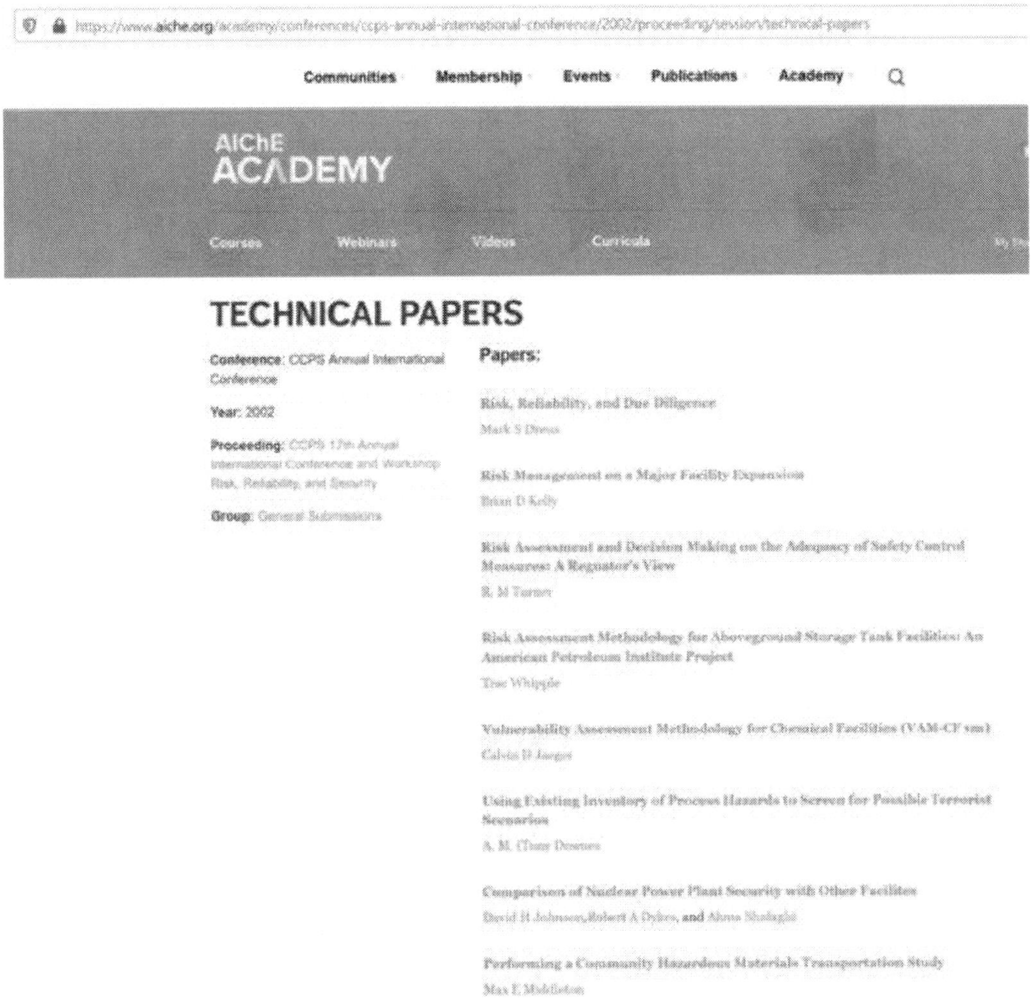

Figure 5 List of papers available for the CCPS Annual International Conference, 2002.

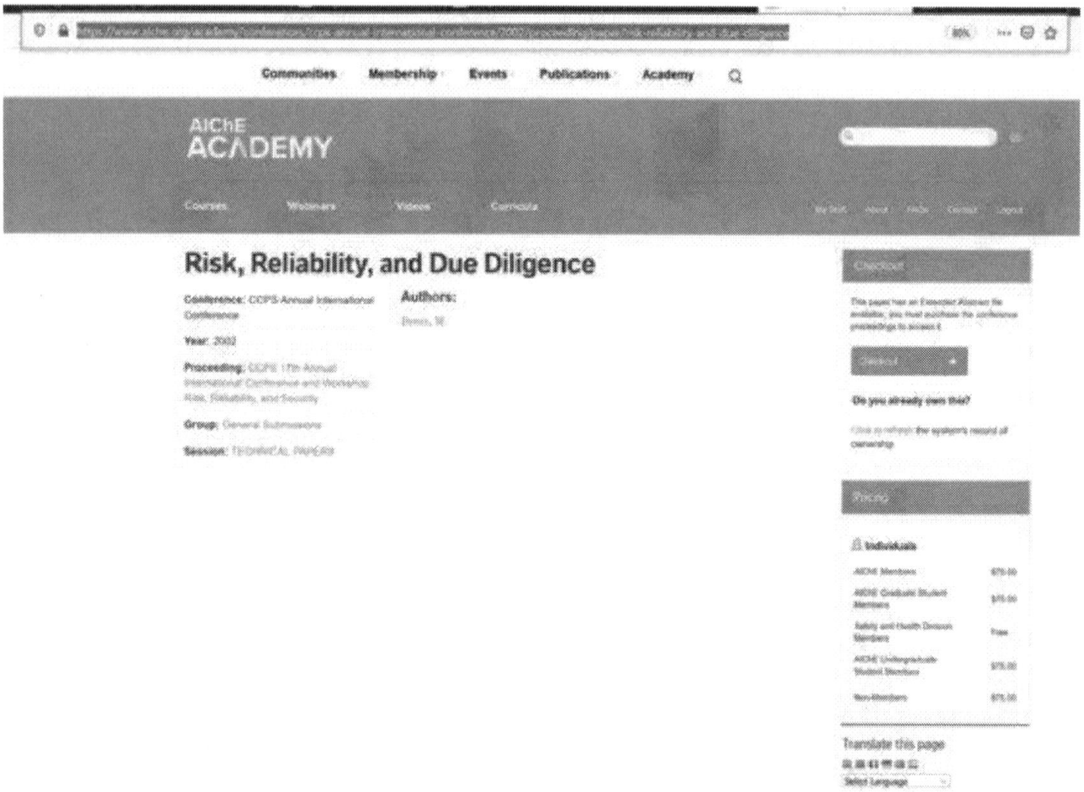

Figure 6 AIChE Academy representation of Risk, Reliability, and Due Diligence, by M. Dreux.

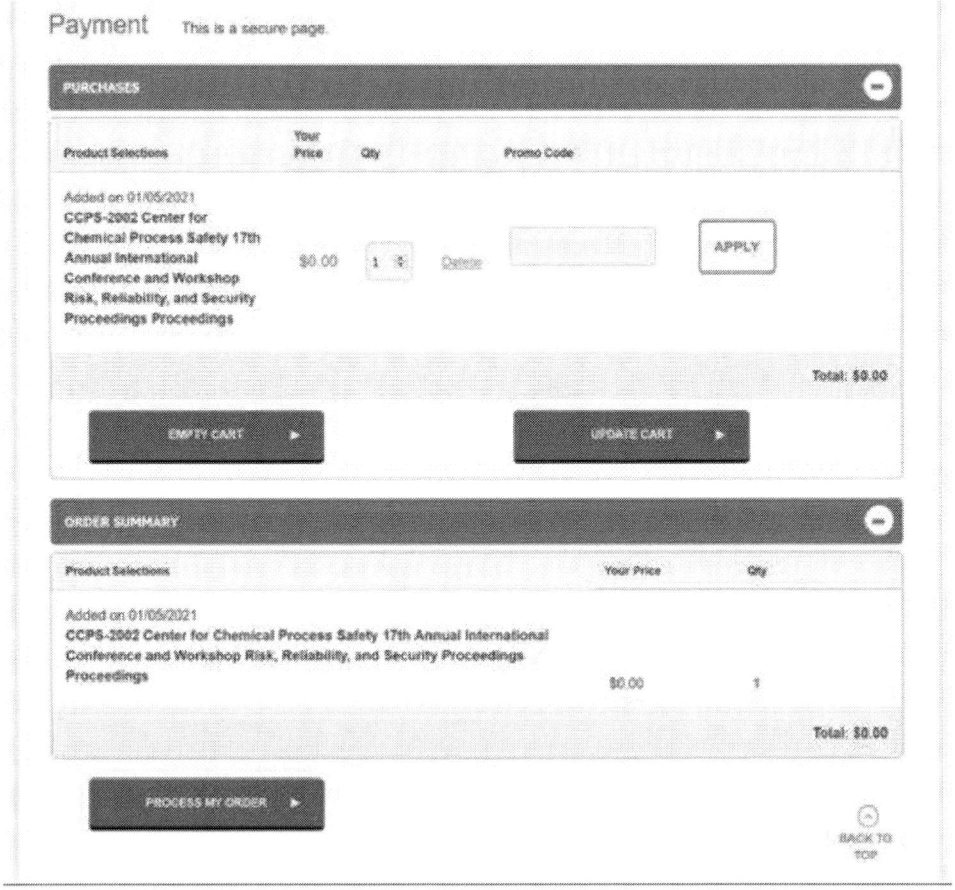

Figure 7. Purchase confirmation page.

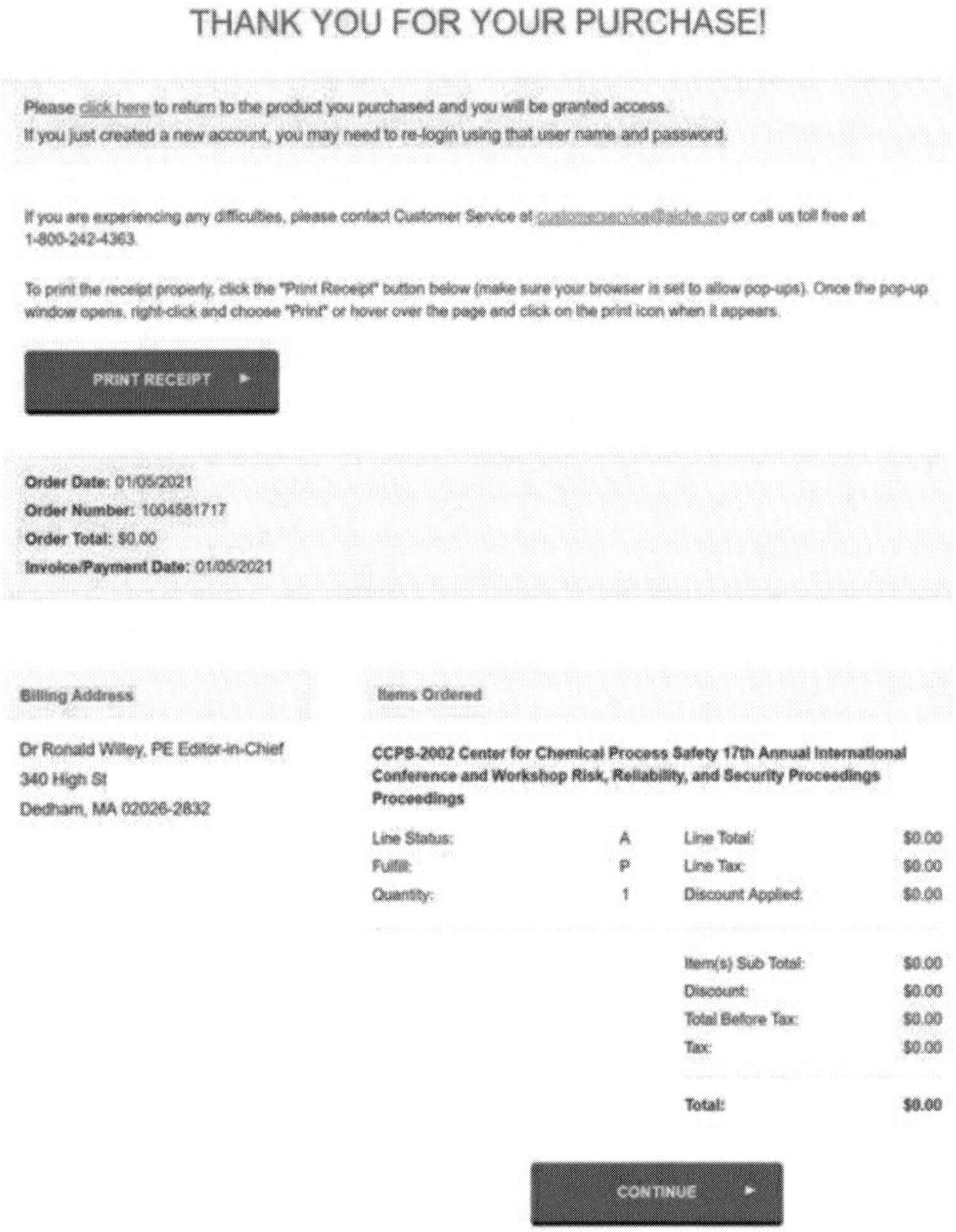

THANK YOU FOR YOUR PURCHASE!

Please click here to return to the product you purchased and you will be granted access.
If you just created a new account, you may need to re-login using that user name and password.

If you are experiencing any difficulties, please contact Customer Service at customerservice@aiche.org or call us toll free at 1-800-242-4363.

To print the receipt properly, click the "Print Receipt" button below (make sure your browser is set to allow pop-ups). Once the pop-up window opens, right-click and choose "Print" or hover over the page and click on the print icon when it appears.

PRINT RECEIPT ►

Order Date: 01/05/2021
Order Number: 1004581717
Order Total: $0.00
Invoice/Payment Date: 01/05/2021

Billing Address

Dr Ronald Willey, PE Editor-in-Chief
340 High St
Dedham, MA 02026-2832

Items Ordered

CCPS-2002 Center for Chemical Process Safety 17th Annual International Conference and Workshop Risk, Reliability, and Security Proceedings Proceedings

Line Status:	A	Line Total:	$0.00
Fulfill:	P	Line Tax:	$0.00
Quantity:	1	Discount Applied:	$0.00

Item(s) Sub Total:	$0.00
Discount:	$0.00
Total Before Tax:	$0.00
Tax:	$0.00
Total:	**$0.00**

CONTINUE ►

Figure 8. Receipt page

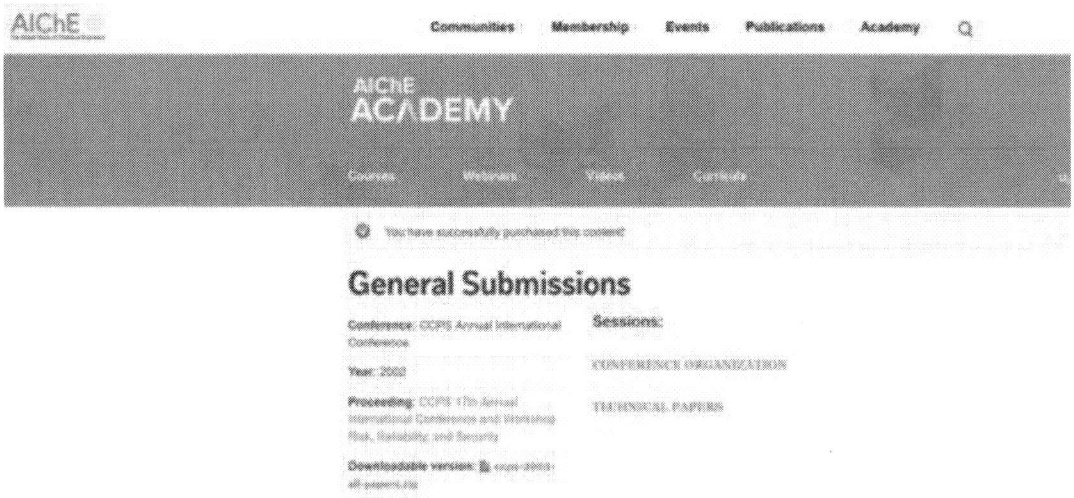

Figure 9. New View of Conference Proceedings

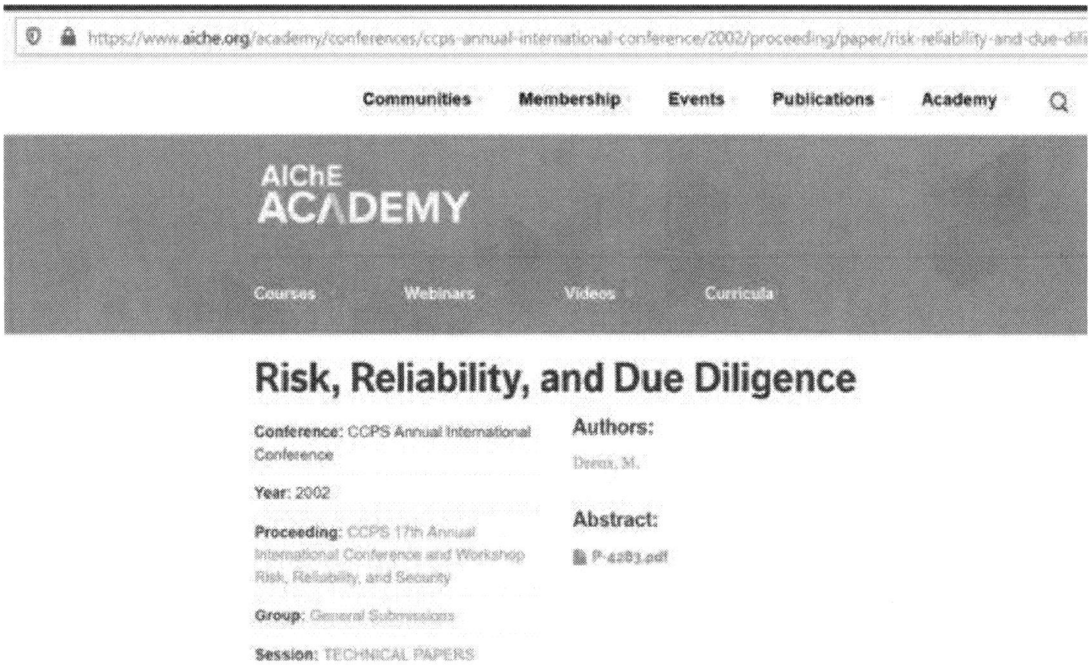

Figure 10. Paper Title Page.

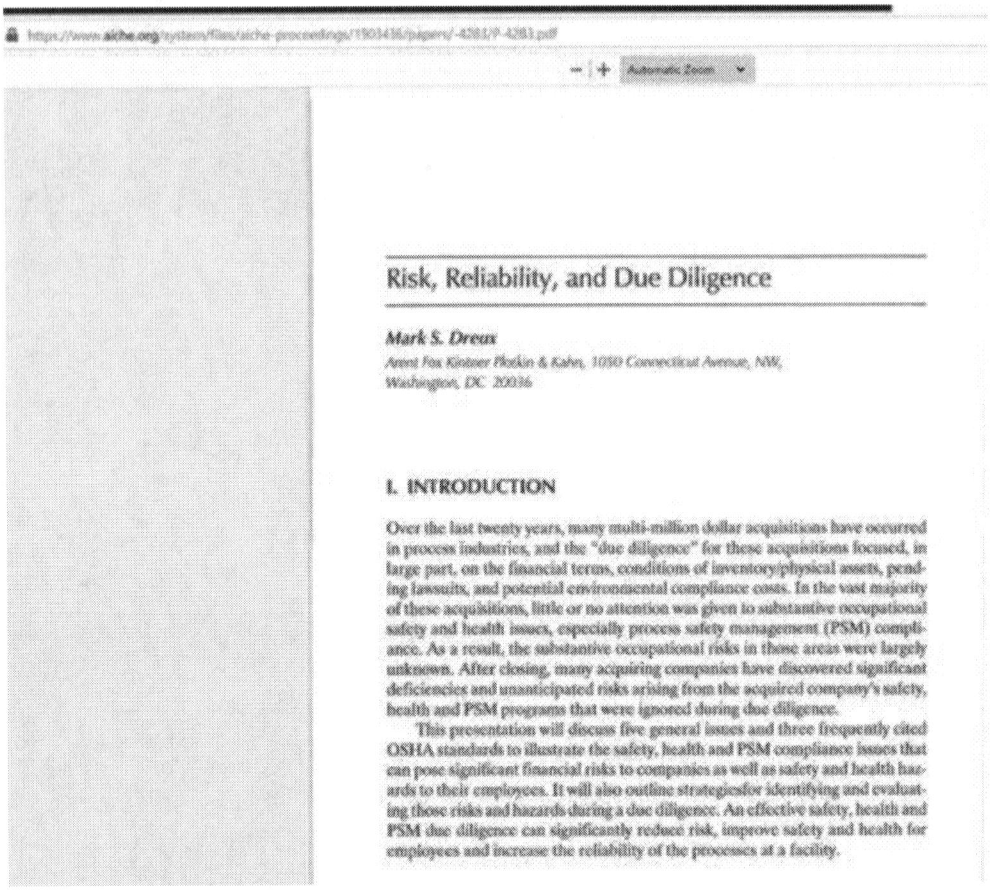

Risk, Reliability, and Due Diligence

Mark S. Dreux
Arent Fox Kintner Plotkin & Kahn, 1050 Connecticut Avenue, NW, Washington, DC 20036

I. INTRODUCTION

Over the last twenty years, many multi-million dollar acquisitions have occurred in process industries, and the "due diligence" for these acquisitions focused, in large part, on the financial terms, conditions of inventory/physical assets, pending lawsuits, and potential environmental compliance costs. In the vast majority of these acquisitions, little or no attention was given to substantive occupational safety and health issues, especially process safety management (PSM) compliance. As a result, the substantive occupational risks in those areas were largely unknown. After closing, many acquiring companies have discovered significant deficiencies and unanticipated risks arising from the acquired company's safety, health and PSM programs that were ignored during due diligence.

This presentation will discuss five general issues and three frequently cited OSHA standards to illustrate the safety, health and PSM compliance issues that can pose significant financial risks to companies as well as safety and health hazards to their employees. It will also outline strategiesfor identifying and evaluating those risks and hazards during a due diligence. An effective safety, health and PSM due diligence can significantly reduce risk, improve safety and health for employees and increase the reliability of the processes at a facility.

Figure 11. Paper finally appears.

Figure 12. Conference Download Option.

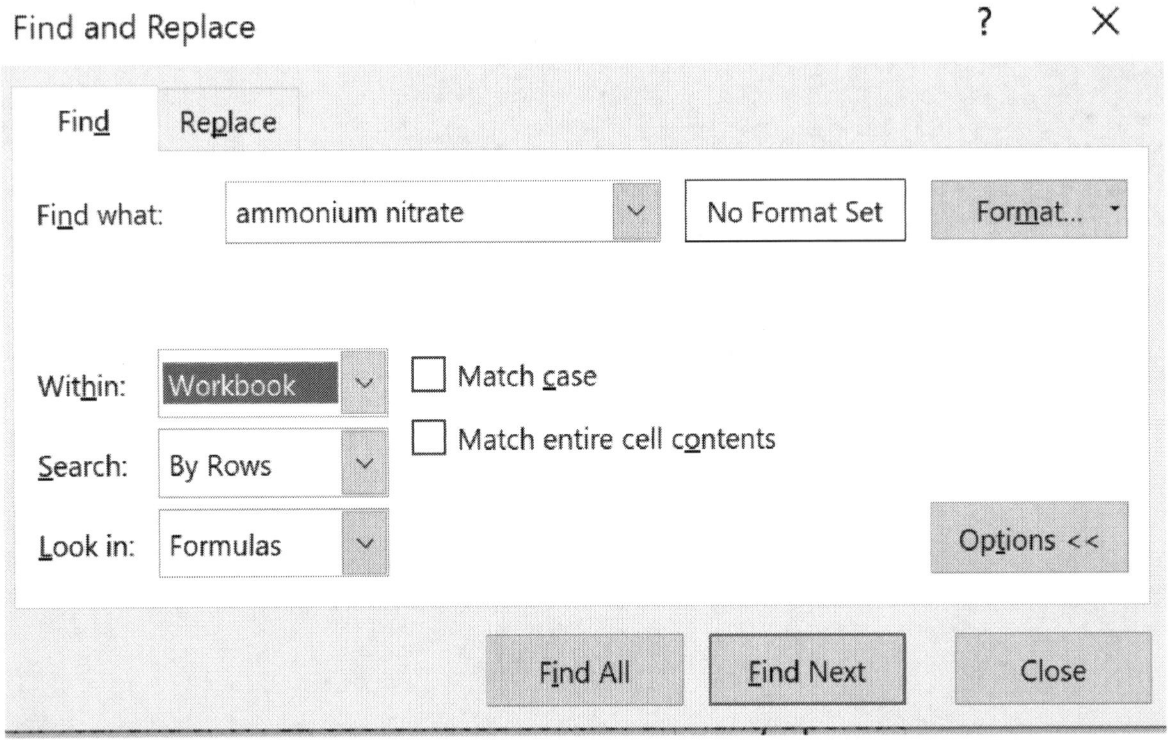

Figure 13. Setting up an Excel Search for the key word combination "ammonium nitrate".

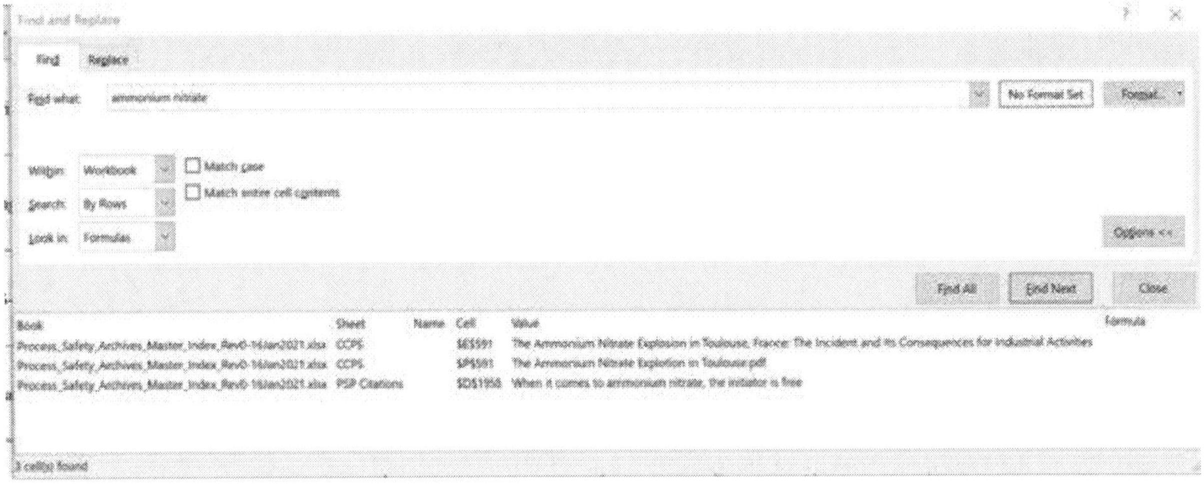

Figure 14. Search results using the term "ammonium nitrate" in the Excel Spreadsheet.

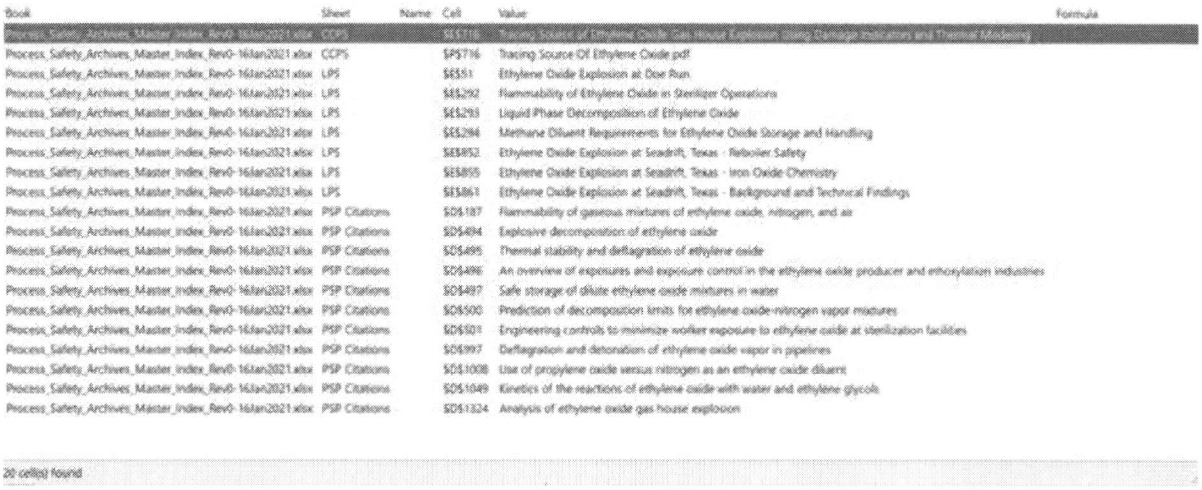

Figure 15. Results using "ethylene oxide" as a search term in the Excel Spread Sheet.

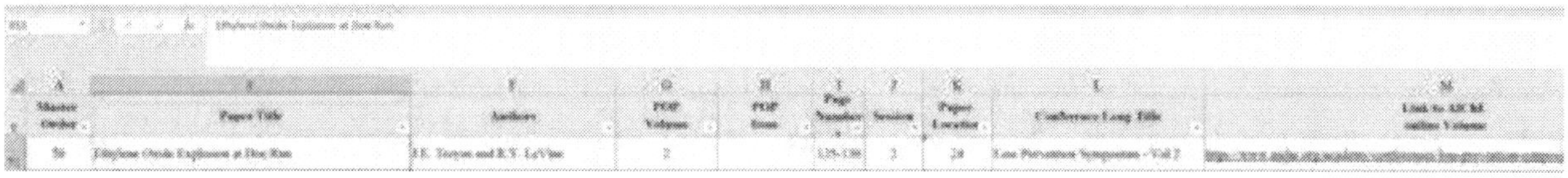

Figure 16. Clicking on the third entry in Figure 15 search takes one to Cell E51 of the LPS tab.

Figure 17. Column M contains a direct link to all proceedings presented in that conference.

Figure 18. Column S contains a direct link to the actual paper.

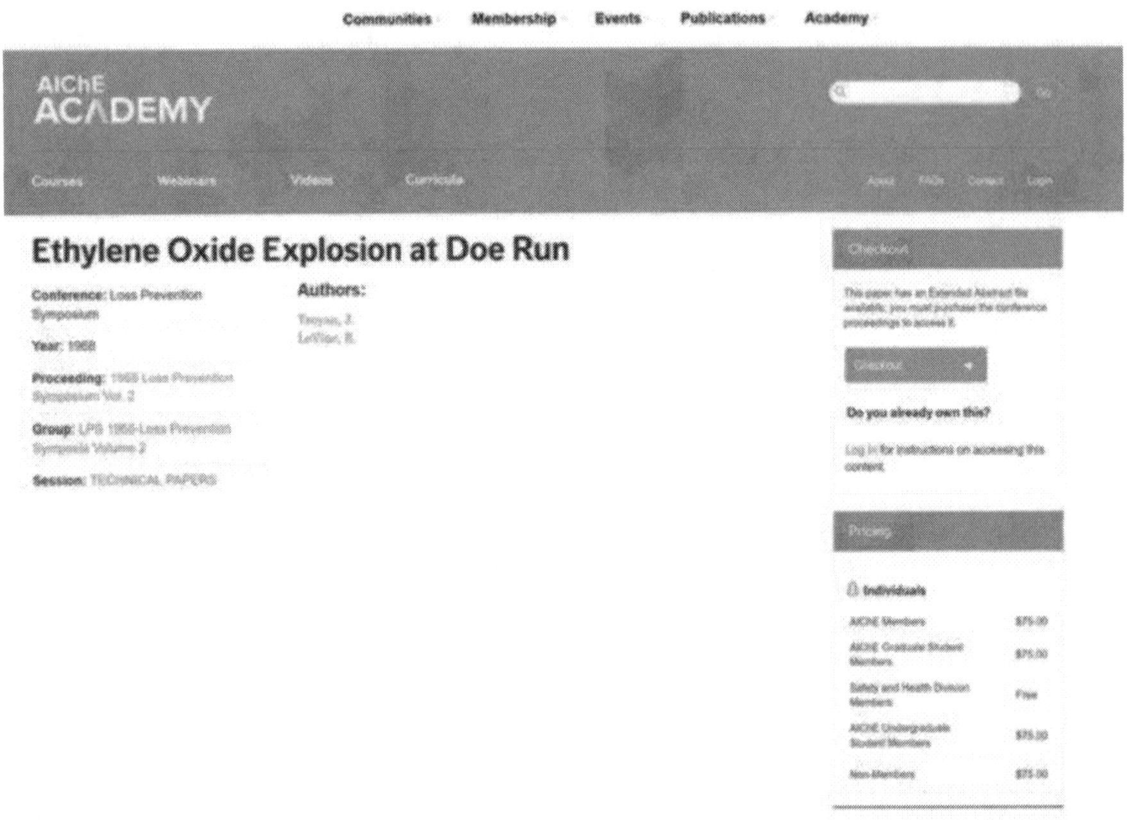

Figure 19. Paper Title Page

619027 Millennial Strategies to Capture Baby-Boomer Knowledge

Abstract:

Risk data is some of the most difficult data in the world to read, digest and understand. Focusing on Process Hazards Analysis (PHA) reports specifically, although the framework and guidewords are followed, for the most part, universally it is still difficult for someone with fresh eyes to pick up a report, read it and know what is important. Never mind having the time to wade through 500-5,000 pages of risk assessment information or multiple risk assessments looking for the specific piece of equipment that they need to know about.

Wouldn't it be great if a new operator or engineer, starts on Monday morning and by noon hour knew the most important risk information relevant to the facility or project they're working on? Why does this matter? Thinking about critical defeats processes, understanding critical procedures and which alarms matter most and need to be prioritized there are many times and processes where this information will be important.

A solution will be presented to make your risk data user friendly for millennials and all generations who are familiar with basic search engine functionality. Being able to search risk data for specific equipment tags and seeing that data in easy to understand, interactive visualizations puts it in context for what they need when they need it. Understanding the relationships between initiating events, safeguards/ IPLs (mitigative/ preventative) and consequences using bowties to tie it all together.

In addition, gamification of risk data will be presented with online, modern and interactive puzzles that can help people understand in minutes the strategies used to safeguard facilities and process units. It also adds a fun factor, competition and other nuances to engage and excite people about safety.

The future workforce needs better solutions to help enable them to leverage the knowledge of the baby boomers and generations that went before. Tens of thousands of experienced workers, who are often the SME in PHA sessions, are retiring on a weekly basis each with decades of knowledge. Although significant knowledge is being lost, the information captured in documents like PHA reports can help the next generations to learn, in second and minutes, what others took a lifetime to build. This knowledge transfer is only possible when presented in an accessible and consumable format.

619104 Lessons Learned for Nextgen PHA Facilitators

Abstract:

There are many training sessions available to train the next generation of PHA facilitators on how to execute a PHA; however, as seasoned facilitators look back on their experiences, they realize there are many details that did not get transferred in their classroom sessions. They learned these details on the job through years of valuable experience and through face-to-face and virtually-led workshops. These learnings can improve the quality and efficiency of the PHA workshop, the documentation of the workshop and the PHA experience. In 2020, this knowledge transfer became even more necessary during the COVID-19 pandemic. Organizations had to move toward conducting workshops virtually while still striving to achieve the same standards needed to ensure a safe facility and satisfy OSHA PSM requirements.

In this paper, seasoned facilitators capture both the small and the significant findings they have learned in their years of PHA facilitation from the very beginning of the PHA (at the preparation phase) to the reporting and follow-up phase. These learnings touch on setting expectations in the terms of reference, maintaining consistent documentation throughout the workshop to aid in effective reporting, maintaining consistency when the team members change (which happens in longer-term workshops), increasing engagement, especially in virtual workshop settings, and increasing use of tools (such as PHA Pro) to improve consistency and efficiency throughout the documentation process.

It is our desire to pass these lessons learned to the next generation of PHA leaders to supplement their classroom learnings and help them succeed in different settings and environments.

618410 "Begin at the Beginning and Go on till You Come to the End: Then Stop" Lifecycle of a Safety Instrumented System

Abstract:

Safety Instrumented Systems have specific steps in their lifespan. Many engineers are responsible for managing these safety systems at a specific point in the lifecycle. This seminar will describe all of the phases of the SIF lifecycle, beginning at the beginning and going through to the end.

617758 Adopting an Integrated Process Maturity Risk Model Approach to Process Safety

Abstract:

Adopting an Integrated Risk Maturity Model for Process Safety

Process Safety can mean different things to different people but one thing is for sure, that when it comes to automation we need to address the three disciplines of: alarm management cybersecurity and functional safety. The reason is that there are consequences of not addressing these from a risk point of view, such as:

- the consequence of inaction due to operators being flooded with alarms;
- the consequence of compromise due to malware or viruses that disable and/or take over the Safety Instrumented System (SIS) or Basic Process Control System;
- the consequence of failing on demand due to the SIS not being maintained properly.

The key point here is that all three disciplines are managed via performance-based standards that have a lifecycle approach with the same three phases: Analysis, Design/Implementation and Operation and Maintenance. By adopting an integrated risk maturity model, companies can assess how well they are doing at implementing these three disciplines and managing the risks associated with them, to meet tolerable levels. This paper outlines an approach and methodology to help companies establish where they are from a maturity point of view, help identify where they want to get to and how to address any gaps. The paper will also present some examples of where companies have benefited from this type of approach and that it is not just large companies that can benefit but companies of all sizes within the process industries.

617249 Reimaging Management of Change

Abstract:

For many companies, Management of Change opportunities ranks among the higher number of findings and observations of all OE processes. Additional experience has found that MOC is one of the more difficult process safety management elements to implement and to get and keep it right. But how can we improve MOC, improve leadership appreciation for what the MOC process is trying to prevent and help eliminate the "get 'er done" culture to make the change and document when we have time? Reimagine an improved overall MOC system, including people, tools, competency and efficiency.

Reimaging a MOC system would include:

- MOC system(s) are more fault tolerant and resistant to circumvention or human error
- Competent, knowledgeable personnel are involved upfront to decide what changes need to be put into the MOC process
- Qualified personnel, normally independent of the MOC originator, review the request to identify any potentially adverse impacts
- Achieve better results with fewer, qualified resources
- Monitor MOC performance and efficiency in a practical way

618537 Using Data Analytics to Generate PHA Checklists – Saving SME Time and Supporting SME Knowledge

Abstract:

The majority of those that have been exposed to Process Safety are aware of Hazard and Operability Studies (HAZOPs); a Process Hazard Analysis (PHA) methodology that is regularly used in industry because of its scalability and relative ease of execution when analyzing the risks of a process unit. Those who have been exposed to HAZOPs are also aware that these studies are qualitative, subjective and incredibly dependent on the people involved in the analysis session. Everyone who participates in a HAZOP does the best they can to do a complete and comprehensive analysis, however their contributions can be limited and biased based on their human knowledge and personal experience. This can lead to inconsistencies in identifying and capturing risks in comparable facilities/units, which can result in increased risks at site.

In order to strive towards higher quality, more comprehensive and more consistent HAZOPs, operation companies are starting to create HAZOP/PHA Checklists to help their teams better prepare for their sessions and be aware of potential high risk scenarios.

Unfortunately, HAZOP/PHA Checklists often take a large amount of time and effort for a Subject Matter Expert (SME) to create, and this amount of SME time is not always available in smaller operating companies.

So, what to do? There is obviously a benefit in having HAZOP/PHA Checklists, so how can the time of an SME required to generate one of these checklists be reduced while still creating a high quality reference tool for HAZOP teams? Also, if the insights of a few SMEs are so beneficial, what about the knowledge of ten or a hundred SMEs?

From these wants, a new method has been found to accomplish this task of generating HAZOP/PHA Checklists by leaning on the power of automatic data processing and data analysis of already completed PHAs. By combining the SME knowledge of tens to hundreds of HAZOPs of the same technology process together, a complete HAZOP/PHA Checklist can be generated with minimum manual SME input.

This paper shows a case study comparison between an SME generated HAZOP Checklist and a Big Data generated HAZOP Checklist, and highlights the high similarity/accuracy when compared, as well as the time savings of the a Big Data generated HAZOP Checklist methodology.

617639 How Effective Is Your Fire and Gas System?

Abstract:

A well-conceived fire and gas system enables the early detection of loss of containment incidents, and the resultant actions can significantly reduce the magnitude and duration of potential consequences. With a wide range of variables affecting fire and gas hazard behavior, and an unlimited number of potential fire and gas detector configurations, how does one ensure that the system is effective and fit-for-purpose?

This presentation will outline an adaptable and consistent approach to assist with the strategic development of a fire and gas detection philosophy that is applicable to both onshore and offshore environments. Emphasis is placed on the importance of hazard assessment to gain an understanding of the behavior of the specific hazards and how this can be used to inform the detection strategy, in conjunction with relevant standards and codes. The presentation goes beyond what may be included in a typical fire and gas philosophy document and covers detector layout strategies, describes modern verification methods to assess detector coverage (e.g. 3D detector mapping), and provides approaches for the management of ongoing inspection and testing of the fire and gas detection system.

618809 Factors Contributing to US Chemical Plant Process Safety Incidents from 2010 – 2020

Abstract:

Process safety incidents occur more frequently than desired – they can result in injuries, fatalities, environmental impacts, facility damage, downtime & lost production, as well as impacts on a company's and industry's reputation. While such incidents occur across many industries, this study is focused on an analysis of the most commonly reported contributing factors to process safety incidents in the US chemical industry.

The database for the study contained 79 incidents from 2010-2019, investigated by the Chemical Safety Board (CSB) and/or by companies. To be included in the database, the incident had to occur at a company classified as 325 in the North American Industry Classification System (NAICS), assigned to businesses who participate in chemical manufacturing. Unlike other summary studies of process safety incidents, the subject analysis is specific to the US chemical industry and based on contributing factors to the incidents reported post investigation.

All of the incidents included in the study were classified as Tier 1 incidents according to the American Petroleum Institute (API) RP 754 definition. Using this methodology for determining incident severity, each incident in the database included a severity score, and was placed into one of four severity categories, identified as low, medium, high and very high severity. All incidents investigated by the CSB were deemed very high severity. The list of contributing factors included in this analysis also provided more granularity and specificity than typically in other studies, and included factors such as preventive maintenance; safeguards, controls & layers of protection; and hazard awareness and identification.

Reported contributing factors were catalogued for each incident by severity category, and then a prioritized (weighted) list of contributing factors was developed for the database of incidents. The top three weighted contributing factors for the chemical industry were: design; preventive maintenance; and safeguards, controls & layers of protection. The finding that the top contributing factors to process safety incidents included factors such as operating procedures, preventative maintenance, and emergency preparedness and response, areas that are traditionally covered under the occupational health and safety administration (OSHA) regulations, prompted a review of the federal OSHA citations database. The investigation and citation history for NAICS 325 companies was analyzed to assess whether there was any overlap between the top reported contributing factors to process safety events and the top OSHA citations recorded for the industry.

A database consisting of the inspection and citation history for the chemical industry identified by NAICS code 325 was assembled for inspections occurring between 2010-2020 (August). Analysis of the thousands of entries provided an understanding of which regulations were most commonly cited, and which were most commonly cited following fatalities. Findings were presented for NAICS code 325, as well as for each sub grouping, including NAICS 3251, 3252, 3253, 3254, 3255, 3256, 3259.

The analysis of the citation history for the chemical industry specifically, identified that the key factors reported as contributing to incidents overlapped with the most common OSHA violations. The overlap of contributing factors for process safety incidents with the most common OSHA violations suggests that addressing regulatory issues may also pay dividends in reducing process safety incidents. These findings will likely be most beneficial if shared among those with the ability to create influence within the chemical industry, assess practices in these key areas, and identify improvement opportunities seeking industry best practices.

618590 Optimal Machine Learning Models for Liquid Aerosolization Safety Contributors

Abstract:

To date, many attentions have been paid to the new types of marine fuels to meet the requirement of the long-term sustainable strategy issued by the International Maritime Organization. Speaking to marine fuel selection, the safety factors should always be top prioritized. Flammable and explosive hazards, which have been well studied, are the major concerns of safety aspects of fuel application. However, the liquid aerosolization issue, making bulk liquids more hazardous on combustion and explosion, has not widely recognized in industry or academia.

This study is aimed at identifying the contributors of liquid aerosol formulation and solving their data deficiencies by developing the corresponding quantitative structure–property relationship models. 1215 liquid chemical substances and 14 predictors have been input to train the machine learning models via k-fold cross validation with the consideration of principal component analysis. Three rounds of comparisons were executed to find the final optimal models for liquid dynamic viscosity (LDV), surface tension (ST) and liquid vapor pressure (LVP). The most persuasive model for LDV is obtained by exponential Gaussian process regression (GPR) approach with seven principal components while the Matern 5/2 GPR algorithm is the most robust one for ST and LVP to formulate the optimal models. Since the reasonably good interpretation and prediction performance provided by the developed models, they can be served as effective tools to expand the database for the liquid aerosol formulation properties of organic compounds.

SPRING21
+17TH GCPS
A Joint AIChE and CCPS Meeting

Human Factors Implementation – for Plant Workers

William Bridges
President
Process Improvement Institute, Inc.
1321 Waterside Lane,
Knoxville, TN 37922 USA
wbridges@piii.com

William C. Rhodes II, P.E.
Process Improvement Institute, Inc. (PII)
82 Via Santa Maria
San Clemente, CA 92672
wrhodes@piii.com

PROCESS IMPROVEMENT INSTITUTE

Prepared for Presentation at
American Institute of Chemical Engineers
2021 Spring Meeting and 17th Global Congress on Process Safety
Virtual
April 18 - 22, 2021

Human Factors Implementation – for Plant Workers

William Bridges
President
Process Improvement Institute, Inc.
1321 Waterside Lane,
Knoxville, TN 37922 USA
wbridges@piii.com

William C. Rhodes II, P.E.
Process Improvement Institute, Inc. (PII)

Keywords: human factors, process safety, PSM, human error, risk-based process safety, RBPS, organizational culture

Abstract

Process safety is about controlling risk of failures and errors; controlling risk is primarily about reducing the risk of human error. Often it is believed that human errors committed by plant workers are the cause of most process safety accidents. However, the entire organization contributes to these human errors. Therefore, actions are needed by the organization to reduce plant worker human errors by improving Human Factors that contribute to the plant worker human errors, and to build an organizational culture that seeks to learn from plant worker human errors, rather than to assign blame.

This paper introduces the multiple types and categories of human error and the Human Factors that influence the rate at which human errors are made. It establishes the need for creating management systems for these Human Factors, and how to implement them in a manner to reduce plant worker human errors. Finally, it describes a Safety-Principled Organizational Culture that allows an organization to create the proper environment to enable these critical improvements to reduce plant worker human error. This paper builds on earlier papers, starting from 2010, on the same topic.

The data presented is from basic research by the authors on the root causes of more than 3000 accidents and near misses; and also based on the review of hundreds of accidents analyzed by others and summary data from many companies. This Video Presentation and the related slides are even more keenly focused on the selected human factors for which the frontline workers should take the lead so that the base human error rate at a site is as low as possible. Case studies and examples are used to illustrate key points.

1. Introduction

All accidents (or nearly all, if you consider that there are some natural phenomena that we either cannot guard against or choose not to guard against) result from human error. Quite often, the erroneous conclusion is drawn that the error was "caused" by the actions of the individual directly involved in the incident. However, this view not only does a disservice to the affected individual, but it prevents proper corrective actions because **humans govern and accomplish all of the activities necessary to control the risk** of accidents. In other words, humans influence other humans in the process. Not only do humans cause accidents (unintentionally) by making errors directly related to the process itself, but they also cause errors by creating deficiencies in the design and the implementation of management systems. Specifically, we make errors in authorities, accountabilities, procedures, feedback, proof documents, continual improvement provisions, etc. Ultimately, these management systems govern the human error rate of those directly contacting or influencing the process. Thus, the entire organization (plant workers, technical staff and management) contribute to the human errors resulting from the actions of the plant worker.

99% of accidental losses (except for natural disasters) begin with a human error *(supported by data from more than 1500 investigations)*

Root causes of accidents are management system weaknesses
(Center for Chemical Process Safety, American Institute for Chemical Engineers, "Guidelines for Investigating Chemical Process Incidents", 2003) – OSHA agrees

Weak Management Systems → Human Error → Accidents

Figure 1: Relationship between Accidents and Management

The optimum approach to reduce plant worker human errors therefore is not to "fix" the plant worker but rather to strengthen management systems so that it is less likely that the human errors will occur (and that when they inevitably do, their consequence is controlled to the greatest extent possible).

This paper outlines this approach by first providing context to the term "human error" so that the means to reduce and mitigate it can be better understood. It then establishes that the most practical and efficient means to minimize plant worker human errors is through properly designed and implemented management systems. Next, it establishes the key Human Factors that contribute to human errors. Conceptually, plant worker human errors are controlled by understanding the Human Factors related to them and their work to the degree that effective management systems for each of them can be developed. Therefore, the paper then details how these Human Factors can be improved for plant workers. Finally, it explains that this entire process is only possible with the right type of organizational culture.

This paper introduces the concept of a Safety-Principled Organizational Culture as one that enables the proper application of Human Factors to plant workers to reduce their human errors. Further, with this culture, when human errors are made, the appropriate lessons are learned so that those errors, and even seemingly unrelated errors, are not repeated. In this way, a plant's overall safety performance will improve.

Definitions for Use in This Paper

- This paper will use the term **Human Error** to mean the errors that are made during direct interface or direct influence of the process.
- **Human Factors** are those aspects of the process and related systems that make it more likely for the human to make a mistake that in turn causes or could cause a deviation in the process or could in some indirect way lead to the increased probability of an accidental loss.
- **Management Systems** are the administrative controls an organization puts in place to manage the people and workflow related to the process under consideration, and so these inherently attempt to control human factors.

2. Types and Categories of Human Error

Quite often people will analyze a plant incident and draw the (typically correct) conclusion that it was due in large part to a plant worker human error. However, they end their analysis at that point, and are therefore unable to draw conclusions as to what could or should be done to prevent its recurrence. To fully protect against future harm, however, plant worker human errors must be fully analyzed, and the first step in this analysis is to understand the types and categories of human error.

In simplest terms, there are only two types of human error: *Errors of Omission* (someone skips a required or necessary step) and *Errors of Commission* (someone performs the step wrong). But in addition, these errors occur either inadvertently (unintentional error) or they occur because the plant worker believes his or her way is a better way (intentional error, but not intentional harm). Intentional errors can usually be thought of as errors in judgment. Some believe a "lack of awareness of the risk" causes these errors, but in actual practice, the plant worker who commits an intentional error may well be aware of the risk. They instead believe they know a better way to accomplish a task or they believe there are already too many layers of protection (so bypassing one layer will not cause any harm).

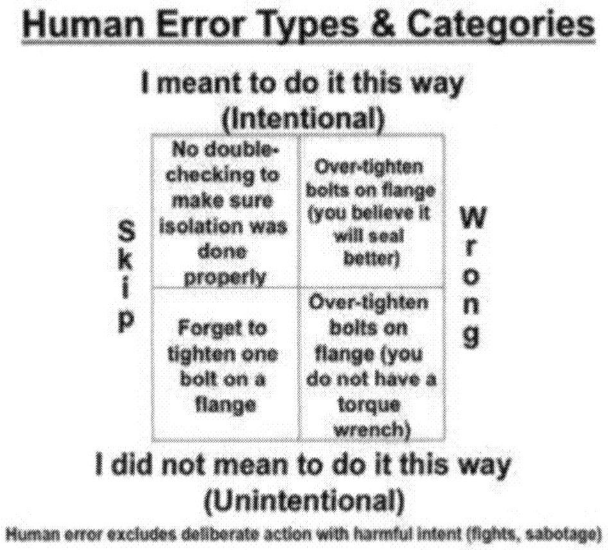

Figure 2: Types and Categories of Human Error

Regardless of type or category of human error, the organization can and should exert considerable control of the errors through properly implemented management systems.

3. Controlling Human Error through Management Systems

A process is a combination of the utilities, raw materials, and human actions (direct actions and those actions involving programming the process to accomplish automatic functions). If anything goes wrong with these Inputs, or if there are basic design flaws or basic fabrication flaws in the process, then the Outputs will not be desired. The desired output is acceptable (or high) production rates at acceptable or higher quality factors with no harm to humans (long term or short term), no harm to the environment, and with acceptable (or higher) life of the process components. The negative outcomes resulting from plant workers failing to control the raw material quality, failing to control the utility levels consistently, making errors directly related to the operation of the process, or making errors in the care of the process (such as maintenance) will result in lower production, lower quality, higher number and severity of safety-related accidents, and more negative impact on the environment. The potential of the negative outcomes is collectively referred to as ***business risk*** – more precisely, the risk is a product of the likelihood of one or more of these negative outcomes and the severity of each outcome.

A critical concept is therefore:

If an organization does not directly control risk, the organization cannot directly control quality, safety, environmental impact, or production to acceptable levels. An organization must sustainably control human error to manage the risk of accidental losses that impact quality, safety, the environment, production, or assets.

In order to sustainably control the risk of a complex process (such as an oil/gas operation, refinery, chemical plant, steel plant, automobile manufacturing, aircraft manufacturing, etc.), the organization must design and implement management systems to:

- **Understand the Risk** – This involves predicting problems; which in turn includes predicting the risk of possible accident/loss scenarios, establishing the appropriate design and the right layers of protection to control risk to a tolerable level
- **Control Risk Factors Day-to-Day** – This involves controlling the original design by maintaining the established layers of protection and managing changes to the design using integrated management systems
- **Analyze Actual Problems and Determine Weaknesses in the System** – This involves identifying weaknesses in designs and management systems and weaknesses in risk understanding through root cause analysis of actual problems (losses and near-losses)

These three elements are illustrated in the following figure:

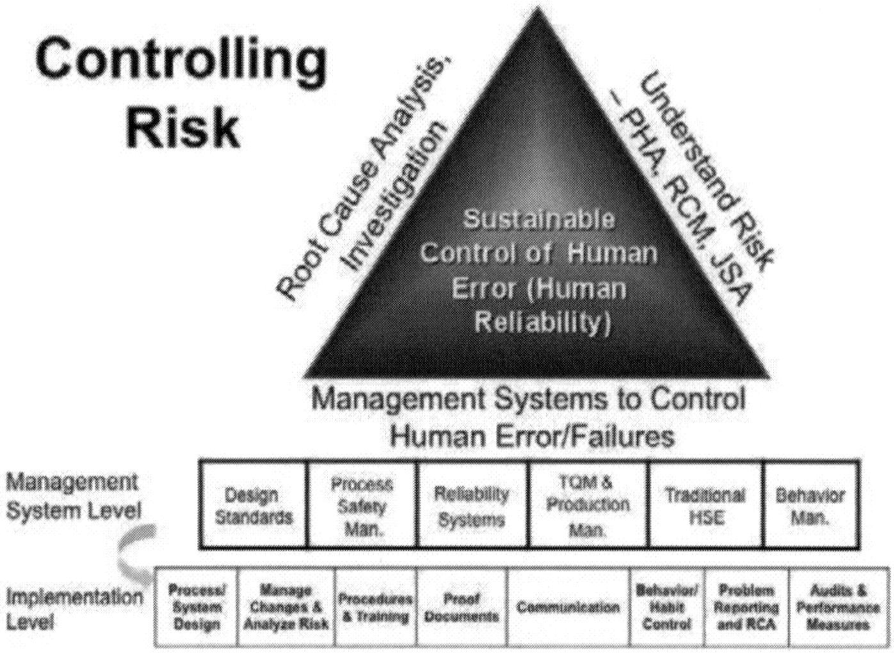

© Process Improvement Institute, Inc. (2004-2021)
Figure 3: Controlling Risk through Human Reliability

Management systems control the interaction of people with each other and with processes. They are the high-level procedures we use to control major activities like conducting PHAs, management of change, writing operating procedures, training employees, evaluating fitness for duty, conducting incident investigations, etc. If management systems are weak, then layers of protection will fail, and accidents will happen.

To reiterate, accidents are caused by human error. In general, Process Safety Management (PSM) is focused on maintaining these human errors at a tolerable level because:

- All accidents happen due to errors made by humans; including premature failure of equipment. There is a myriad of management systems to control these human errors and to limit their impact on safety, environment, and quality/production
- When these management systems have weaknesses, near misses occur
- When enough near misses occur, accidents/losses occur

4. Human Factors that Contribute to Human Errors

To minimize plant worker human error, process safety systems should address the **Human Factors Categories** (see various US NRC and US DOE standards from 1980s and 1990s)[1,2]. Table 1 on the next page lists the key human factor categories along with multiplication factors that poor human factors can have on the base human error rates.

TABLE 1: SUMMARY of 10 HUMAN FACTOR CATEGORIES

Based in part on: Gertman, D.; et. al., *The SPAR-H Human Reliability Analysis Method*, NUREG/CR-6883, U.S. Nuclear Regulatory Commission, Office of Nuclear Regulatory Research, Washington, DC, August 2005. PII has modified the list slightly to account for general industry data and terminology and to incorporate PII internal data.

Human Factor Category	Human Factor Issue/Level	Multiplier for Cognitive & Diagnosis Errors
Available Time	Inadequate time	P(failure)=100%
(includes staffing Issues) – *for responses only*	Barely adequate time (≈2/3 x nominal)	10
	Nominal time (1x what is expected)	1
	Extra time (at least 2x nominal and >20 min)	0.1
	Expansive time (> 4 x nominal and > 20 min)	0.01
Stress/Stressors	Extreme (threat stress)	5
(includes staffing issues)	High (time pressures such as during a maintenance outage; issues at home, etc.)	2
	Nominal	1
Complexity & Task Design	Highly complex	5
	Moderately complex (requires more than one staff)	2
	Nominal	1
	Obvious diagnosis	0.2
Experience/Training	Low	10
	Nominal	1
	High	0.5
Procedures	Not available in the field as a reference, but should be	20
	Incomplete; missing this task or these steps	8
	Available and >90% accurate, but does not follow format rules ***(normal value for process industry)***	3
	Good, 95% accurate, follows >90% of format rules	1
	Diagnostic/symptom oriented	1
Human-Machine Interface (includes tools)	Missing/Misleading (violates populational stereotype; including round valve handle is facing away from worker)	20
	Poor or hard to find the right device; in the head calc	10
	Some unclear labels or displays	2
	Good	1
Fitness for Duty	Unfit (high fatigue level (>80 hrs/wk or >20 hr/day, no day off in 7-day period; or illness, etc.)	20
	Highly degraded fitness (high fatigue such as >15 hr/day, illness, injury, etc.)	10
	Degraded Fitness (>12 hr day and >72 hr/wk)	5
	Slight fatigue (>8 hr per day; ***normal value for process industry***)	2
	Nominal	1
Work Processes & Supervision	Poor	2
	Nominal	1
	Good	0.8
Work Environment	Extreme	5
	Good	1
Communcation	No communication or system interference/damage	10
	No standard for verbal communication rules ***(normal value for process industry)***	3
	Well implemented and practiced standard	1

If all of the human error factors shown above in Table 1 are controlled very well, then the optimized human error rates listed in Section 5 of this document are achievable. Alternatively, if one or more of these factors are compromised, the human error rate will increase by the values shown in this table. As an example, an individual whose fitness for duty rating is unfit due to the excessive work hours shown in the table will make errors at a rate 20 times greater than an individual whose fitness for duty is normal.

With excellent control of each of the human factors listed above, a company can begin to approach the lower limits that have been observed for plant worker human error. These lower limits are about:

- **1 mistake in 100 steps for most procedures-based tasks** (such as starting up a process unit), a little less for a routine (daily) task that becomes almost a reflex
- **1 in 10 chance or a little better for diagnosis and response to a critical alarm**

Excellent control requires superior design and implementation of management systems, which is enabled through a thorough understanding of these factors, as outlined below.

5. Strengthening Human Factors for Plant Workers

In order to achieve the lowest practical plant worker human error rates noted above, comprehensive management systems must be developed for each of the Human Factors provided in Table 1. Key elements for each of these Human Factors are provided below.

A. <u>AVAILABLE TIME</u> and <u>STRESS / STRESSORS</u>

The **Available Time** for the task refers to the time the task is expected to be required for completion. In many cases, the time determined for a process is determined using the theoretical times where the conditions are "ideal" and not necessarily realistic. When task and process duration are not realistic plant workers tend to "find the best and easy way" to get it done, ultimately creating latent conditions for error occurrence. **Work Stress** refers to the emotional response that arises when work demands exceed the plant worker's capacity and capability to cope. Some processes have been designed paying little attention to job design, work organization and management systems. Failing to fulfill this requirement eventually results in work stress.

Staffing levels directly impact both factors. Staffing is the process of assessing, maintaining and scheduling personnel resources to accomplish work. An adequately staffed organization ensures that personnel are available with the proper qualifications for both planned and foreseeable unplanned activities. Staffing is a dynamic process in which plant management monitors plant worker performance to ensure that overall organizational performance goals are met or exceeded. The result of an effective staffing process is a balance between personnel costs and the achievement of broader organizational goals.

Three key issues must be considered when staffing decisions are made; the first of which is selecting the right staff for a job. Each organization requires the proper amount and type of expertise to operate the plant safely and competently under a variety of conditions. The term "expertise" includes the attributes of talent, effectiveness, knowledge, skills, abilities, and experience necessary to operate and maintain plant systems, structures and components.

The second key issue is avoiding staff overload. Surges in workload, such as during outages, typically require staff augmentation as well as longer work hours for permanent staff. The introduction of contractor personnel or company personnel from other sites may increase the likelihood of errors due to unfamiliarity with the plant, its procedures and hardware, for example. Longer work hours have the potential to increase fatigue, which also contributes to the likelihood of error.

The third key factor is rotating staff every 1 hour or less for tasks that require high vigilance. Humans are inherently unable to remain alert for signals that seldom, if ever, occur. Even a sailor whose life is at stake cannot maintain an effective watch (look-out) for hostile submarines for more than 30 minutes or so. The following figure illustrates the rapid decrease of vigilance with time.

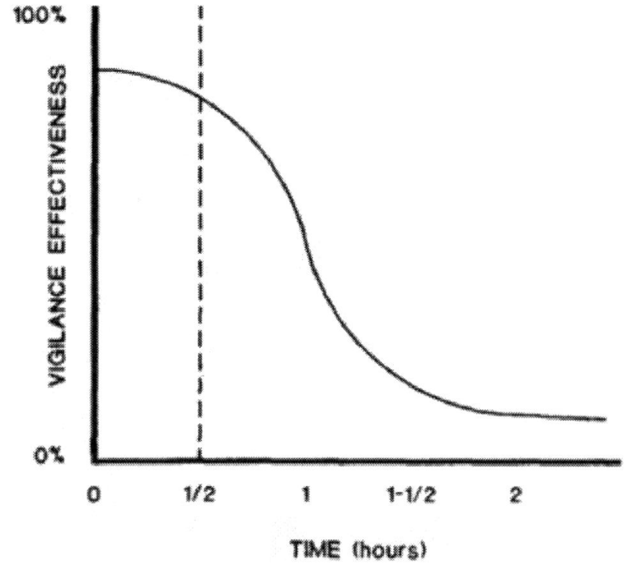

Figure 4: Vigilance as a Function of Time

Placing a worker in situations requiring extended, uneventful vigilance may lead to accidents. Therefore, it is important to design control systems in a manner that requires regular operator interaction so that the operator will remain attentive.

B. COMPLEXITY & TASK DESIGN

A task that is designed with the human limits in mind is much more likely to work effectively than one that assumes humans can and will "always" do what is written. The task must consider that humans think and remember and factor in prior data and prior experiences. Factors to consider in assessing **task design** adequacy (i.e., minimizing the likelihood of plant worker human error during its completion) include:

- Complexity of the task
- Probability of repeat errors (coupled error) on redundant aspects of the design
- Error Detection and Error Recovery

Complexity of Task (procedure-based or a call for action) – If the task is too complex, then plant workers can forget their place in the task, fail to understand the goal of each step or sub-step, or fail to notice when something isn't going right.

Task complexity is a function of:

- Number of choices available for making a wrong selection of similar items (such as number of similar switches, number of similar valves, number of similar size and shaped cans)
- Number of parallel tasks that may distract the worker from the task at hand (leading to either an initiating event or failure of a protection layer)
- Number of staff involved (more staff = more complex)
- Number of adjustments necessary to achieve the goal
- Amount of mental math required (as a rule, NO math should be required in anyone's head when accomplishing a standardized task)
- How much judgment is required to know when you have accomplished each goal within the task

For most chemical process environments, the complexity of the task is relatively low (one action per step), but for response actions (in which plant workers serve as the Independent Protection Layer (IPL)), there are almost always other tasks underway when the out-of-bounds reading or the alarm is activated. Complexity is difficult to predict (since it is not known when a human intervention will be needed), but higher complexity can increase error rates by 2 to 10 times.

Probability of Repeat Errors (Coupled errors) – For many maintenance tasks, making a repeat error or "common cause error" or "dependent error" can lead to failures of all backup systems. Such errors have led to MANY airplane crashes and major process safety accidents as well.

Coupling represents the probability of repeating an error (or repeating success) on a second identical task, given that an error was made on the first task. The increased probability of failure on subsequent tasks given that an error has already been made is known as dependence. The list below provides some starting point guidance on values to use:

- 1/20 to1/90 – if the same tasks are separated in time and if visual cues are not present to re-enforce the mistake path. This error rate assumes a baseline error rate of 1/100 with excellent human factors. If the baseline error is higher, then this rate will increase as well.
- 1/2 – if the same two tasks are performed back-to-back, and if a mistake is made on the first step of two. This error rate assumes a baseline error of 1/100 with excellent human factors. If there the baseline error is higher, then this rate will increase as well.
- 8/10 to 10/10 – if the same three tasks are performed back-to-back and strong visual cue is present (if you can clearly see the first devices you worked on), if a mistake is made on the first step of the two or more

- Two or more plant workers become the same as one person (with respect to counting of errors from the group), if people are working together for more than three days; this is due to the trust that can rapidly build.

These factors are based on the relationships provided in NUREG-1278[1] and the related definitions of weak and strong coupling provided in the training course by Swain (1993)[1] on the same topic, as shown below in Table 2.

The following relationship is for errors of omission, such as failing to reopen a root valve or failing to return a safety instrumented function (SIF) to operation, after bypassing the SIF. The qualitative values in Table 2 are based jointly on Swain[1] (1993) and Gertman (SPAR-H, 2005 which is NUREG/CR-6883)[2].

One can readily conclude that staggering of maintenance tasks for different channels of the same SIF or for related SIFs will greatly reduce the level of dependent errors. Unfortunately, most sites PII visits do not stagger the inspection, test, or calibration of redundant channels of the same SIF or of similar SIFs; the reason they cite is the cost of staggering the staff. While there is a perceived short-term higher cost, the answer may be different when lifecycle costs are analyzed.

Simple Rule: Staggering of maintenance can prevent a significant number of plant worker human errors in redundant channels. In the airline industry, the US Federal Aviation Administration (FAA) requires staggering of maintenance for aircraft with multiple engines or multiple control systems (i.e., hydraulics) (*FAA Advisory Circular 120-42A*, as part of Extended Operations (ETOPS) approval).[6]

Table 2: Guideline for Assessing Dependence for a within-SIF Set of Identical Tasks (based partially on SPAR-H, 2005 [1,2], and partially on field observations by PII) Courtesy Process Improvement Institute, Inc., All Rights Reserved

Level of Dependence	Same Person	Actions Close in time	Same Visual Frame of Reference (can see end point of prior task)	Worker Required to Write Something for Each Component
Zero (ZD)	No; the similar tasks are performed by different person/group	Either yes or no	Either yes or no	Either yes or no
Zero (ZD)	Yes	No; separated by several days	Either yes or no	Either yes or no
Low (LD)	Yes	Low; the similar tasks are performed on sequential days	No	Yes
Moderate (MD)	Yes	Moderate; the similar tasks are performed more than 4 hours apart	No	No
High (HD)	Yes	Yes; the similar tasks are performed within 2 hours	No	No
Complete (CD)	Yes	Yes; the similar tasks are performed within 2 hours	Yes	Either yes or no

Once the level of dependence is known, the probability of either repeat success or repeating errors on identical tasks can be estimated. For these probabilities, we use Table 3, which is a re- typing of Table 20-17 from NUREG-1278[1] (and the similar table in SPAR-H [Gertman, 2005][2]).

Table 3. Equations for Conditional Probabilities of Human Success or Failure on Task N, given probability of Success (x) or Failure (X) on Task N-1, for Different Levels of Dependence *Courtesy Process Improvement Institute, Inc., All Rights Reserved*

Level of Dependence	Repeating Success Equations (but shown as error probability)	Repeating Failure Equations
Zero (ZD)	$P_{Success@N} = x$	$P_{Failures@N} = X$
Low (LD)	$P_{Success@N} = (1+19x)/20$	$P_{Failures@N} = (1+19X)/20$
Moderate (MD)	$P_{Success@N} = (1+6x)/7$	$P_{Failures@N} = (1+6X)/7$
High (HD)	$P_{Success@N} = (1+x)/2$	$P_{Failures@N} = (1+X)/2$
Complete (CD)	$P_{Success@N} = 1.0$	$P_{Failures@N} = 1.0$

Error Detection and Error Recovery – Is there enough feedback in the process to allow the plant worker to realize (in time) that they made a mistake? Have they been trained on how to reason through how to recover from mistakes they or others make? (Sometimes, doing a step too late is far worse than NOT doing the step at all.) Is there enough time available for the type of intervention necessary?

C. EXPERIENCE/TRAINING

Training is necessary for general functioning of management systems and for task-specific skills such as how to start up a compressor, repair a pump, lead a root cause analysis, perform a proper Lock-Out/Tag-Out, etc. Organizations normally do a good job of training, including hands-on training. Training systems can be weak (inadequate), however, if the training system does not adequately:

- Address how to troubleshoot the process (i.e., handle process deviations or upsets).
- Provide plant workers with a mental model(s) of the process so they can perform the assigned tasks correctly, diagnose process upsets properly and quickly, and understand the consequences of their actions.
- Provide plant workers with enough practice of critical tasks.

Ironically, training is many times listed as the "cause" or "root cause" of an accident, when in fact the training is adequate, and some other Human Factor is the major cause.

D. PROCEDURES

Many procedures do not follow best practices for controlling human error, and so the written procedure actually "contributes" to increased plant worker error rates. Further, many organizations are missing guides on how to troubleshoot (what to do when process deviations occur). The best practice rules for writing and validating procedures have been published for many years ***(see Bridges, 1997-2010)***[3, 7, 8]. Below is a checklist based on the current set of best practice rules for developing operating, maintenance and other work-instructions (procedures):

TABLE 4: PROCEDURE QUALITY CHECKLIST (courtesy PII, 2008)

#	Issue	Response
	Procedure Content Checklist	
1	Is the procedure drafted by a future user of the written procedure? (Engineers should not author procedures to be used by operators or maintenance staff.)	
2	Is the procedure validated by a walk-down in the field by another future user of the procedures?	
3	Is the procedure reviewed and commented on by technical staff (engineers or vendors)?	
4	Is the procedure checked versus the Page and Step format rules below?	
5	Is a hazard review of step-by-step procedures performed to make sure there are sufficient safeguards (IPLs) against the errors that will occur eventually (when a step is skipped or performed wrong)?	
6	Is the content measured using "newly trained operators" to judge the % of steps that are missing, steps that are confusing or wrong, and steps that are out-of-sequence? (A score of 95% accuracy of content is good.)	
	Page Format Checklist	
1	Is the title of the procedure the largest item on the page?	
2	Is the procedure title clear and consistent with other titles, and does it uniquely describe the topic?	
3	Are the document control features the smallest items on the page?	
4	Are temporary procedures clearly identified?	

#	Issue	Response
5	Is white space used effectively? • Is there one blank line between each step? • Does indentation help the user keep their place? • Are the margins large enough to reduce page congestion?	
6	Is type size 12 point font or larger?	
7	Is serif type used (rather than sans-serif)?	
8	Is mixed case used for words of steps, with ALL CAPS used only for special cases (such as IF, THEN, AUTO, and WARNING)?	
9	Is the step number very simple (single level of number used)? Only an integer?	
10	Have sections or information not necessary to performing the steps been moved to the back or to another part of the manual or training guide?	
11	Are section titles bold or larger than the text font? Do sections have clear endings?	
12	Is the decision on electronic presentation versus hard copy correct? Are electronic linkages to procedures clear and accurate and easy to use? If paper is chosen, is the color of the paper appropriate?	
13	Is the overall page format (such as Outline format or T-Bar format) appropriate to the use of the procedure?	
14	Are play script features added for tasks that must be coordinated between two or more users? • Play script is normally used when there are two or more hand-offs of responsibility for steps.	
15	Are rules followed for formatting of Warnings, Cautions, and Notes? (See annotated rules, such as Warnings are for worker safety and Warnings must clearly standout from rest of page.)	
	Step Writing Checklist	
1	Is each step written as a command?	
2	Is the proper level of detail used throughout? This is judged based on: • *Who* will use the procedures • Same level of detail used in similar procedure steps	
3	On average, is there only one implied action per instruction? Best practice is to average 1.2.	
4	Does the procedure indicate when sequence is important? • If sequence matters, each step should be numbered (with an integer or letter) • If sequence does not matter, bullet lists should be used	
5	Are only common words used? Apply "education" level test (5 grade reading level is best)	
6	Do all <u>acronyms</u>, <u>abbreviations</u>, and <u>jargon</u> aid understanding? • Develop a list of such terms for use in procedures *and* communication. • Use terms that users use (within reason)	
7	Is each step <u>specific</u> enough? No room left to guess/interpret: • The meaning of a word or phrase (Check vs. Make sure) • The intent of a step or series of steps • A desired quantity or value • To what equipment the step applies	
8	Is the procedure free of steps that require in-your-head <u>calculations</u>? • Values expressed as ranges rather than targets with error bands • Conversion tables, worksheets, or graphs provided where needed	
9	Are graphics to the user's advantage? • No explanatory paragraphs or lengthy instructions that could be replaced by a picture • No impressive graphics that provide no real advantage	
10	Are references to the user's advantage? • No lengthy explanations or instructions that could be replaced by branching to a reference • No references to a procedure that references still another • No gaps or overlaps between this procedure and a referenced document • If branching, must branch to a procedure, not to a specific step in a procedure	

#	Issue	Response
11	Are rules followed for writing warnings, cautions, and conditional steps? • No more than 2 per page • No actions within a warning or caution (actions must always be numbered steps) • Warnings and Cautions contain descriptions of potential consequences	

For procedures to be effective in ensuring that tasks are performed correctly, they must be used. There are many reasons that plant workers may not use procedures. ***Deficient Procedures*** are the most prevalent problem in process industries since procedures have not traditionally been developed from the perspective of optimizing human factors; instead, procedures have been traditionally developed to meet a compliance requirement to have written procedures. Examples of procedure deficiencies (inaccuracies) include:

- Incorrect/incomplete/nonexistent. Most procedures we have audited have been only 70 – 85% accurate; the inaccuracies include missing critical steps, steps as written are not what needs to be done, or the steps are out of sequence
- No/misplaced warnings. For example, a warning should never **contain** the action to take; it should instead **emphasize** the action to take
- Poor format and presentation rules

Other reasons plant workers may not use procedures include:

- Procedures are out of date
- No procedure has been written for the task
- Plant workers cannot find the procedure they want to use
- Plant workers don't need a procedure because the task is simple
- Plant workers need more information than the procedures contain
- Plant workers see procedures as an affront to their skill
- Procedures are difficult to use in the work environment
- Procedures are difficult to understand

So, in addition to the rules for writing procedures that are shown in the table above, the organization must also address the reasons that cause the plant worker not to use the written procedure.

E. HUMAN-MACHINE INTERFACE (INCLUDING TOOLS)

The **Human-Machine Interface** (also known as the Human-System Interface, or HSI) is defined as the technology through which plant workers interact with plant systems to perform their functions and tasks. The major types of HSIs include alarms, information systems, and control systems. Each type of HSI is made up of hardware and software components that provide information displays, which are the means for user-system interaction, and controls for executing these interactions.

Plant worker use of HSIs is influenced directly by (1) the organization of HSIs into workstations (e.g., consoles and panels); (2) the arrangement of workstations and supporting equipment into facilities, such as a main control room, remote shutdown station, local control station, technical support center, and emergency operations facility; and (3) the environmental conditions in which the HSIs are used, including temperature, humidity, ventilation, illumination, and noise.

There are three important goals to be achieved in the design and implementation of the HSI. These are:

- **Design for operability** refers to designing the HSI to be consistent with the abilities and limitations of the plant workers who will be operating it. Weaknesses in the design processes can result in an HSI that is not well suited to the tasks that they must perform to ensure plant safety, resulting in increased workload, decreased performance by personnel, and an increased likelihood of errors.

  The wide-spread adoption of distributed control systems has provided many advances in process control. However, if implemented improperly, these systems can greatly reduce the operability of the facility. Although they were unable to provide as much information, old analog panels were typically laid out in a manner such that experienced plant workers were able to monitor the entire process and effectively identify conditions meriting attention. Due to space limitations, and their cost (e.g., hardware and installation), alarms were typically limited to those deemed to be associated with a significant process upset.

  The advent of distributed control systems has eliminated the space limitations associated with alarms, and greatly reduced their costs (in many cases, alarms can be added merely through programming). Without proper alarm prioritization, configuration structure and discipline, this can lead to "alarm flooding" (an excessive number of simultaneous alarms) during abnormal operating conditions. The human error of failing to respond properly to an alarm increases by a factor of over 1,000 if ten alarms activate simultaneously compared to a single alarm activation.

  Therefore, it is critical that control and alarm system designers in facilities with distributed control systems properly prioritize information so that plant workers can respond appropriately to abnormal conditions. Further, trouble-shooting guides should be readily accessible via the distributed control system to reduce human error in alarm response.

  Beyond the topic of alarms, the proper design of the visual displays can also reduce human error. Certain color combinations, such as a light gray background with darker gray equipment and black lines coupled with a flashing red indication for those components in an alarm state effectively draw the plant workers' attention to the relevant part of the process requiring attention. Further, an effective display should include:

  - Safety systems that are bypassed, impaired, or out of service
  - Process alarms that are disabled
  - Process conditions that are not within established limits
  - Backup systems that are unavailable
  - Special maintenance or testing activities that are currently in progress

- **Design for maintainability** refers to designing the HSI and associated plant equipment to ensure that plant workers are able to perform necessary maintenance activities efficiently. Weaknesses in the design process can result in systems that impose excessive demands on plant workers for maintenance and, therefore, are prone to maintenance errors or problems with reliability and availability.

- **Design for flexibility** refers to the way that changes, such as upgrades to the HSI, are planned, and put into use. A new HSI component may require the plant worker to perform functions and tasks in new ways. Skills that the plant worker developed for managing workload when using the former design, such as ways for scanning information or executing control actions, may no longer be compatible with the new design. The new HSIs must also be compatible with the remaining HSIs so that plant workers can use them together with limited possibilities for human error. Also, HSI modifications may not be installed or put into service all at one time, causing the plant worker to adapt to temporary configurations that are different from both the original and final configurations. Weaknesses in HSI implementation can increase plant worker workload and the likelihood of errors.

Tools are a special category of HSI, which typically refers to hand tools or devices that are generally designed with the plant worker in mind. A big part of human factor consideration is how to make the equipment and process operation mistake-proof (to prevent errors as much as possible). The following are some factors to consider when mistake-proofing in designs:

- **Design for unambiguous assembly.** Design the product or device such that the assembly process is unambiguous (by designing components so that they can only be assembled one way); i.e., design matching parts that are easy to insert and align. For example, use notches, asymmetrical holes, and stops to mistake-proof the assembly process. Products that go together in only one way require less plant worker training, perform more reliably, and can be repaired more quickly.
- **Consult plant workers.** Operators, technicians, and maintenance personnel can pinpoint the most troublesome areas.
- **Avoid symmetry.** When a particular orientation is critical to the design, avoid symmetry. For example, use nonsymmetrical hole patterns.
- **Use labels sparingly.** Labels tend to come off equipment too easily and often are wordy.
- **Review the environment.** Environmental problems that encourage mistakes include poor lighting; high/low heat; excess humidity, dust, and noise — anything that distracts plant workers.

The following are some factors to consider when mistake-proofing in process/operations:

- **Error-proof Mechanisms.** Error-proof mechanisms are very powerful in improving system reliability when incorporated into the design. These mechanisms, by design, will not allow a plant worker to perform an illegal operation. For example, if a plant worker enters a value that is outside the accepted range of operation, the control logic will not accept the value.

- **Automatic Alerts.** Automatic alerts immediately inform the plant worker of an illegal operation to prompt corrective action. These alerts are particularly useful in critical operations that allow time for corrective action before some adverse consequence occurs. For example, a piping manifold may be designed such that a warning alarm sounds if valves are not opened/closed in the proper, critical sequence, or if a critical valve is left opened/closed.
- **Automatic System Shutdown.** Automatic system shutdown should be incorporated into the design when an illegal action is performed during a critical operation and no time is available for corrective action. For example, if a plant worker uses the wrong sequence in charging a reactor, the reactor will shut down before other materials are charged that may lead to critical temperatures and pressures.

F. FITNESS FOR DUTY

Successful task performance requires that the capabilities that plant workers bring to the task fall within an expected range. **Fitness for Duty** issues include reduction in a plant worker's mental or physical capabilities due to substance abuse, fatigue, illness, or stress, which increases the likelihood of errors. Types of possible impairments include:

- Physical attributes – strength, reach, eyesight and color acuity, hearing,
- Mental attributes – drug and alcohol (abuse), mental stress (on and off the job);
- Fatigue – issues from on the job and off the job (especially control of hours per work-day and per work-week)

Safeguards to prevent fitness for duty related errors include company programs for the detection and prevention of potential or actual impairment, as well as the individual responsibility of plant workers to decline assignments if they are impaired for any reason. The latter safeguard is a weak one, however, because humans are generally over-confident of their capabilities when under the influence of drugs or alcohol, or are stressed, fatigued or ill. Other factors that may discourage self-reporting include the fear of poor performance reports from supervision or not receiving pay for extra overtime.

Company programs that may be implicated in errors caused by plant worker impairment include:

- **Fitness for Duty Program** – Company fitness-for-duty programs are primarily responsible for detecting and preventing impaired plant workers from performing tasks that may affect public health and safety. Medical evaluations of personnel, behavioral observation programs, employee assistance programs and drug and alcohol testing are used to detect impairment. Impaired plant workers can be prevented from performing tasks by establishing protocols for instances in which a plant worker is believed to be unfit for duty, training for supervisors on detection of and response to fitness for duty issues, and employment guidelines for personnel who voluntarily enter drug or alcohol treatment programs. Weaknesses in a Fitness for Duty program may allow impaired personnel to have access to vital areas in a plant where they could commit errors. One excellent starting point for a Fitness for Duty sub-element is the guidance provided in ***US NRC's 10 CFR 26 (2005)***[15].

- **Overtime Policies and Practices** – Most companies establish limits for work hours to reduce on-the-job fatigue. It has been shown that 17 hours of work without a break results in the same error rate as being legally drunk. And, at 10 days straight of 12 hours work-days, the error rates for non-routine tasks such as startup of a continuous unit can increase to 1 mistake in 5 to 10 steps (as opposed to the target of 1 mistake in 100 steps). Routine authorization for work hours in excess of those recommended may result in fatigued plant workers. Further, a practice of excluding training or meetings that occur outside of a plant worker's normal work schedule from work-hour limitations will also contribute to fatigue. In ***US NRC's 10 CFR 26 (2005)***[15], the guidance given for control of overtime hours is no more 72 hours of work per 6-day period, no more than 16 hours of work in one day within that period, and a minimum of 24 hours contiguous hours away from work within a 7-day period. The US DOT has even more stringent rules on limiting work hours and establishing required hours for recovery from fatigue.
- **Shift Scheduling** – Shift scheduling may also affect the likelihood that plant workers will show performance decrements due to fatigue. A change in the assigned shift or a rotating shift schedule will disrupt circadian rhythms and may increase the likelihood of errors.
- **Safety Culture** – The effectiveness of self-reporting and behavioral observation programs depends greatly upon the safety culture at a site.

G. WORK PROCESSES AND SUPERVISION

Supervision is the process by which work is directed and overseen by first-line management. Successful supervision requires a combination of leadership skills and technical competence. Supervision differs from peer checking or quality control because a supervisor has line management responsibility for the plant worker(s) as well as responsibility for the work activity.

Supervision is more than the moment-to-moment direction of a work activity. Successful supervision requires the assessment and shaping of plant worker attitudes and motivation, communication and implementation of management expectations for performing work, the assignment of the best-qualified workers to various tasks, as well as the technical competence to identify incorrect actions and stop improper activities before an error is committed. Effective supervision involves directing the work, overseeing how it is performed, and leadership.

Organizations are typically structured to have sufficient supervision of the job, but many times the delineation between the trainer and the supervisor roles is blurred. Supervision can and normally does play a key role in selecting of the right worker for the job, scheduling of workers to match the required tasks for the day/week, and generally overseeing the task execution to ensure policies and procedures are followed. Supervisors are not always trained on all of their key roles in support of control of human factors, such as detecting issues in plant workers related to fitness for duty or fatigue.

H. WORK ENVIRONMENT

The **Work Environment** refers to the physical conditions in which work is performed. Environmental conditions that can affect performance include excessive vibration and noise, temperature extremes, and insufficient lighting, as summarized below:

- There are two types of **vibration** that may cause errors. The first is whole-body vibration, in which vibration is transferred to the plant worker from standing or sitting on a vibrating surface. The second is object vibration, in which a stationary worker interacts with a vibrating object in some fashion. The effects of vibration depend upon its frequency and acceleration. Frequency is the number of oscillations (cycles) that occur in one second. Acceleration is the force, or intensity, of the vibration.
- **Noise** is unwanted sound. Noise can cause errors in several ways. It may disrupt communications, affect the ability to perform tasks and annoy plant workers. The effects of noise on communications are complex. Even relatively low levels of noise can mask speech, but only under some circumstances. For example, speakers naturally raise their voices when there is background noise and may be able to overcome some of its effects on communication. Being able to see the speaker's face or using standardized phrases also improves communication in a noisy environment. The type of background noise also affects communication. It is easier to communicate over noise that is steady and uniform than noise that includes sharp tonal peaks, such as background speech.
- **Heat exposure** is a common problem in many areas of a plant, such as the turbine building when the plant is operating. The extent to which plant workers will be affected by heat depends on many factors. These include their physical characteristics, such as age, weight, acclimation to heat, physical fitness and dehydration. Other factors that determine the effects of heat on performance include airflow, humidity, clothing, and level of physical activity.
- **Exposure to cold** affects the performance of manual tasks. Decreases in the ability to control hand movements begin at an air temperature of approximately 54° F. The fingers may become numb to pain at this temperature and touch sensitivity is reduced. Performance of gross manual tasks, such as those involving the arms and legs is also degraded at 54° F. The speed at which manual tasks can be performed is affected by the rate of cooling. Slow temperature drops have a greater negative impact on manual dexterity than rapid temperature decreases, during the initial exposure period.
- **Adequate lighting** is required for accurate performance of nearly every task in a unit operation.

The organization must have engineering controls to help control each factor, but sometimes there is no other choice but to rely upon administrative controls.

I. COMMUNICATION (most importantly, verbal communication)

Communication is the exchange of information while preparing for or performing work. Verbal communication occurs face-to-face, by telephone, sound-powered phones, or walkie-talkies, as well as over public address systems. Written communication occurs, for example, through policies, standards, work packages, training materials, and e-mail.

Communication involves two sets of behaviors: (1) creating and sending messages and (2) receiving and interpreting them. Communication always involves at least two individuals, the sender and the receiver, and occurs:

- Between individuals
- Within and among work groups
- In meetings
- In pre-job or pre-evolution briefings
- During shift turnover

Successful communication requires several steps. The sender first develops the intention to communicate **either** verbally or in writing. The sender then composes a message that presents the meaning as clearly as possible. The receiver must pay attention to the message and then interpret its meaning. If the communication is successful, the receiver interprets the message consistently with the sender's intended meaning. However, there are many potential errors in both sending and receiving the communication:

- Sending Errors
    - *Content wrong*
    - *Content inconsistent*
    - *Content inappropriate for the job*
    - *Content inappropriate for the receiver*
    - *Standard terminology not used*
    - *Familiar terminology not used*
    - *Message production inadequate or interfered with*
    - *Necessary information not sent*
    - *Wrong place or person*
    - *Wrong time*
    - *Sending verification failure*
- Receiving Errors
    - *Information not sought*
    - *Information not found*
    - *Information not used*
    - *Receiving verification failure*
    - *Message misunderstood.*

Communication errors may be reduced in the following ways:

- Structuring messages in a standard format, which alerts the receiver if important information has been skipped
- Providing written instructions or procedures for complicated tasks
- Establishing protocols to repeat key parts of verbal communications
- Preceding special or nonroutine activities with a pre-job briefing
- Using structured protocols, checklists and logs to supplement written instructions

One key opportunity for communication errors is when a new shift takes over responsibilities from the previous shift. An effective tool to reduce these errors is a properly designed shift handover process that includes appropriate checklists and logs of activities, and a structured and consistent method for verbally reviewing this information. Additionally, the logs and discussions need to include:

- Routine operations
- Transition operations (e.g., changes to produce a different product or switch to a different raw material source)
- Nonroutine operations (e.g., maintenance activities or temporary operating conditions)

6. Establishing a Safety-Principled Organizational Culture

To ensure the success of the management systems created to improve Human Factors for plant workers, the proper culture of the organization must be developed so that this can be done effectively. Merriam Webster defines culture as "the set of shared attitudes, values, goals, and practices that characterizes an institution or organization". An organization's culture ultimately reflects the collective values of individuals and groups and can be influenced by the personalities of charismatic leaders, by shifting priorities based on the perceived scarcity of resources or actions necessary for survival, or by a sustained history of successes, behaviors, and beliefs. In an especially sound culture, deeply held values are reflected in the group's actions.

Organizational Culture is a somewhat intangible attribute of a company that is evident by the words and actions of management and all employees – including the plant workers. It is further defined informally as "how we do things around here", "what we expect here", or "how we behave when no one is watching". Reducing human error by plant workers therefore requires that the proper organizational culture be built and maintained.

A Safety-Principled Organizational Culture is one in which the actions of the entire organization are driven by the genuine desire to complete every task and achieve every objective in a manner that protects employees, plant equipment, and the community in which the plant is located to the greatest extent possible. While profitability and production targets are also important goals for the organization, the entire organization understands that these are enabled by continuous safe operations, rather than despite them.

Establishing a strong Safety-Principled Organizational Culture is a long-term process that requires dedicated resources for as long as the organization exists. Additionally, it is the result of the quality and quantity of effort put into its implementation. A solid, sensible effort by management and technical staff to establish a culture of this nature will be noticed by the plant workers and they will want to help. This will have the effect of catalyzing the program, thereby helping to ensure its success. Alternatively, merely stating the objective of building this culture will have little effect on creating excellent organizational safety culture if the implementation and control of safety is weak.

Establishing a strong Safety-Principled Organizational Culture will minimize safety incidents, and this will save lives. Since essentially all accidents begin with a human error, the actions or inactions of the plant worker can be the limiting factor in determining overall safety performance.

The proper culture helps plant workers understand why strict adherence to procedures is the right thing to do. Further, it helps to enable thoughtful compliance and to alert management when procedures are wrong. In this way, a sound culture is essential to maximizing the results of ALL safety activities.

Establishing a strong Safety-Principled Organizational Culture can be achieved by successfully pursuing 5 Key Strategies – each of which are described below.

Strategy 1: Establish Safety as a Core Value

Attaining the goal of superior plant worker performance (i.e., low human error rates) begins by providing strong and consistent communication about the importance of safety **from the very top of the organization**. This is achieved by setting an appropriate tone that safety is equal to or greater than production and by not diluting the message with a lot of other high priority core values. The first step in setting the tone is by clearly stating the importance of safety in a high-level mission statement. Then, this mission statement must be communicated to the organization, along with the related goals and plans for achieving it. An important aspect that must be communicated is the key role of controlling plant worker human error by focusing on Human Factors. Additionally, it is critical that this effort not be reduced to a "program of the month" – it must be spoken about frequently and consistently to all levels of the organization.

Of course, words are hollow without consistent corresponding actions. Therefore, strong leadership must be provided by demonstrating the commitment to safety excellence through actions and decisions. Replace slogans and banners with high standards, shared values, and core beliefs which sustain the organization regardless of who is in charge. To help ensure these values propagate throughout the entire organization, managers must be educated in this culture, vision, and related standards, as well as the expectations for their roles and responsibilities in this critical effort.

A strong Safety-Principled Organizational Structure relies upon high standards of performance that are consistently enforced. This is achieved by first defining appropriate safety goals – namely, those that provide an indication of the organization's capability to perform their work safely. Too often, organizations tie the success of their safety programs on the lagging indicator of incident rates. While this is of course the ultimate goal of the organization, this isn't necessarily a reliable indicator of safety for the organization. Consider how many ways a plant worker can be hurt completing a single task, then multiply that by the number of tasks the individual completes per day, and then by the number of plant workers within the organization. Clearly there are thousands of opportunities for injury each day. Fortunately, however, for nearly all organizations the number of incidents that occur is very small compared to the number of opportunities for an incident. Therefore, unless incident rates are abnormally high, a change in an organization's injury rate is typically due to factors outside of their control. It is much more valuable, then, to establish goals centered around the organization's work in improving their safety-related management systems.

Another aspect of establishing safety as a core value is to establish responsibilities and accountabilities for process safety roles. Every individual within the organization should be held accountable for steadfastly adhering to the newly established expectations. While it may be tempting to overlook seemingly minor deviations from these expectations, this can lead to normalizing deviations which can potentially lead to serious incidents due to a relaxed response to these warning signs of potential hazards. Therefore, the leaders of the organizations must hold direct reports accountable for demonstrating behavior consistent with this value and set expectations for each level of the organization to hold their direct reports to this same behavior.

Finally, it must be recognized that improving the organizational culture involve costs and benefits that are both long-term in nature and sometimes difficult to measure. By establishing appropriate goals as noted above that focus on management system health, however, the organization has the means to continuously improve its performance.

Completing all of the above will help ensure that the core value of safety is consistently embraced throughout the entire organization, including the plant workers.

Strategy 2: Provide the Necessary Resources to Ensure Success

Providing resources to transform the organizational culture is the most tangible way to demonstrate to the plant workers that safety is indeed an organization's core value. In this context, resources constitute monetary commitments, as well as those of personnel.

The place to begin is to fund and allocate personnel to enable good safety performance by creating and effectively implementing the Human Factor management systems described above. Done properly, these will close any gaps in the control of Human Factors as well as those evident in existing systems that don't meet a critical need such as performing a PHA of procedures to find scenarios unique to non-continuous modes of operation as described by Bridges and Marshall[26]. Adequate personnel must also be allocated for completing safety-related tasks to support these management systems (inspections, maintenance, etc.), for maintaining these management systems as changes or newly identified requirements occur, and for training and educating personnel.

It is also important to designate a high-ranking leader for safety who can help ensure that major decisions are based upon a proper consideration of safety. But it is also important to provide the positional support necessary to promote strong safety performance throughout the organization. This is achieved by identifying resources on paper and filling with effective and competent people. For example, if a trainer is shown on an organization chart, it should be ensured that this is their full-time job and that they aren't responsible for other unrelated tasks. This demonstrates that the organization places a high value on training and development of people and groups. This can be further enhanced by ensuring that training materials are maintained current, and by maximizing the use of instructor-led training, rather relying than too much on the use of Computer-Based Training.

Furthermore, a properly designed and implemented training program will allow the organization to demonstrate that it places a high value on safety knowledge and expertise. This can be enhanced by identifying and providing the proper mix and amount of expertise for its most important tasks (for example, by pairing an individual with a high level of expertise with one who has the potential, but not yet the experience, to develop this same level of expertise). To properly ensure optimum prevention of human error, the organization should focus on building expertise in Human Factors and seek out expertise for critical decisions.

Finally, since excessive overtime impacts employees' abilities to perform their jobs safely and increases safety risks, the organization must devote sufficient manpower resources to properly control overtime.

In summary, the overarching approach here is to ensure that a structured, strategic approach is developed to properly allocate funds and personnel to create, implement and maintain the management systems needed for safety enhancement opportunities.

Strategy 3: Establish a Blame-Free and Learning Environment

It is critically important that the organization commit to a blame-free culture for all plant worker human errors, including for near misses reported and investigation findings. At an intellectual level, everyone understands that no one is perfect and that zero human errors are not possible. Plant workers are no exception. A key driver for the creation of a blame-free environment then is driven by how the organization responds to the inevitable human errors that plant workers will make.

The response to plant worker errors can be thought of as a continuum that spans two extremes; on the left side the response is one of blaming or fault-finding, while on the right side the response is one of trying to understand or learn why the error was made. Quite often people will automatically default to the "blame" side of the continuum for plant worker human errors, saying something to the effect that if only the plant worker had handled the situation differently the error would not have occurred. While this is true at face value, it is not a helpful response. After the fact, it is easy to see what should have been done differently because the outcome is now known. But when the plant worker was taking actions, this outcome was not obvious to him or her. Significantly, people do not come to work planning to make errors; instead, they come to work to complete their tasks in the most efficient manner possible. It is only after an incident occurs that the benefit of alternate actions is clear. Thus, the more productive response is to try to understand the context of the plant worker's action – what was it that made this particular action appear to be correct? It is only by taking this approach that the lessons from the incident can be fully understood.

Each person in the organization needs to understand where on the human error response continuum they reside and make every effort to move as far to the right side as possible. As shown in Figure 5, there are three aspects to an individual's response to a plant worker human error: the **internal response** (what we say to ourselves when we are made aware of a human error), the **external response** (what

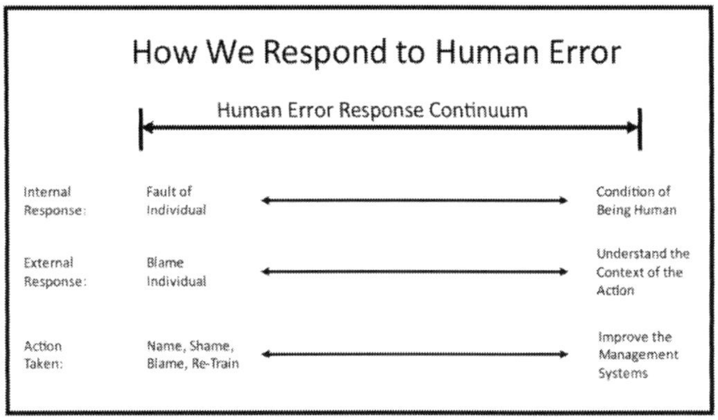

Figure 5: How We Respond to Human Error

we say out loud when we are made aware of a human error), and the **action taken** (what we do when we are made aware of a human error).

The internal response continuum ranges from thinking that the error was the fault of the plant worker, to recognizing that the error was a condition of that plant worker being human. The best approach for an individual to move to the more productive right side of this spectrum is to consider experiences in their own life in which they have made an error, which in retrospect seems to have been easily avoided. With this perspective, it is easier to understand that the plant worker who made the error in question was doing the best job possible but did not fully understand a specific hazard associated with it.

The external response continuum ranges from blaming the plant worker, to understanding the context of the action (i.e., why it made sense for the plant worker to take the action he or she took). Only by understanding the context of the action is it possible to develop measures so that another plant worker faced with the same choice does not repeat the error. As an example, a major chemical company experienced an acid spill from a tank truck one very cold evening. The truck driver was directed by the chemical company's guard to the wrong operating unit and began unloading the acid into that unit's tank. The operator from the unit soon emerged from the local control room and informed the driver that no acid was needed there. After a few phone calls, it was learned that the delivery was for a different unit which was located about ½ mile away. Accordingly, the driver disconnected his truck but failed to properly tighten the unloading hatch on top of the tank truck (most likely due to the cold ambient temperature). Therefore, as the truck was driven to the other unit, acid was spilled along the entire route. The chemical company's response to this incident was to inform the delivery company that that driver was no longer allowed on company property – in essence, blaming the driver for the spill. But this decision left many questions unanswered. Why was the driver directed to the wrong unit by the company's guard? Why was the driver allowed to unload directly into a company tank without checking in to the local control room (which is a requirement even for company employees visiting from a different plant area)? Only by answering these questions – understanding the context of the action – could assurance be provided that future spills of this nature would not occur.

The action taken response continuum ranges from a combined action of naming the plant worker responsible, shaming the person for making the error, blaming them for the consequences, and then re-training them, to making improvements to the underlying management systems. Using the acid spill example, the driver most certainly knows how to correctly complete all of the tasks needed to safely unload product; therefore, training would provide no real benefit other than to make him feel worse than he already did. However, by improving the management systems (strengthening the process for directing drivers to the proper locations, enforcing the rule to check in to the local control room before unloading) a better assurance that the human error could be prevented from recurring is possible.

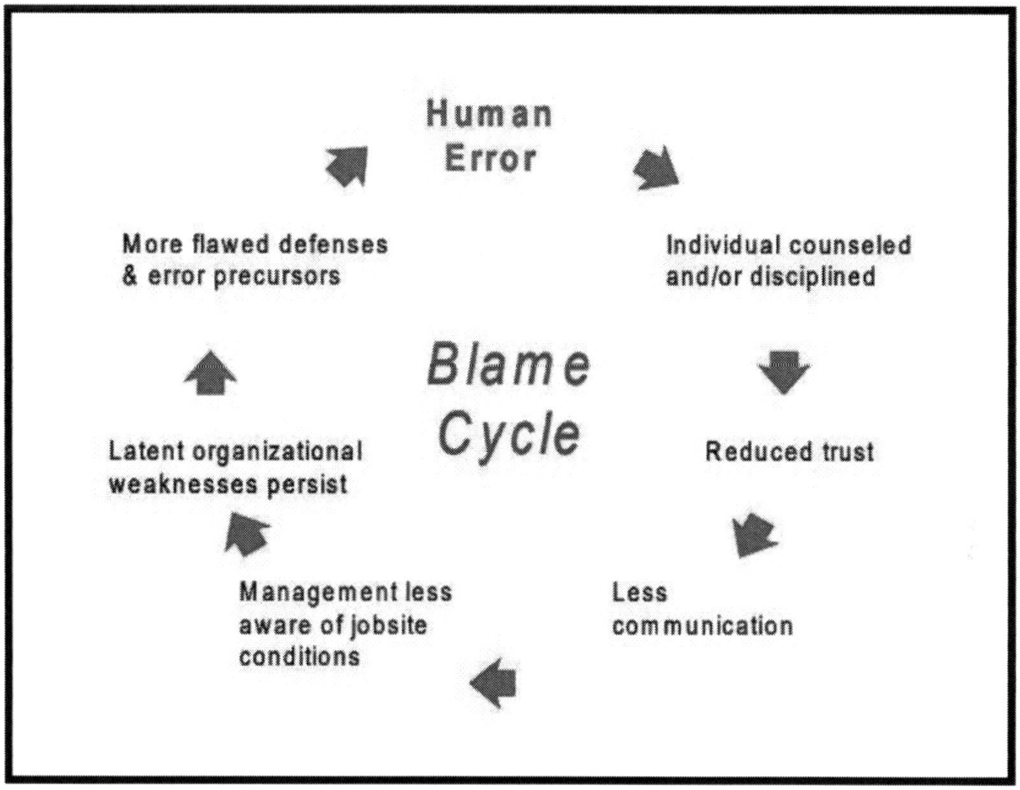

Figure 6: The Blame Cycle

The above discussion highlights the fact that residing on the "blame" end of the human error response continuum prevents an organization from truly learning about the root causes of incidents, thereby potentially leading to a recurrence of the error. However, blaming plant workers pose further risks to the organization. This is illustrated by the "Blame Cycle" that was developed by the US Department of Energy[27]. This cycle establishes a phenomenon that when a plant worker is blamed for a human error, this results in this individual ultimately not reporting future errors, which the organization is therefore unaware of and is unable to address. This then leads to future human errors due to these latent weaknesses, which are unreported, and the cycle repeats itself potentially leading to very serious consequences.

With this perspective then, a blame-free and learning environment can be achieved by ensuring that the first questions following an incident are what and why, not who. Those are followed by asking how we can improve the procedures, task design, practices, equipment, and other management systems to prevent this from happening again. With this environment, all incidents (especially near misses) are reported and investigated without blame, which results in root causes (management system failures) being identified, and corrective measures being taken to implement sustainable and permanent solutions. Trust and collaboration are high because the culture is blame-free when it comes to mistakes and people ask for help when they need it and give help without being asked.

Some key approaches to establishing a questioning/learning environment include properly performing risk reviews, conducting thorough incident investigations, performing table-top drills, and sharing and seeking lessons inside of and outside of facility. Significantly, plant workers should be highly involved in each of these activities. It is also critical to maintain a high awareness of hazards and consequences and remain vigilant for indications of system weaknesses. This can be achieved by ensuring that all plant workers are educated on the hazards of the operation, that they understand how safety hinges on control on Human Factors, that they all are educated on the consequences of deviation from safe practices, and that they share lessons learned.

A final aspect is to provide timely response to, and communication of, safety issues and concerns and lessons learned from incident investigations, audits, risk assessments, employee concerns and discrepancies between practices and procedures.

In summary, an organization must ensure that a blame-free and learning environment is consistently embraced throughout the entire organization. This can be achieved by reinforcing this at all opportunities through words and actions by personally seeking solutions instead of blame. Furthermore, top management must hold direct reports accountable for demonstrating behavior consistent with this value and set expectations for each level of the organization to hold their direct reports to this same behavior.

Strategy 4: Enhance Employee Empowerment and Involvement in Safety

Plant workers can best contribute to the Safety-Principled Organizational Culture and thus successfully fulfill their safety responsibilities by having the necessary authority and accountability for safety responsibilities. Leadership can enable this by reinforcing that all employees have responsibilities to themselves, fellow employees, company, and community and by involving the plant workers in safety management systems to ensure that they are widely understood and supported. This approach has the added benefit of recruiting additional resources to support the many tasks required.

A prerequisite to this is for the organization to ensure open and effective communication – both vertically and horizontally. Strategically, this can be achieved by emphasizing the need to promptly observe and report non-standard conditions. A key management system to enable this is a properly deployed near miss reporting program. Bridges[28] outlines an approach for creating a program of this nature, as well as specific, common hurdles that must be overcome to ensure its success. Key items in a properly designed program include first providing means for all plant workers to communicate hazards to management (in a blame-free environment) and for management to consistently act on these stated concerns.

Related to this, plant workers should be involved in identifying hazards and deciding how they should be addressed. In certain situations, plant workers should be authorized to take action to address hazards without management involvement. A very important example of this is that plant workers should have the authority to question decisions which are being made about process safety, including a "Stop Work Authority".

Ensuring optimum plant worker involvement in the safety management systems must be intentional. Therefore, the organization should develop an Employee Participation Plan, which lists key components of the management system that plant workers can best contribute to. A few examples to include identifying and controlling deficiencies in Human Factors, writing and reviewing operating and maintenance procedures, and participating in Process Hazard Analyses, Root Cause Analyses, and audits. Other opportunities can be developed by critically reviewing components of all safety management systems (including those developed for Human Factors) that will benefit by having the specialized knowledge of plant workers.

Strategy 5: Drive Continuous Improvement in the Safety Management Systems

Even the best designed safety management systems are not perfect (because they are developed and managed by humans). Therefore, the best Safety-Principled Organizational Cultures undertake specific activities to understand gaps in the systems and then take actions to close these gaps. This is achieved by establishing relevant metrics on the health of the management systems, which as noted previously, should predominately be leading metrics. Typical examples include the timeliness of correcting deficiencies, adherence to ITPM schedules/procedures, and the expansion of problems detected in one area to investigate similar applications in other areas.

Over time, trends should be developed of these metrics, presented to management who should then direct actions to improve these Key Performance Indicators. Further, these metrics should be shared within the organization so that all employees – including plant workers – have a clear sense of how well their efforts to contribute to the success of these programs are working. It is important to remember that many of these safety-related metrics have long-term results, rather than an immediate impact. Therefore, continued diligence in these continuous improvement efforts are needed to ensure a sustained Safety-Principled Organization Culture, and ultimately outstanding safety performance for the organization.

7. Closing

Human error is inevitable. And the human errors that are typically thought of as the ones that lead to serious accidents are often attributable to those made by plant workers. Although this assessment is unfair (as other members of the entire organization contribute in meaningful ways to these incidents), it is nonetheless very important for organizations to undertake measures to reduce plant worker human errors.

The means to do this are outlined in this paper. It relies first on the organization having the proper culture – one that doesn't react to plant worker errors by blaming them; rather, it seeks to understand why the error was made and taking actions to fix the issue instead of the plant worker. It recognizes that to do this requires that the error must be traced to the underlying Human Factors that increased the likelihood of this error's occurrence, and that the best way to improve these Human Factors (and thus reduce plant worker human errors) is to develop and implement robust management systems for them.

The information provided in this paper to strengthen the plant worker Human Factor management systems is the most important method to reduce plant worker human errors. However, even with perfectly implemented management systems human errors will still occur – albeit at a lower frequency and likely with a smaller adverse impact than before. Therefore, in cases in which the consequence of a human error is unacceptably high, the organization should implement solutions that will lower the residual risk further. These solutions can be designed to either help prevent the error or compensate for the error, but **they must be independent of the human error initiating event**. Bridges and Bridges [25] have compiled a list of the best remedies for human error for these situations. In a similar vein, plant worker human error can be further reduced by performing the previously mentioned PHA of procedures to find scenarios unique to non-continuous modes of operation as described by Bridges and Marshall[26].

Finally, a small fraction (about 15%) of plant worker human error is attributable to acquired habits, or behaviors; the remedy for which is a properly designed and implemented plant worker observation program. The reader is encouraged to review these other important documents and methodologies to obtain a fully comprehensive understanding of the best ways to reduce plant worker human errors.

8. References:

1. Swain, A. D., Guttmann, H. E., *The Human Reliability Handbook - with Emphasis on Nuclear Power Plant Applications*; NUREG/CR-1278, U.S. Nuclear Regulatory Commission, Division of Facility Operations, 1983.

2. Gertman, D.; Blackman, H.; Marble, J.; Byers, J. and Smith, C.; *The SPAR-H Human Reliability Analysis Method,* NUREG/CR-6883, U.S. Nuclear Regulatory Commission, Office of Nuclear Regulatory Research, Washington, DC, August 2005.

3. Bridges, W., and Tew, R., *Human Factors Elements Missing from Process Safety Management (PSM),* 6th GCPS, AIChE, March 2010.

4. *Risk-Based Process Safety*, CCPS/AIChE, John Wiley & Sons, Inc.; Hoboken, NJ, 2007.

5. *Conduct of Operations and Operational Discipline*, CCPS/AIChE, John Wiley & Sons, Inc.; Hoboken, NJ, 2011.

6. *US FAA Advisory Circular* 120-42A (as part of ETOPS approval)

7. Bridges, W. G., Williams, T. R., "Effective Procedure Writing – A Two-Part Series (Part 1: Reducing Risk of Worker Errors; Part 2: Addressing Process Safety Management Requirements for Operating Procedures)," *Chemical Engineering Progress*, 1997.

8. Bridges, W., and Collazo-Ramos, G., *Human Factors and their Optimization,* 8th GCPS, AIChE, April 2012.

9. *Guidelines for Hazard Evaluation Procedures*, 3rd Edition, CCPS/AIChE, John Wiley & Sons, Inc.; Hoboken, NJ, 2008.

10. Bridges, W., *Optimizing Hazard Evaluations,* LPS/AIChE, April 2009.

11. Bridges, W., Lorenzo, D., and Kirkman, J., *Addressing Human Errors During Process Hazard Analyses,* CCPS Conference, AIChE, 1991 (also published in Chemical Engineering Progress May 1994).

12. *Guidelines for Investigating Chemical Process Incidents, 2nd Edition,* CCPS/AICHE, John Wiley & Sons, Inc.; Hoboken, NJ, 2003.

13. Bridges, W., *Gains in Getting Near Misses Reported,* 8th Conference, ASSE-MEC, Bahrain, 2008.

14. Della Rocco, P.S., et al., *The Effects of Napping on Night Shift Performance,* Federal Aviation Administration, U.S. Department of Transportation, 2000.

15. *Fitness for Duty Programs, Preamble, 10 CFR 26, U.S. Nuclear Regulatory Commission, Revised 2005.*

16. US Chemical Safety and Hazard Investigation Board (CSB), *Anatomy of a Disaster: Explosion at BP Texas City Refinery,* published 2006.

17. University of South Australia (UniSA), Adelaide Centre for Sleep Research, *SDA Fact Sheet - AT03*, 2006.

18. Hallbert, B.; Whaley, A.; Boring, R.; MCabe, P.; and Chang, Y.; *Human Event Repository and Analysis (HERA)*, NUREG/CR-6903, U.S. Nuclear Regulatory Commission, Office of Nuclear Regulatory Research, Washington, DC, 2007.

19. Bridges, W., and Clark, T., *LOPA and Human Reliability - Human Errors and Human IPLs,* 7th GCPS, AIChE, March 2011.

20. Bridges, W., and Thomas, H., *Accounting for Human Error Probability in SIL Verification Calculations,* 8th GCPS, AIChE, April 2012.

21. Bridges, W., *The Impact of Human Factors on LOPA,* 9th GCPS, AIChE, April 2013.

22. Bridges, W., and Tew, R., *Best Practices for Writing Operating Procedures and Trouble-shooting Guides,* 13th GCPS, AIChE, March 2017.

23. Bridges, W., *Proven Approaches to Ensuring Operators Can Respond to Critical Process Deviations in Time (Human Response IPL),* 13th GCPS, AIChE, March 2017.

24. Bridges, W. and Rhodes, W., *Everything you need to Know about Human Reliability and Process Safety,* 17th GCPS, AIChE, March 2019.

25. Bridges, S. and Bridges, W., *Abnormal Modes of Operation – A risk-based focus,* 16th GCPS, AIChE, March 2020

26. Bridges, W. and Marshall, M., *Necessity of Performing Hazard Evaluations (PHAs) of Non-normal Modes of Operation (Startup, Shutdown, & Online Maintenance)*, 12th GCPS, AIChE, April 2016

27. DOE Standard – *Human Performance Improvement Handbook, Volume 1: Concepts and Principles*

28. Bridges, W., *Gains from Getting Near Misses Reported*, 8th GCPS, AIChE, April 2012

619491 Lessons-Learned from Remote HAZOP/LOPA Applications in 2020

Abstract:

Going into Year 2020, basic brainstorming team session approaches for the performance of a Hazard and Operability (HAZOP) Study or a Layer of Protection Analysis (LOPA) had not changed significantly over the years; however, the infection prevention measures necessary to address COVID-19 resulted in a rapid transition to widespread application of telecollaboration tools that previously had only been applied sporadically. Prior to 2020, telecollaboration tools for Process Hazard Analysis (PHA) had been used to short-circuit the need for pulling together international teams, for very short PHAs or MOCs, and other select applications. As 2020 drew to a close, numerous applications of HAZOP/LOPA Team Exercises using telecollaboration tools resulted in valuable insights into tools and techniques that were productive for different PHA applications. Properly applied, applications ranging from shorter Process Hazard Analyses (PHAs) and Management-of-Change (MOC) evaluations to longer studies addressing Five-Year PHA Updates and Capital Projects can be implemented productively using telecollaboration approaches.

Post COVID-19, many PHA applications will still be best addressed with the HAZOP/LOPA Team being co-located, but the savings in travel costs, the logistics associated with personnel travel, and the ability to quickly convene and implement a HAZOP/LOPA review can allow remote HAZOP/LOPA, through the use of contemporary technologies, to be an attractive option that can increase overall efficiency, without compromising quality. Experiences in the successful application telecollaboration tools in 2020 mean that Remote PHAs are here to stay.

A core objective of this paper is to provide insights into good applications and tips that can directly be used by the HAZOP/LOPA Facilitator to optimize the logistics to get the most out of a HAZOP/LOPA, where some or all team members are geographically distant.

SPRING21
+17TH GCPS
A Joint AIChE and CCPS Meeting

Research on Human Resource Development Exercises for Resilience to Cyber Incidents in Critical Infrastructure

Yuitaka Ota
Nagoya Institute of Technology,
System Management and Engineering
Showaku, Gokiso, 4668555 Aichi, Japan
cjr17009@nitech.jp

Erika Mizuno, Asai Haruna, Tomomi Aoyama, Yoshihiro Hashimoto, Ichiro Koshijima
Nagoya Institute of Technology,
System Management and Engineering
Showaku, Gokiso, 4668555 Aichi, Japan
ckv17032@nitech.jp, h.asai.852@nitech.jp, aoyama.tomomi@nitech.ac.jp,
hashimoto@nitech.ac.jp, koshijima.ichiro@nitech.ac.jp

Prepared for Presentation at
American Institute of Chemical Engineers
2020 Spring Meeting and 17th Global Congress on Process Safety
Virtual
April 18 - 22, 2021

Research on Human Resource Development Exercises for Resilience to Cyber Incidents in Critical Infrastructure

Yuitaka Ota
Nagoya Institute of Technology,
System Management and Engineering
Showaku, Gokiso, 4668555 Aichi, Japan
cjr17009@nitech.jp

Erika Mizuno, Haruna Asai, Tomomi Aoyama, Yoshihiro Hashimoto, Ichiro Koshijima
Nagoya Institute of Technology,

Keywords: OT-Security, cyber incident exercise, business continuity management

Abstract

Current cyber-attacks target IT systems and OT (Operational Technology) systems that are indispensable for the automatic operation of critical infrastructure industries. There have been various reports of cyber incidents targeting the OT system. In 2019, Norsk Hydor was infected with particular ransomware worldwide through its corporate network, forcing the manufacturing site to operate manually. In May 2020, Taiwan's CPU, responsible for Taiwan's oil supply, was targeted by a cyber-attack.

COTS (Commercial Off The Shelf, such as Windows OS, Intel PC, and open source applications) devices have been proposed to reduce the deploying OT system budget. Also, DX efforts are being made to shift from physical operation to virtual operation by using virtualization with IoT, AI, and Cloud. Due to this technological trend, the expanded attack surface is forced to manage the OT system as a "Zero Trust System" that assumes cyber-attacks. In IT systems, countermeasures such as whitelisting and firewalls prevent attackers from outside the system. For OT system, Network Security Zoning described in IEC62443 is recommended to mitigate the effects of cyber-attacks.

However, attackers exist all over the world, and they are continually developing new attack methods. For this reason, the authors believe that it is impossible to prevent all attacks, no matter what measures the defenders take. Therefore, when responding to cyber-attacks against the OT system, it is essential to consider countermeasures by focusing on plant accidents caused by control system failures, most of which are assumed as safety measures.

To ensure the plant's safety and continue the business operation even if it has a cyber-attack on the OT system, it is indispensable to develop human resources to respond to the incident flexibly. For this reason, we had been developing cybersecurity exercises to acquire the meta-knowledge necessary to respond to cyber-attacks targeting the OT system.

In this paper, the authors present the design framework, the implementation method, and OT expert's evaluation results of a proposed exercise.

1 Introduction

1.1 Background

An industrial control system (ICS) is a device or group of devices for managing and controlling other devices and systems. ICS is widely used in the Critical Infrastructure (such as electric power, city gas, and transportation) and the manufacturing industry that supports our daily lives, from those that are familiar to us. The conventional ICS has an isolated network as an independent system built with its hardware, software, communication protocol and has been considered resilient to cyber-attacks. However, in recent ICS, the adoption of COTS (Commercial Off The Shelf, such as Windows OS, Intel PC, and open source applications) devices is progressing due to versatility and budget. Also, DX (Digital Transformation) is being promoted for the purpose of plant operation support and operation optimization using IoT (Internet of Things) and AI (Artificial Intelligence) through open networks with the outside. As a result, ICS is exposed to the threat of cyber-attacks, and companies using ICS need to recognize not only the risk that has been assumed as the risk of plant shutdown but also cyber-attacks as one of the risks.

Cyber-attacks targeting ICS have been increasing in recent years. In 2010, malware called Stuxnet attacked Iran's uranium enrichment centrifuge [1]. Hackers caused power outages in Ukraine in 2015[2] and 2016[3]. In 2019, Norsk Hydros in Norway, the world's leading manufacturer of rolled aluminum products, was affected by cyber-attacks in plant operations [4]. In addition, the Information Security Agency (BSI) of the German Federal Government has published "Industrial Control System Security: Top 10 Threats and Countermeasures" [5].

In this way, cyber-attacks are a risk that threatens the safety of the plant, and companies must respond as a risk that threatens their business.

1.2 Response to a cyber incident

1.2.1 Cyber-attacks targeting Industrial Control System

Fig. 1 shows a cybersecurity testbed and a virtual ICS network. The testbed (Fig. 1) heats water with a heater in the lower tank and circulates the heated water through a simple pipeline to the upper tank (using a pump). The testbed is also equipped with actual industrial control devices and is controlled automatically[6]. In this ICS network, OLE for Process Control (OPC) Servers collect and exchange process data and monitor them by using Supervisory Control And Data Acquisition (SCADA) function included in the OPC Servers. OPC Servers receive various parameters from the testbed's sensor, and SCADA gives instructions to the control equipment using the parameters' data.

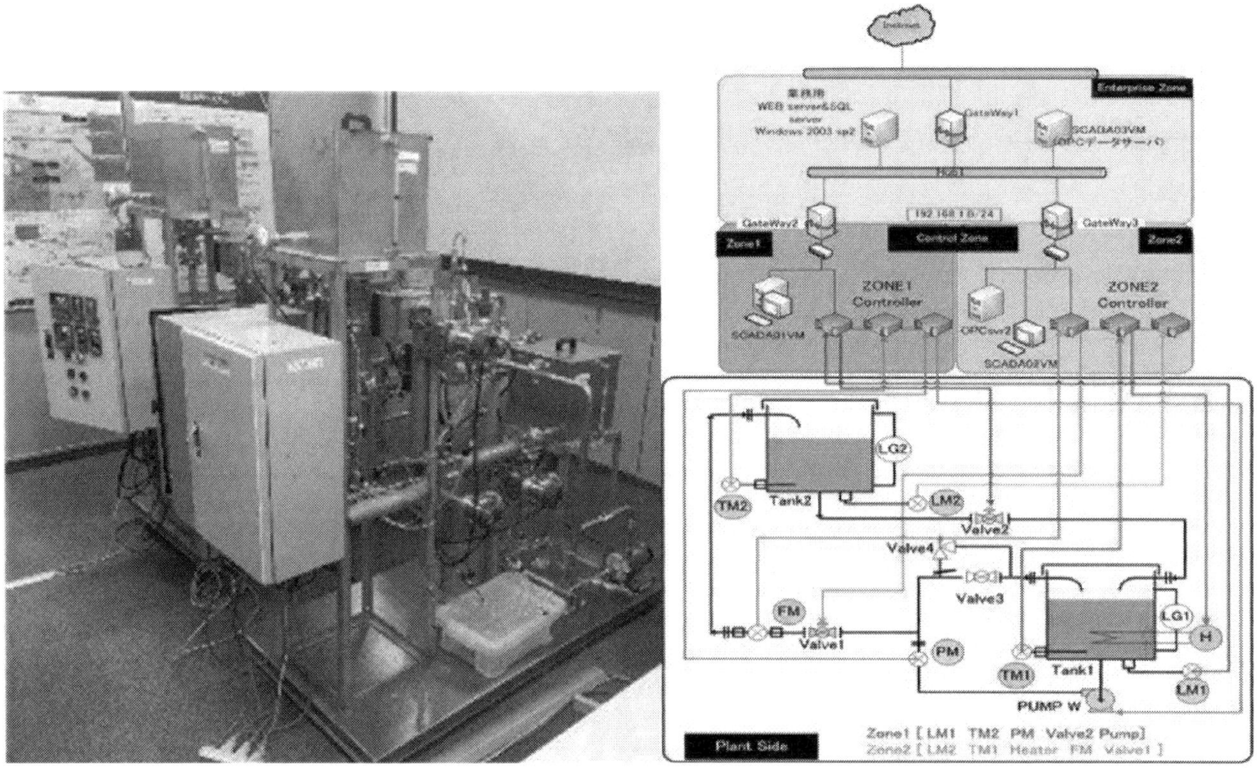

Figure 1. Cyber Security Testbed and ICS network

The authors have demonstrated cyber-attacks against the testbed, according to the steps of the anatomy of a cyber-attack. The attacking procedure is determined as follows.

1) Attack Step 1: Reconnoitering and enumeration

The attacker executes the Armitage to launch the Metasploit framework. Then, the attacker executes the Nmap, which scans a target network (ICS-2 Network 192.168.12.0/24). After completing the scan, the attacker gets target information.

2) Attack Step 2: Intrusion and advanced attacks

According to the functions of the Armitage, the attacker ascertains an attack method and vulnerability in the Metasploit framework. The attacker exploits payloads, which execute inappropriate code in the use of the vulnerability. After sending the payloads, the attacker obtains the administrator's permission for the OPC2 illegally.

3) Attack Step 3: Malware insertion

The attacker uploads a malware execution file to an OPC2 background process to execute a malware file. The configuration file of OPC2 was rewritten, and then the OPC2 was restarted. They were resulting from the attack, the valve controlled by the OPC2 shutdown.

4) Attack Step 4: Clean-up

The attacker deletes the malware file and the configuration file from OPC2 storage to not leave the cyber-attacks evidence and finish the cyber-attack.

As the valve is opened, the stage of the testbed changes. Usually, the testbed's state transition appears on the monitor of the operator (Fig. 2). The changing graph shows the liquid level of the upper tank. The liquid level is changing due to the malfunction of the valve. The operator can notice the testbed abnormality due to the unusual process deviations from the steady-state condition through the monitor. However, the attacker conceals the information about cyber-attack in order not to let anyone notice the cyber-attack. The testbed monitoring screen is concealed, as shown in Fig. 3, while the cyber-attack causes severe accidents at the testbed.

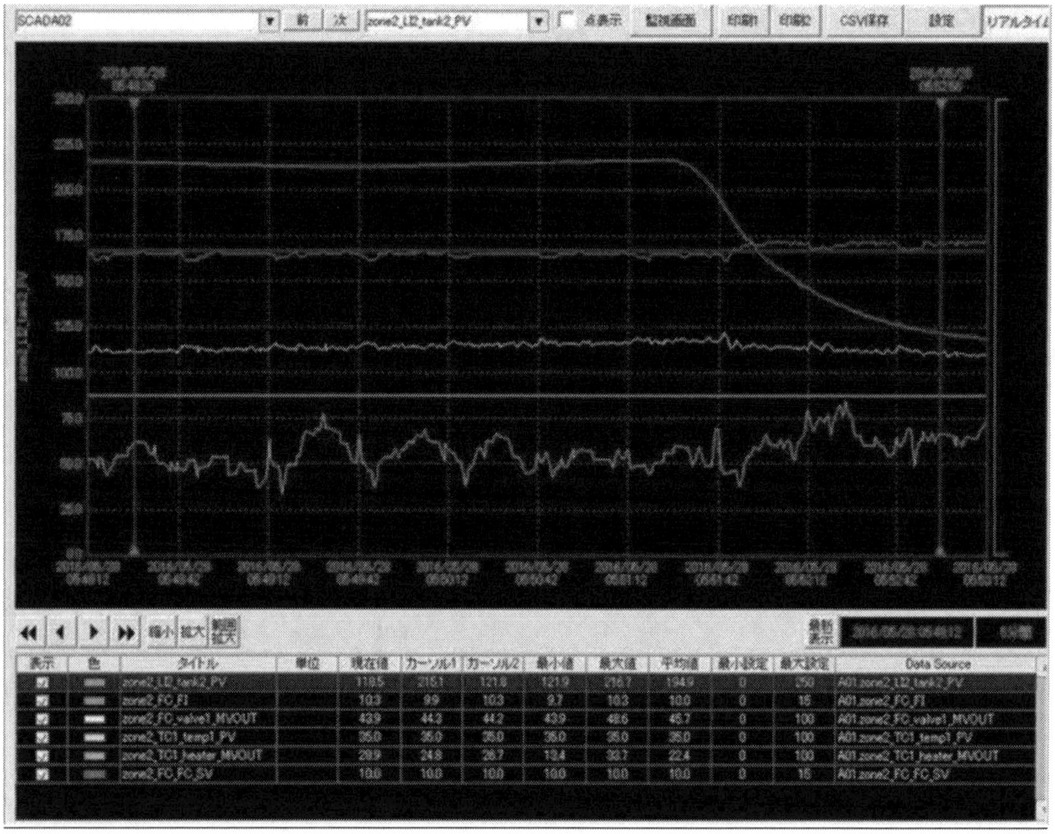

Figure 2. HMI Screen Shot in Operation (Safety Incident)

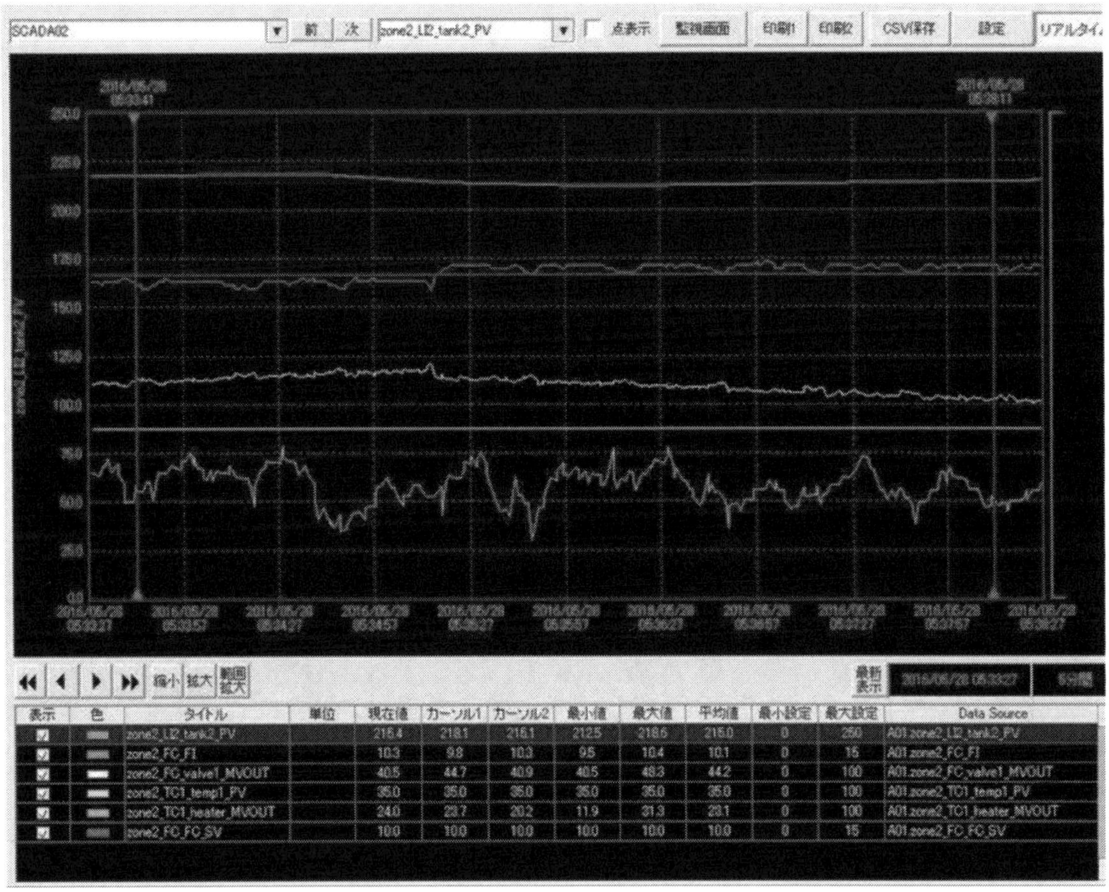

Figure 3. HMI Screen Shot in Operation (concealed by the attacker)

This demonstration shows a possibility of safety incidents caused by cyber-attacks to ICS.

1.2.2 Response to Cyber Incidents

The consequences of a cyber-attack depend on the characteristics of the plant. For example, an attacker hijacks a data server (DCS, OLE for Process Control server, etc.) that gives parameters and commands to control devices such as a PLC (Programmable Logic Controller) and gives the control devices malicious instructions. Such an attack can lead to abnormal operation and an incident, similar to a data server or control equipment failure or operator error with traditional equipment. That is, incidents caused by cyber-attacks in ICS systems are incidents that have already been considered in HAZOP scenarios with different initiating events. Operating companies have taken countermeasures to ensure the safety of the plant for many years. Companies are preparing countermeasures to ensure safety (Safety Response) under an incident at their plant. Human resource development is also carried out based on the countermeasures to respond when an incident occurs. In other words, even if an abnormality occurs due to a cyber-attack, it is possible to ensure the safety of the plant if the operation staff takes a proper Safety Response. However, problems peculiar to cyber-attacks, such as rewriting server configuration files and rewriting control device programs, may hinder the implemented safety-based Safety Response. Figure 4 shows the incident response structure that takes into account the cyber-attacks proposed here. If the plant is operating abnormally caused by the cause considered so far, it is possible to restore plant safety by performing a safety response, as shown in the left figure. However, there is a possibility that cyber-attacks will interfere with safety response (e.g., concealment of a plant's information on the monitor). Companies must consider where cyber-attacks interfere with safety response. Moreover, it is necessary to take the responsibility to eliminate the effect of cyber-attacks (security response) to take safety response as shown in the right figure. Therefore, if a cyber-attack causes a plant's abnormal operation, a response to rectify the situation is required in combination with a structure to add a security response to the safety response.

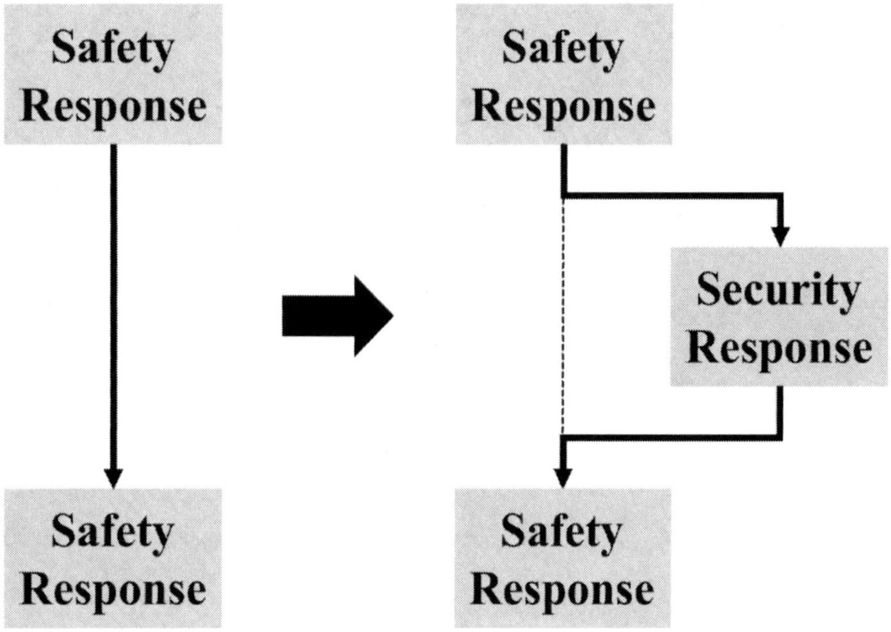

Figure 4. Incident Response Structure[7]

Responding to cyber-attacks targeting ICS systems, it is necessary to integrate a Safety Response to ensure the plant's safety and a Security Response to eliminate the effects of cyber-attacks. It is essential to create a plan that considers security incidents in addition to the existing safety response plan, and make it known to each operations staff.

1.3 Required organizational cooperation

The right staff needs to take action at the right time. To respond to incidents competently, it is necessary to properly share the current situation and communicate it among the right staff.

When an incident occurs due to cyber-attacks at a plant, two responses, a Safety Response to ensure the plant's safety and a Security Response to remove the cyber-attack effects, proceed simultaneously. Since the Safety Response responds to ensure the plant's safety, the response's responsibility is on the OT side (operators working at the plant). Since the Security Response is a response to remove the effects of cyber-attacks, the response's responsibility is on the IT side (IT engineer). Also, considering a cyber-attack on ICS, the result of the cyber-attack appears in the plant, and the cause is the IT network. In other words, it is a structure in which the cause-effects occur in different places. Also, in the incidents that we have experienced so far, there was no problem in dealing with the cause, but in cyber incidents, the same incident may occur in all parts connected to the Internet. Therefore, information sharing is vital for appropriately responding to the above-mentioned new type of incident, "Simultaneous Multiple Accidents."

2 How to develop Security Personnel

2.1 Preparation required as a company

・Formulation of ICS-BCP

Many companies have created a business continuity plan (BCP) in advance so that their business can continue even during natural disasters such as earthquakes and typhoons. It is important to develop a similar program for cyber incidents. It is necessary to analyze the risks in the event of a cyber incident and formulate an ICS-BCP that includes countermeasures for reducing, holding, avoiding, and mitigating the effects of the risks.

・Human Resource development

Even in unprecedented situations such as cyber incidents, companies must respond to continue their business. And it is the employees who respond to the incident. If the personnel's ability to respond to the incident is not sufficient, the company's damage will increase.
In other words, as a company, it is required to develop human resources who will respond.

2.2 Importance of virtual experience

As mentioned in 2.1, companies must prepare a BCP in advance for cyber incidents and develop human resources to carry it out. However, many companies have no cyber-attacks experience with ICS, and it is not easy to imagine how cyber-attacks may affect companies.

A mechanism is needed to experience cyber incidents and discuss necessary countermeasures.

As an approach to that mechanism, the authors chose an exercise. Figure 5 shows the flow for developing a training exercise. By analyzing incidents in the real world, we can extract the causes of incidents and the commonality of cyber-attacks targeting ICS. By giving the extracted information rules and game characteristics, the exercise participants will learn systematically. Thus, participants experience cyber-attacks on ICS through exercises.

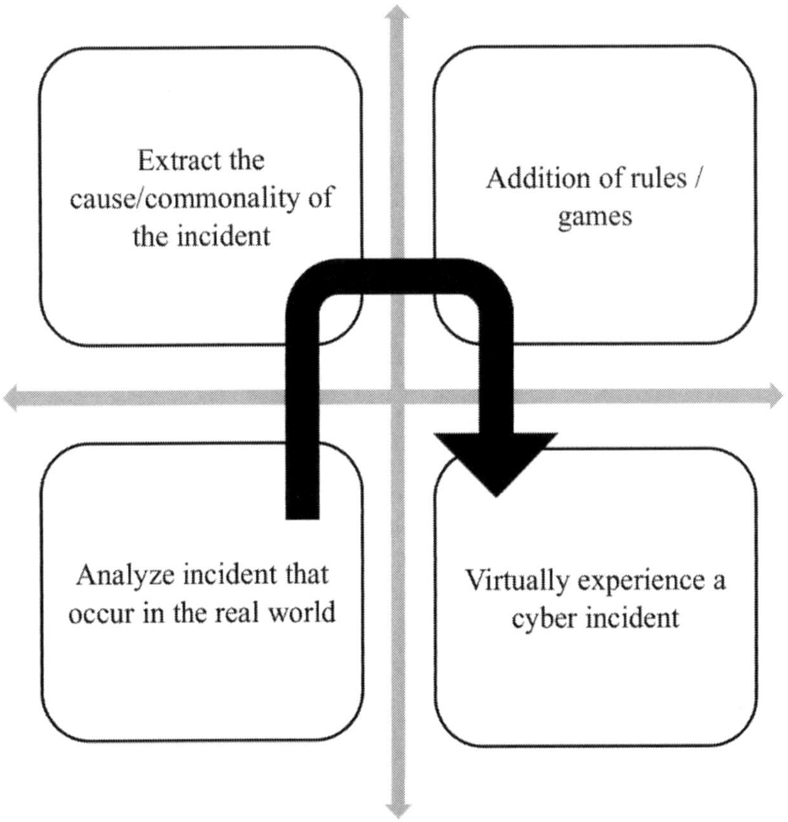

Figure 5. Exercise development process

Even staff who have never experienced a cyber incident can increase their experience with cyber incidents through exercises. By expanding this experience value, staff will more likely think and act even if people have a similar incident or an incident that people have never experienced before.

2.2 Human resource development method using virtual experience (exercise)

By taking the following four steps, we aim to improve the incident response capability of the company.

1. Information input
   Many companies have no experience in responding to cyber incidents, so it is necessary to understand the impact of cyber incidents on companies and plants. It is also required for the parties concerned to understand that cyber-attacks on plants are problems on the IT side and OT side.

2. Create ICS-BCP from a management perspective
   Companies are required to generate stable profits. When an incident occurs, companies need to think about what to prioritize to keep the business running, when to recover, and alternatives if there is no recovery. Each company must create an ICS-BCP assuming a cyber incident and communicate it to employees.

3. ICS-BCP with added on-site perspective
   It is necessary to respond to incidents based on the ICS-BCP defined as a company created in step 2. However, this BCP only includes the perspective of how to respond as an organization.

   By checking the BCP from the viewpoint of the operational site, it can be made into an ICS-BCP that can be operated.

4. Share the idea of all layers
   Employees must understand ICS-BCP to respond even if a never-experienced-before incident occurs. Therefore, by conducting training centered on the planned ICS-BCP, we will deepen the understanding of employees and objectively evaluate the ICS-BCP.
   By repeating this training multiple times, it is possible to confirm omissions and deficiencies in BCP and increase ICS-BCP accuracy.

The activities mentioned above don't do only once. Each company can improve its resilience ability by making assumptions based on various preconditions (scenarios), rotating the PDCA cycle, and continuing discussions.

3 Exercises for improving resilience developed

Based on the recognition that not only hard skills but also soft skills are important for responding to cyber incidents, we have developed discussion-based ICS security exercises.

3.1 Theme of the exercise

As mentioned above, it is very important to prepare a response plan to respond to an incident with hard and soft skills. Therefore, the purpose of the exercise is to understand the required actions to be taken by cyber-attacks and the importance of inter-organizational collaboration needed to take those actions.
The final goal is to be able to create an ICS-BCP using the knowledge gained from the exercises.

In order to make participants understand the purpose of the exercise, the exercise was divided into three phases according to the purpose of correspondence. (Fig.6)
When an incident occurs in a plant, whether it is a cyber incident or not, it is necessary to take measures to ensure the safety of the plant. Then, it is necessary to investigate the cause after ensuring the safety of the plant. If the cause is a cyber-attack, it is necessary to investigate the extent to which the impact of the cyber-attack has spread.
Initially, the safety workload is large, but the security workload gradually increases.
Then, there will be a response to confirm the extent of the impact of cyber-attacks not only on plant safety but also on business continuity.

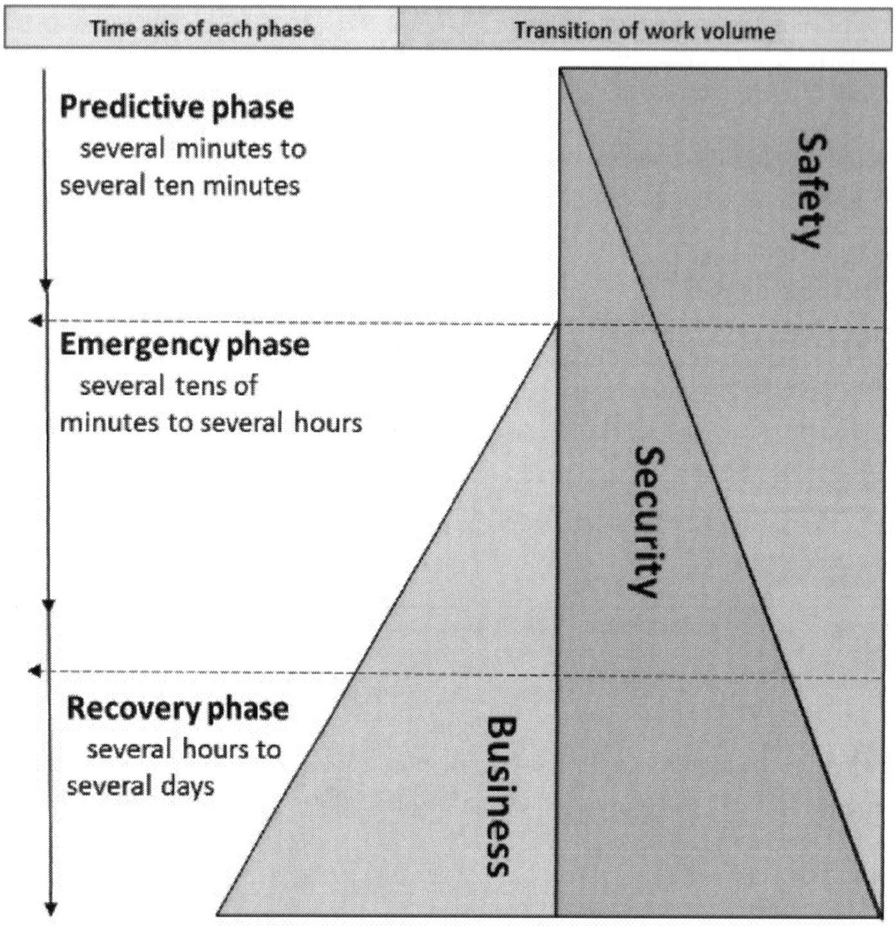

Figure 6. Structure of Exercise with Workload Image among Safety, Security, and Business Continuity Activities[7]

The authors visualized the state that this working volume change with the progress of time. (Fig6) Fig.6 expresses the work volume at a certain spot. The work volume is divided into safety, security, and business. The participant understands a change of this working volume through exercise.

1) Predictive Phase

The purpose of this phase is to ensure the safety of plants threatened by cyber-attacks. In this phase, the participants examine the safety response obstructed by a cyber-attack.

After detecting suspicious communication in a section that monitors the network, the exercise begins with an alert for the status of the liquid in the plant. In this phase, the participants must ensure the safety of the plant. Once safety is ensured, participants must investigate the abnormal operation. In this phase, participants can recognize how different causes can influence the safety response.

2) Emergency Phase

The purpose of this phase is to continue the business. The participants consider the security of the plant and how the exercise implicates a real cyber-attack. They must also take into account

how their business will be affected by the cyber-attack. In this phase, the participants consider practical measures taken to ensure business continuity throughout the company.
This phase begins after the abnormal operation of the plant reveals a cyber-attack. In this phase, participants can recognize that the entire company must be mobilized to treat a situation caused by a cyber-attack.

3) Recovery phase
The purpose of this phase is to restore own business to normal operation rapidly. The participants must find a way to nullify the influence of the cyber-attack so that the plant can return to normal operation. In addition, they must consider how business is affected by the temporary plant shut down. In this phase, the participants examine the start-up procedure of a plant crippled by a cyber-attack. In addition, the participants examine methods to prevent the recurrence of similar cyber-attacks.
This phase begins after the plant is stopped by a cyber-attack. In this phase, participants think about how to restart the plant. In this phase, participants can see how different causes affect the restoration of normal plant operations.

The following three types of training exercises have been developed.

- Card-Type Incident Response Exercise
- BCP Creation Exercise
- Simulation Exercise called "IMANE" (Incident MANagement Exercise)

3.2 Core scenarios for exercise

The developed exercises are based on the same scenario with different methods.
The base scenario is required to be able to consider the ideal organizational cooperation and necessary soft/hard skills for the organization to respond quickly in the event of a cyber-attack through exercises. The created scenario was composed of three scenarios to think about the three aspects of safety, security, and business.

Virtual Company:
Participants from various companies, departments have different backgrounds. A common scenario is indispensable for the participants to share the exercise together. In addition, using a simple and realistic virtual company makes it easier to apply the awareness of the exercise to the business. Figure 7 is an organizational chart of the created virtual company. Practical prerequisites, such as the company's business profile and its organization's intercommunication rules, have been set to achieve the exercise's objectives. Participants can discuss how information should be shared and how an organization should respond to cyber-attacks. Figure 8 is the network structure of a virtual enterprise. The attacker will attack using the computer that was hijacked by the cyber-attack as a stepping stone. The attacker steps on the hijacked computer and moves to the computer he wants to attack. In other words, the effects of cyber-attacks are spreading not only to the attack target but also to the attack process. For mitigating the impacts of cyber-attacks, it is necessary to consider countermeasures within the range of the effects of cyber-attacks. By conducting exercises based on this network structure, participants can discuss the Security Response required to respond to cyber incidents.

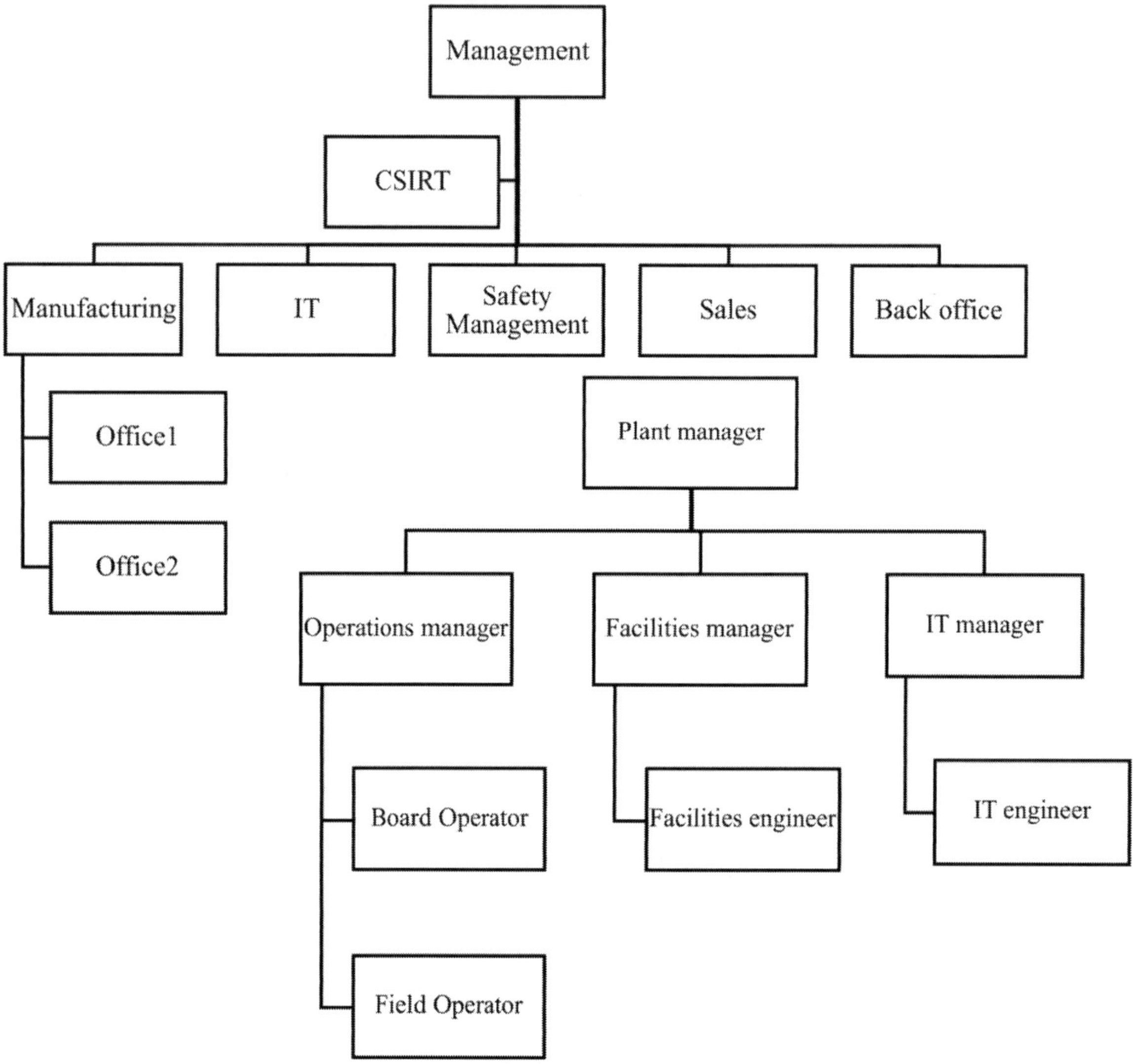

Figure 7. The organization chart of a virtual company

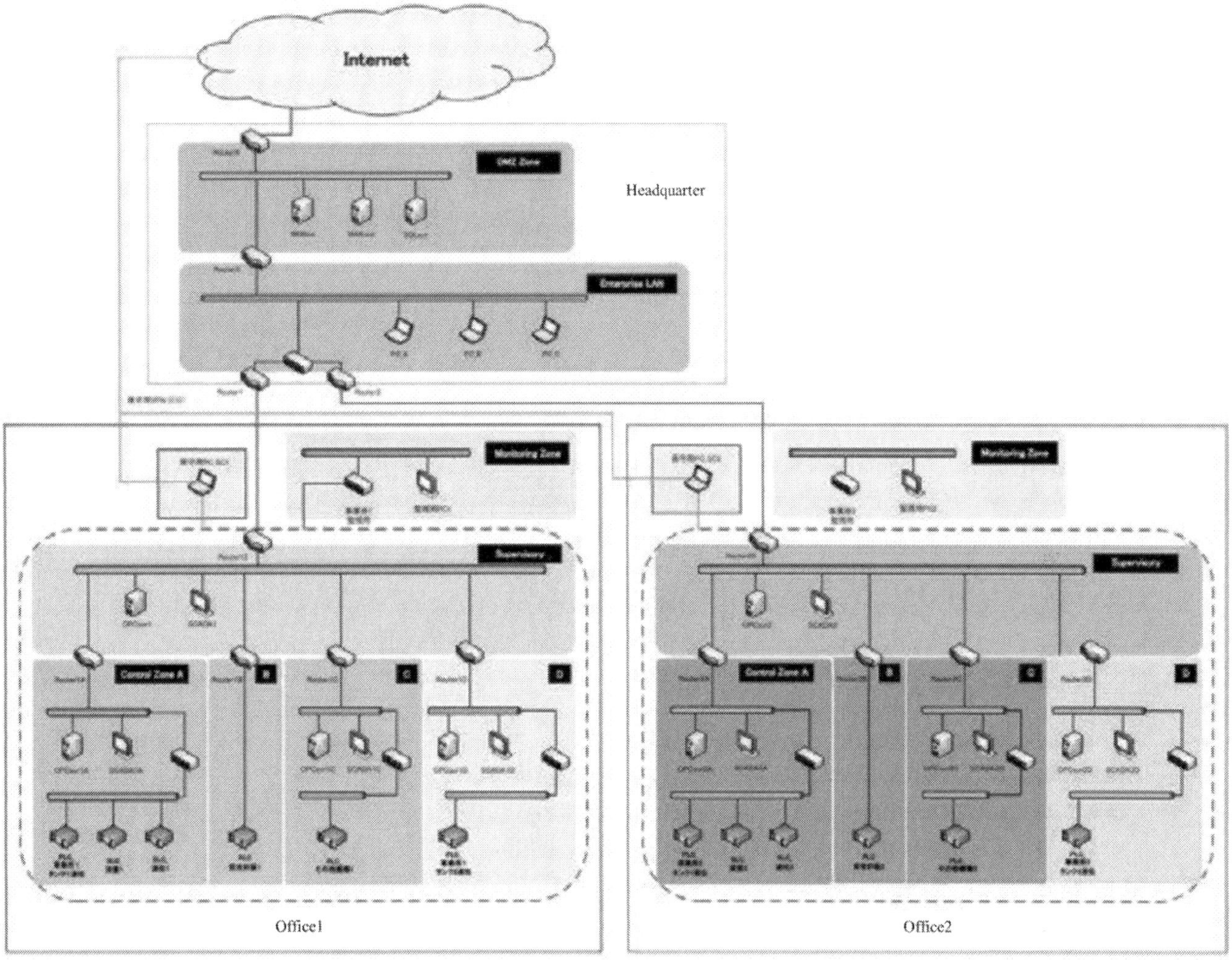

Figure 8. The network structure of the virtual company

Plant Operation Status Scenario:
A real-world plant consists of many devices, pipes, valves, and so on. If the plant is used as the subject of the exercise, the discussion may be complicated, and the purpose of the exercise may not be achieved. The plant used as the subject of the exercise should have a simple structure. By using a plant with a simple structure as the subject, it is possible to facilitate discussions focusing on the cyber incident that is the purpose of the exercise. In addition, it becomes easier to obtain meta-knowledge about cyber incidents, and it becomes easier to apply the knowledge gained through exercises to actual plants.

This scenario is based on our cybersecurity testbed. And we have introduced the control equipment used in the actual plant. This testbed aims to test and develop ICS security solutions. Based on the cyber-attack demonstration conducted as part of the research, the plant operation information caused by the cyber-attack is used for the exercise.

Cyber-attack Scenario:
The flow of cyber-attacks on plants is almost the same as cyber-attacks on IT systems.
To imagine how to respond to a cyber incident, it is important to deepen your cyber-attacks knowledge. There are many cyber-attacks methods, but attackers usually attack based on the cyber kill chain [8]. The cyber kill chain [9] classifies cyber-attacks into seven phases (Reconnaissance, Weaponization, Delivery, Exploitation, Installation, Command & Control, Actions on objectives). The idea is that appropriate measures can be taken in each phase, and cyber-attacks can be prevented if the attack does not proceed to the next phase.
This scenario is based on a cyber-attack demo conducted in our laboratory. It also includes the idea of a cyber kill chain. Throughout the exercises, participants can discuss the measures to protect the company by understanding cyber-attacks workflow and intention.

3.1 Card-Type Incident Response Exercise

The first exercise is Card-Type Incident Response Exercise. This exercise aims to gain an awareness of what a cyberattack on a plant is and what actions will be required.
This exercise uses a prepared action card to create a workflow for responding to cyber-attacks. Participants will achieve the purpose of the exercise by considering the response to cyber incidents while understanding the given action card's contents.

3.3.1 Exercise steps

These exercises consist of five steps: introduction, explanation of prerequisites, group work, presentations, and open discussions. Participants guess the company's situation from the given preconditions and create a workflow while thinking about the actions and information sharing necessary to make the plant safe and carry out business continuity.

Introduction:
The facilitator explains the purpose of the exercise and how to proceed. The general scenarios of the exercise are introduced to the participants. At this step, participants understand the importance of the exercise and, therefore, be motivated to apply themselves fully. By doing so, participants can increase the benefit of the exercise.

Description of Prerequisites:
Participants perform group work as a virtual company that owns a plant and provides products. The facilitator explains the preconditions about the virtual company at the beginning of the exercise. Prerequisites include the plant's status, the state of the IT network, and the company situation. Participants watch a video to understand the preconditions (In this phase, we use the videos we create to help participants understand the prerequisites.).

Participants envision the outcome based on the prerequisites.

Group Work:
The group task is to create a workflow by envisioning the outcome from the given information and considering the possible situation's response.
Originally, this exercise was conducted in group work in person. With COVID-19, we improved this exercise so that we could perform it using web conferencing.

In group work, the Excel file shown in Fig. 9 is used. Worksheets and action cards are pre-stored in this Excel file.

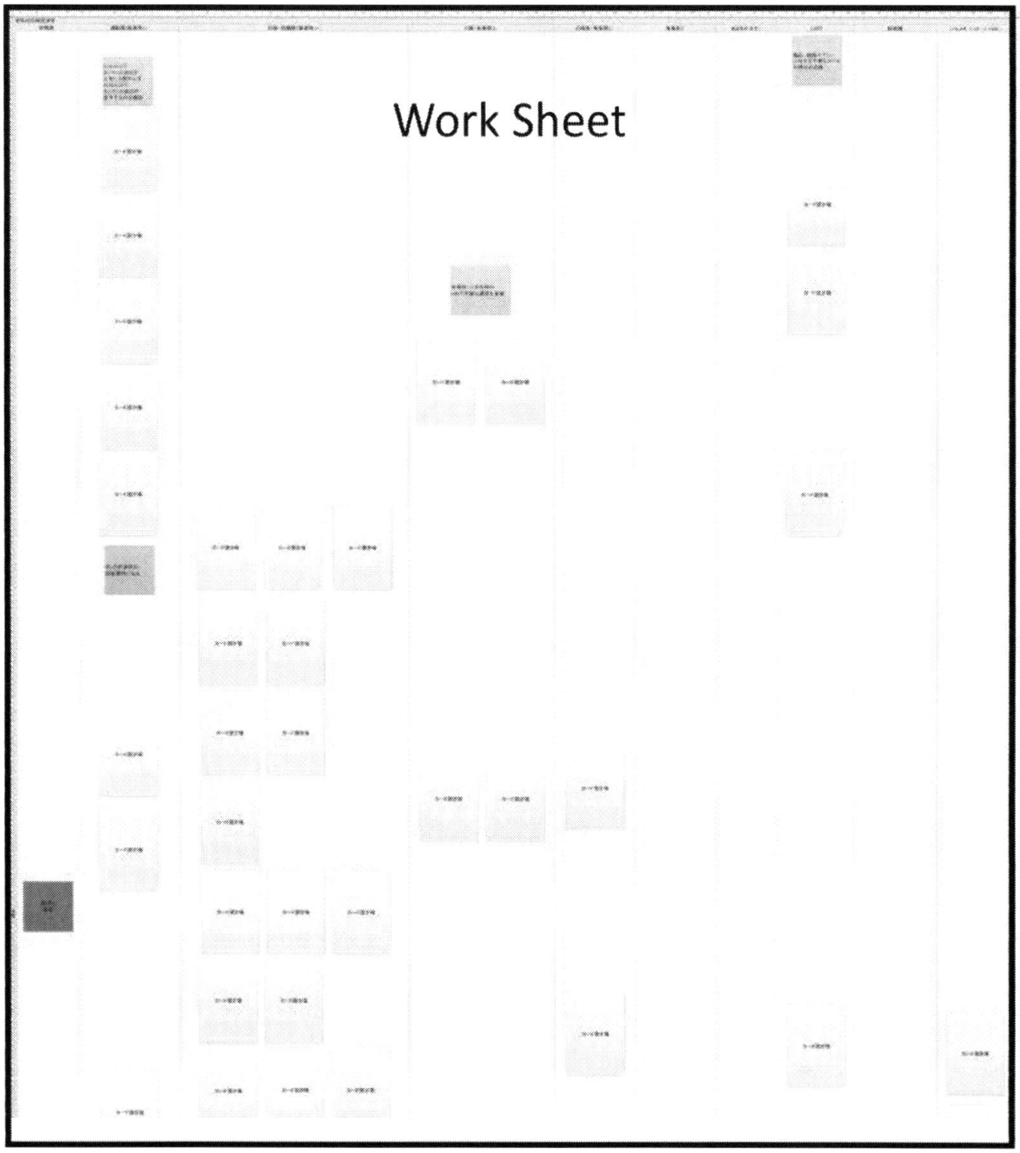

Figure 9. Group worksheet (Excel Screen)

The worksheet contains columns with the virtual company's departments, prerequisites, and action card storage roles. Action cards are classified into two types: normal cards and special cards (the upper part of Fig. 10 is the normal card and the lower part is the special card).

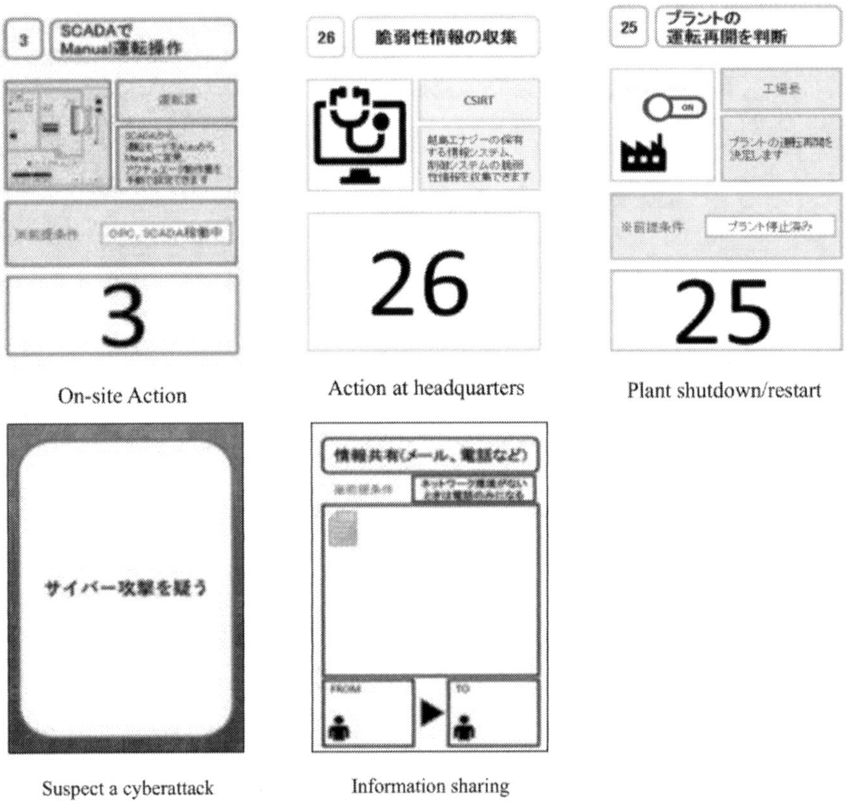

Figure 10. Type of action cards

The normal card is an incident response card, and there are a total of 38 cards of three types:

- "action at the site"
- "action at the head office"
- "shutdown/restart of the plant"

On the normal card, the result after taking action is written on the backside, and the participant thinks about the next action based on that information.

There are two types of special cards: "suspect/validation of cyber-attack" and "information sharing." The information-sharing card is used to share the information obtained by each department with other departments. In this exercise, participants discuss and strategize countermeasures from a bird's-eye-view, that is, without playing a specific role. However, in reality, information cannot be shared with others unless it is shared by the person who has the information. Participants use the information-sharing card to share the information held by each department. This makes it possible to express the flow of information sharing on the workflow. An important action in this exercise is to suspect/validate a cyberattack. No security response will be given unless the cause of the plant malfunction is suspected to be a cyber incident. There is information that can be a trigger to suspect that it is a cyber incident, and someone must judge that it is a cyber incident based on that information. By describing it on the workflow, participants identify the departments that will be the key persons in the cyber incident response. Participants will use these action cards to create a workflow, as shown in Figure 11.

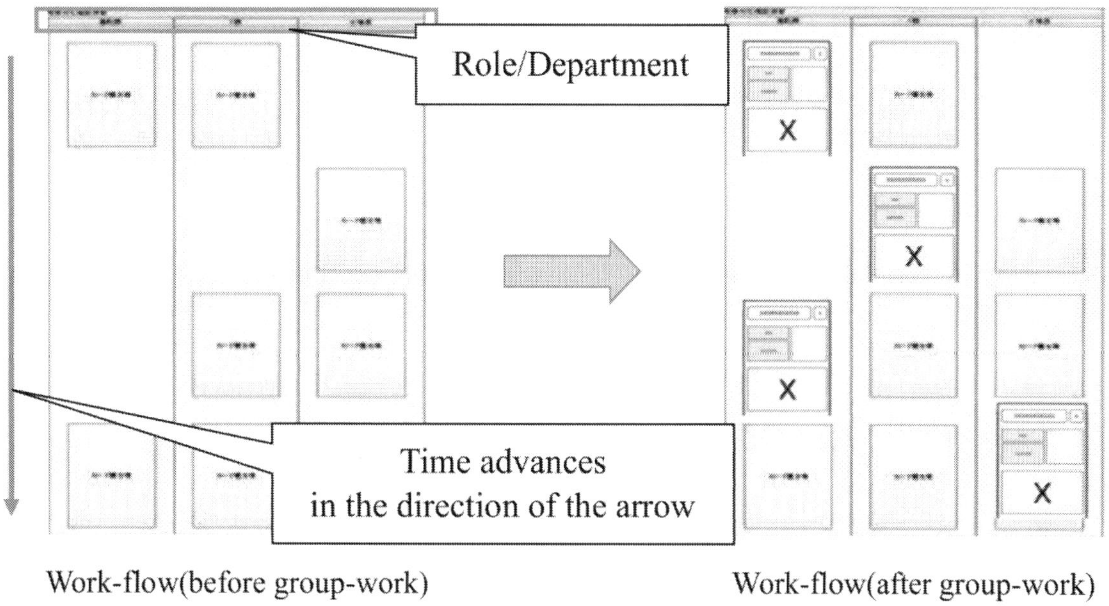

Figure 11. Workflow creation (image)

Presentation:
For groups with a lot of work experience in plants, create workflows that focus on plant safety. For groups with a lot of work experience in the IT department, create workflows that mainly support IT systems. Workflow, which is the product of group work, depends on group members' background and experience. By sharing the workflow and the workflow creation process, the exercise participants confirm the similarities and differences between the ideas of other groups and their own group's ideas. This step can enhance the effectiveness of the exercise.

When each group makes a presentation, ask them to focus on the following three points.

- How did you think about "who (which department), when, why" and placed the action card?
- What did you imagine when you created the workflow from the information given?
- How did you feel through the exercises what kind of human resources, information sharing, and organization-related systems should respond to cyber incidents?

Participants can understand the ideas of other groups and gain new knowledge by listening to other groups' presentations and asking and answering questions.

Open-Discussion:
After presenting the group's plans, the facilitator will show an example of the exerciser's workflow. This workflow presented is the optimal solution that the exercise developer thinks. The facilitator will give the same presentation as the participants in the exercise. After the

facilitator's presentation is over, the ideal inter-organizational collaboration for cyber incidents, information sharing, and necessary human resources will be discussed.

Through the exercises, exercise participants will experience the response to cyber incidents and understand the impact of cyber-attacks on ICS. And it is possible to gain insights on ideal organizational cooperation and necessary soft skills and hard skills so that the organization can respond quickly in the event of a cyber-attack.

3.4 ICS-BCP Creation Exercise

The second exercise is a BCP creation exercise. Participants create an incident response exercise based on a given scenario (such as a virtual business scenario). Participants will consider how to attack the plant from the attacker's point of view, consider actions to protect the company from cyber-attacks, and create an ICS-BCP. The purpose of the exercise is to understand the perspective of things needed to create an ICS-BCP.

3.4.1 Exercise steps

This exercise consists of five steps: scenario explanation, group work, demonstration, exercise review, and open discussion. Participants in the exercise identify the risks that the company faces from the given prerequisites. Next, consider an example of a cyber-attack that may cause that risk. Then, think about how to respond to the incident caused by the cyber-attack and create an ideal ICS-BCP.

Description of the Scenario:
Virtual enterprise scenarios, plant scenarios, and cyberattack scenarios are required to create incident response exercises. In this exercise, we want participants to discuss how to create an ICS-BCP rather than how to create a scenario. Therefore, the scenario is provided by the facilitator in advance. In addition, since there id an infinite number of attack scenarios, some scenarios are presented, and participants select from those scenarios. In this step, those scenarios will be explained.

Group-Work:
This exercise also creates deliverables through group work.
The deliverables to be created are the worksheets and action cards required for the card-type incident response exercise.

- First, participants analyze the risks of a virtual company from a given virtual company and plant scenario.
- Next, participants create a cyber-attack scenario that causes the analyzed risk. Participants create a scenario of the impact on the company and the plant's operating status due to the created cyber-attack scenario.
- Finally, participants create an ideal workflow for the company to perform incident response exercises and create actions that use that workflow as action cards.

Figure 12. An example of Team Discussion in Group Work

Demonstration:
In this step, participants perform their own exercises on other groups. Exercises will be conducted for different groups using workflows and action cards created in group work. Through this step, participants confirmed that they could convey the message through the exercise (ideal response / inter-organizational collaboration) to other groups. Participants can also gain more cyber-incident experience by taking exercises in various scenarios.

Group-work (Review of the exercise):
There are many things that can be gained by doing exercises.
Participants review the exercises they have created based on the results of the exercises conducted by other groups. The group once again discusses whether the ideal workflow that they have considered is the optimal solution, what factors hinder the ideal workflow that they have created, and so on.

Open-Discussion:
Each group will share what they have reviewed their exercises and what they have learned through the exercises in this step. By listening to various people's opinions through this step, it is possible to discuss how to view the things necessary for creating an ICS-BCP.

3.5 IMANE (Simulation exercise)

This exercise aims to confirm the effectiveness of the created ICS-BCP by simulating the incident response. This exercise is a simulation exercise that uses a personal computer to create a game based on a scenario, input chat, and respond to an incident by selecting a response.

3.5.1 Exercise steps

This exercise consists of five steps: identifying stress factors such as no one to deal with or no tools to deal with, adding storylines, simulating incident response, reviewing the exercise, and open discussion. Participants will play the role given and aim for business continuity while responding to incidents as virtual company members.

Identifying Stress Factors:
The ICS-BCP created by ICS-BCP creation exercise is an ideal workflow. In other words, the plan is based on the premise that all the assumed actions can be performed. However, in reality, it may not be possible to perform actions that should be possible due to various factors, such as lack of resources such as people and tools required to perform actions. Participants will identify stress factors that can hinder their actions in operating the ICS-BCP they have created. Stress factors are the risks involved in conducting ICS-BCP, and it is important to have advanced preparations and alternatives to eliminate the stress factors.

Adding storylines:
The created ICS-BCP only identifies the actions and the shared information, instruction contents, and information on the operation transition of the plant due to the actions are hidden between the actions. Participants identify the hidden information and necessary information communication paths. Then, create a game scenario by incorporating the stress factors created in the previous step.

Simulating Incident Response:
This step simulates an incident response. Figure 13 shows the structure of this exercise. In previous exercises, exercise participants were required to create a workflow in groups without being assigned to a specific role. In this exercise, roles are assigned to each exercise participant, and incident response is performed by taking actions for each role. It is an exercise that is closer to the actual incident response.

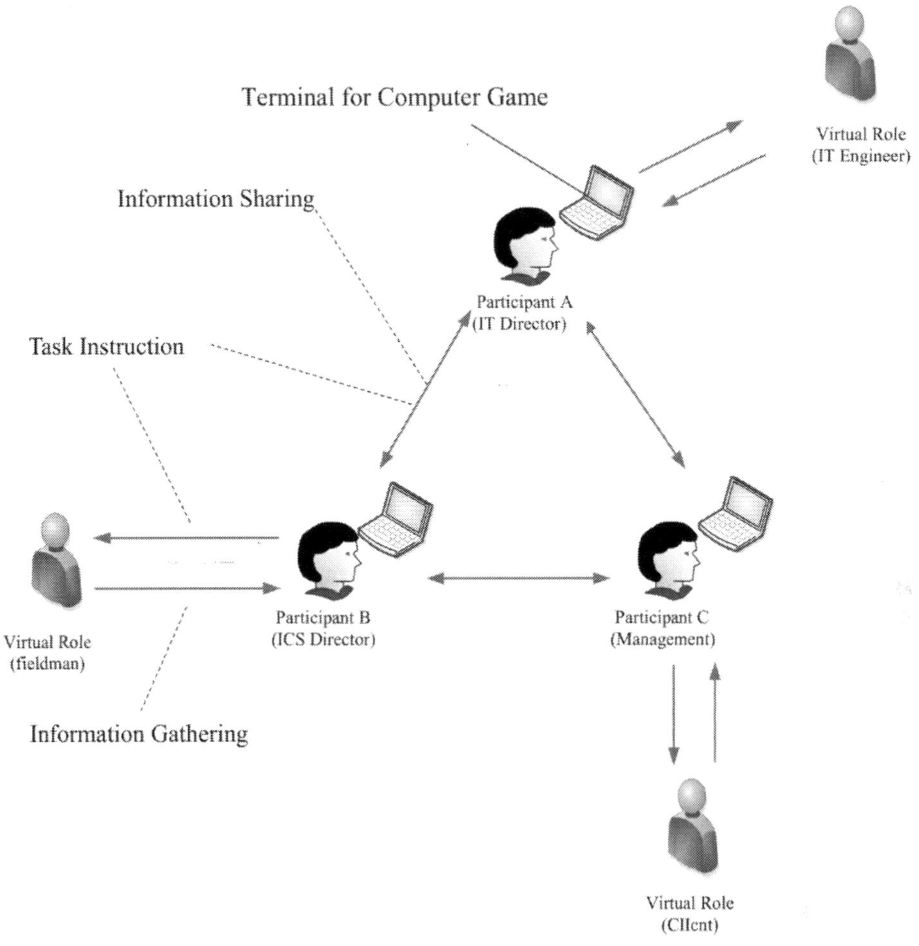

Figure 13. IMANE structure

Figure14 is a console screen operated by the participants. Participants share information with other players and take actions based on the information flowing in the tray. Each participant is required to understand the situation and take action based on the information displayed on their console screen. In addition, since the actions that can be performed for each role and the information is given are different, it is also required to share the information necessary for taking action to converge the situation with other players. Participants use the interface in Figure 14 to share information and select actions. The information displayed in the tray in Figure 15 is linked to the story created in the previous step, and the information displayed differs depending on the action performed so that the incident response can be made more realistic.

Figure 14. The user interface of IMANE exercise

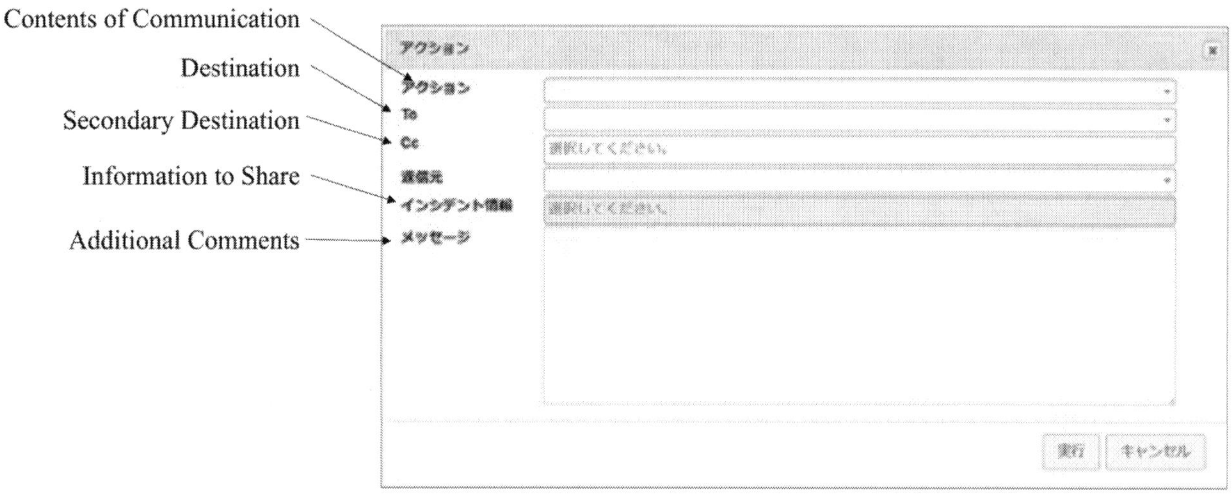

Figure 15. The action user interface of IMANE Exercise

Review of the Exercise:

After the end of this exercise, it can be output as a product of the exercise, as shown in Figure 16. This workflow includes the information given to each role, the actions taken, and the flow of information-sharing and participants can use this workflow to review the incident response after the exercise.

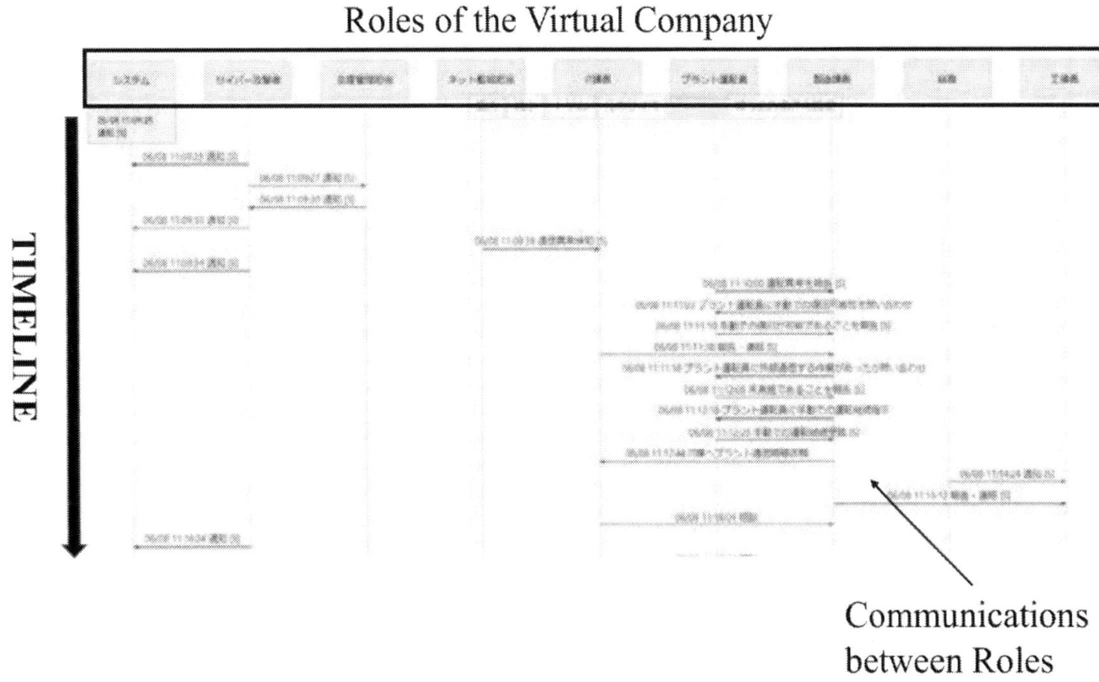

Figure 16. Example deliverable of the exercise

Using this deliverable, exercise participants can objectively review ICS-BCP by looking back on their response. It is possible to improve the accuracy of ICS-BCP by checking for missing or defective ICS-BCP.

Open-Discussion:
In this step, each participant shares the impressions gained through the exercise. We discuss the ideas needed to create an ICS-BCP and human resource development using this exercise.

By performing these exercises in various scenarios, the exercise participants can increase their imagination and resilience to cyber incidents.

4 Pilot Exercise

While improving the developed exercises step by step, many security and operation experts participated in the Nagoya Institute of Technology security workshops from 2018. To get an opinion to improve the exercises, we asked the participants to complete the questionnaire. The questionnaire was answered with a maximum of 5 points. In the exercise conducted on December 11, 2020, 28 people participated, and the following questionnaire results were obtained (Max point is 140). From the results of this questionnaire, it can be seen that the people who participated in the exercise understood the purpose of the exercise that we set and received high praise for the exercise.

Questionnaire content for Card-Type Incident Response Exercise and ICS-BCP Creation Exercise:

1. Was the goal set for the exercise, and was it possible to attain it?
2. Do you think you learned about organizational collaboration in the event of a cyber incident?
3. Do you think you learned about the organizations needed to respond to cyber incidents?
4. Was the information necessary to proceed with the exercise given?

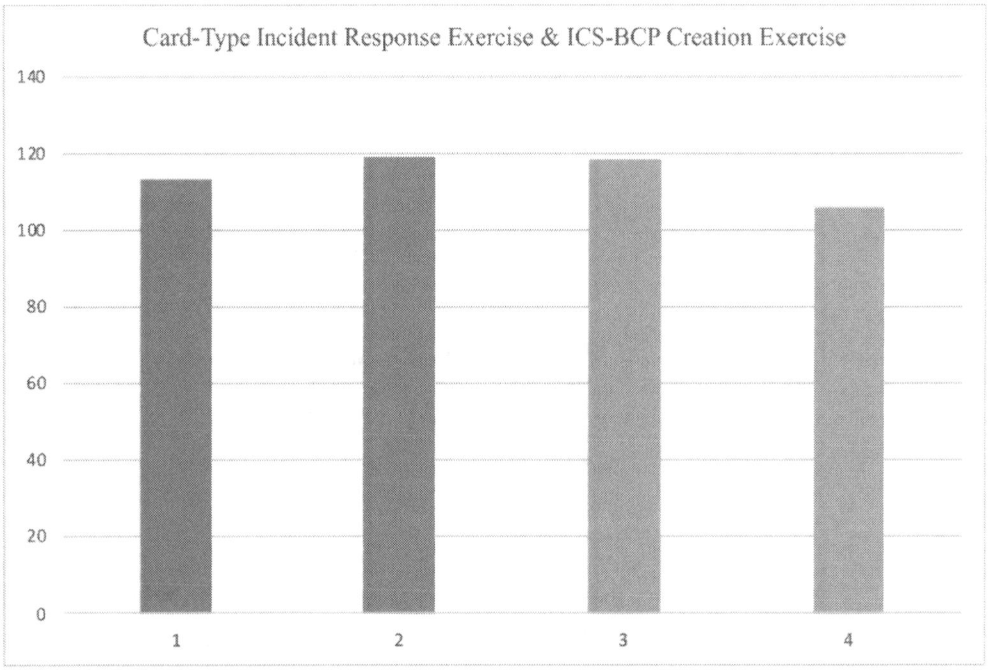

Figure 17. Questionnaire results of Card-Type Incident Response Exercise & ICS-BCP Creation Exercise

Questionnaire content for IMANE:

1. Was the goal set for the exercise, and was it possible to get it?
2. Do you think you learned about organizational collaboration in the event of a cyber incident?
3. Do you think you learned about the organizations needed to respond to cyber incidents?
4. Is it designed so that necessary information is communicated and given to each role?
5. Do you feel that you were able to experience information gathering and decision-making in an emergency?

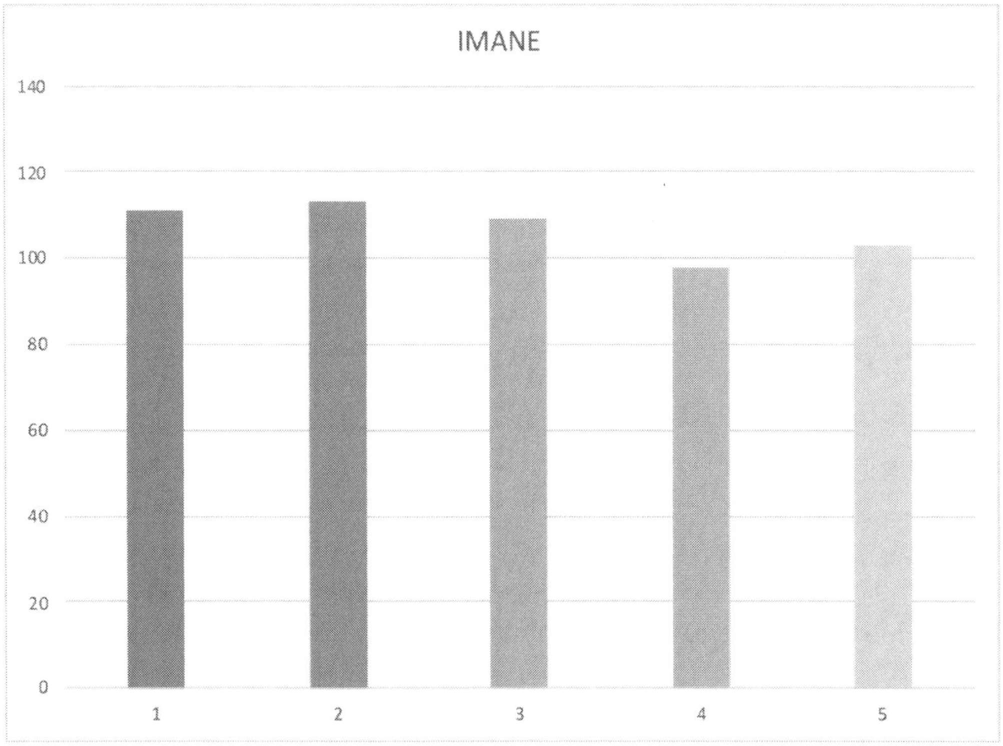

Figure 18. Questionnaire results of IMANE

In addition, the following was found by analyzing the deliverables of the exercise.

・ Plant safety first

Many workflows can keep the plant safe, even if the backgrounds of the participants in the exercises vary. Even if the cause of the plant abnormality is a cyber-attack, it is easy to imagine the visible situation, and it can be seen that appropriate measures can be taken.

・ No action has been taken to identify a cyber-attack

Analyzing the workflow, it is possible to select situation sharing and actions for performing Safety Response. However, information sharing for performing Security Response, such as identifying the range of influence of cyber-attacks is often not successful. This is thought to be due to the fact that they have never experienced cyber-attacks and lack the knowledge to consider actions. In addition, Safety Response is easy to imagine because it corresponds to the visible physical impact on the plant, but Security Response is difficult to imagine because it corresponds to the invisible network.

・ Organization that is the center of the response

To respond to cyber incidents to ICS, it is necessary to fluidly change the organization that is the core of the response.

In the initial response, it is necessary to respond to plant operation abnormalities caused by cyber incidents. The person working at the plant takes action. After ensuring the plant's safety, it is necessary to eliminate the effects of cyber incidents. IT department takes this action. Finally, it is necessary to continue the business considering the impact of the cyber incident on the business.

Management considers actions for the business gains.
In this way, the work to be done changes over time, and it is necessary to respond while changing the organizational response that is the core of the response accordingly.

5 Concluding Remarks

In this paper, the authors presented the factors required of a company against the threat of cyber-attacks targeting ICS and explained the exercises developed to increase the company's resilience.

Through the workshop, we received positive opinions from the participants that they could gain various awareness of cyber-attacks. However, the effect of the exercise could not be confirmed quantitatively, and it could not be confirmed that the response-ability was improved.

In the future, we will conduct exercises for more people and obtain deliverables. Then, by analyzing the deliverables, we will consider the method of planning ICS-BCP considering cyber-attacks and the conditions necessary to confirm the validity of the planned ICS-BCP.

6 References

[1] Institute for Science and International Security, "Did Stuxnet Take Out 1,000 Centrifuges at the Natanz Enrichment Plant?", https://isis-online.org/isis-reports/detail/did-stuxnet-take-out-1000-centrifuges-at-the-Natanz-enrichment-plant/8 (accessed Jan. 25, 2021)

[2] ICS-CERT,
ICS Alert(IR-ALERT-H-16-056-01), "Cyber-Attack Against Ukrainian Critical Infrastructure", https://us-cert.cisa.gov/ics/alerts/IR-ALERT-H-16-056-01 (accessed Jan. 25, 2021)

[3] SANS Institute, ICS Defense Use Case No. 6, "Modular ICS Malware", https://ics.sans.org/media/E-ISAC_SANS_Ukraine_DUC_6.pdf (accessed Jan. 25, 2021)

[4] Norsk Hydro, "Cyber-attack on Hydro", https://www.hydro.com/en/media/on-the-agenda/cyber-attack/ (accessed Jan. 25, 2021)

[5] BSI, "Industrial Control System Security Top 10 Threats and Countermeasures2019", https://www.allianz-fuer-cybersicherheit.de/ACS/DE/_/downloads/BSI-CS/BSI-CS_005E.html#download=1 (accessed Jan.25, 2021)

[6] T. Aoyama, M. Koike, I. Koshijima, Y. Hashimoto, "A unified framework for safety and security assessment in critical infrastructures". In: Safety and Security Engineering V. WITPRESS LTD. (Sep 2013)

[7] Y. Ota, T. Aoyama, D. Nyambayar, I. Koshijima, "Cyber Incident Exercise for Safety Protection in Critical Infrastructure," Int J. Saf. Secur. Eng. 8,246-257 (2018)

[8] J. Sakhnini, H. Karimipour, A. Dehghantanha, R.M. Parizi, G. Srivastava, Security aspects of internet of things aided smart grids: a bibliometric survey. Internet Things (2019). https://doi.org/10.1016/j.iot.2019.100111

[9] A. Shostack, *Threat Modeling: Designing for Security* (Wiley, Indianapolis, In, 2014)

SPRING21 +17TH GCPS
A Joint AIChE and CCPS Meeting

Hazard Identification during Process Related Maintenance and Operations Tasks

T Michael O'Connor
Mary Kay O'Connor Process Safety Center
Texas A&M University, College Station, TX
mikeocon@gmail.com

Dr Zohra Salim
Mary Kay O'Connor Process Safety Center
Texas A&M University, College Station, TX
zohrahalim@tamu.edu

Prepared for Presentation at
American Institute of Chemical Engineers
2021 Spring Meeting and 17th Global Congress on Process Safety
April 18 - 23, 2021

Hazard Identification during Process Related Maintenance and Operations Tasks

T Michael O'Connor
Mary Kay O'Connor Process Safety Center
Texas A&M University
College Station, TX
mikeocon@gmail.com

Dr Zohra Halim
Mary Kay O'Connor Process Safety Center

Keywords: serious injury, fatality, chemical manufacturing, petroleum refining

Abstract

Trends in process related fatalities are presented for chemical manufacturing and petroleum refining. The trend analysis revealed the more than half of fatalities occur in single fatality incidents. This observation led to study of the nature of these low fatality incidents. Three common tasks were identified that are associated with a large portion of process related fatalities and severe injuries. These 3 tasks are clearing, cleaning and purging lines and vessels; opening lines and vessels; and transferring fluids. The specific hazards encountered in incidents associated with these tasks are identified. The goal is to provide the basis to improve system design, training, procedures, job hazard analysis, PPE requirements and other factors based on hazards actually associated with major harm to individuals performing these tasks.

Introduction

There are 3 primary components of this study:

- Trends in fatalities from 1984-2019 in chemical manufacturing (CM) and petroleum refining (PR)
- Tasks associated with severe injuries and fatalities in CM and PR
- Nature of serious injury incidents in CM only

Data is extracted from two US OSHA databases:

The first is the Fatality and Catastrophe Inspection Search (FCIS) database which includes all incidents with a fatality or 3 or more injuries. Since incidents without a fatality and less than 3 injuries are not included in the data injuries from this database are not included in the study. This database covers the years 1984 to the present whichallows long-term trending including the implementation of OSHA PSM in 1992. This analysis is limited to the states in OSHA Region 6. These states are TX, LA, AR, OK and NM. The data was accessed through OSHA's online query system.

The second database is the Severe Injury Reports (SIR) database. This database is limited to the 28 States with federal, not state, OSHA programs. This database includes all the Region 6 states. It covers the years 2015 to July 2020 and includes cases with hospitalization or amputation. This entire dataset can be downloaded as a spreadsheet.

For both databases the authors have interpreted the accident description to determine whether it is a process related incident and the task being undertaken. The serious injury database contains the hazardous substance, nature of injury, affected body part and event type. In some cases, the description is not sufficient to extract the needed data and these items are classified as unknown. Some incidents have been investigated by the Chemical Safety and Hazard Investigation Board (CSB) or other agencies. In those cases, information may have been extracted from those sources.

This paper focuses on both the Chemical Manufacturing (CM) and Petroleum Refining (PR) industry as determined from their NAICS and SIC codes. Both industries were analyzed for fatality trends from 1984-2020. The tasks associated with both severe injuries and fatalities were determined for both CM and PR. A more detailed analysis of selected tasks was performed only for CM. Detailed analysis of PR incidents are not yet completed.

Scope of Incidents in Study

The following definitions were utilized to determine if an incident was process related and was to be included in the study:

Hazardous Material – Any substance the is toxic, reactive, flammable, asphyxiating or that presents a hazard to people, the environment, or property due to pressure or temperature.

Process Safety Incident – The sudden, unintended release of, or exposure to, a hazardous material, associated with an industrial process, that results in or might reasonably have resulted in death, injury, significant environmental or property damage, evacuation or sheltering in place.

These are generic definitions used by the MKOPSC. In this paper all of the incidents involve a fatality or injury since they are from the OSHA databases.

Trends in Fatalities in Chemical Manufacturing and Petroleum Refining

Long term trends with time are available from the Inspection Database for fatalities. Serious injuries are included in the database if there are 3 or more injuries or they occur with a fatality. However, incidents with one or two injuries are not included making any trending information questionable. The Serious Injury database only covers the years 2015 through 2020, making any trends of doubtful significance, so only fatalities were used for trending.

Plots are shown for CM and PR. Fatalities and fatal incidents over the years 1984 to 2020 are illustrated. There are a number of high fatality incidents in the latter 1980s. This is followed by a dramatic decrease in fatalities around the time of the implementation of OSHA PSM in 1992. It seems likely this decrease was due to both the increased focus on process safety due to the new regulation as well as a direct response to these high fatality incidents. The trends are somewhat similar for CM and PR with the exception of the BP Texas City explosion in 2005.

Figures 1 through 4 show the large number of incidents with a low number of fatalities. After 1992, for CM and PR, greater than 80% of fatalities occurred in single fatality incidents. This statistic in particular motivated a more detailed examination of these numerous incidents with mostly low numbers of fatalities as well as incidents with serious injuries.

In this plot (see Figure 1), the total height of the bars represents the total fatalities in a year. The colored bands making up the bars show individuals incidents and the number of fatalities in that event. The red arrow denotes the implementation of OSHA PSM in 1992. Clearly there is a sharp decrease in fatalities after that time. In addition, there are no incidents with more than 4 fatalities after PSM began.

Figure 1 - Fatalities by Event in Chemical Manufacturing Region 6

This plot (see Figure 2) shows that not only the annual number of fatalities dropped in 1992 but also the number of fatal events.

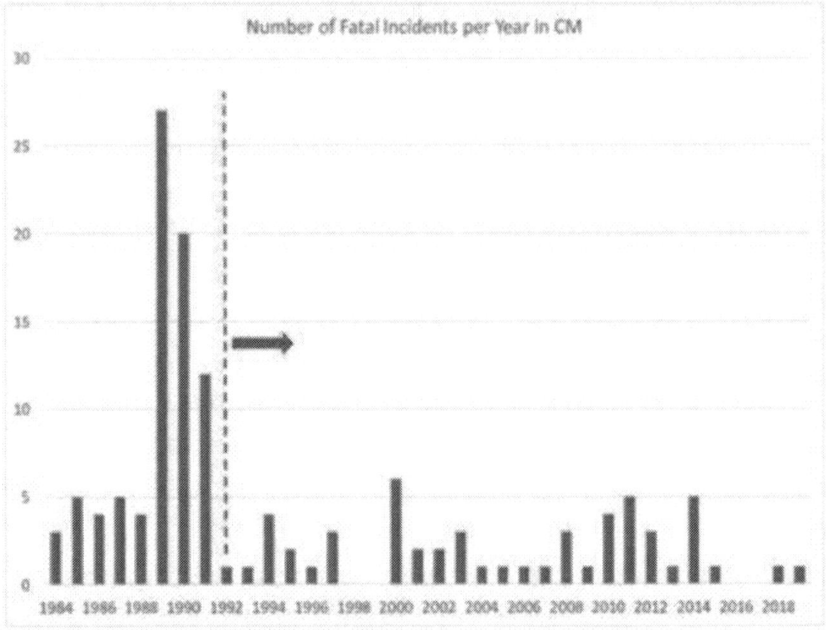

Figure 2 - Fatal Events in Chemical Manufacturing

Like CM there is a distinct reduction in fatalities in PR about the time of PSM implementation. This reduction appears to have occurred in 1994, a few years after PSM. There is another reduction about the time of the event at BP Texas City (see Figure 3). We theorize that this high-profile event may have re-focused the PR industry on process safety. This reduction has persisted for 14 years.

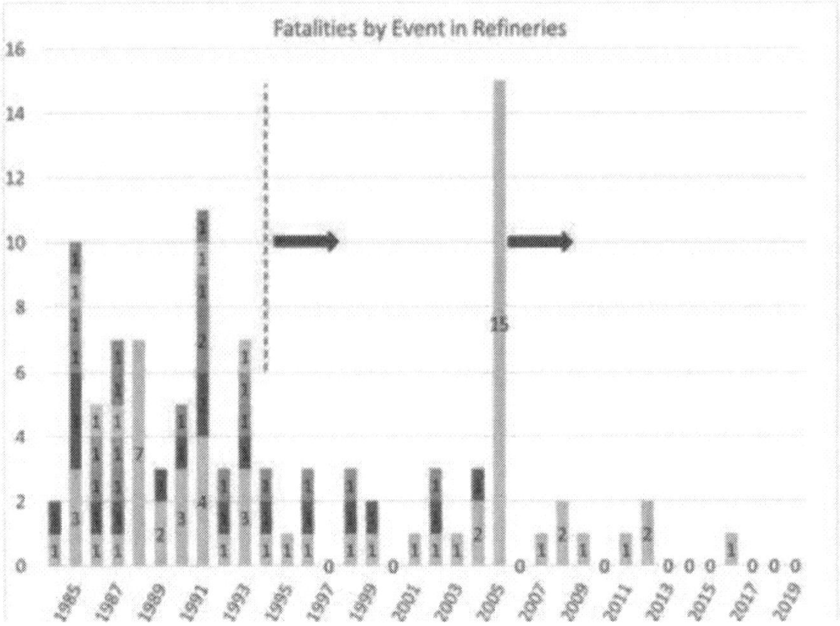

Figure 3 - Fatalities by Event in Petroleum Refining

It is clear that not only the number of fatalities was reduced around 1994 but also the number of events. Since 2002 there has been either zero or one fatal event per year (see Figure 4).

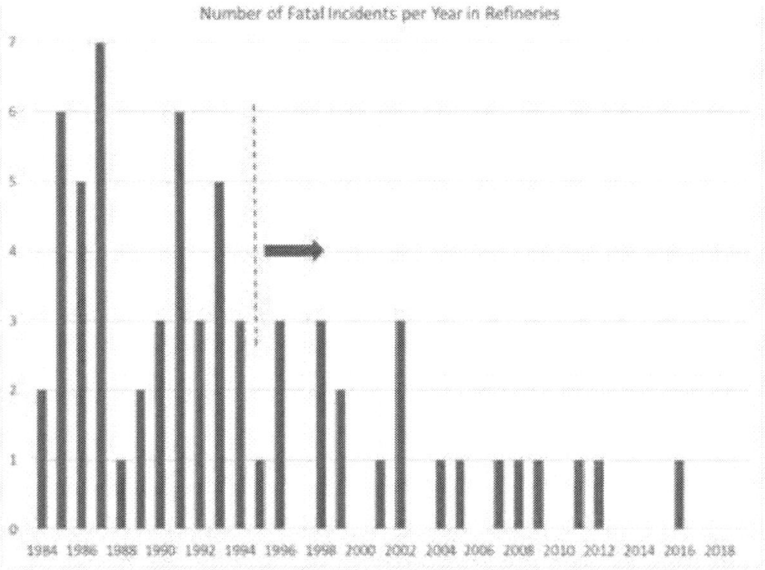

Figure 4 - Fatal Incidents in Petroleum Refining

Table 1

The fatality and fatal event annual rates are shown for CM and PR in Table 1. For CM there is a 50% reduction in fatal events per year after 1992. The fatalities per year dropped dramatically by a factor of almost 5. For PR there is a 60% reduction in fatal events in 1994 and an overall reduction of 75% after 2005. Fatalities in PR were reduced by nearly 50% in 1994 and a remarkable 10 to 1 after 2005. The relative magnitudes of the PR and CM rates are not comparable since they are for different numbers of facilities and employees.

Event and Fatality Rates for CM and PR - Region 6			
Chemical Manufacturing			
	1984-1992	1993-2020	
Fatal Events/yr	4.1	2.0	
Fatalities/yr	10.0	2.2	
Petroleum Refining			
	1984-1994	1995-2005	2006-2020
Fatal Events/yr	4.0	1.6	1.0
Fatalities/yr	6.0	3.2	0.6

Tasks Associated with Serious Injuries and Fatalities

As the descriptions of each incident were analyzed it was soon apparent that most of the incidents involved a task being performed by a small number of people and the consequences typically suffered by those workers. Both maintenance and operational activities are involved.

Table 2 lists the tasks for serious injury and fatal incidents, from both databases, in both CM and PR. The term "unrelated" is used to denote incidents that are not related to the activity of the victim. Most of these "unrelated" cases are mechanical failures, leaks or explosions and the victim is by chance in the vicinity.

Table 2

Severe Injury and Fatal Incidents by Task									
	Chemical Manufacturing				Petroleum Manufacturing				
	Severe Injury Incidents	%	Fatal Incidents	%	Severe Injury Incidents	%	Fatal Incidents	%	% Averaged for Sort
Cleaning, clearing, purging lines and vessels	35	16	16	20	10	29	11	11	19
Opening lines and vessels	47	21	8	10	3	9	15	15	14
Transferring fluids	43	20	8	10	3	9	7	7	11
TOTAL % for 3 Tasks		57		39		47		32	
Unrelated*	10	5	11	13			16	16	8
Unknown	20	9	3	4	2	6	8	8	7
Maintenance	18	8	3	4	5	15	4	4	8
Operating	6	3	6	7	3	9	4	4	6
Starting equipment	6	3	5	6	2	6	5	5	5
Confined Space Entry			7	9			8	8	4
Mixing chemicals	8	4	3	4			7	7	4
Troubleshooting	7	3			2	6	4	4	3
Fall or step into hazard	3	1			2	6	1	1	2
Other	5	2	4	5			1	1	2
Sampling	3	1	3	4	1	3			2
Handling explosives			3	4					1
Lighting heater							5	5	1
Transferring solid	7	3							1
Reactor startup			2	2					1
Starting process					1	3	2	2	1
Hot Work							3	3	1
Testing	2	1					1	1	0
Inspecting							1	1	0
Total Incidents	220	100	82	100	34	100	103	100	100

*Unrelated indicates the employee was not involved in a task telated to the event.

Tasks Associated with Serious Injury and Fatal Incidents in Chemical Manufacturing.

In both CM and PR, there are 3 activities that account for a large fraction of the fatalities. These tasks are:

- Cleaning, clearing, or purging lines and vessels
- Opening lines and vessels
- Transferring fluids and solids

These 3 types of tasks usually involve the intentional interruption of a pressure containing equipment component. That is of course the case with opening lines and vessels. For cleaning, clearing, or purging, it is necessary to open the pressure containment before proceeding. Transferring fluids often involves making or breaking connections as well as venting and draining lines and vessels

Cleaning, clearing, and purging by their nature involve removal of a substance, usually from the interior of the containment. Difficulties are encountered in numerous ways. These include:

- Substance is not what was expected
- Quantity of substance is greater than expected
- Unexpected pressure behind a blockage
- Blockages are propelled out of the containment by pressure
- Tools, covers and hatches are propelled by unexpected pressure
- Containment is over-pressurized when clearing an obstruction

Opening lines and vessels often involve problems similar to those above. These include:

- Unexpected pressure
- Unexpected substance

  Unexpected quantity

Transferring fluids involve many different situations based on the source of the fluid, receiving container, and type of transfer line or hose.

- Hoses often disconnect unexpectedly
- Hoses are subject to failure
- Disconnected hoses may contain an unexpected substance
- Hard pipes often contain vents and drain lines that are open
- Transfers may result in mixing incompatible substances
- Rail cars and tank trucks may have complex sets of valves for venting, draining, multiple compartments and other confusing situations
- Totes, drums, buckets, and other containers can be dropped or damaged, sometimes by forklifts
- Static electricity may be an issue in many of these operations

Nature of the Serious Injury Incidents in CM

The OSHA database, SIR, contains four factors that describe the incident and the resulting injuries. The four factors are the following:

- Type of Event
- Hazardous Substance
- Nature of Injury
- Affected Body Part

The results for all tasks in CM are shown in four tables. These four factors are the results of the incident and infer a limited amount about the causes. Hopefully information about the most common hazards will assist in focusing efforts for improved safety. Data for all incidents are included, not just those for the 3 most common tasks.

Table 3

Chemical Manufacturing		
Serious Injury - All Tasks		
Event Type	#	%
Exposure through intact skin, eyes, or other exposed tissue	82	37
Contact with hot objects or substances	74	33
Inhalation of harmful substance-single episode	21	9
Ignition of vapors, gases, or liquids	14	6
Struck by dislodged flying object, particle	9	4
Explosion of pressure vessel, piping, or tire	8	4
Explosion, n.e.c.	6	3
Other	9	4
TOTAL	223	100

Event Type

Two types of events account for 70% of all events in roughly equal frequency. These two are exposure of external tissue to a hazardous substance and contact with hot objects and surfaces. These are such broad categories that it does little to identify a particular type of event which deserves more attention than others. Two other event types also occur in a significant portion of incidents. These are inhalation of a substance and ignition of a substance. In planning for and executing a task, it is important to consider if any of these events could occur. This can lead to identifying procedures to avoid the event and to utilize proper PPE and emergency procedures.

Table 4

Hazardous Substance

Perhaps surprisingly, water is the most frequent hazardous substance involved in serious injuries. Of course, this is due in part to its widespread presence in many processes. Often the water is condensate and results in burns. Nearly as common are acids followed by resins. Resins are often associated with processes such as extrusion were contact burns are common. While the table shows a number of common substances there are an additional 40 substances that were only involved in one incident. In addition, there are 30 objects. In OSHA's classification system, objects, rather than a hazardous substance, are identified as the source of injury. These are usually cases where an object propelled by pressure or an explosion strikes the victim. These objects are typically shrapnel, pipe plugs, covers and hatches or hoses.

Chemical Manufacturing		
Serious Injuries - All tasks		
Substances	**#**	**%**
Water	**27**	**12**
Unknown, not classified	27	12
Acids	**25**	**11**
Resins	**17**	**7**
Multiple	12	5
Chlorine, chlorine bleach, compounds	**9**	**4**
Steam and vapors	**9**	**4**
Sodium and potassium hydroxide, potassium carbonate	**8**	**4**
Ammonia, anhydrous ammonia	**6**	**3**
Chemicals, n.e.c.	6	3
Phenol	**4**	**2**
Liquids, unspecified	4	2
Aromatics and hydrocarbon derivatives, except halogenated, n.e.c.	**3**	**1**
Other Substances	40	18
Physical Objects*	30	13
TOTAL	**227**	**100**
* OSHA lists objects such as those propelled by pressure		

Table 5

Nature of Injury

The nature of injury generally reflects the type of event shown in the previous table. Thermal burns are naturally associated with contact with hot objects and substances and chemical burns with exposure of tissue. Traumatic injuries result from being struck by object and explosions. Poisoning, toxic, noxious, or allergenic reactions are typically due to inhalation. While there is little information in these databases about PPE it appears that it is often inadequate for the actual circumstances.

Chemical Maufacturing		
Serious Injuries - All Tasks		
Nature of Injury	**#**	**%**
Thermal Burns	**93**	**40**
Chemical Burns	**79**	**34**
Traumatic	**26**	**11**
Poisoning, toxic, noxious, or allergenic effect	**26**	**11**
Other	**9**	**4**
TOTAL	**233**	**100**

Table 6

Affected Body Parts

One third of the incidents involve multiple body parts. This may reflect the serious nature of the studied incidents. Severe injuries may be the result of releases of larger quantities of hazardous substance, higher pressures and other factors that lead to more widespread injuries. This contrasts with less serious injuries associated with more common cases involving days away from work. In a study of these injuries (Mannan), there were typically specific body parts injured. The most often injured body part was the face. The second most affected body part for severe injuries is body systems. This means internal systems implying inhalation or ingestion of a substance. The remaining body parts are more typical parts of face, extremities, trunk, and head. The adequacy of PPE may be a factor in the severity of these injuries.

Chemical Manufacturing		
Serious Injuries - All Tasks		
Body Parts	#	%
Multiple body parts, n.e.c.	77	33
Body Systems	28	12
Nonclassifiable	28	12
Face	25	11
Lower Extremities	25	11
Upper Extremities	24	10
Trunk	15	7
Trunk & Head	8	3
TOTAL	230	100

Incidents with Detailed Investigations

The CSB and the US Department of Labor investigated several of the CM incidents in this study. These incidents were Phillips 66 1989, Honeywell 2003 (3 incidents) and DuPont 2014. These incidents generally involve one of the 3 common tasks. Phillips 66 involved the removal of a plug in the settling leg of a reactor. This led to a release of isobutane resulting in 23 fatalities and more than 100 injuries. Honeywell had 3 incidents involving 3 different substances. The first was a chlorine release due to a mechanical failure. The second was an antimony pentachloride release which resulted in a fatality. The release was the result of removing a plug from a cylinder. The third release was hydrogen fluoride during the transfer of fluid using a venturi stick, resulting in hospitalization of one worker. The DuPont incident resulted in 4 fatalities during the removal of a hydrate plug, releasing mercaptan. These three investigation reports show that larger scale incidents may be the result of the 3 common tasks.

Observations

The trends in fatalities before and after the implementation of PSM in 1992 shows a remarkable improvement in fatalities and fatal events. Following the reduction in 1992, most fatalities occur in single fatality incidents. These trends are counter to what one might conclude by observing high profile incidents in the media. This is true in part because of incidents in storage, transportation and at user facilities which are not included in this study.

The nature of these incidents seems to straddle the line between process safety and occupational safety. It may be that there is a need to better coordinate the safety efforts of what are in many companies two distinct organizations. The same may be true of the educational system. For example at Texas A&M University, occupational safety is taught in the School of Public Health while process safety is addressed in the College of Engineering.

Based on the incident description, the vast majority of fatalities appear to occur with relatively small releases affecting workers in close proximity to the release. At least 80% of the events are associated with what seem to be routine maintenance and operations tasks. They do not appear to be the result of complex technical failures such as impaired control systems.

There has been great success in limiting fatalities since the introduction of OSHA PSM. In order to further reduce fatalities, it appears that additional focus on less complex small scale releases is warranted.

The nature of these incidents suggest that improvements may be needed in training for hazard recognition, procedures, and job safety analysis.

The incidents are often caused at least in part by the dynamic nature of the tasks being performed, where the operation is changing over time. Procedures are needed that recognize and adapt to these changes.

In many cases, there are only two physical barriers to protect the workers, the pressure containment and PPE. Since many of the tasks involve opening the containment, adequate PPE is especially important. PPE requirements should be based on all possible release scenarios.

Recommendations for Industry and Academia

Process and Occupational Safety specialists should work together to understand and address small scale process incidents associated with common tasks. Some specific recommendations:

- Utilize this insight to better educate operations and maintenance personnel.
- Design process systems to minimize the need for hazardous tasks or reduce the severity of possible releases.
- Incorporate these findings in junior college process operator curriculum.
- Consider whether current procedures adequately cover the type of tasks most often associated with severe injuries and fatalities.

Further Investigation by the Authors

This study focused on CM and PR. Fatalities were addressed only for OSHA Region 6. The other 9 regions should be studied to confirm the trends and patterns and increase the confidence in the results. The severe injury data covered the 22 federal OSHA states. The scope for severe injuries and fatalities should be expanded to Oil and Gas and other industries.

The descriptions of the incidents in this study were extracted by the authors. To identify and interpret the entire dataset, machine learning techniques should be applied.

Develop systematic methods for identifying hazards for the most common tasks which were identified in this study.

Investigate the need for and characteristics of procedures that would prevent or mitigate the consequences of incidents from the common tasks.

References

https://www.osha.gov/data, This is the site for both data sources in this paper, "Fatality and Catastrophe Investigation Search" and "Severe Injury Reports".

Mannan etal, "Patterns and Trends in injuries due to chemicals based on OSHA occupational injuries and illness statistics", Journal of Hazardous Materials, 2008, p 349-356.

Chemical Safety and Hazard Investigation Board (CSB), Honeywell Incidents, Report No. 2003-13-I-LA, August 2005

Chemical Safety and Hazard Investigation Board (CSB), Toxic Chemical Release at the DuPont La Porte Chemical Facility, La Porte, Texas, Incident Date: November 15, 2014, No. 2015-01-I-TX

US Department of Labor, Phillips 66 Company's Houston Chemical Complex (HCC), Report to the President, April 1990.

618389 Incident Investigations to Enhance Safety Culture: Refinery Examples

Abstract:

The investigation of small incidents and near misses provides an opportunity for any company to change its safety culture at the grass roots level. This change is due in part to features unique to small incidents and near misses. The purpose of this paper is to outline the benefits of using small incidents and near misses to increase situational awareness and enhance safety culture. The paper will also include a refinery case study demonstrating how these small steps can make a big difference for process safety management.

Scope: The primary purpose of the investigation is to understand the underlying causes of the incident and to prevent it from happening over and over again. Events are not just human error or a result of a physical failure, each event is indicative of a fault in the system. The investigations are an important part of safety culture of company. They increase the awareness as well.

Methods, procedures, Process: The important elements of the event research are the timely research and report creation, the effective documentation of the findings and the sharing of the results. The method of the investigation is selected due to the potential risk of Incident/Near Miss/Safety Observation. In addition to the method, the assignment of team members is made by different levels of the organization depending on the level of potential risk assessment of the event.

Results, Observations, Conclusions: The current and potential evaluation of the impact of the incident to environment, human, reputation and to asset, the establishment of research teams, the appointment of root causes and actions, the automatic creation of the corrective and preventive actions or maintenance requests are made on specific investigation tab of the SAP (SAP EHSM) system. The publication of the accident research report are carried out by SAP EHSM as well.

In most of the applications Tier 1 and 2 Process Safety incidents are deeply investigated but sometimes Tier 3 and 4 incidents left behind in terms of investigation and actions completion speed. However, if you check all the Tier 3 and 4 incidents root causes comprehensively, you may see that they already call for the Tier 1 and 2 incidents. For example, you may find some corrosion reports as Tier 3 since it did not caused any Tier 1 or 2 incident. As you have investigation you will come up with some defaults due to crude oil quality or process technologies applied. Having such an investigation in the phase of Tier 3, you will be able to prevent such a big leakage. Moreover emphasizing on Tier 3 and 4 incidents investigation, gives the message that you are more proactive since you are not only dealing with the incidents only caused harm to people or asset but also potential risks. That will also help management to communicate to employees that, they are one-step ahead the incidents, which will lead employees to think potential risks in the workplace.

619133 Effective Implementation of Dhas – Trends One Year Past the NFPA Deadline

Abstract:

The deadline incorporated into NFPA 652 for existing facilities handling combustible dusts to complete a dust hazard analysis (DHA) has come and gone as of September 2020. How are companies doing with incorporating the findings of their DHAs into capital expense and operational expense budgets? Are authorities having jurisdiction (AHJs) even aware of the requirement? This presentation will review key findings from surveys of our clients and AHJs.

American Petroleum Institute
What Don't You Know about Your PSM Program?: How the Process Safety Site Assessment Program (PSSAP®) Can Help Sites Find and Address These Blind-Spots.
March 22, 2021

Often companies struggle to understand what process safety management areas need their focus. What specific topics am I neglecting? Which of my programs or procedures could use some help to be truly effective? How does my holistic program compare with my peers in the industry? The Process Safety Site Assessment Program (PSSAP®) is an industry-created, industry-led program that can help answer these questions.

The American Petroleum Institute (API) and the American Fuel and Petrochemical Manufacturers (AFPM) kicked-off the PSSAP program in 2012. Since then, API has administered over 130 of these assessments across the globe. These assessments cover the main focus areas of process safety management: Process Safety Leadership, Operating Practices, Mechanical Integrity (focused on fixed equipment), Safe Work Practices, Management of Change (MOC), Process Hazards Analysis (PHA), Facility Siting, Product Storage & Transfer, and Incident Learning. In addition to these traditional areas, the PSSAP program can also assess your Hydrofluoric Acid (HF) Alkylation/API RP-751 processes, as well as the Safeguards a site has in place to address risks, our newest offering.

But how can an assessment help my site? Won't I just receive a report? The PSSAP program is not a normal assessment. While our assessments are based on satisfying industry-created good practice protocols, an operator gets much more from our assessors than a report following their time on-site. API sends a full team of vetted, qualified assessors that are leaders in their field, with the average experience of one of our assessors being 40+ years. Our assessors spend time on site, and in addition to evaluating your programs for your blind spots, they also are constantly sharing the good-practices they have seen across the industry as part of their participation in this program. Not only does a site receive an assessment of their programs, they also can find out what the rest of the industry is doing to address these areas. Not only does a site get this through good-practice sharing from our assessors, but sites also get to see the benchmarking data from our previous assessments. With over 130 worldwide assessments, a site can statistically gauge how they compare with their peers. Sites also receive a detailed examination of the specific areas that could use more of their attention.

In addition to the different focus areas and our assessor pool, it is important to note that these assessments are scalable. Are you a small site? Is your operation too small for you to think such an assessment would be worthwhile? API created a shorter program for these sites, enabling them to access our industry-leading expert assessors so that all may benefit from their wealth of knowledge, continuing to drive refining, petrochemical and chemical operators toward our shared goal of zero process safety incidents.

We will explore our newest offerings in this forum, and will also show examples of what a site stands to gain from their participation in this industry-led and driven program.

619049 Bouncing Back from a Heart-Wrenching Process Safety Incident

Abstract:

During startup of Urea Fertilizer Plant after a brief shutdown, due to recurrence of a leakage, the field maintenance team was inspecting top head gasket of condenser on a 31-ft high platform when Carbamate solution splashed on the workers. They tried to escape via monkey ladder but fell down one after the other resulting in loss of life of one of the employees and hospitalization of other four.

Detailed incident investigation by a multi-disciplinary team highlighted number of gaps related to human elements, job handover/execution and crisis management. The incident occurred due to failure of multiple layers of protection including management of change, mechanical integrity, risk assessment, PTW, job supervision, fatigue element, over-occupancy, communication, training & development, emergency response, PPEs etc.

Interestingly, this happened in less than 3 years after successful compliance certification of Process Safety Management System but raw material unavailability and resulting financial crunch hugely jeopardized organization's process safety management system and safety culture.

The paper will cover this eye-opening case of huge PSM deterioration in very short time due to financial factors and how did the organization bounce back from that desperate situation. It will shed light on causes of incident, lessons learning and results achieved after revamp of the process safety management program. It will enable other participants to learn from story of our setbacks & revival, and implement similar cultural shift programs in their respective organizations.

619358 Reactive Metal Waste Particulate Combustibility

Abstract:

Reactive Metal Waste Particulate Combustibility

Besides the usual particulate combustibility sensitivity to composition and particle size, metal waste particulate combustibility also depends on processing, handling, and additive complications. These other complications render it almost impossible to generalize waste particulate combustibility properties, and require careful sample selection and testing to provide accurate property data. Examples of these considerations are given with regard to the testing of aluminum, manganese, titanium, and zirconium particulate samples.

Explosibility test data for aluminum swarf and grinding samples showed the expected variation in test results depending on particle size, but particulate with more than 90 weight percent greater than 420 micrometer, and less than 0.5% smaller than 75 micrometers were combustible with P_{max} values in the range 2.1 to 5.3 bar-g. Finer aluminum particulate samples were tested alone and mixed with titanium grind dust and carbon fiber composite grind and sanding dust. The aluminum-composite grind dust had the highest P_{max} and K_{st} values (over 9 bar-g and 200 bar-m/s, respectively) even though the composite itself had only moderate values of these parameters that were less than those for the metals. The composite-aluminum mix samples also had very low Minimum Ignition Energies.

Manganese is a metal whose combustibility properties are especially sensitive to the presence and thickness of a surface oxide passivating layer. Waste particulate that has been exposed to air for days and weeks often will not ignite at all in the UN layer burn rate test apparatus, and suspended dust clouds will have very low P_{max} and K_{st} values, often under 5 bar-g and 35 bar-m/s, respectively. On the other hand, manganese particulate that has been transported and processed in a near-zero oxygen atmosphere have very low MIE values, and are easily ignited by friction and weak electrostatic discharges soon after exposure to air. Therefore, manganese particulate waste-handling precautions need to be based on detailed knowledge of the particulate processing conditions and site-specific combustibility test data for the applicable particulate waste.

Large particulate titanium swarf typically has moderate K_{st} values (under 100 bar-m/s). High moisture contents can further reduce the measured P_{max} and K_{st} values, but the threshold moisture content to render samples noncombustible are higher than previously reported. Safe handling of used titanium powder is a major concern in additive powder based fusion manufacturing equipment. Titanium filter particulate from AM inerted equipment is often pyrophoric and has been responsible for several fire and explosion incidents. The observed flame and ejected sparks from tested AM dust samples demonstrate the enhanced particle explosibility hazard compared to most combustible organic dusts.

Zirconium swarf particulate is even more sensitive to handling conditions than most other reactive metal particulate. Data for large particulate samples exposed to air for several weeks indicate marginally combustible dust cloud test results. UN burn rate swarf layer test results range from non-ignitable to readily and violently ignitable (with a single weak spark) depending on the moisture level and presence of residual cutting oil and possible rust contamination from swarf drums.

617989 Dangerous When Wet: Exploring Important Differences in Common Water Reactivity Testing Methodologies

Abstract:

It is important to accurately characterize a substance or mixture for water reactivity during processing, storage and transportation. There are several methods available to evaluate this behavior. This paper will provide an overview of the various methods including the United Nations (UN) Recommendations on the Transport of Dangerous Goods, Model Regulations, Rev. 21 (2019), U.S. Department of Transportation (DOT) Code of Federal Regulations: 49 CFR 173.124, National Fire Protection Association (NFPA), and International Fire/Building Code (IFC/IBC). Each method has an application associated with the test design and measured parameters. The test methods will be compared and contrasted to identify their applicability and limitations. Examples of test results from specific methods will be reviewed to identify key learnings that could lead to incomplete or incorrect classification of substances. The objective is to understand the limitations of the test methods and explore additional characterization to understand the behavior of the material as it relates to the usage need of process, storage or transportation.

The following learning objectives will be achieved:

1. Introduce test methods to characterize water reactivity
2. Understand applicability and limitations of specific test methods.
3. Identify additional test methods to characterize water reactivity for safe processing, storage and transportation of materials.

617877 Novel Approach to Developing Hazardous Area Classification Boundaries

Abstract:

This paper addresses the need to refine the methodology to develop documented reliable and cost effective HAC contours. The paper will provide an alternate approach for developing HAC boundaries at a fraction of the cost of the previous methodologies described. A cost and time estimate comparison will be provided for the different approaches, as well as an estimate on the capital cost that can be saved for operators who choose a different approach to HAC.

Performing a hazardous area classification (HAC) analysis, also known as electrical area classification, has long been a crucial part of hazard prevention practices across facilities that handle flammable/ combustible material. These terms represent the same work process, but are often are confusing depending on the industry. Various organizations such as NFPA and API in the United States and ATEX and IEC in Europe and Asia provide guidance on how a HAC analysis should be performed and the basic distances to define the HAC boundaries. In the United States, OSHA mandates a documented HAC analysis be performed to classify locations depending on the properties of the flammable gas, flammable liquid–produced vapor, combustible liquid–produced vapors (aka flammable/combustible material) that could be present, and the likelihood that a flammable or combustible concentration or quantity is present at these facilities.

Currently there are various approaches to define HAC boundaries based on the guidance provided in documents such as NFPA 497 and API RP 500. The common approach taken by many includes simply labelling an entire process block with a conservative classification rating based on the process conditions and materials handled in the block. This quick and broad-brush approach to defining the hazardous area boundaries and classifications often results in an overly conservative boundary and classification. As a result, operators may then purchase and install more costly electrical/electronic equipment, which can add to maintenance costs over the life of the facility and possible hazards for maintenance personnel.

A more recent approach to defining HAC boundaries and classifications that could remove some of the conservatism of the previous method described above is the point-source method. To develop a hazardous area classification around the process equipment, this method reviews:

- the process streams to identify the flammable/combustible material(s), and their process pressures and flows;
- selected process source equipment that handles the flammable/combustible materials, and the
- use of expert judgment and the guidance provided in recommended practice documents.

Point-source analysis leads to a more accurate HAC boundary that is shaped by the specific process equipment and equipment layout rather than encompassing an entire process block. Considering all equipment that handles a flammable/combustible material is a costly endeavor and in some cases such analysis is not needed to produce a reliable HAC contour to reflect appropriate potential ignition mitigations and controls addressed by Recognized and Generally Accepted Good Engineering Practices (RAGAGEP).

Mobile Computing for Procedure Execution Facilitation to Reduce Risk

While large sums of many have, and will continue to be, spent on engineered safeguard equipment to protect chemical process facilities from catastrophic accidents, the accidents still continue to happen. Furthermore, many of the accidents that occur on a regular basis are simple human errors such as moving the wrong valve or working on the wrong piece of equipment. In order to reduce risk, tools that can assist plan operators and maintenance technicians to perform the correct actions on the correct pieces of equipment are critical. Currently available technology such as internet connected mobile phones with scanning devices can provide this critical and potentially lifesaving assistance. When procedures are executed with the assistance of software tools that connect the operators, maintenance technicians, shift supervisors, PSM professionals, and management using a single dashboard that allows everyone to know everything that is going on, risk can be reduced, and efficiency and profitability can be increased. This paper will describe a process for using mobile computing tools to increase safety by allowing the persons that are executing procedures to get information about each procedure step including location of the action (with GPS mapping), videos of instruction on how to perform the step, and scanning of the equipment via QR code to confirm correct locations. Furthermore, supervisors can also track the location and status of the multiple procedures that are being executed during complex operational modes such as startup and shutdown.

617862 Control Room Design – Could We Have Anticipated a Pandemic?

Abstract:

Unusual times raise awareness of risks to our operations that are often ignored, or at best, considered of lesser importance. The current pandemic is a significant example of this, and we are seeing companies taking unusual steps to ensure the safety of their employees while struggling with the need to keep plants running. To meet these goals in a timely manner, we have seen some inventive and unique solutions and even the questioning of standard practices, such as centralized control rooms. However, an important question should be 'could we have been ready for this?'. An obvious focus is on the control room, although this is only part of the solution. This paper discusses a best practice approach to the design of new, or redesign of existing facilities. If followed, we believe this process can effectively identify, consider and mitigate plausible risks at all stages in the design process and address not only the physical aspects of the design, but also supporting work practices.

619058 Human Factors for Process Safety: A Handbook By the Centre for Chemical Process Safety and the Energy Institute

Abstract:

Human performance is a factor in almost every major process incident. The costs of major process incidents are well known. Death. Catastrophic destruction of facilities. Immense monitory costs. Reputational loss. Closure.

An understanding of human errors and mistakes makes it possible to identify how to reduce the possibility that they occur. Human Factors is more than common sense. There are many reasons why people may make mistakes. There are many factors that influence how people perform. A Human Factors view states that it is vital to ask why and how errors and mistakes might occur. This includes asking, for example:

- How an individual's performance is influenced by the conditions they work in;
- Whether the information and equipment they have been given are suitable and sufficient;
- Whether the training they have been given is sufficient; and
- How an individual's performance is influenced by the prevailing culture.

Understanding human performance is also critical to support people to successfully complete tasks. Human Factors helps to achieve high performing teams. It helps to achieve reliable teams who can be trusted to perform safety critical tasks, to be diligent, to learn, and to improve.

Process operations can be complex and involve many difficult tasks. In a process environment, with many complex tasks and safety critical operations, Human Factors methods need to be implemented systematically. The highest standard of human performance is necessary.

The Centre for Chemical Process Safety and the Energy Institute have produced (written by a Human Factors consultancy, Greenstreet Berman Ltd) a comprehensive but simple to understand handbook for people responsible for the 'sharp-end' of process plant operations and maintenance. The handbook provides practical advice, examples of good practice as well as principles, and tools. It covers how to support human performance through procedures and job aids, training and learning, effective task planning, high reliability communications, fatigue risk management, development of error management skills, and preparing people to perform emergency response tasks.

The handbook is intended for people who understand process operations, but have limited or no Human Factors knowledge. This includes Frontline supervisors, operations and maintenance managers, plant superintendents, process engineers and process safety and health and safety personnel.

The handbook expressed the view that errors and mistakes happen because of a combination of problems in the working environment and due to the support, or lack of it, offered by the organization. Directors, managers and supervisors should create the right environment for successful task performance. This is sometimes called the "systems" approach to Human Factors. This means that its vital to ensure that, for example, the system of training, the system of supervision, the rota system, the system of communication and so on, are all designed to positively support successful task performance.

In order to support human performance and reduce the potential for error and mistakes, it is important to understand the nature of the tasks, the type of human performance required, and the causes of possible error and mistakes. Also, people have a range of capabilities. A general understanding of human performance can be used when designing training programme, setting staffing levels, selecting a type of job aid and many other Human Factors matters. By understanding people's capabilities, ways of thinking and limitations, it is possible to design work to maximize human performance and minimize the potential for errors and mistakes. It may also be possible to train people to be more aware of how mistakes might be made. This would help people to recognize the risks or early warning signs of a mistake and give people the time and ability to prevent these mistakes from happening.

This paper will summarize how the handbook draws together latest good practice on key Human Factors topics. Each chapter can be read by itself. Altogether they provide a comprehensive aid to optimising the human factor of chemical process operations.

After providing overarching Human Factors concepts and knowledge, the first chapters cover core topics of job aids, equipment design, competence, fatigue management, staffing levels and task support.

It then moves onto error management in task planning, preparation, and control; capturing, challenging, and correcting error, communicating information and instructions. For example, it includes the role that "error assessment" can play in foreseeing and preventing human error, by engaging in effective task planning and preparation. Many tasks are complex, with many task steps, creating the potential for errors and mistakes. Motivated staff can be very "task focused" and intent on completing the task and solving the problems. This can create a risk of losing awareness of the situation, improvising unsafe ways of completing a task, and overlooking unexpected events or conditions that require a change in their actions. When people are task focused, they can miss "weak signals" around them that the situation is unsafe or is changing.

Recognising Error-likely situations can help to spot, capture and correct errors before and during operations. The handbook notes tactics for minimizing distraction and interruptions, such as "heads down" announcements and cordoning off activities. An open team culture that invites, values, and accepts challenge is also vital. When challenges are made, they should be explored and understood. People need to be able to trust that their challenges will be welcomed by their colleagues.

It then moves onto other essential non-technical skills, including situational awareness and agile thinking. Human perception and memory possess limitations. A strong focus on one part of the situation, or element of the environment, can lead to other sources of information being neglected or missed. High workload, stress, and extensive task and time focus can impair their cognition. For any team to work effectively they must have "shared situational awareness" – that is, a common understanding of the event.

Non-technical skills training can increase awareness and knowledge of situational awareness and agile thinking, and allow people to practice using these skills. This is sometimes known as Crew Resource Management. For example, how to gather and interpret information, how to foresee future implications of current events, and how to recognize a mismatch between personal situational awareness and situational awareness of others, and act accordingly. Tools and job aids such as Dynamic Risk Assessment and Stop-Think-Act-Review (STAR) assessment can help people to pause and maintain awareness of the situation.

The next chapters cover the human factors of emergency response. This again focuses on non-technical skills, such as stress management, psychological resilience and challenge behaviours, as well as ensuring people have the technical knowledge and job aids such as decision trees to understand what is happening in an emergency situation and to devise appropriate actions.

Cutting across all of these, the handbook concludes with the use indicators to recognize human performance issues, and how to learn from error and performance. Every chapter of this handbook applies equally to contractors and subcontractors. Finally, a chapter covers the Human Factors of operational level change, whether this is new plant, changed procedures or organisational change. It advises on how to recognise the impact on people, assess and manage these impacts.

The scope of the handbook and the guidance, examples and tools has been specified by representatives of chemical process companies, ensuring that it is directly applicable and practical. It focuses on existing process operations. The handbook is in press and is expected to be published in late 2021.

SPRING21 +17TH GCPS
A Joint AIChE and CCPS Meeting

Too Many or Not Enough? A Methodology for Evaluating Fire and Gas Detector Layouts at LNG Facilities

Bryant Hendrickson, PE, CFEI
bhendrickson@blueeandc.com

Filippo Gavelli, Ph.D., PE
Jake Piekarz, PE

Blue Engineering and Consulting
6011 University Blvd, Suite 220
Ellicott City, MD 21043

Prepared for Presentation at
American Institute of Chemical Engineers
2021 Spring Meeting and 17th Global Congress on Process Safety
Virtual
April 18 - 22, 2021

Too Many or Not Enough? A Methodology for Evaluating Fire and Gas Detector Layouts at LNG Facilities

Bryant Hendrickson, PE, CFEI
bhendrickson@blueeandc.com

Filippo Gavelli, Ph.D., PE
Jake Piekarz, PE

Blue Engineering and Consulting
6011 University Blvd, Suite 220
Ellicott City, MD 21043

Keywords: detector mapping, detector coverage, hazard analysis, siting study, vapor dispersion

Abstract

NFPA 59A *Standard for the Production, Storage, and Handling of Liquefied Natural Gas (LNG)* requires facilities to have the equipment necessary for the detection and control of fires, leaks and spills of hazardous materials, yet provides no requirements or guidance on the location of the detectors. Literature available in the public domain outlines key factors to consider in developing layouts for flame and gas detectors, but a methodology to use this information in evaluating a hazard detector layout is not clearly defined; instead, generic and non-quantifiable terminology such as "quick and reliable" is often used. As a result, there is no consistent approach to developing flame and gas detector layouts for land-based LNG facilities and there is no systematic method for regulators to evaluate these designs. The ability to quantify hazard detector performance and define performance targets are of particular interest to the Department of Transportation (DOT) Pipeline and Hazardous Materials Safety Administration (PHMSA), which regulates numerous LNG facilities in the United States.

Blue Engineering and Consulting is working on a DOT-PHMSA sponsored project to develop a risk-based approach and criteria for hazard detector layouts at LNG facilities. This project builds upon performance-based design principles outlined in NFPA 72 *National Fire Alarm and Signaling Code* and the International Society of Automation (ISA) technical report 84.00.07 *Guidance on the Evaluation of Fire, Combustible Gas, and Toxic Gas System Effectiveness*. The methodology divides the LNG facility into Detection Areas based on the hazards present and the plant layout, identifies the appropriate hazard scenarios to evaluate the detector layout, establishes performance targets for each Detection Area, and quantifies hazard detector coverage. This paper will describe and demonstrate the proposed methodology. It should be noted that, while performance targets must be chosen for the purposes of the demonstrative included in this paper, these targets should not be interpreted as requirements of DOT-PHMSA nor as acceptable to them.

1 Introduction

Fire and gas detection systems are required to be installed at LNG facilities by NFPA 59A *Standard for the Production, Storage, and Handling of Liquefied Natural Gas (LNG)*, however, no requirements or guidance for the location of hazard detection devices is provided [1]. Literature available in the public domain outlines key factors to consider in developing layouts for flame detectors (e.g. size of fire to be detected, fuel involved, sensitivity and field of view of detector, including obstructions, etc.) and gas detectors (e.g. pressure, temperature, density of the released material, topography, air movement and temperature effects, location of potential ignition sources, etc.), but a methodology to use this information in developing a detector layout is not clearly defined; instead, generic and non-quantifiable terminology such as "quick and reliable" is often used. Further, available guidance repeatedly calls for gas detectors to be placed in proximity to potential leak sources, which is often interpreted as placing point gas detectors right next to likely leak locations. This attempt at detecting the leak instead of the resultant gas cloud is known to be flawed and could result in a failure to detect the release in many cases, depending on factors such as release orientation and wind direction. It is also difficult to detect pressurized releases close to the source, where the cross-sectional area of the jet is quite small.

The lack of a consistent approach to developing hazard detector layouts for land-based LNG facilities and of a systematic method for regulators to evaluate these designs are of particular interest to the Department of Transportation (DOT) Pipeline and Hazardous Materials Safety Administration (PHMSA), which regulates numerous LNG facilities in the United States. PHMSA permits applicants to use a 10-minute design spill duration in their hazard analyses if the process design includes acceptable detection, isolation, and shutdown, however, applicants are also permitted to evaluate a release duration shorter than 10 minutes based on demonstrable surveillance, shutdown and isolation design. Since there is currently no standard for the design of hazard detection systems at LNG facilities, it follows that there is no consistent methodology for demonstrating or evaluating successful detection, isolation, and shutdown provisions.

Blue Engineering and Consulting is working on a DOT-PHMSA sponsored project to develop a risk-based approach and criteria for hazard detector layouts at LNG facilities. This project builds upon performance-based design principles outlined in NFPA 72 *National Fire Alarm and Signaling Code* [2] and the International Society of Automation (ISA) technical report 84.00.07 *Guidance on the Evaluation of Fire, Combustible Gas, and Toxic Gas System Effectiveness* [3]. The ISA guidance describes how the total effectiveness of the fire and gas detection system is based on three factors: safety availability, mitigation effectiveness and detector coverage. While consideration should be given to all factors in LNG facility design, the current analysis is focused on detector coverage.

The proposed methodology closely follows the performance-based design process outlined in ISA TR84.00.07 and is summarized below in Table 1-1. The methodology divides the LNG facility into Detection Areas based on the hazards present and the plant layout, identifies the appropriate hazard scenarios to evaluate the detector layout, establishes performance targets for each Detection Area, and quantifies hazard detector coverage. It should be noted that, while performance targets must be chosen for the purposes of the demonstrative included in this paper, these targets should not be interpreted as requirements of DOT-PHMSA nor as acceptable to them.

Table 1-1: Fire and Gas Detector Evaluation Methodology

Step	Title	Description
1	Fire and Gas Detection Philosophy	Set detector coverage targets and voting criteria
2	Designate Detection Areas	Group areas of the facility that contain similar hazards and perform Detection Area Evaluations
3	Identify Hazard Scenarios	Perform a fire and explosion risk assessment and determine a reduced set of scenarios for detector evaluation
4	Assign Performance Targets	Compare the results of steps 2 and 3 to criteria from step 1
5	Flame Detector Mapping	Evaluate flame detectors according to performance targets
6	Gas Detector Mapping	Evaluate gas detectors according to performance targets

2 Fire and Gas Detection Philosophy

The Fire and Gas Detection Philosophy should be developed prior to assessing the facility risk or performing the hazard detector layout so that objective and unbiased performance criteria are clearly established, and all forthcoming decision-making is based on the risk tolerability of the project stakeholders.

NFPA 59A requires hazard detection systems to be designed, installed and maintained in accordance with NFPA 72. NFPA 72 offers prescriptive methods for determining the location of initiating devices such as heat-sensing and smoke-sensing fire detectors for indoor applications, however, it does not offer prescriptive methods for locating gas and flame detectors outdoors. Such detectors need to be located using a performance-based design approach, which requires the designer to document each performance objective and applicable scenarios, along with calculations, modeling or other technical substantiation that justifies adequate protection. Therefore, flame and gas detector layouts developed for LNG facilities need to meet these requirements for performance-based design analysis and documentation.

2.1 Detector Coverage Approaches

Successful detection is based on two leading parameters:

1. Detector coverage criteria – based on risk presented and how critical detection is.
2. Detector voting criteria – based on what actions need to be taken upon detection.

Two approaches for achieving effective detector coverage are defined in the ISA technical report 84.00.07: geographic coverage and scenario coverage.

Geographic coverage is defined as the fraction of an area or volume that is covered by the detectors. Calculating geographic coverage is a straight-forward process with flame detection, where each detector has a well-defined field of view. With respect to gas detection, this method requires first establishing a dimensioning cloud size, which represents the threshold for unacceptable damage. As the dimensioning cloud is moved around the detection area, the detector layout must be such that at least the minimum required number of detectors is always within the dimensioning cloud;

the most efficient way to comply with this requirement is with an evenly spaced grid of point gas detectors.

Scenario coverage is defined as the ratio of detected releases to total credible releases. Unlike geographic coverage, this method requires the evaluation of each credible release scenario. The scenario coverage approach provides the opportunity for the hazard detection layout to be optimized for the specific scenarios, weather conditions and configuration of a facility, however, it requires a much more detailed analysis.

The detector coverage targets in this study will be assigned based on the three risk classifications below in Table 2-1, which are adapted from the ISA TR 84.00.07 to LNG facilities designed and built to current PHMSA regulations. This ranking allows for hazard detection resources to be focused on the scenarios that present the greatest risk, without placing overly burdensome requirements on low-risk scenarios. The criteria will be applied separately to flame detection and flammable gas detection, as outlined in the following subsections.

Table 2-1: Detector Coverage Basis

Risk Classification	Minimum Detector Coverage
High	90%
Normal	80%
Low	60%

While consideration must also be given to sensor reliability and other factors, detector voting is largely based on what actions need to be taken upon detection. Typically, one detector activation (1ooN) will initiate operator notification and two detectors are required to activate (2ooN) to confirm the hazard before automated action is taken. Therefore, when evaluating detector coverage, a scenario that presents a high risk and/or warrants immediate emergency shutdown should require coverage by 2ooN detectors, whereas coverage by 1ooN (which requires less detectors to satisfy) is sufficient for a scenario that does not warrant immediate emergency shutdown (i.e. the delay and reduced reliability associated with operator intervention is acceptable). Note that the voting criteria used to evaluate the detector layout is a minimum threshold and does not limit the functionality of the fire and gas detection system. For example, the minimum detection coverage target may be specified at 1ooN, however, 2ooN detector activation can still initiate emergency shutdown.

2.2 Flame Detection

Flame detectors are used at LNG facilities to detect hydrocarbon flames, such as pool fires and jet fires, as well as major equipment fires. Flame detectors generally have a field of view that extends up to 45 degrees off the center axis, resulting in what is referred to as a "cone of vision". The listed maximum range of a flame detector is on-axis, fuel specific, and determined for an unobstructed source in accordance with FM 3260 *Radiant Energy-Sensing Fire Detectors for Automatic Fire Alarm Signaling* [4]. Multi-Spectral Infrared (MSIR) flame detectors have the highest immunity to false stimuli and are listed for use up to 125 ft (38 m) for a 32-in methane or propane plume. MSIR flame detectors have different sensitivity settings which adjust the maximum detection range, in order to avoid nuisance sources of radiation such as a flare.

Given the well-defined field-of-view of flame detectors, evaluating a layout by geographic coverage does not require hazard modeling, and a more detailed scenario-based approach is not necessary. Flame detector layouts are often developed on a 2D plot plan of the facility showing the cone of vision for each detector based on the sensitivity setting. However, flame detectors are based on line-of-sight, therefore the effective coverage area is impacted by obstructions that may cause blind spots. To detail the impact of obstructions, flame detector layouts can be performed using software capable of 3D ray tracing, which visualizes the field of view in the presence of equipment and other obstructions.

Flame detection is often utilized to prevent cascading damage from thermal effects by initiating inventory isolation and/or active mitigation measures. NFPA 59A requires the following equipment to be considered for cascading damage hazards: LNG storage containers, LNG marine carriers, refrigerant storage vessels, buildings, or equipment required for the safe shutdown and control of the hazard. Such cases should require a high detector coverage target and 2ooN voting for executive action. Conversely, other fire hazards may include delayed consequences where operator intervention (initiated by 1ooN voting) is sufficient, such as jet fire impacts on structural steel, and performance targets should be reduced accordingly. In any case, a site-specific analysis is necessary to determine acceptable exposure intensity and duration. The flame detector coverage criteria used in this analysis are summarized in Table 2-2.

Table 2-2: Flame Detector Coverage Criteria

Risk Classification	Geographic Area Description	Minimum Detector Coverage
High	Equipment coverage where cascading damage hazard due to fire is identified	90% by 2ooN
Normal	Detection Areas with flammable fluid service	80% by 1ooN
Low	Detection Areas with combustible fluid service	60% by 1ooN

2.3 Flammable Gas Detection

The most common combustible gas detector at LNG facilities in the US is the infrared (IR) point gas detector. This detector measures infrared absorption at two wavelengths: one calibrated to the target gas and one reference wavelength. Point IR combustible gas detectors measure the gas concentration at a discrete location with a reading in %-LFL (lower flammability limit). A point IR gas detector activates in a matter of seconds, however, it cannot "see" the gas at a distance; instead, it requires the flammable gas cloud to overlap its specific location. This often results in the need for an array of point gas detectors to monitor a specific region of the facility.

Open path gas detectors are the long-distance version of the point IR gas detectors. With the ability to separate the IR source and receiver by distances greater than 100 meters, a single open path detector can cover a much larger area, but only along a single straight line (e.g., at a single elevation). The IR source and receiver must maintain line-of-sight, therefore they can be difficult to use in some congested areas. Open path gas detectors measure the integral of the gas concentration along the path and output the result in LFL-m (equivalent to a meter-long gas cloud at LFL concentration). These detectors are often used for perimeter monitoring to ensure flammable gas does not extend from one area of a plant to another.

As discussed previously, a geographic approach to gas detection requires establishing a dimensioning cloud size, which represents the onset of unacceptable damage. This approach is based on overpressure hazards and is often implemented by use of spheres ranging from 5-10 meters (16 – 33 feet) in diameter. This concept does not translate well to the onshore LNG industry for several reasons, including but not limited to:

1. Vapor clouds from pressurized releases do not grow as a sphere radiating from the source.
2. Isolatable inventories at LNG facilities can be quite large, therefore, the cloud size will continue to grow after detection (during isolation and de-inventory).
3. The large size of onshore LNG facilities would result in an impractical number of gas detectors using a uniform grid layout.
4. Many areas of LNG facilities contain low levels of congestion, and many facilities use a nitrogen-based liquefaction system and only contain low reactivity fuels, both of which lead to reduced overpressure hazards.

Therefore, a site-specific analysis using a scenario-based approach is better suited to evaluating gas detector layouts at onshore LNG facilities.

LNG facilities under PHMSA jurisdiction are already designed to limit the likelihood of an accidental release and prevent offsite impacts. However, if a project relies on hazard detection to satisfy PHMSA siting requirements, unsuccessful detection could lead to severe consequences. Therefore, a minimum detector coverage of 90% is assigned to releases where the project took credit for a reduced release duration based on detection and isolation[1]. Conversely, releases in open areas with well-controlled ignition sources or low-reactivity vapor releases pose comparatively smaller risk to the facility; given the difficulty to detect such releases, the same detection target would require a disproportionate share of detection resources. Therefore, these releases are considered low risk and assigned a detection target of 60%. For the purposes of this demonstrative, 2ooN voting is assigned uniformly to result in automatic emergency shutdown for all loss of containment scenarios.

Table 2-3: Flammable Gas Detector Coverage Criteria

Risk Classification	Scenario Description	Minimum Detector Coverage
High	Releases where duration less than 10 minutes is assumed based on detection	90% by 2ooN
Normal	All remaining scenarios	80% by 2ooN
Low	Releases in open areas with well-controlled ignition sources and all low reactivity vapor releases	60% by 2ooN

[1] Total release duration must consider the time for the gas to reach the detector, detector activation time, signaling time, valve closure time, and depressurization/de-inventory of the isolated section.

3 Designate Detection Areas

A generic LNG facility developed for PHMSA research projects is used in this demonstration exercise (Figure 3-1) which includes two liquefaction units, each with a dedicated pretreatment area. For brevity, this paper will focus on Refrigerant Storage for flame detection and the Liquefaction Units for flammable gas detection, however, a full Detection Area Evaluation will be completed for both areas in the following subsections to detail the hazards present and inform detector performance targets. An assessment of the fuel reactivity and typical level of congestion is used to determine if VCE hazards should be considered for a given Detection Area; the assessment is based on the Baker-Strehlow-Tang (BST) method for predicting blast loads, which is widely used to evaluate VCE hazards in the petrochemical industry.

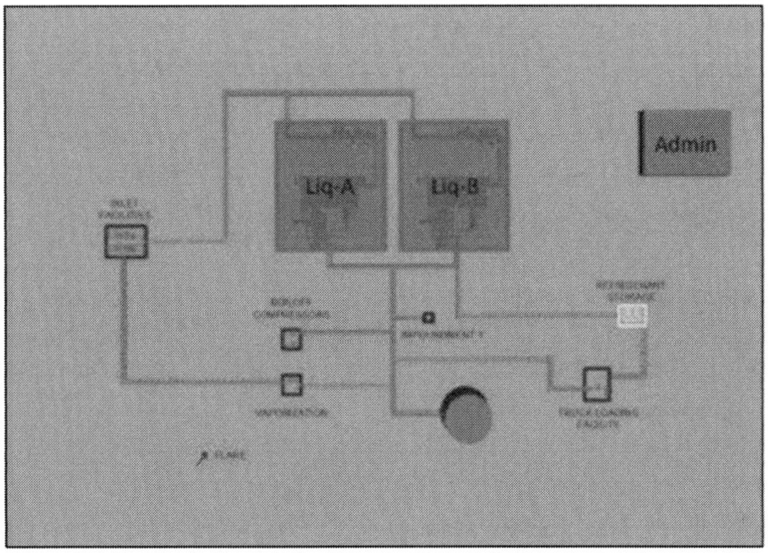

Figure 3-1: Detection Areas for the generic LNG facility used in this study.

3.1 *Refrigerant Storage Evaluation*

Facilities utilizing flammable refrigerants for liquefaction typically have a Refrigerant Storage area remote from the Liquefaction area to limit hazardous exposures. The Refrigerant Storage area evaluation is provided below in Table 3-1.

Table 3-1: Refrigerant Storage Evaluation

Flammable Streams	Refrigerants (e.g., Ethylene, Propane, Butane)
Potential Leak Scenarios	Flashing and jetting releases from liquid inventory piping. Liquid spill from large-bore piping ruptures flowing into the facility trenches and impoundment.
Flash Fire	Yes
VCE	Yes (medium/high reactivity fuels; low/medium congestion)
Jet Fire	Yes
Pool Fire	Yes
Cascading Damage	BLEVE of pressurized storage vessels

3.2 *Liquefaction Unit A/B Evaluation*

There are several liquefaction technologies available in the LNG industry, mainly depending on the desired production rate of the facility. Many small-scale LNG facilities use a closed loop nitrogen refrigerant liquefaction system; though a nitrogen release presents an asphyxiation hazard in the vicinity of equipment, this hazard is often bounded by flammable gas hazards from the natural gas and LNG lines. Facilities needing a larger production rate often utilize a mixed-refrigerant liquefaction process.

This demonstrative considers an LNG facility with a mixed-refrigerant liquefaction process, which introduces hazards from additional flammables such as ethylene, propane, and butane. These refrigerants are more reactive than methane and present increased VCE hazard potential. Though ethylene is considered a 'high' fuel reactivity, it usually makes up less than half of the mixed refrigerant streams, therefore the highest fuel reactivity for the Liquefaction area is considered 'medium'. Additionally, the Liquefaction area tends to be the most congested area of an LNG facility.

Table 3-2: Liquefaction Evaluation

Flammable Streams	Natural Gas and LNG (Methane) Refrigerants (Ethylene, Propane, Butane, and mixtures)
Potential Leak Scenarios	Flammable vapor releases from 'clean gas' lines from the Pretreatment area or mixed-refrigerant vapor lines. Flashing and jetting releases of LNG or liquid mixed-refrigerant inventories. Liquid spills from large-bore line ruptures flowing into the facility trenches and impoundment.
Flash Fire	Yes
VCE	Yes (medium reactivity fuel; medium congestion)
Jet Fire	Yes
Pool Fire	Yes
Cascading Damage	BLEVE of pressurized vessels

4 Identify Hazard Scenarios

A risk-based approach requires developing a list of potential hazard scenarios that can be used to assess detector coverage. Fire and explosion risk assessments typically include a range of potential releases, varying from pinhole leaks to full-bore ruptures. It is not feasible to design a fire and gas detection system capable of detecting every conceivable leak. Therefore, the list of potential hazard scenarios used to evaluate the hazard detection system should exclude scenarios with minor risk contributions, where the cost of detection would be disproportionate to the benefit. The proposed methodology includes performing a hole size sensitivity study to identify the minimum leak size required to be included.

The following figures compare the location specific individual risk (LSIR) contours for the baseline case (which includes all release categories) with the risk contours including only subsets of releases. Specifically, Figure 4-1 only includes releases 6 inches and larger in the comparison subset while Figure 4-2 includes releases 2 inches and larger in the comparison subset.

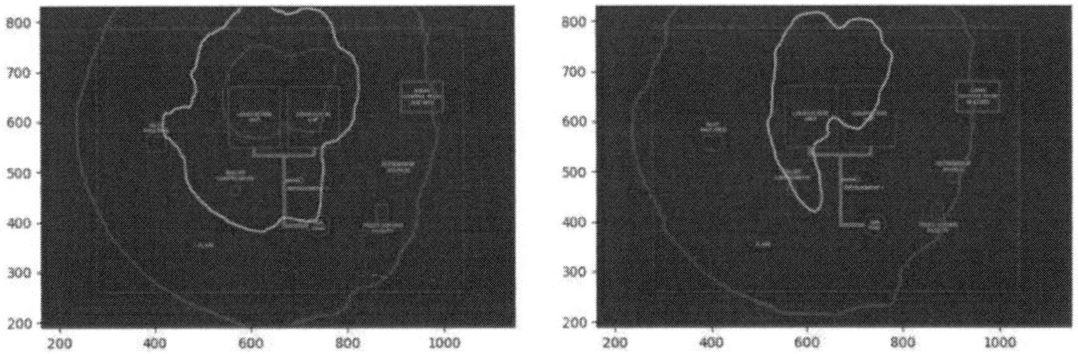

Figure 4-1: LSIR contours for baseline case (left) and releases 6 inches & larger (right).

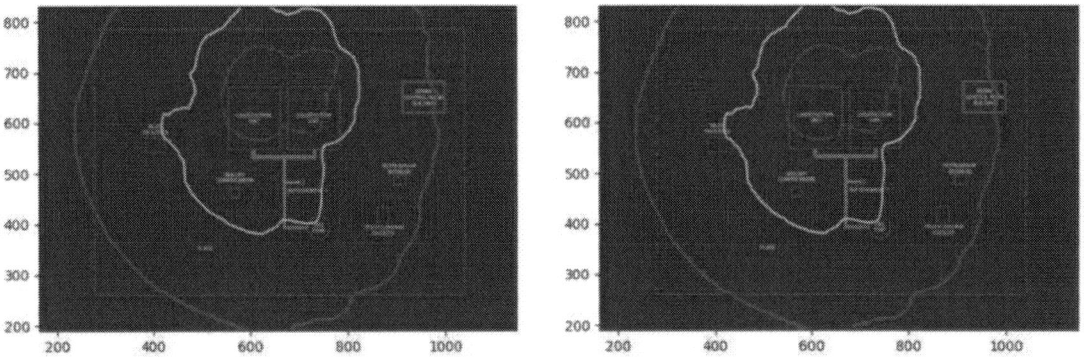

Figure 4-2: LSIR contours for baseline case (left) and releases 2 inches & larger (right).

The comparisons show that, for this specific case, releases smaller than 2 inches diameter do not contribute appreciably to the individual risk at the specified tolerability thresholds. Therefore, the performance-based evaluation of the hazard detection layout for this demonstrative was limited to evaluating detection of releases 2 inches and larger.

It is important to note that the results of the sensitivity study are case-specific and depend on the risk tolerability criteria used; they should not be considered generally applicable to other facilities. For each facility, a hole size sensitivity analysis should be performed to determine the minimum hole size to be considered in the detector evaluation.

5 Assign Performance Targets

Having detailed the hazards present in each Detection Area and established which scenarios should be used to evaluate the hazard detection layout, performance targets for flame and gas detection can be assigned to each Detection Area. This paper will focus on flame detection for the Refrigerant Storage Area and flammable gas detection for the Liquefaction Units A/B, though flame and flammable gas detection should be evaluated for every Detection Area within the facility in practice.

5.1 *Refrigerant Storage Area (Flame Detection Only)*

Cascading damage hazards due to fires were identified for the Refrigerant Storage Area, therefore, flame detection will be evaluated at a minimum detector coverage of 90% and 2ooN voting to initiate executive action (inventory isolation and/or active mitigation systems).

Table 5-1: Flame Detector Coverage Criteria for Refrigerant Storage Area

Geographic Area Description	Minimum Detector Coverage
Refrigerant Storage vessels along with neighboring piping and equipment	90% by 2ooN

5.2 *Liquefaction Units A/B (Flammable Gas Detection Only)*

Based on the results of the hole size sensitivity in section 4, vapor dispersion modeling can be limited to releases 2 inches diameter and larger. Natural gas releases from the pretreatment area will be evaluated with a minimum scenario-based detection coverage of 60%. The remaining scenarios (LNG, liquid and vapor lines for mixed refrigerants) will be evaluated at a minimum scenario-based detection coverage of 80%. Additional layouts will be evaluated to demonstrate the increase in detectors required to achieve 90% coverage for select inventories. Automatic shutdown is assumed for all loss of containment scenarios, requiring 2ooN voting for all inventories.

Table 5-2: Flammable Gas Detector Coverage Criteria for Liquefaction Units A/B

Scenario Description	Minimum Detector Coverage
All remaining scenarios	80% by 2ooN
Natural gas releases	60% by 2ooN

6 Flame Detector Mapping

Obstructions in congested areas can cause blind spots for flame detectors which can be difficult to quantify in 2D layouts. Therefore, example flame detector mapping will be shown using the 3D software Detect3D by Insight Numerics. Detect3D allows the detector coverage to be quantified by two methods as shown in Figure 6-1: a Point Cloud Method that establishes a 3D grid of points and determines how many can be seen by each detector, and a Fire Area Method that defines a fire size and how much of the fire needs to be seen for a detector to activate, which is then translated throughout the zone of interest.

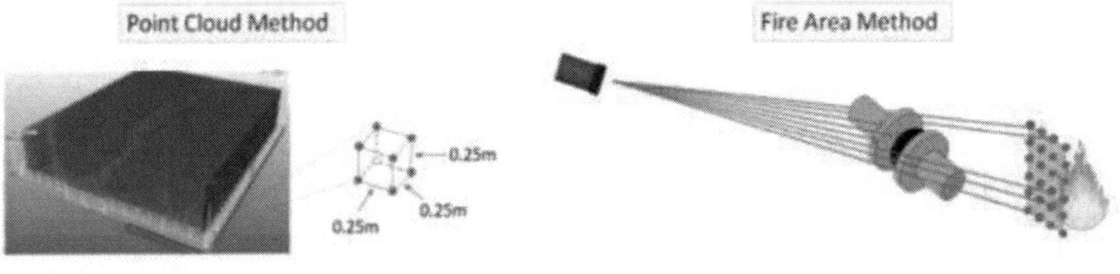

Figure 6-1: Diagrams for Point Cloud Method and Fire Area Method from Detect3D User Manual.

The Point Cloud Method is a strenuous test on flame detector coverage as even small obstructions can block the view of individual points in space, even though these blockages may only result in small shadows that would not negatively impact the performance of the flame detection system. Therefore, the Fire Area Method is more commonly used. Since there is a large range of possible fire scenarios (e.g. jet flames of different sizes and orientations, pool fires of different sizes), rather than calculating how much of a given fire size the detector can see, it is proposed to calculate how much of a given fire size the detector cannot see (i.e. define the size of an 'acceptable shadow'). This can be achieved by defining a fire area with the same dimensions as the acceptable shadow and setting a very low activation threshold (i.e. 1% of surface area).

Requiring a flame detector layout to cover 100% of a protected volume is impractical as well as unnecessary, since very small fires do not have the capability of causing escalation of events and therefore can be detected by slower means. Such an approach has been captured in the NORSOK-S001 standard for the offshore industry that specifies the following minimum requirements for flame detection [5]:

- A flame size of 0.5 meter in diameter and length of 1 meter shall be detected by at least one detector
- A flame size of 1 meter in diameter and length of 3 meters shall be detected by at least two detectors

In effect, the NORSOK criteria require any shadows caused by obstructions to be small enough that a cylinder 0.5 m in diameter and 1 m long cannot be hidden from any flame detector. While pool and jet fire flames at LNG facilities are likely to be larger than the NORSOK criteria listed above, multiple obstructions could exist between a flame detector and the fire event, causing multiple shadows (i.e. unseen portions of the flame). Therefore, a conservatively small acceptable shadow size is necessary to evaluate geographic coverage. Considering all possible orientations of the smaller NORSOK cylinder essentially creates a 1 meter (3.3 foot) diameter sphere, which can be approximated as a 1 meter cube for the purposes of flame detector mapping. By defining an acceptable shadow as a 1 meter cube, this will filter out small blind spots when evaluating flame detector coverage.

A sample refrigerant storage area geometry is shown in Figure 6-2, where the red box denotes the protected volume where flame detector coverage is desired. The protected volume should include areas surrounding the targets subject to cascading damage hazards and sources of thermal radiation (i.e. pool and jet fires) that could initiate said hazards. While hazard modeling is not required to evaluate geographic coverage, it can be useful in defining the protected volume. In the case shown below, the protected volume does not include the piperack as any jet fire hazard from this area would have to extend into the protected volume to impact the storage vessels. In the geographic approach, detector coverage will be evaluated as the percent of the protected volume that meets the performance targets.

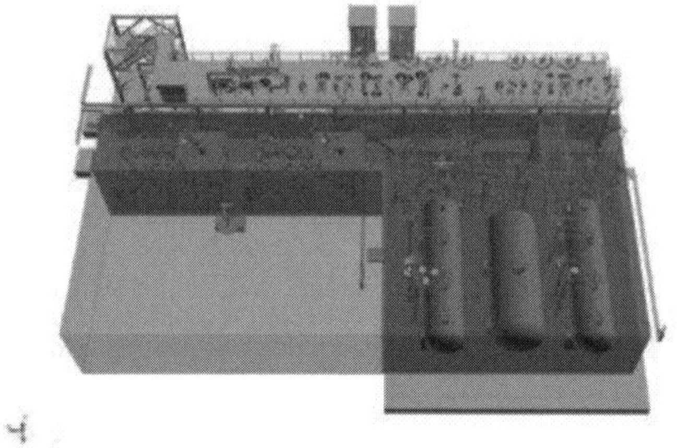

Figure 6-2: Sample refrigerant storage geometry with protected volume shown in red.

The Detect3D software was utilized to determine flame detector locations and orientations to achieve 90% coverage by 2ooN detectors with an acceptable shadow of 1 m x 1m. One possible solution is shown below using six flame detectors and a maximum range of 125 ft for the detection of a 32-inch propane plume (actual coverage 93.3% for 1ooN and 90.1% for 2ooN). Note that all detectors were angled down to prevent concerns with nuisance alarms from distant sources of radiation. Additionally, the detection range for different fuels needs to be confirmed with product listings and coordinated with the manufacturer for fuels not included in the listing.

Figure 6-3: Flame detectors for 90% coverage by 2ooN detectors; detector locations and orientations indicated by orange arrows.

With the minimum detector coverage criteria satisfied, blind spots should still be evaluated. Figure 6-4 shows the volume not seen by any flame detectors in black (concentrated in the lower-right corner). Since there are no jet fire sources in this area, grading could be utilized under the vessels to prevent accumulation of liquids and the presence of a pool fire in the blind spot, in which case the blind spot may be deemed acceptable. A single flame detector could be added to provide 1ooN coverage and operator notification, or two detectors would be required to achieve automatic ESD for a fire event in this blind spot.

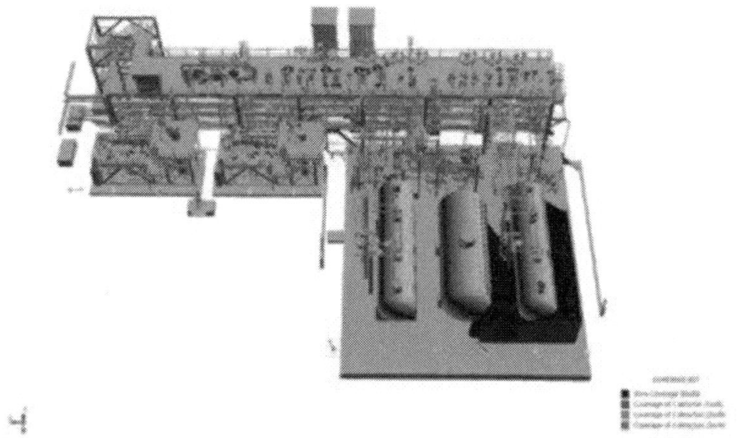

Figure 6-4: Flame detector blind spots shown in black for 6 detector configuration above (i.e. 0ooN coverage).

The above example demonstrates that 90% coverage by 2ooN targets can be achieved with an appropriate number of flame detectors using the acceptable shadow approach. In addition to quantifying flame detector coverage, this methodology also provides a means to optimize the location and orientation of flame detectors in the presence of obstructions.

7 Flammable Gas Detector Mapping

The hole size sensitivity discussed in section 4 revealed that releases 2 inches diameter and larger were driving the risk for the facility, therefore the hazard detection layout should be evaluated against hole sizes 2 inches and larger. The example dispersion modeling used in this paper was performed using the integral model Phast (version 8.22), however, the same methodology could be performed using other integral models as well as computational fluid dynamics (CFD) tools.

The vapor cloud envelope for each release scenario is tested against the gas detector layout to determine if the scenario will result in successful detection (2ooN at 40% LFL or higher) and the time at which this occurs. Based on the total number of release scenarios for each inventory type in the Detection Area, the percentage of successful detection can be calculated (i.e. scenario-based coverage). This demonstration will only report successful detection cases that occur before steady-state conditions are reached (approximately 30 seconds according to Phast).

The cloud footprint will obviously differ between integral and CFD models in the presence of obstructions. Generally speaking, the narrower clouds associated with Phast are likely to present a more rigorous test of the detector layout as these clouds will be harder to detect. However, this is not the case with open-path gas detectors. Since the Phast cloud is allowed to extend freely downwind in the unobstructed field, the effectiveness of open-path gas detectors could be over-predicted. Therefore, open-path gas detectors were included in this analysis, however, a maximum of one open-path gas detector is counted towards successful detection of the cloud. Since CFD modeling can account for obstructions to cloud development, this restriction on open-path gas detectors would not apply.

In the following examples, detailed in Table 7-1 and Figure 7-1, point gas detectors are located at elevations of 0.5 meters and 3 meters, while the open-path gas detectors are located at 1 meter.

Perimeter detection was provided by open path gas detectors around the pretreatment area in the top-right corner of each liquefaction unit, as well as the main liquefaction equipment.

The gas detector layouts were identical for Liquefaction Units A and B. The detector layout was first evaluated with a 10-meter grid of point gas detectors to see how this simple grid layout would perform in a scenario-based coverage calculation (Layout 1). Layout 1 has a total of 81 point gas detectors for Liquefaction Unit A (21 for pretreatment area and 60 for the liquefaction area). With Layout 1, all detection targets are satisfied except the MR-Vapor inventories.

Table 7-1: Scenario-Based Flammable Gas Detector Coverage (2ooN)

Inventory	Target	Layout 1	Layout 2	Layout 3
Natural Gas	**60%**	70%	60%	60%
LNG	**80%**	90%	85%	90%
MR-Vapor	**80%**	74%	80%	80%
MR-Liquid	**80%**	89%	86%	88%

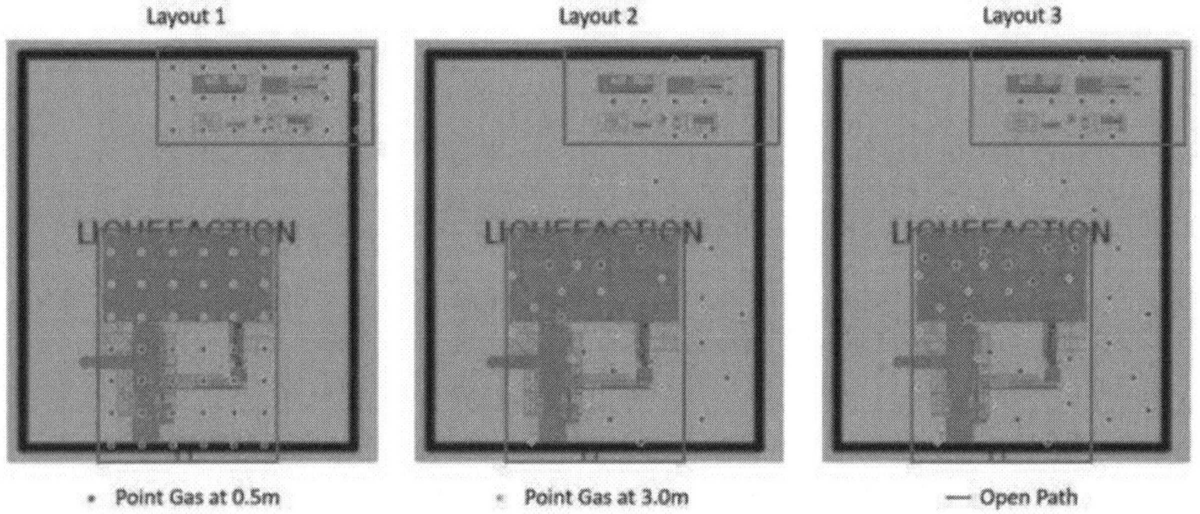

Figure 7-1: Gas detector layouts for Liquefaction Unit A. Point gas detector counts are sequentially 81, 41, and 49.

Layout 2 was defined considering release points and cloud shapes for the various inventories. With a total of 41 point gas detectors (8 in pretreatment and 33 elsewhere) all detector targets were satisfied. This is about half of the point gas detectors needed for the 10-meter grid in Layout 1.

Layout 3 was defined to achieve a reduced release duration for the LNG scenarios. Under the performance targets proposed in the current study, this would require 90% detection by 2ooN gas detectors. This was achieved by using 49 point gas detectors (8 in pretreatment and 41 elsewhere).

It should be noted that scenario-based gas detector coverage is generally sensitive to the number of release points and release directions evaluated. Given the lack of detailed design information for this example case, a conservative number of release directions were assumed. This sensitivity to release points and directions further amplifies the importance of a site-specific analysis for gas detector layouts.

8 Conclusions

Fire and gas detectors at LNG facilities need to be located using a performance-based design approach, which requires the designer to document each performance objective and applicable scenarios, along with calculations, modeling or other technical substantiation that justifies adequate protection. The examples in this paper demonstrate that both fire and gas detector performance can be quantified and compared to detector coverage targets. Performance targets are characterized by detector coverage percentage (based on the risk to the facility) and detector voting criteria (largely based on what actions need to be taken after detection).

Flame detection is recommended to be evaluated by geographic coverage using an acceptable shadow approach. This method does not require detailed hazard modeling and, in addition to quantifying detector coverage, optimizes the location and orientation of flame detectors. Congestion and large obstructions can have significant impacts on flame detector coverage, therefore 3D flame detector mapping is useful in these situations.

Gas detection is recommended to be evaluated by scenario-based coverage. Integral models such as Phast can help with spacing detectors away from release sources in an effort to detect the resultant cloud instead of attempting to detect the leak location, which increases scenario coverage rates and decreases the number of detectors required to do so. However, there are two main limitations of using integral models for detector evaluation: integral models cannot account for obstructions and they cannot fully account for open path detectors. Therefore, while this exercise can be conducted with integral models, CFD modeling would further optimize the detector layout and better utilize open path gas detection.

The detector layouts presented in this paper are generic examples and should not be translated to other facilities. Additionally, it should be noted that other layouts could also satisfy the performance targets set in this paper. The proposed methodology can be carried out using different risk tolerance thresholds, modeling tools, and detector coverage criteria. While performance targets must be chosen for the purposes of the demonstrative included in this paper, these targets should not be interpreted as requirements of DOT-PHMSA nor as acceptable to them.

9 References

[1] National Fire Protection Association, NFPA 59A Standard for the Production, Storage, and Handling of Liquefied Natural Gas (LNG), 2019.

[2] National Fire Protection Association, NFPA 72 National Fire Alarm and Signaling Code, 2019.

[3] International Society of Automation, "TR84.00.07 Guidance on the evaluation of fire, combustible gas, and toxic gas system effectiveness," 2018.

[4] Factory Mutual, FM Approvals 3260 Radiant Energy-Sensing Fire Detectors for Automatic Fire Alarm Signlaing, 2018.

[5] NORSOK Standard, S-001 Technical Safety, 2018.

SPRING21
+17TH GCPS
A Joint AIChE and CCPS Meeting

What Is the Safest Hazard Assessment Approach for Distances to Storage Tanks?

Delmar "Trey" Morrison, Ph.D., P.E., CPSP
Ryan Hart, Ph.D., P.E.
Jessica M. Morris, Ph.D.
Navid Zanganeh, Ph.D.
Exponent, Inc.
4580 Weaver Parkway, Suite 100,
Warrenville, Illinois, USA 60555

Prepared for Presentation at
American Institute of Chemical Engineers
2021 Spring Meeting and 17th Global Congress on Process Safety
Virtual Conference
April 18-22, 2021

What Is the Safest Hazard Assessment Approach for Distances to Storage Tanks?

Delmar "Trey" Morrison, Ph.D., P.E., CPSP
Ryan Hart, Ph.D., P.E.
Jessica M. Morris, Ph.D.
Navid Zanganeh, Ph.D.
Exponent, Inc.
4580 Weaver Parkway, Suite 100,
Warrenville, Illinois, USA 60555

Keywords: Consequence Modeling, Manage Risks, and Understand Hazard and Risk

Abstract

Safety in the liquified natural gas (LNG) industry is achieved by four elements: primary containment, secondary containment, safeguard systems, and separation distance from sensitive receptors. The first two elements can be accomplished by appropriate engineering design and robust maintenance programs to minimize the potential for an LNG release. Determining a safe separation distance from a potential release (including secondary containment location) to personnel in buildings is outlined in API RP 752 and 753. This recommended practice provides guidance for managing the hazards from fires, explosions, and toxic releases to personnel in buildings. In the recommended practice, three assessment approaches may be utilized: 1) consequence-based approach based on maximum credible events, 2) risk-based approach combining consequences and their frequencies, and 3) spacing tables based on literature data. Each of these assessments vary in their methodology, resulting in different distances to the same hazard. Here, all three approaches are applied to an asset that all LNG facilities have, an LNG Storage Tank. LNG Storage tank failure data is not as prevalent as other plant assets failure data - such as pipeline failures. With a lack of LNG storage tank failure rates, what is the most appropriate method to assess the distances to hazards from an LNG storage tank?

1 Introduction

Safety in the liquified natural gas (LNG) industry is achieved by multiple engineered elements of a process safety program: primary containment, secondary containment, safeguard systems, and separation distance from sensitive receptors. These elements are integrated with industry standards, regulatory compliance, and codes. The safety of containment systems can be addressed by appropriate engineering design and robust maintenance programs. The third element, safeguard systems, can elevate the safety level by rapid and reliable detection of a hazardous situation to enable the operators to respond and control the hazard. The last element, separation distance, provides a safe separation distance from the potential release (including secondary containment location) to minimize the potential effects on personnel and the public. [1]

1.1 Primary and Secondary Containment

Primary containment is the first boundary to contain hazardous chemicals at a facility. An example of this would be a storage tank. Secondary containment would be utilized if there was a loss of containment or failure with the primary containment storage tank. An example of secondary containment would be a wall surrounding the storage tank or a sump for the material to be collected in and contained within if a leak were to occur. Appropriate engineering design along with employing appropriate materials to construct the LNG storage tanks are the backbone of primary and secondary containments. Today, there are multiple types of LNG storage tanks, such as single containment, double containment, full containment, membrane, and in-ground tanks used in the LNG industry. Typically, a single containment tank system is a liquid-tight single tank, constructed from low-temperature tolerant metal and surrounded by insulation (with a system to manage the vapor produced). The single containment tank can then be surrounded by a dyke wall, which acts as a secondary containment to hold the liquid in case of leakage in an incident. A double containment system tank consists of a single containment tank system surrounded by a dyke wall. The secondary containment can be made of metal or prestressed concrete and it is intended to contain the liquid spills from a primary tank; however, it is not necessarily intended to contain or manage any vapor leakage. Double containment tank system consists of a primary liquid-tight tank with a secondary liquid-tight tank. The full containment system should be capable to hold liquid and manage vapors from the primary and secondary containment [2]. Figure 1 schematically shows single, double, and full containment tanks.

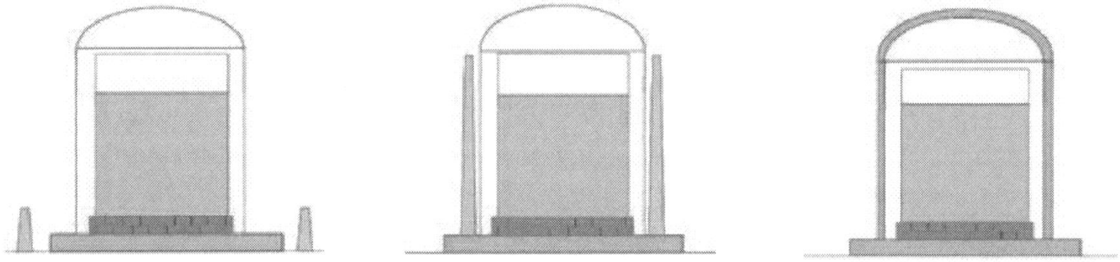

Figure 1. three common types of LNG storage tanks – Single Containment (left), Double Containment (middle), and Full Containment (right).

1.2 Safeguard Systems

The third layer of safety is safeguard systems. Safeguard systems employ rapid and reliable detection to control hazards related to LNG releases. In this layer of protection gas, smoke, and flame detectors along with temperature/pressure sensors and cameras are used in LNG terminals to detect flammable gas releases and/or fires. Emergency shutdown systems are also in place to minimize the loss of LNG in the case of a release. Fire control systems, vapor reduction systems, and trained operators are also utilized in this layer of protection to prevent hazards.

1.3 Safe Distances

The final layer of protection presented here is providing a safe distance between the hazard and personnel to minimize the risk in case a hazard occurs. Determining a safe separation distance from a potential release (including secondary containment location) to personnel in buildings is outlined in API RP 752 and 753. These recommended practices provide guidance for managing the hazards from fires, explosions, and toxic releases to personnel in buildings [3-4]. In the recommended practices, three assessment approaches are presented: 1) spacing tables based on literature data, 2) consequence-based approach based on maximum credible events and 3) risk-based approach combining consequences and their frequencies. The safe distances from LNG storage tank hazards can be determined through each of these assessment approaches. This paper analyzes the different hazard analysis methods to determine the most appropriate approach to determining the safe distance to an LNG storage tank.

2 Methodology

API RP 752 and 753 are the recommended practices that deal with the management of hazards associated with permanent and portable buildings in a process plant, respectively. The objective of both practices is to ensure that the location of permanent and/or portable buildings in an LNG process plant will protect the occupants against all potential hazards. In API RP 752 and 753, three assessment approaches are provided in which each of them will be discussed in the following sections:

2.1 Spacing Tables Approach

The Spacing Tables Approach is based on using data tables to determine the minimum safe distance between potential hazard and the hazard receptor (i.e. personnel located in buildings or other hazard receptors). As stated in API RP 752, "*Industry groups, insurance associations, regulators, and owner/operator companies have developed experienced-based spacing tables*" to estimate the minimum safe distance. Spacing table data are usually only applicable for fire hazards and they are typically not sufficient to address the safe distance for explosion and toxic release hazards. Due to this limitation, API RP 752 only permits the use of spacing tables for fire hazards. In case of investigating the explosion and/or toxic hazards, alternative methods such as consequence-approach and risk-based approach must be used. The spacing tables data can be found in various resources such as the NFPA Standard for the Production, Storage, and Handling of Liquefied Natural Gas (LNG) [2] or CCPS Guidelines for Facility Siting and Layout [5]. In a building siting evaluation utilizing the spacing tables approach, the safe separation distance is the one that meets or exceeds the given distance in the spacing table for the specified scenario and hazard receptor.

2.2 Consequence-based approach

The Consequence Based Approach investigates the effects from fires, explosions, and toxic releases. This approach is based on the concept of defining maximum credible events (MCE) and the hazard(s) that accompany these scenarios. Three types of hazards are evaluated based on API RP 752 and 753 guidance: explosion overpressure, thermal radiation, and toxic gas exposure. In this approach, a series of MCE scenarios are evaluated for each of these hazards to determine the appropriate distance populations in buildings should be located. A summary of building exposure criteria and consequence criteria is given in Table 1.

Table 1. Summary of building exposure and consequence criteria.

Building exposure criteria	Consequence criteria
Blast load	Occupant vulnerability
Thermal flux and exposure time	Potential building damage
Flammable gas and/or toxic concentration and exposure time	Building internal environment degradation

After determining the appropriate distance to the specified consequence, the buildings located within those regions need to be identified. If a building sustains exposure criteria greater than what is specified, additional protection may need to be provided to the building and its occupants. These protection measures can include relocating the building at a safe distance that does not exceed exposure criteria or modifying the building to be able to withstand the appropriate building exposure criteria (i.e. adding a blast wall for overpressure hazards).

2.3 Risk-based Approach

Risk is the probability of a specific event having an adverse effect on human health, environment, or economic loss. Risk is a function of the estimated consequence(s) of a hazard and the estimated frequency of occurrence. The risk-based approach combines the likelihood of the consequences of hazardous releases. The first step of this approach is to determine the hazardous scenarios (similar to the consequence-based approach). The second step is to apply failure rates to the hazardous scenarios to determine the likelihood of these scenarios occurring. To determine likelihood, established failure rates for parts are utilized. The scenario information, likelihood, and site conditions (weather, population, ignition sources, etc.) are all used to calculate the risk. Given the complexity, risk assessment software is used to perform the risk calculations.

3 Analysis

An analysis was performed to determine an appropriate distance to an LNG storage tank. In this analysis, a theoretical LNG peak shaving facility was used. LNG peak shaving facilities are the facilities where surplus LNG is stored to be used when the LNG demand is at its peak. When needed, the LNG is gasified through a vaporization process and distributed to the users via pipeline. These facilities are important to ensure adequate natural gas storage and supply in regions where the LNG demand may be seasonally higher than usual and there is a lack of pipeline or natural reservoir storage inventory. These facilities can either (1) produce LNG onsite or (2) receive LNG from trucks (where the LNG is produced at another facility). For this study, we consider type (2), where the LNG is received by truck and is then stored in LNG storage tank(s) awaiting regasification. Figure 2 schematically shows an LNG peak shaving facility with truck unloading, piping from the truck unloading to the LNG storage tank, the LNG storage tank, piping from the LNG storage tank to the vaporizers, and then piping from the vaporizers to the pipeline.

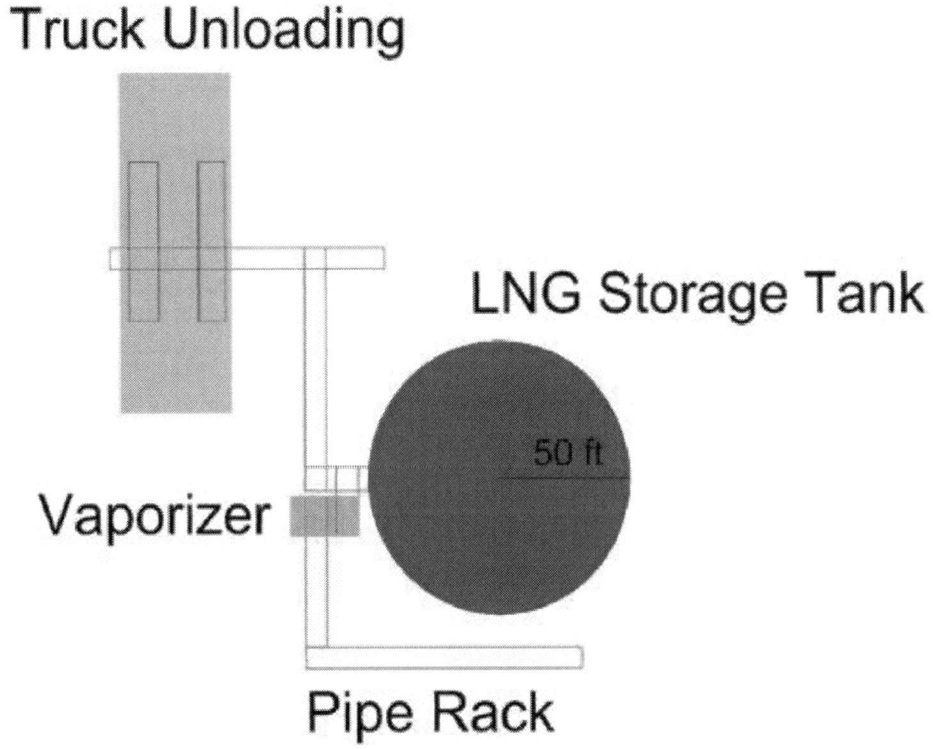

Figure 2. Schematic of an LNG peak shaving facility.

A comparison of the different hazard assessment approaches described in API RP 752 and 753 was performed for an LNG peak shaving facility to determine the most appropriate distance to an LNG storage tank.

3.1 Spacing Tables approach

Spacing tables are based on experience to provide an estimated minimum safe distance from the hazard to the hazard receptor. The minimum safe distance is based on hazards from a fire or thermal radiation. Spacing tables do not take into consideration overpressure or toxic hazard distances due to their complexity. In this analysis, the minimum safe distance from an LNG storage tank is determined with NFPA 59A and CCPS guidance.

NFPA 59A has a minimum separation distance associated with any type of container or tank that contains LNG or flammable refrigerant. These distances are specified below in Table 2 (or NFPA 59A Table 6.3.1). Additional methods are recommended in NFPA 59A for performing an analysis on the fire hazards from LNG.

Table 2. Distances from Containers and Exposures from NFPA 59A (Table 6.3.1). [2]

Container Water Capacity		Minimum Distance from Edge of Impoundment or Container Drainage System to Property Lines That Can Be Built Upon		Minimum Distance Between Storage Containers	
gal	m^3	ft	m	ft	m
<125*	<0.5	0	0	0	0
125–500	≥0.5–1.9	10	3	3	1
501–2,000	≥1.9–7.6	15	4.6	5	1.5
2,001–18,000	≥7.6–63	25	7.6	5	1.5
18,001–30,000	≥63–114	50	15	5	1.5
30,001–70,000	≥114–265	75	23		
>70,000	>265	0.7 times the container diameter but not less than 100 ft (30 m)		¼ of the sum of the diameters of adjacent containers [5 ft (1.5 m) minimum]	

*If the aggregate water capacity of a multiple container installation is 501 gal (1.9 m^3) or greater, the minimum distance must comply with the appropriate portion of this table, applying the aggregate capacity rather than the capacity per container. If more than one installation is made, each installation must be separated from any other installation by at least 25 ft (7.6 m). Do not apply minimum distances between adjacent containers to such installation.

Another source for spacing tables is provided in CCPS "Guideline for Facility Siting and Layout" [5]. The values from this reference are located in Table 3. Distances to overpressure and toxic releases require additional analysis for appropriate distances. Depending on the hazard receptor (property boundary, roads, off-site population, etc) the distances from the storage tanks to the hazard receptor varies.

Table 3. Typical Flammable and Combustible Liquids and Liquefied Flammable Gases Tank Spacing to Other Areas and Equipment for Fire Consequences. [5]

	Text References	Atmospheric and Low Pressure Storage (non-boil over) less than 10,000 gals	Atmospheric and Low Pressure Storage (non-boil-over) greater than 10,000 gals	Atmospheric Storage (boil over potential)	Pressurized Storage	Refrigerated Flammable Storage
Text Reference		5.9 and 6.8.20				
Property Boundary or Adjacent Industry	5.2.5	25	100	200	200	200
Public Access Right of Way (road, rail lines, parks)	5.2.5	50	100	200	250	250
Off-site populations: housing, offices	5.2.5	50	250	500	350	250
Process Area battery limits (excepting portable containers related to the process)	5.7.3 6.8.1	25	100	100	250	100
Utility Battery Limits (excepting portable containers related to the utility)	5.5	25	100	100	250	100
Fire water pumps (excepting the fuel source for the pump)	5.8.2	50	200	200	250	200
ESD and mitigation system activation points (activation point must be outside of tank diked area.)	6.8.15	50	50	50	50	50

3.2 *Consequence-based Approach*

The consequence-based approach takes into consideration the different hazards from the different scenarios that can occur. Consequence-based approaches commonly provide the distances to specified hazard thresholds or will provide the observed hazard threshold at the specified distance.

3.2.1. Consequence Analysis Software – Phast v6.7

The consequence analysis performed in this study utilized an industry-proven commercial computational model developed and distributed by Det Norske Veritas (DNV) called Phast v6.7. Phast is widely used for the calculation of hazard distances from the release of several hazardous substances, including LNG. The Phast software uses the Unified Dispersion Model (UDM), which is an integral equation model to calculate the dispersion from pressurized and unpressurized releases. Phast is able to model discharge dynamics, dispersion dynamics, and hazard consequences for a multitude of scenarios such as near-field jet dispersion, non-equilibrium droplet evaporation and rainout, touchdown, pool spread, and vaporization, heavy gas dispersion, and far-field passive dispersion to name a few of its capabilities. The UDM also allows for possible plume lift-off when a grounded plume becomes buoyant. The UDM has been validated extensively with experimental data and is the subject of several peer-reviewed scientific papers [6]. Phast v6.7 UDM has also been approved by the U.S. PHMSA for analyzing LNG vapor dispersion exclusion zones [7].

3.2.2. Scenario Description – Consequence Analysis

Two hypothetical scenarios for the LNG peak shaving facility were modeled using Phast v6.7. Model parameters followed guidance provided in 49 CFR § 193.2059. A summary of the weather input parameters is shown below in Table 4.

Table 4. Weather Parameters used in the Consequence Analysis.

Weather	
Ambient Temperature (°F)	50
Surface Temperature (°F)	50
Relative Humidity	50%
Wind Speed (mph)	4.5
Pasquill-Gifford Atmospheric Stability	Category F
Surface Roughness Factor (m)	0.03

LNG transport lines at the hypothetical peak shaving facility were modeled as potential release sources going into and out of the LNG storage tank. LNG is transported to the facility via truck in ISO containers. The "Truck Unloading" line transports the LNG from the portable container to the LNG storage tank. When natural gas is needed, LNG is transported via pump from the LNG storage tank to the vaporizer in the "Send-out Line." Relevant information for both of these lines (defined here as MCEs) are provided below in Table 5.

Table 5. Input Conditions in Phast Software for Consequence Analysis.

- LNG Release Scenarios -		
Segment	**Truck Unloading**	**Send-out Line**
Component	Methane	Methane
LFL	5%	5%
Release Type	Leak	Leak
Pressure (psig)	80	500
Temperature (°F)	-265	-265
Phase	Liquid	Liquid
Pipe Size (in)	4	4
Leak Size (in)	Varies	Varies
Piping Inventory (gal)	10,000	300
Duration of Release (s)	600	600
Leak Elevation (ft)	20	20

In the consequence model, distances to the thermal hazards are calculated. For the two scenarios at the hypothetical peak shaving facility in this study, distances to the flash fire and jet fire hazards were all reported in Table 6 and Table 7. While overpressure hazards are normally reported, in these two scenarios, overpressures were not experienced and therefore not reported.

Table 6. Distances to <u>Flash Fire Hazards</u> from Truck Unloading and Send-out Line.

Scenario	Elevation (ft)	Wind Speed (mph)	Leak Dia. (in)	Distance to LFL (ft)	Distance to 1/2-LFL (ft)
Sendout Line	20	4.5	0.118	25.7	39.7
			0.394	75.5	123
			1.97	479	679
			4.0	644	837
Truck Unloading			0.118	19.8	27.4
			0.394	51.3	88.9
			1.97	496	1,150
			4.0	1,200	1,910

Table 7. Distances to Jet Fire Hazards from Truck Unloading and Send-out Line.

Scenario	Elevation (ft)	Wind Speed (mph)	Leak Dia. (in)	Distance to 0.35 BTUs/ft^2 (ft)	Distance to 1.10 BTUs/ft^2 (ft)	Distance to 3.30 BTUs/ft^2 (ft)
Sendout Line	20	4.5	0.118	31.8	n/a	n/a
			0.394	137	106	84.7
			1.97	608	490	416
			4.0	1,150	919	779
Truck Unloading			0.118	n/a	n/a	n/a
			0.394	98.0	71.1	n/a
			1.97	447	358	299
			4.0	842	671	561

3.3 Risk-based Approach

The risk-based approach is a quantitative analysis taking into consideration both the consequence and the likelihood of the scenario occurring. The calculated risk, from likelihood and severity, is then compared to a benchmark risk which is considered by stakeholders as acceptable or not acceptable. Distances to different levels of risk are reported in this study for the same two scenarios described in the consequence-based approach.

3.3.1. QRA Software – Phast Risk

The quantitative risk assessment (QRA) performed in this analysis utilized an industry-proven commercial computational model developed and distributed by Det Norske Veritas (DNV) called Phast Risk (v6.7). This software takes into consideration the discharge dynamics, dispersion dynamics, consequences, population, weather, and likelihood to quantity the associated risk for each hazardous release scenario. In this analysis, the distance to the associate risk is reported.

3.3.2. Scenario Description – Risk

Risk takes into consideration the consequence and the likelihood, or frequency, of the consequence occurring. To perform a risk-based approach to the hypothetical peak shaving facility, the frequencies associated with loss of containment for different pipeline hole sizes need to be accounted for. Industry pipe failure rates reported in DNV's Failure Frequency Guide were used in this analysis [8]. The failure rates for process pipe and valves are shown below for the 4" pipe used in the two scenarios.

Table 8. Process Equipment Leak Frequencies for 4" pipe [8].

Equipment	Leak Size	Total Failure Frequency
Process Pipe	1 – 3 mm	5.36 E-05
	3 – 10 mm	2.01 E-05
	10 – 50 mm	7.46 E-06
	50 – 150 mm	6.61 E-06
	Total	**8.78 E-05**
Flange	1 – 3 mm	5.75 E-05
	3 – 10 mm	2.10 E-05
	10 – 50 mm	7.61 E-06
	50 – 150 mm	8.06 E-06
	Total	**9.42 E-05**
Actuated Valve	1 – 3 mm	5.66 E-04
	3 – 10 mm	1.79 E-04
	10 – 50 mm	5.45 E-05
	50 – 150 mm	2.14 E-05
	Total	**8.20 E-04**

A hypothetical quantity of process equipment parts and length of pipe for the Truck Unloading Line and Send-out Line is provided below.

Table 9. Parts Count and Pipe Length used in Risk Analysis.

Scenario	Truck Unloading	Send-out Line
Pipe Length (ft)	250	300
Valves	20	10
Flanges	4	10

The failure rates for the Truck Unloading Line and Send-out Line were determined using the parts count, pipe length, and DNV Failure Frequency values specified in Table 9. Phast Risk applies the probability of failure to the consequences. In addition to the failure rate of each line, the probability of each hazard occurring is able to be taken into account in the Phast Risk QRA software. Default probabilities were used in this analysis.

The quantitative risk assessment calculated the risk from the different hazard consequences. In this assessment, the 3D Baker-Strehlow-Tang (BST) overpressure model was used for explosion overpressure hazards. The assumptions made for the BST input parameters are shown below.

Table 10. BST Explosion Assumptions for QRA Model.

BST Input Parameters	LNG / Methane
Material Reactivity	Low
Flame Expansion	3
Obstacle Density	Low
Ground Reflection Factor	2
Congested Areas	100%

The distances to the specified risk values from the two scenarios are shown below.

Table 11. Distance to Hazard Zones.

Risk	Distance from Truck Unloading Line (ft)	Distance from Send-out Line (ft)
1E-05	475	380
1E-06	626	487
1E-07	673	545
1E-08	680	584

4 Results and Discussion

In this study, spacing tables, consequence-based, and risk-based approaches were applied to determine a safe distance to an LNG storage tank at a hypothetical peak shaving facility to analyze the different hazard approaches. The results for each approach are compared to determine the most appropriate method to assess the distances to hazards from an LNG storage tank. All analysis was performed in accordance with API RP 752 and 753. The maximum distances from each hazard comparing the three methods are shown below in Figure 3.

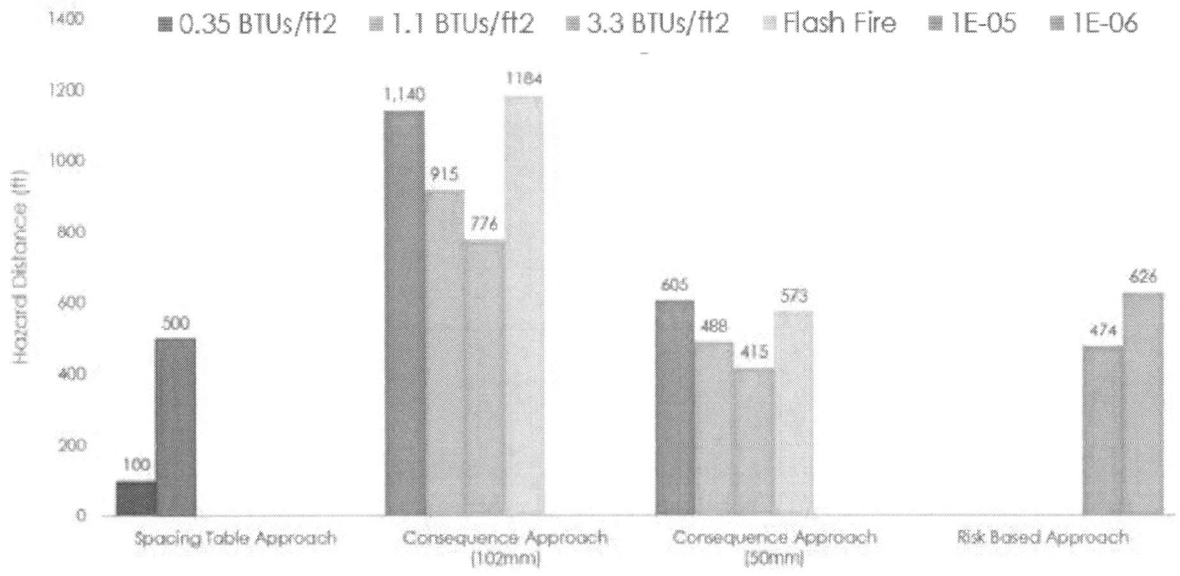

Figure 3. Minimum Safe Distances to LNG Storage Tank form different hazard analysis approaches.

4.1 Spacing Tables Approach

As shown in Figure 3, the minimum safe distance obtained from the spacing table approach is between 100-500 ft, depending on the appropriate hazard receptor. This value indicates that for a fire hazard, an LNG storage tank should be located 100-500 ft away from the appropriate hazard receptor. The spacing table approach is the quickest and easiest method to determining the minimum safe distance from a hazard. Although quick and easy, this method is very limited in the additional parameters that affect the hazard such as weather conditions, leak size, material properties, leak elevation, and more. Other methods are much more considerate of the parameters that affect the release in addition to the different hazards that may be present (such as overpressure or toxic releases). While spacing tables only provide the distance to a fire hazard, and are generally not used for overpressure or toxic release, they may be used as a starting point. Alternative methods such as consequence and risk-based approaches may be necessitated in order to assess other hazards not encompassed in the spacing table method and to yield a higher degree of certainty in the spacing estimate.

4.2 Consequence-based Approach

The consequence-based approach gives more detailed results for the maximum credible events that may occur for the design conditions of the analyzed process. The largest distances to the different credible hazards are reported in the consequence-based approach. In this study, two scenarios 1) Send-out Line and 2) Truck Unloading Line were modeled based on our hypothetical plant design. Each scenario contained details such as weather conditions, material composition, type of hazard, phase, temperature, and pressure of stored materials, etc. While this method is able to take into consideration multiple parameters in the analysis, the consequence-based approach is more time-consuming than the spacing tables approach and requires specialized software and training. The output of our PHAST simulations are the sets of minimum safe distances based on the defined fire hazards (see Table 6 and Table 7).

As shown in Figure 3, the minimum safe distances for flash fire and jet fire hazards are obtained with a leak size of 4 in (102 mm). While this scenario provides the largest distance to the maximum credible events, this event may not be the most likely scenario. Based on the failure rate data in Table 8, a 4 in (102mm) leak is less likely to occur compared to a 2 in (50mm) leak. While a 4 in leak gives the largest hazard distances, the 2 in release scenarios are also reported. This approach takes into consideration all hazard types and different release parameters but does not take into consideration the likelihood of the leak sizes or different scenarios occurring.

4.2 Risk-based Approach

The risk-based approach takes into consideration the consequences and the likelihood of their occurrence. The likelihood was determined through established failure rates provided in DNV's Failure Frequency Guide [8] applied to the hypothetical peak shaving facility. Total failure rates for each scenario were determined and incorporated into the QRA software. The output of the simulation is the series of minimum safe distances based on the likelihood of defined scenarios occurrence (see Table 11). Risks in the magnitudes of 1E-5 and 1E-6 are typically used for LNG facilities to provide the minimum safe distances from all hazards. The risk-based approach provides the most comprehensive method incorporating the consequences with their likelihood of occurring. Although it is the most comprehensive method, it is the most time-consuming, requires the most training, and most expensive software compared to the other methods.

5 Summary

Three different approaches, spacing tables approach, consequence-based approach, and risk-based approach were applied to investigate the minimum safe distance of an LNG tank storage in a hypothetical peak shaving facility in compliance with API RP 752 and 753. The objective of this study was to determine the most appropriate method for determining the safe distance from an LNG Storage tank. Although the spacing table approach is a quick and easy method to determine the minimum safe distance from a fire hazard, it is unable to incorporate parameters that affect discharge and dispersion and is not appropriate for modeling overpressure or toxic releases. The consequence-based approach and risk-based approach both require specific software and training to perform. The consequence-based approach is able to take into consideration additional parameters and provide hazard details for multiple different types of hazards but it unable to take into consideration the likelihood of occurrence. The risk-based approach is able to incorporate consequence and likelihood of occurring, resulting in the most comprehensive model. Although most comprehensive, it is the most time-consuming, expensive, and specialized approach that a consequence-based approach may be able to provide distances that prove adequate. Each facility poses different scenarios and parameters that need to be taken into consideration when determining the best hazard assessment approach.

6 Reference

[1] Foss, Michelle Michot, et al. "LNG safety and security." Center for Energy Economics (CEE), 2003.

[2] National Fire Protection Association. Standard for the Production, Storage, and Handling of Liquefied Natural Gas (LNG). National Fire Protection Association, 2001.

[3] API RP 752, 3rd Edition, December 2009 - Management of Hazards Associated with Location of Process Plant Permanent Buildings, Reaffirmed, August 2020.

[4] API RP 753, 1st Edition, (R 2020) – Management of Hazards Associated with Location of Process Plant Portable Buildings; First Edition, Reaffirmed, August 2020.

[5] Center for Chemical Process Safety. "Guidelines for Facility Siting and Layout"; August 2003.

[6] Witlox, H.W.M. and Holt, A., 1999, A unified model for jet, heavy and passive dispersion including droplet rainout and re-evaporation, International Conference and Workshop on Modeling the Consequences of accidental Releases of Hazardous Materials, CCPS, San Francisco, California, September 28–October 1, pages 315–344.

[7] PHMSA Docket No. 2011-0075, October 11, 2011.

[8] DNV Failure Frequency Guide. "Process Equipment Leak Frequency Data for use in QRA", April 2013.

617846 Develop a CFD Heat Transfer Model for Using High Expansion Foam in an LNG Spillage

Abstract:

To effectively store and transport natural gas, liquefied natural gas (LNG) is widely used in the industry. However, a LNG spill can result in a vapor cloud, which can potentially cause fire and explosion. The National Fire Protection Association (NFPA) 11 recommends using high expansion (HEX) foam to effectively mitigating the vapor hazard and controlling LNG pool fire. In the experiments, the liquid nitrogen pool was used to simulate LNG leakage, because the boiling point of liquid nitrogen is close to methane, which is the main component of LNG. A lab-scale foam generator was used to produce HEX foam and carry out experiments to evaluate the temperature profile in the foam. Based on these experimental results, the heat balance among HEX foam and vaporized gas was discussed, and a heat transfer model using ANSYS Fluent® was developed to estimate minimum HEX foam height that should be applied in the LNG spillage risk mitigation. Finally, we performed a sensitivity analysis on the effect of vaporization rate, the diameter of the vapor channel, and the heat transfer coefficient on the required minimum height of the HEX foam. It can be observed that at least 1.2 m HEX foam in height is needed to ensure the risk mitigation under the common situation. The simulation result can be used not only for understanding the heat transfer mechanism when applying HEX foam but also for effectively formulating HEX foam related standard by the policymaker.

SPRING21
+17TH GCPS
A Joint AIChE and CCPS Meeting

It Was a Dark and Stormy Night – Investigation of Acrylate Storage Tank Explosions

Kathleen Kas, Dow Chemical Company, Washington Crossing, PA
kkas@dow.com

Delmar "Trey" Morrison, Exponent Inc, Warrenville, IL
tmorrison@exponent.com

Prepared for Presentation at
American Institute of Chemical Engineers
2020 Spring Meeting and 17th Global Congress on Process Safety
Virtual
April 18 - 22, 2021

It Was a Dark and Stormy Night – Investigation of Acrylate Storage Tank Explosions

Kathleen Kas, Dow Chemical Company, Washington Crossing, PA
kkas@dow.com

Delmar "Trey" Morrison, Exponent Inc, Warrenville, IL
tmorrison@exponent.com

Keywords: Lightning, monomer, explosion, ignition

Abstract

It was in the early morning hours of May 16, 2012 when a thunderstorm rolled into the Bristol, Pennsylvania plant where The Dow Chemical Company produced acrylic polymers. As the weather turned stormy, lightning struck in the tank farm area containing inhibited ethyl acrylate, butyl acrylate, and methyl methacrylate. Within seconds, an ethyl acrylate tank exploded and was followed a few minutes later by a butyl acrylate tank explosion. The explosions and ensuing fire damaged the tank farm and led to a lengthy shutdown; fortunately, there were no injuries. This paper details the investigation into the complex sequence of events that led to the two explosions. The investigation team evaluated data gathered from eyewitnesses, security system cameras, weather services, and the process control system. The team also analyzed physical evidence including electrical bonding and grounding systems, deflagration arresters, and various tank components. Although the adjacent ethyl and butyl acrylate tanks were involved in the event, the investigation did find that flame isolation devices functioned as intended and prevented the propagation of flames to surrounding processes via the interconnected venting system. However, there was a puzzle left to unravel. The tank installations were determined to be in compliance with the industry lightning standards such as NFPA 780. If the tanks met industry standards, how did the explosion happen? While the exact mechanism could not be determined with certainty, the ignition of the atmosphere in the ethyl acrylate tank most likely occurred from a spark discharge inside the tank's headspace, likely between internal components at different electrical voltage potentials. This type of mechanism could invalidate the Faraday Cage assumption inherent in current industry guidance and may represent a hazard that is not currently well-recognized in industry. It should also be noted that, while lightning can't be prevented, the outcome of a lightning strike can also be mitigated by the use of safeguards such as inertion, flame arresters, diking, and foam systems.

1 Introduction

On May 16, 2012, at approximately 3:34 am, a lightning strike occurred at the Dow Bristol Plant in Bristol, Pennsylvania (Figure 1). Immediately after a lightning flash was observed, flames were observed on top of the Ethyl Acrylate Monomer (EA) tank about one second before it exploded. Liquid EA spilled into the dike shared with the Butyl Acrylate Monomer (BA) tank, and the BA tank subsequently exploded a few minutes later. The explosions and ensuing fire damaged the tank farm and led to a lengthy shutdown; fortunately, there were no injuries.

This paper details the investigation into the complex sequence of events that led to the two explosions. Key questions that the investigation team needed to answer included:

1. Was lightning the true initiating event?
2. Why did the contents of the EA tank ignite if the tank was grounded?
3. Why did the BA tank explode, since the headspace was not in the flammable region prior to the event?
4. What learnings can we take away from this incident?

This paper summarizes the event, the evidence collection and analyses, the testing of hypotheses, the conclusions from the event, and key learnings.

Figure 1 The dike fire and aftermath of the explosions.

2 Description of the Storage Tank System & Safeguards

The Dow Bristol Plant creates products for the Dow Coating Materials business. Acrylate-based coatings are produced at the facility. The facility receives, stores, and processes acrylate monomers including Ethyl Acrylate ("EA," CAS No. 140-88-5) and Butyl Acrylate ("BA," CAS No. 141-32-2).

2.1 Acrylate Properties

At room temperature, EA monomer is a colorless liquid with a pungent odor and a low odor threshold of approximately 1.2 ppb [1]. BA monomer is also a colorless liquid, with a fruity odor and an odor threshold of about 35 ppb.

The acrylate monomers are inhibited with the methyl ester of hydroquinone (MEHQ) to prevent unintended polymerization during storage. This inhibitor requires the presence of oxygen to retain activity. While systems exist to provide a mixed nitrogen/oxygen gas with a low enough oxygen level concentration to maintain both inhibitor activity and inertion, this type of mixed gas system has the potential to fail to an unsafe state. The limiting oxygen concentration for EA is 9.0 volume percent; thus, the control system to manage the atmospheric concentration would require a high level of reliability. The failure of the atmospheric control system can, and has, resulted in unintended polymerization in industry. In this installation, the risk of a runaway polymerization due to mixed gas failure was determined to be a greater risk than that of headspace ignition, so the tank was blanketed with air to maintain inhibitor efficacy.

The flash point of EA was below the ambient temperature of 19°C at the time of the incident, so the vapor headspace of the tank was flammable. The nearby BA tank's headspace, however, was not in the flammable region. The initial headspace vapor concentrations are shown on a vapor pressure curve for EA and BA in Figure 2.

Table 1. Physical properties of the acrylate monomers.

Property	Ethyl Acrylate	Butyl Acrylate
Flash Point	8°C	37°C
Lower Explosion Limit (LEL)	1.40%	1.30%
Upper Explosion Limit (UEL)	14.0%	9.90%
Maximum Experimental Safe Gap (MESG)	0.86 mm	0.88 mm
Odor Threshold	1.2 ppb	35 ppb

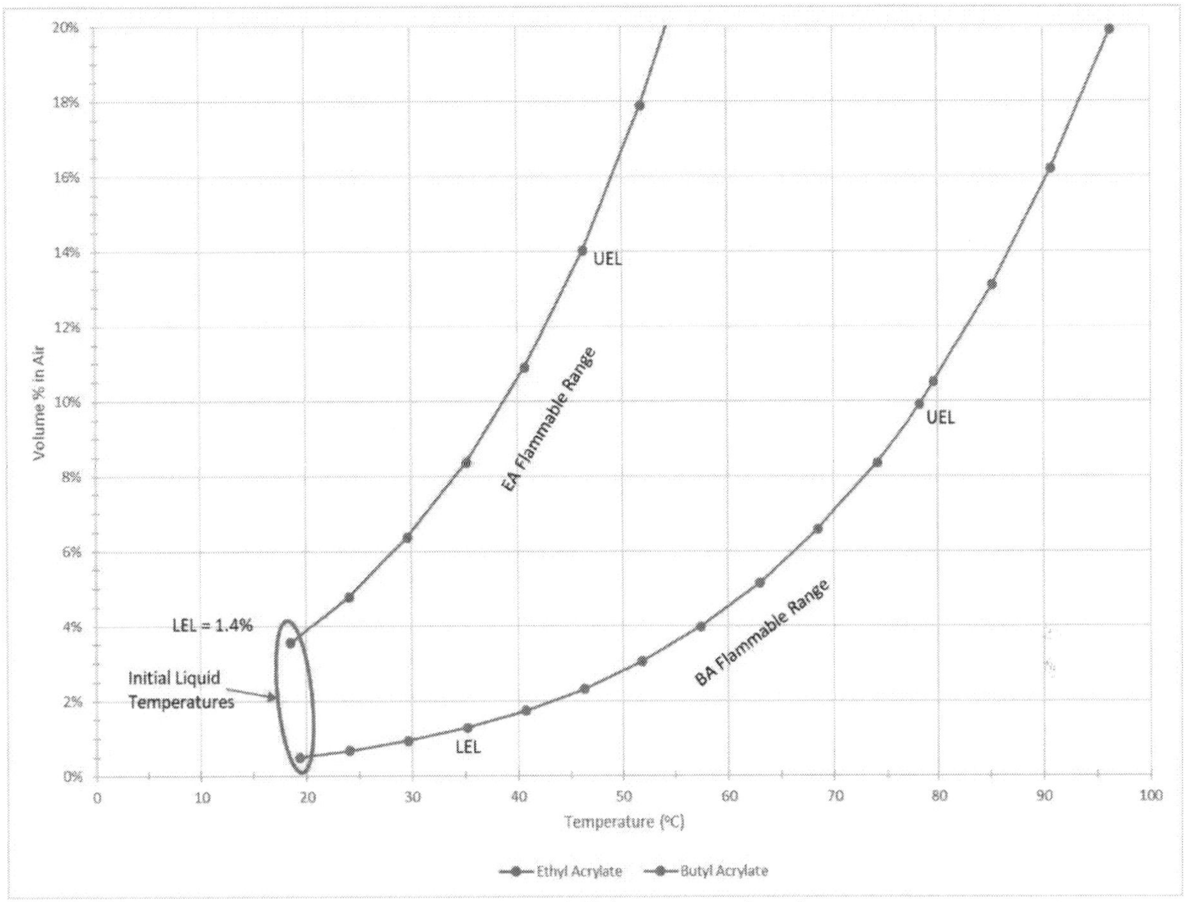

Figure 2 Volume percent of EA and BA vapors in air versus temperature.

2.2 Description of the Tank Farm

EA and BA monomers were stored in Tanks 120 and 119, respectively, in Tank Farm 30A.

Figure 3 is an aerial photograph of the area taken circa 2007. Liquid monomers were pumped into the EA and BA tanks from trucks and rail cars and transferred to the downstream manufacturing processes as needed. Truck and rail unloading stations are located around the perimeter of the tank farm dike walls. Figure 4 is a view of the tanks after the incident.

Figure 3 Aerial image of EA Tank (south) and BA Tank (north) in Tank Farm circa 2007.

Figure 4 View of tank farm after fire suppression efforts. Tanker truck in foreground was unloading MMA at the time of the incident.

The two atmospheric storage tanks involved in the incident contained inhibited BA liquid and inhibited EA liquid. The EA tank inventory was 673,482 lb and the BA tank held approximately 536,220 lb. At the time of the event, no transfers were occurring in or out of the EA or BA tanks. Truck unloading was occurring to a Methyl Methacrylate (MMA) tank in a separate tank farm area; however, the investigation concluded that MMA did not play a role in the event.

As shown in Figure 5, each storage tank was filled through an overhead dip tube and was equipped with a vapor header connection to the thermal oxidizer, an atmospheric conservation vent, and an atmospheric emergency pressure relief vent. The vapor header connection and atmospheric conservation vent were isolated from the tank using inline deflagration arrestors rated for this service.

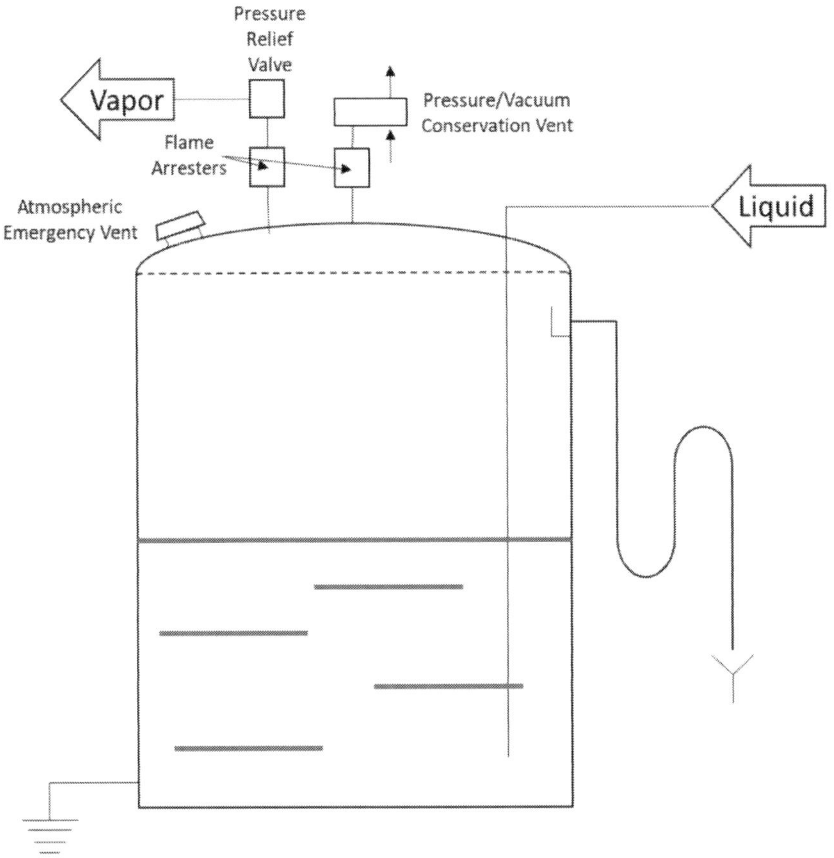

Figure 5 General tank appurtenance configuration and relationship.

2.3 Tank Construction

Tank 120 was in dedicated inhibited Ethyl Acrylate (EA) monomer service, and Tank 119 was in dedicated inhibited Butyl Acrylate (BA) monomer service. The carbon steel tanks were operated as atmospheric storage tanks and were originally constructed in the 1940s. The tanks were 30-feet in diameter by 29-feet tall with dome roofs. The dome roofs were originally constructed from petal sections with riveted overlap joints. At some point in the past, the rivets had been removed and the overlapped joints had been welded. The tanks were both 30 ft in diameter by 29 ft tall carbon steel cylindrical storage tanks with domes roofs. The steel thickness at the roofs and walls was approximately 0.3 inches. The tanks were equipped with a static grounding system and lightning protection consistent with guidelines in NFPA 780.

Table 1. Tank appurtenances.

Vent Device	**Liquid Line**	**Instrumentation**
Pressure/vacuum conservation vent to atmosphere	Liquid fill line	Capacitance level probe
Pressure relief valve to vent system (oxidizer)	Liquid withdrawal line	Liquid temperature probe
Pressure relief hatch to atmosphere	Overflow line with loop seal	Liquid temperature gauge
	Fire foam nozzle	

2.4 Tank Farm Grounding Installation

At the time of the event, National Fire Protection Association (NFPA) 780-2011, Standard for the Installation of Lightning Protection Systems, was a recognized industry standard. Section 7.2.2. of NFPA 780 advises: [2]

"Metallic structures that are electrically continuous; tightly sealed to prevent the escape of liquids, vapors, or gases; and of 3/16 in. (4.8mm) thickness or greater to withstand direct strikes... shall be considered to be inherently self-protecting."

American Petroleum Institute (API) Recommended Practice 2003, seventh edition (2008), also provided guidance. Section 5.4.1 stated: [3]

"Metallic tanks, equipment, and structures commonly found in the petroleum industry that are in direct contact with the ground (i.e. no nonconducting membranes) have proved to be sufficiently well grounded to provide for safe propagation to ground of lightning strokes. Supplemental grounding by means of driven ground rode neither decreases nor increases the probability of being struck, nor does it reduce the possibility of ignition of the contents."

The EA and BA tank installations at Bristol were found to be in compliance with these provisions. The vessels were steel tanks exceeding 3/16" thickness that were mounted on concrete pads and attached to grounding rods. Metallic piping and conduit were appropriately bonded to the tanks. There was an insulating membrane installed in the containment area; however, it did not isolate the tank and its associated pad from the connection to earth. Routine maintenance and testing were performed on the grounding system to maintain it in good working order. Therefore, no gap between industry guidance and the tank installations was identified.

2.5 *Deflagration Arresters*

Deflagration (flame) arresters were installed on the storage tanks to prevent the propagation of deflagrations from an external source into a tank storing combustible or flammable material (Figure 6). Flame arresters typically contain small, high surface area passages through which momentum and heat of a flame front passing through a flammable vapor/air mixture in piping are quenched. Flame arresters are sized according to the chemicals involved. NFPA 497 reports maximum experimental safe gap (MESG) values between the plates of 0.86 mm for ethyl acrylate (inhibited) and 0.88 mm for n-butyl acrylate (inhibited).[4] These gap sizes correspond to Class I Group D materials, or those materials with an MESG greater than 0.8 mm.[5]

Figure 6 Example pressure/vacuum relief device (VPRV) mounted above a deflagration arrester.

3 Investigation

The investigation was initiated once the fire was extinguished and the scene was rendered safe. The focus of the investigation was to gather data to understand the sequence of events, identify the ignition mechanism for the EA tank, and determine the means by which the BA tank also ignited.

3.1 Understanding the Sequence of Events

3.1.1 Eyewitness accounts

Two witnesses were located in the area of the tank farm at the time of the lightning strike and explosion, with positions shown in Figure 7. None of the witnesses saw the location of the lightning strike. One witness (Witness #1) was sitting in a truck by the MMA unloading station completing paperwork. He saw the flash and heard the lightning strike, which was shortly followed by the EA tank explosion. A second witness (Witness #2) was located in a nearby building, sitting at a desk on the east side. He looked up immediately after the strike and noted flames on top of the EA tank, which preceded the explosion.

Figure 7 Witness locations to the lightning strike and ensuing events.

The lightning discharge itself was not captured by surveillance video; however, a flash associated with the lightning strike was recorded by a camera pointing to the northwest.

3.1.2 Evidence collection

Immediately after the emergency response activities were completed, the scene was secured. Photographs were taken of the tank farm and surrounding areas. A chain-of-custody process was established for evidence. Debris from the explosions was located, tagged, photographed, logged, bagged, and retained in a locked area. Security camera footage from all site cameras was secured. The Distributed Control System (DCS) data was retrieved and archived so that it would not be overwritten.

3.1.3 Verification of lightning strike

At the time identified in security video and supported by witness testimony, a lightning strike to ground was recorded by two lightning verification services: StrikeNet and TOA. Both organizations recorded a strike to ground within the Bristol site with a magnitude between -36 kA and -38kA. Although the sensors used by the two organizations placed the lightning strike within the Bristol site at slightly different locations, the tank farm fell within the confidence zones for both sets of data.

The equipment, debris, and surrounding areas were inspected after the event to determine if the locations of one or more lightning attachment points could be located; however, no burn marks or other indications of a lightning strike were evident on the steel piping or equipment in the tank farm or connected to equipment in the tank farm. Attempts to detect erratic magnetic fields on steel equipment were also inconclusive. However, based on the eyewitness testimony, security camera footage, and lightning strike data, it was concluded that a lightning strike had occurred and initiated the event.

3.2 Evaluation of Physical Evidence

3.2.1 Review of tank damage

About a second after the lightning strike, the EA tank exploded (Figure 8). The explosion displaced the dome roof, separated piping systems, and damaged the lower tank shell. The explosion yielded an immediate fireball followed by a tank fire and pool fire in the surrounding dike area from burning EA monomer liquid. A post-incident photograph of the EA tank is shown in Figure 9. Flames were observed at dome level immediately after the lightning strike by an employee in a building about 100 feet southwest of the tank. The roof of the EA tank fractured into multiple pieces, and a portion was projected outward and hit a nearby trailer. A large section of the roof, including the deflagration arrester and vacuum/pressure relief assembly, fell into the tank. The northwestern and western portions of the upper-most segment of the sidewall of the tank, as well as the external ladder of the tank, were bent inward. The flat tank bottom bowed due to the internal pressure prior to rupture of the tank roof.

Figure 8 **First frame of surveillance video showing the EA tank explosion. This frame is time-stamped 1 second after the frame capturing the lightning strike.**

Figure 9 **Ethyl acrylate tank after the incident. The upper portion of the tank was curled into the tank from when the roof fell back into the tank after the explosion. Oxidation and soot deposition are evident on the outside of the tank wall.**

Approximately two minutes after the EA tank exploded, the BA tank also ruptured (Figure 10). The roof was found on a nearby loading rack with apparent soot deposition but with paint intact. The dome was curved inward and several of the radial seams of the roof were separated. A photograph of the dome appears in Figure 11.

Figure 10 BA tank from the west.

Figure 11 BA tank dome, which landed on a nearby loading rack.

3.2.2 Deflagration Arresters

The EA tank conservation vent device was found to be damaged but still attached to the tank's dome. The overpressure relief component of the conservation vent was separated from the rest of the assembly. The pressure relief valve pallet had been displaced upward, and the weather hood studs had been pulled out of the assembly flange. There was evidence that the guide cap impacted on the top weight, and the guide cap rivet was pulled out. The fluorinated ethylene propylene (FEP) polymer seal on the pallet appeared heat-affected.

As shown in Figure 12, the EA tank's conservation vent flame arrester internal cartridge was damaged mechanically – the plates were displaced. Examination of the plates in the cartridge revealed that the tank headspace side of the cartridge displayed a circular pattern of erosion indicative of an unknown duration exposure to flames venting from the tank into the flame arrester. The atmospheric side of the cartridge did not exhibit a similar heat-affected damage pattern. The presence of erosion on the tank side of the cartridge but not on the opposing side indicated that flame travel through the arrester was directed from the inside of the tank to the outside. This confirmed that the flame arrester operated as intended and did not allow flames to vent from the tank. This also supports a conclusion that the flame arrester cartridge would have prevented an external fire from passing into the tank headspace.

Figure 12 EA tank conservation vent deflagration arrester cartridge. Tank side on the left and the outside is on the right. Flame impingement pattern highlighted on inside of cartridge.

As shown in Figure 13, the BA tank conservation vent flame arrester showed similar flame impingement patterns from the inside of the tank as the EA tank.

Figure 13 BA conservation vent deflagration arrester cartridge. Tank side on left and outside on right. Flame impingement pattern highlighted on inside of cartridge.

The BA tank's vent line inline deflagration arrester was ejected from the tank and later found inside the tank after being subjected to the pool fire inside the tank. The deflagration arrester was too heavily damaged to determine if it was attacked from an internal or an external fire. However, the EA tank vent line flame arrester was intact enough to conclude that the EA tank fire was unlikely to communicate with the BA tank through the vent header. The EA tank vent line was equipped with a 3-inch inline flame arrester, shown in Figure 14. The flame arrester was located between the tank and the vent header, which went to the thermal oxidizer. The flame arrester assembly was fragmented and pieces were projected in multiple directions away from the tank. The impact caused collapse and deformation of the corrugated aluminum plates in the internal cartridge based upon the lack of fire/explosion damage to the edges of the plates. The plate array also displayed a circular pattern of erosion indicative of an unknown duration of exposure to flames venting from the tank into the flame arrester. The atmospheric side of the cartridge did not exhibit a similar heat-affected damage pattern. The plate surfaces were also free of heat-affected damage patterns. The presence of erosion on the tank side of the cartridge but not on the opposing side indicates that flame travel through the arrester was directed from the inside of the tank to the outside. As shown in Figure **15**, the clean plate surfaces indicate that the flame penetrated a minimal depth into the cartridge. This indicates that the flame arrester operated as intended and did not allow flames to vent from the tank. This also indicates that an external fire, for example due to the ignition of a release of flammable vapor through the shared vent line, did not pass into the tank through the arrester.

Figure 14 EA vent line inline deflagration arrester cartridge. Tank side on left and outside on right. Flame impingement pattern highlighted on inside of cartridge.

Figure 15 Image of clean plate surface inside cartridge indicating that flame penetration depth into the cartridge was minimal.

3.2.3 Location of EA tank headspace ignition

After the lightning strike, there was a time delay before the EA tank exploded. Based on the witness statements, security camera footage, and recorded process data, the following sequence of events was identified for the explosion. Table 3 provides the sequence of events.

Table 3. Sequence of events.

3:47:08	Lightning strike
	Flames visible on roof of EA tank
3:47:09	Explosion of EA tank
	EA tank fire and pool fire in dike
	BA tank explosion

The duration of the delay was approximately 1 to 1.2 seconds. During this interval, Witness #2 reported seeing flames on top of the EA tank but could not be more specific about the nature of those flames. Possible explanations include ignition of a pre-existing vapor release from one of the roof appurtenances or a crack in the roof, burning of a combustible solid material such as a plastic component, or venting of flames from the emergency vent. There are many parameters that may have affected this timing, including the flame speed in EA vapor/air mixtures and the effects of pressure relief devices. The approximate method presented in the SFPE Handbook was used to estimate the pressure rise with time inside the EA tank head space by assuming propane combustion properties for the EA vapors (Equation 1).[6] Given these calculations, the estimated burst pressure for the vessel of approximately 2 psig would be reached in about 1 second, as shown in the graph in Figure 16. Thus, the 1 to 1.2 second delay between the lightning strike (ignition) and explosion is reasonable and consistent with the delay expected for an internal deflagration to cause the tank to rupture.

Equation 1:

$$\frac{P - P_0}{P_0} = \left(\frac{1}{\gamma} + \frac{P_0}{P_m - P_0}\right)^2 \left(\frac{P_m - P_0}{P_0}\right)^3 \left(\frac{\chi S_u t}{a}\right)^3 \ll 1 \quad (7)$$

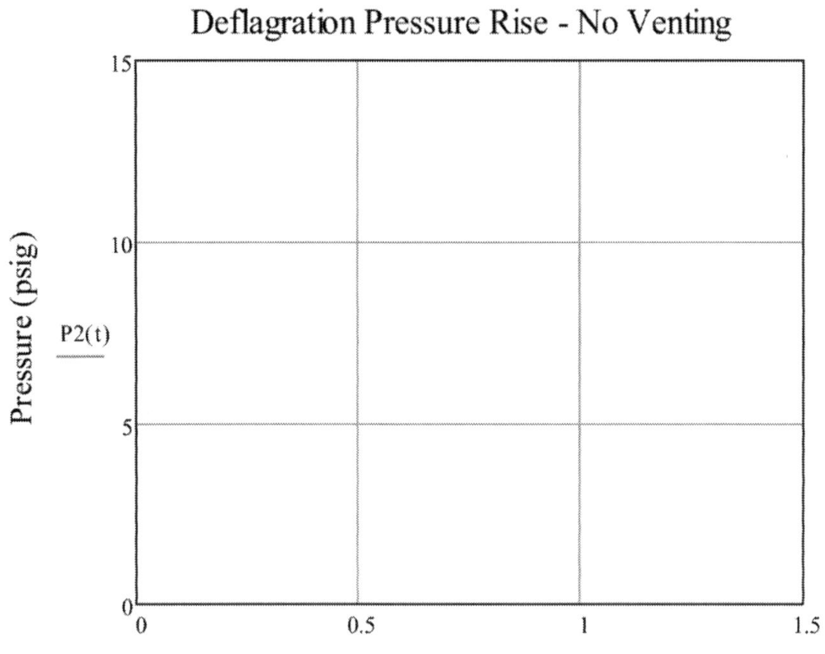

Figure 16 Example pressure rise curve versus time for propane combustion properties assigned to EA tank head space.

Based on this calculation and logical deduction, the point of ignition inside the headspace was likely near the 12-inch emergency vent such that flames would be issuing from the vent for a period of time prior to tank rupture pressure being reached. The emergency vent would be anticipated to open quickly after an internal pressure rise begins, which would allow flames to vent. If the internal deflagration had originated away from this area, the witness would have only observed flames close to or coincident with the explosion.

3.3 Testing Hypotheses

For a fire to occur, it is necessary to have a fuel, an oxidant, and an ignition source. In this case, the flashpoint of EA was 8°C – well below the 19°C temperature of the liquid. The tank contents were stored under an air atmosphere. Since fuel and oxygen were both present in the EA tank, the investigation focused on determining potential ignition mechanisms for the EA tank.

While the headspace of the EA tank was flammable, the flash point of BA is 37°C – above the ambient temperature at the time of the event. As a result, hypotheses were developed to determine not only the ignition source for the BA tank but also how the tank temperature increased to the point where the contents became flammable.

3.4 Hypotheses: Ignition of the EA Tank

Three potential sources of the EA tank ignition were evaluated:

1. Ignition source entered the EA tank via the vent system
2. Ignition source entered the EA tank while it was breathing/venting to atmosphere
3. Ignition source internal to the tank.

3.4.1 EA Hypothesis 1: Ignition source entered the tank via the vent system

The vent header could possibly have contained a flammable mixture due to the MMA truck unloading activities that would displace the MMA tank flammable headspace into the vent header. However, ignition in the vent header system outside of the tanks was eliminated as a potential cause based upon several observations. The fire pattern present on the EA tank vent header deflagration arrester was located on the inside (i.e., tank side) of the deflagration arrester, indicating that flames were venting into the arrester cartridge for some period of time prior to the explosion. Additional evidence that discounted external ignition in the vent system included the lack of soot in examined piping, the lack of overpressure damage in piping and connected tank components, and the lack of indication of failure of the PRV isolating the tank from the vent header.

3.4.2 EA Hypothesis 2: External ignition source while EA tank was breathing/venting

Although the EA tank was not directly associated with the filling activities ongoing at the time of the incident, the possibility for flammable EA vapors to leak from the tank was evaluated as a potential path to internal ignition. Since the EA tank was anticipated to act as a Faraday-like cage and prevent penetration of an electrical discharge, the possibility of an external discharge igniting a vapor leak was considered. Four primary leak locations were identified and evaluated:

1. Release from a crack in the shell or dome roof
2. Release from an incompletely filled overflow line loop seal

3. Release from EA conservation vent
4. Release from EA 12-inch emergency vent

EA vapors have a very low detection threshold (0.5 ppb), and from the investigation team's experience, a leak in the tank would have been detected during normal operator inspections if it was near ground level. A small leak at roof level could not be as easily detected. A recent tank inspection did not detect cracks in the tank shell or vapor leaks in the tank. If such a small crack existed, it would be unlikely to be larger than the minimum experimental safe gap (MESG) of 0.86 mm; thus, it was unlikely that ignition could propagate back into the tank through such a crack. Therefore, a leak from a crack in the shell or dome roof was refuted as a potential ignition pathway.

The loop seal fluid had been replaced in 2010, and a leak past the seal fluid would be anticipated to be detected by field operators since the opening is near ground level adjacent to the tank. Thus, a leak from the loop seal was refuted as a potential source for vapors.

An ignition pathway through the EA conservation vent was also considered. However, the deflagration arrester did not exhibit signs of impingement by an external flame. Additionally, the deflagration arrester appeared to extinguish the internal flames as intended before they could pass through the cartridge. The deflagration arrester cartridge was packed with parallel plates and suitable for suppressing flames in either flow direction. Thus, an external ignition at the VPRV that flashed back into the tank to ignite the tank headspace vapors was refuted as a possible cause.

The larger 12-inch diameter emergency vent was not equipped with a deflagration arrester. The pallet assembly had a rigid FEP film diaphragm seal, which was an electrical insulator. The device was destroyed in the incident, but most major components were recovered. From the recovered components, no evidence was found of electrical bonding (e.g., bonding strap) between the pallet and the frame or tank. The listed set point for the emergency vent was the same (1.5 psig) as that for the conservation vent; thus, the tank could have breathed out through either vent. An ignition at this emergency pressure vent assembly could not be refuted during the investigation.

3.4.3 EA Hypothesis 3: Ignition source generated from an internal spark

As indicated in NFPA 780-2011 and API 2003 (2008), metallic tanks that are electrically continuous, sealed, and in direct contact with the ground are considered to be protected from lightning. This is based on the assumption that the steel tank system acts like a Faraday cage. A Faraday cage is an electrically conductive enclosure that protects the interior of the enclosure from the effects of external electrical fields. Electrical charges are distributed and flow to ground, cancelling the field's effect on the interior.

However, the protection offered by the Faraday cage can be negated by the presence of an unbonded internal component. As API 2003 (2008) pointed out:

"Metal tanks with fixed metal roofs and horizontal metal tanks that are maintained in good condition are generally protected from damage by direct-stroke lightning and ignition of its contents, if all metal components are in electrical contact (i.e. bonded)."

It was therefore possible that direct internal ignition of the EA vapors inside the tank could have occurred due to a spark discharge across a gap between two internal tank components that were not electrically bonded with each other. A detailed analysis of the tank instrumentation, equipment, and appurtenances yielded a list of several potential ignition locations. Piping penetrations into the headspace were either refuted as non-credible or assigned a low likelihood given the nature of mechanical contact between piping and the tank shell.

Conservation and emergency relief vents

The conservation vent and the emergency relief vent both utilized weighted floating pallets with FEP diaphragm seals. The pallets were not electrically bonded to the device bodies or the tank shell. Given the discontinuous electrical nature, it is possible that a spark discharge occurred inside either assembly. However, the conservation vent was separated from the tank headspace by the deflagration arrester. Thus, the conservation vent was refuted as a potential ignition source.

The pallet manway was assumed to behave as a capacitor in the event of a lightning strike, and calculations were performed to estimate the energy of a spark that could have occurred between the pallet and the tank. The calculations indicated that the energy generated would only reach about half the minimum ignition energy (MIE) of ethyl acrylate. However, due to uncertainty in the estimation, the emergency relief vent could not be refuted based on the available information.

High level capacitance probe

The tank was also equipped with a high-level capacitance probe. It was determined that the electrical ground lead inside the sensor housing was not connected to an earth ground, which may have affected the device's ability to dissipate electrical charge. It was noted, however, that the circuit breaker remained active and did not trip during the event. An examination of the level switch showed that the fuse was intact and there were no signs of overvoltage/overcurrent. However, it appeared that the diodes between probe and shield inputs may have been shorted, which could have occurred if the arc were limited to a few microseconds of exposure. Although evidence was not found to indicate that a spark actually occurred between the high-level probe and the tank, the potential for a spark discharge from the probe to a surface inside the tank could not be entirely refuted.

Surface charge

The team also considered the possibility of a liquid surface charge creating a spark discharge. However, this was deemed unlikely based on the conductivity of EA.

3.4.4 Plausible hypotheses: EA tank ignition

Based on this evaluation, the most likely ignition pathway was an electrical discharge from one of two electrically isolated equipment components internal to the EA tank – the high-level probe or the pallet manway.

3.5 Hypotheses: Ignition of the BA Tank

Although the headspace in the EA tank was flammable, the BA liquid was below its flashpoint and was not in a flammable region. Thus, only one side of the fire triangle was present. For ignition to have occurred within the BA tank approximately two minutes after the lightning strike, a source of ignition energy must have made its way into the tank head space, and the tank head space must have had a flammable concentration of vapors. Hypotheses were therefore generated to explain both the generation of an ignitable mixture of BA in air and the source of ignition.

3.5.1 Creating a flammable atmosphere in the BA tank

The vapor space of the BA tank was not flammable under normal operating conditions. The measured temperature of 19.4°C prior to the incident was well below the flash point of 37°C reported in the Safety Data Sheet. However, there must have been a flammable composition in the vapor space of the BA tank at the time when the tank exploded approximately two minutes after the EA tank explosion. The BA tank was in close proximity to the burning EA tank, and the two tanks shared a common dike area. Heat output from the EA fire in the dike, and possibly heat from the EA tank itself, could have raised the temperature of the BA tank, enriching the vapor space in BA vapors. The temperature probe in the BA tank was located in the lower part of the tank and ceased its output to the control system a little more than a minute after the EA tank erupted. Prior to its failure, the thermocouple had registered no appreciable change in temperature. However, given the size of the tank and proximity to the fire, it is very likely that a portion of the tank contents was raised significantly in temperature from the radiant heat of the EA tank fire and potentially direct flame impingement, possibly creating boiling at the shell interface.

One potential outcome of boiling inside a tank due to pool fire exposure is a Boiling Liquid Expanding Vapor Explosion (BLEVE). This type of explosion is possible if the tank is a rated pressure vessel and pressure relief systems are unable to maintain a safe pressure during fire exposure. Since the BA tank was an atmospheric storage tank and the shell did not display bulges, a BLEVE was refuted as a potential cause of explosion.

Three hypothetical paths to ignition of the BA tank vapor space were identified: ignition in the vent pipe which entered the tank, ignition inside of the tank, and ignition outside of the tank which propagated inside.

3.5.2 BA Hypothesis 1: Communication through the vent header

At the time of explosion of the BA tank, the vent header system had been compromised at the EA tank. Despite this damage, ignition in the vent header system outside of the BA tank was eliminated as a potential cause based upon observations of the deflagration arresters. Although the vent line deflagration arrester was heavily damaged and conclusions could not be drawn, the counterpart deflagration arrester beneath the conservation vent displayed internal fire patterns, not external patterns. This model of flame arrester appeared to successfully extinguish flames in EA and BA vapor streams based upon analysis of the surviving components.

3.5.3 BA Hypothesis 2: Ignition through the conservation vent

Because of the fire damage to the conservation vent flame arrester on the BA tank, no direct evidence of functionality could be determined. However, there were no apparent gaps in the plates in the cartridge. Given its rating for this service, it is unlikely that BA vapors venting through the conservation vent could flash back through the device into the tank headspace.

3.5.4 BA Hypothesis 3: Hot surface ignition by the exterior pool fire

Ignition inside of the tank could have occurred due to hot surface ignition within the vapor space. The low BA auto-ignition temperature of 292°C could possibly have been achieved on the inner tank shell surface from radiant heat or direct flame impingement at or near the liquid level where BA vapors would be anticipated to be within the flammable range. The BA tank shell was heavily heat-affected by the fire, and all paint had been heavily damaged. The tank roof was still painted, but the side facing the EA tank was soot covered indicating fire exposure. Thus, this pathway of hot surface ignition inside the tank head space could not be eliminated as a hypothetical cause.

3.5.5 BA Hypothesis 4: Direct ignition by the surrounding fire

Given exposure of the tank to the surrounding fire and the evidence of an internal deflagration, it is apparent that the liquid inside the BA tank was generating vapors. The BA tank was equipped with both a conservation vent and an emergency pressure vent. Since the conservation vent and the 12-inch emergency pressure vent pallet manway were set at the same relief pressure; it is likely that both vents were relieving BA vapors during the incident. Under this condition, the nearby EA tank and EA liquid pool fires could have directly ignited the vented vapors if flames contacted the vapors. The resulting ignition at the pallet manway would have allowed a flash back into the tank head space. This potential ignition pathway was a likely cause for the BA tank explosion.

3.5.6 Plausible hypotheses: Ignition of the BA tank

Although autoignition of the BA as a result of direct flame impingement could not be ruled out, the more likely ignition pathway was ignition of BA vapors venting from the emergency pressure vent pallet manway and flashing back into the tank.

4 Analysis and Conclusions

While the location of the lighting strike could not be precisely determined, the investigation concluded that the lightning strike caused ignition of the flammable head space vapors in the EA tank. The resulting internal deflagration caused the tank's internal pressure to quickly rise to a level causing the tank dome roof to rupture. The subsequent EA liquid pool fire in the containment dike and in the ruptured EA tank caused the BA tank, which did not normally contain a flammable vapor head space, to raise the head space concentration into the flammable range such that internal deflagration could occur.

The most likely cause of ignition of the EA vapors was determined to be an internal spark discharge from a non-electrically bonded component. The two most likely candidate components were the high-level capacitance probe and the 12-inch diameter emergency pressure vent pallet manway. The most likely cause of ignition of the BA vapors was determined to be either hot surface ignition by the internal shell above the liquid level or direct ignition of vapors vented from the 12-inch diameter emergency pressure vent pallet manway by the surrounding fire.

Following the event, Dow created a corporate learning experience report that was shared throughout the organization. It described the event, detailed the root and contributing causes, and discussed key learnings. A number of follow-up actions were also created. These included:

- Review operations policies and procedures concerning suspending manual loading and unloading of flammable materials when there is nearby lightning. Lightning guidance from the National Weather Service [7] recommends stopping outdoor activities from the time when lightning is first seen/heard until 30 minutes after the last clap of thunder.
- Review flame arrester maintenance procedures to ensure that they include plate spacing observations and proper methods for cleaning. Although the flame arresters were found to have worked as designed in this event, it was a good opportunity for the organization to reaffirm that all flames arresters are being properly serviced.
- Review maintenance procedures for inspection and testing of tank grounds and lightning apparatus to confirm that all are being maintained and work as designed.
- For storage tanks that can have a flammable headspace, complete an assessment of the tank, which included:
    - Ensuring that hatches are closed when not in use
    - Ensuring that roofs are in good condition, with no non-conductive patches
    - Evaluating the installation for the appropriate use of vacuum vent valves, flame arresters, gas padding, and other means to prevent flame propagation into the tank.
    - Assessing the equipment for tank penetrations that are not bonded to the tank shell or are not electrically continuous throughout their construction. These could include dip pipes, float level indicators, etc.

Action plans were developed based on the opportunities that were identified from this review.

5 Lessons Learned

Key takeaway messages from this investigation include:

- Lighting strikes are powerful natural events that can't be prevented. It has been estimated that lightning strikes would be expected to increase by about 12 percent per degree Celsius of warming [8]. It is critical to design our equipment and facilities to withstand extreme weather events like lightning.

- Following industry standards on lighting will not prevent all ignition events.

  As stated in API Recommended Practice 2003 (2008), Protection Against Ignitions Arising Out of Static, Lightning, and Stray Currents Section 5.1, it should be noted that: *"Even when all known precautions are employed, prevention or safe dissipation of direct-stroke lightning cannot be absolutely assured. In the case of indirect lightning currents, incendive sparks may occur in some segments of a system that use the best-known precautionary methods and devices."*

- Current standards may not protect against the potential hazard of unbonded components.

  The Faraday cage assumption inherent in the assumption of grounding in the event of a lightning strike assumes that there are no unbonded components. Should that assumption be invalidated, a system that otherwise complies with industry lightning guidance and best practices may not actually be protective. An evaluation of uninerted tanks in flammable service is recommended to identify and remediate any unbonded components.

- To mitigate the risk of lightning, it is prudent to employ good industry practices against ignition, such as flame arresters, closed operation and venting, and grounding. Also, when installing new tanks, consider the proximity to existing tanks. The BA tank likely would not have ignited had it not been in close proximity to the EA tank and had not shared a common dike.

- Consider inerting flammable tanks in situations where the risk of ignition outweighs the risk of failure of the inerting or mixed gas system.

6 References

[1] ACRYLATE ESTERS A SUMMARY OF SAFETY AND HANDLING 3RD EDITION Acrylate Esters, A Summary of Safety and Handling, 3rd edition, 2002. Available at ICSHAM Acrylic Esters Safe Handling Guide (dow.com).

[2] National Fire Protection Association (NFPA) 780, Standard for the Installation of Lightning Protection Systems, 2020 edition.

[3] American Petroleum Institute (API), Recommended Practice 2003, seventh edition (2008).

[4] NFPA 497, Recommended Practices for the Classification of Flammable Liquids, Gases, or Vapors and of Hazardous (Classified) Locations for Electrical Installations in Chemical Process Areas, 2012 edition.

[5] Grossel, S.S., *Deflagration and Detonation Flame Arrestors*, CCPS (2002).

[6] Society of Fire Protection Engineers (SFPE), Handbook of Fire Protection Engineering, 3rd Edition, section "Closed Vessel Deflagrations," pages 3-406 to 3-410 (2002).

[7] National Weather Service, https://www.weather.gov/safety/lightning-safety-overview, accessed March 14, 2021.

[8] Scientific American, https://www.scientificamerican.com/article/lightning-may-increase-with-global-warming/, accessed March 14, 2021.

618785 Dam(n) Safety Lessons: Risk Based Process Safety Principles Applied to a 2020 Dam Failure

Abstract:

On May 19, 2020, over 10,000 Mid-Michigan residents were ordered to evacuate their homes during a pandemic because of two impending dam failures. Swift emergency response coordination enabled people to evacuate in time and no one lost their lives. Unfortunately, nearly every business in the village of Sanford, MI was destroyed; more than 2,000 homes were damaged or condemned; roadways and bridges were swept away; and two lakes disappeared within hours, resulting in permanent disruption to the local ecosystem.

This disaster could have been avoided: The dam system had experienced three near misses in its 95-year history prior to failure, and deficiencies were discussed between the dam owners, government, and community for over 11 years. In addition, dam operators financially struggled in a significantly changing business model for small-scale hydroelectric production. Interestingly, this experience is not unique to dams and infrastructure; it can be applied to companies that manufacture, use, transport and store hazardous materials.

Process safety is the study of applying management systems to prevent rare, catastrophic events such as fires, explosions and toxic releases. The probability of these catastrophic events is rare, but when they do occur, they can destroy the lives of workers, the community, and the reputation of the organization.

We can apply the elements of Risk Based Process Safety (RBPS) to the 2020 Mid-Michigan dam failures as a case study to demonstrate where systems failed and understand how these risks could be averted in the future. This case study will show the impact of using RBPS to prevent rare, catastrophic events at many types of manufacturing and warehousing facilities.

This paper will walk through the four pillars of the CCPS Risk Based Process Safety Model, using publicly available data from the 2020 Michigan dam failures as a case study to illustrate how impactful the RBPS pillars can be for many non-chemical industry segments:

1. **Commit to Process Safety**, including the importance of leaders giving authority to competent personnel to escalate concerns so that concerns can be identified and resolved swiftly, and to perform due-diligence audits prior to acquiring a high-risk operation.
2. **Understand Hazards and Risks**, including the success story where Emergency Responders used knowledge of the projected worst-case flood analysis to quickly evacuate at-risk member of the community.
3. **Manage Risk**, including maintaining equipment to meet Recognized and Generally Accepted Good Engineering Practices (RAGAGEP), and to swiftly resolve deficiencies once discovered.
4. **Learn from Experience**, including ensuring someone with authority reviews facility incident data; Industry incident data; and third-party audit inspection results to shift resources to fix a systems issue before disaster strikes.

Which eye are you looking through? The impact of cognitive bias on process safety

Trish Kerin
IChemE Safety Centre
Melbourne, Australia
tkerin@icheme.org

Prepared for Presentation at
American Institute of Chemical Engineers
2021 Spring Meeting and 17th Global Congress on Process Safety
Virtual
April 18 - 22, 2021

Which eye are you looking through? The impact of cognitive bias on process safety

Trish Kerin
IChemE Safety Centre
Melbourne, Australia
tkerin@icheme.org

Keywords: bias, culture, leadership, lessons learned

Abstract

Bias surrounds us every day. Everything we look at has a lens of our experience coloring our perception, and everyone has their own lens. This impacts how we share and receive information. So why is this important for process safety? Cognitive biases cover a range of aspects, each of which can have an impact on process safety, from what we remember, to coping with too much information, to a lack of context and meaning or a need to act with great haste. There are many identified specific cognitive biases, not every one of them impacts on process safety, but many can and have. This paper will review several process safety type incidents that had cognitive bias as a factor and discuss ideas on how to manage the bias, so we do not fall into its trap.

1 Cognitive bias

Cognitive bias is defined as a systematic deviation from norms or rational judgement, a way of creating a subjective based reality based on an individual's perceptions [1]. While on the surface this sounds like a pitfall in human decision making, our cognitive biases allow us to make quick decisions. This can be advantageous when time is of the essence, provided the bias does not lead us in the wrong direction. Making decisions based on biases, or heuristics to estimate outcomes is a faster way to make decision than rational decision making, and it also fits in with the idea that we have two mind systems, system 1 and system 2. "System 1 operates automatically and quickly, with little or no effort and no sense of voluntary control. System 2 allocates attention to the effortful mental activities that demand it, including complex computations." [2, p. 20]. The problem with system 1 is that it is sometimes wrong, and unless we use system 2 to override the wrong answer, we can end up with the wrong decision. Take for example Figure 1. Quickly glance at it an decide which figure is the largest. Your system 1 would have said the one on the right.

Figure 1 – Which figure is larger? [2, p. 100]

If you were to engage your system 2, and take a ruler and measure the figures you would see they are all the same size. System 1 was misled by a perspective perception. The challenge for managing bias is to engage system 2 prior to making critical decisions.

Another concept to describe this System 1 thinking is called "thin slicing". This "refers to the ability of our unconscious to find patterns in situations and behavior based on very narrow slices of experience." [3, p. 23] It can be summarized as that initial gut decision we may feel in an instant.

1.1 Most common biases

There are 188 identified different cognitive biases, though they can be grouped into the 4 categories of:

- What should we remember?
- We need to act fast
- Too much information
- Not enough meaning [4]

Common biases that can have an impact on safety are:

- Anchoring
- Confirmation bias
- Framing
- Hindsight bias
- Status quo bias
- Sunk-cost bias [5]

In addition to these six, Halo bias can lead to giving more weight to information from certain people than others.

Table 1 shows the linkages between the groups, sub-groups and biases as categorized by Benson (2017). Note none of the biases discussed in this paper fit into the "What should we remember?" group.

Table 1. Groups, sub-groups and biases [4]

Group	Sub-group	Bias
We need to act fast	To avoid mistakes we aim to preserve autonomy and group status to avoid irreversible decisions	Status quo bias
	To get things done, we tend to complete things we've invested time & energy in	Sunk cost fallacy
Too much information	We notice when something has changed	Framing
		Anchoring
	We are drawn to details that confirm our own existing beliefs	Confirmation bias
Not enough meaning	We project our current mindset and assumptions onto the past and future	Hindsight bias
	We imagine things and people we're familiar with or fond of as better	Halo effect

2 How biases impact decision making

System 1 is the part of our mind that makes decisions quickly, and without active thought. It is doing this based on our past experiences and various cognitive biases that exist for each of us. This means that our biases have an effect on our decision making when our system 1 is in control.

2.1 Status quo bias

This is a tendency to want things to remain the same as they have been, an avoidance of change [5]. An everyday example here may be remaining with your current cellular phone provider even though a competitor may offer better rates or service. This type of bias can become an issue if the current way is not working or has unmanaged hazards in it.

2.2 Sunk cost fallacy

This is a tendency to want to continue with a course of action, even when evidence suggest otherwise, in an attempt to recoup losses [5]. It can occur when large sums of money or effort have been invested in something and it has not shown the expected rewards. There can be a tendency to invest more, even though evidence suggest that it will not be successful or indeed hazardous to continue. An everyday example here could be to 'double or nothing' in a game of chance. The odds of winning have not got better, but an attempt is made to win back previous losses by betting higher amounts. This type of bias can become an issue where it is potentially not safe to continue.

2.3 Framing

How information is presented can change how the information is interpreted [5]. An everyday example here may be that a yoghurt brand markets it product as 90% fat free, the corollary to this is that the product is 10% fat. If you are conscious about the amount of fat in your food you would more likely choose the 90% fat free rather than the 10% fat product. This type of bias can become an issue where data is presented in a misleading way. It may also have an opposite effect when some data is presented in a way to prime the mind so that it is more likely to make a favorable decision when caught up in the "thin slicing" stage of thinking [3].

2.4 Anchoring

This is based on placing more emphasis on the first information received, even if subsequent information shows it to be false [5]. An everyday example here might be when shopping and seeing a marked down price – 'was $200 now $150'. Seeing the first price of $200 may make you think that $150 is a good deal, because your expectation is anchored at $200. This type of bias can become an issue where the first piece of data is incorrect, and should be discounted from the decision making process.

2.5 Confirmation bias

This relates to placing more emphasis on data that supports your initial conclusion [5]. It can lead to the discounting of contrary data or searching for data that confirms your thoughts. An everyday example here is related to social media algorithms. We are presented with streams that typically confirm our own beliefs and we do not see counter information, which can lead us to believe that our thoughts are the same as the broader community. This type of bias can lead us to make the wrong decision because we do not accept data that disagrees with our initial thoughts.

2.6 *Hindsight bias*

This bias "is defined as the belief that an event is more predictable after it becomes known than it was before it became known" [6, p. 411]. Hindsight makes it difficult to understand why a person took the action they did when it led to an incident occurring. We are left asking questions like 'why did they do that; they must have known what would happen?' An everyday example here is preparing to go away for a camping trip, and on arrival realizing that an important piece of equipment was left behind, only to exclaim 'I knew I would forget something!' If this was actually known ahead of time it could have been prevented. This type of bias can lead us to believe that an incident could not happen to us, because would not make that kind of mistake that led to an incident.

2.7 *Halo effect*

This is a tendency to like everything about a particular person, even aspects of them you have not observed [2]. It can lead to following the advice of a person because you do not consider that they may be wrong. The corollary of this is the horns effect, where you do not believe the advice of a person because you do not believe they could be right. An everyday example of this may be thinking that your favorite actor must be a very nice person based on their attractive physical appearance, even though you have never met them. This type of bias can lead us to make an incorrect decision based on the opinion of others.

3 Biases in process safety incidents

There are several well know incidents where cognitive biases contributed to the event occurring. An incident has been selected for each bias, with the exception of hindsight bias, as the typical concern with hindsight bias is that it can inhibit learning from incidents after they have occurred and while this can result in a repeat incident, that has not been explored here.

The summary of incidents and their associate biases are contained in Table 2. This paper does not go into detail about the incidents, only mentioning the bias impacts. For more information on each incident refer to the references.

Table 2. Incident examples and their associated bias.

Bias	Incident	Example
Status quo bias	Columbia	Over the years that the space shuttle program had been running, NASA has normalized foam strikes as a part of launch. This because the status quo, and it was determined that they were not a safety of flight issue, and therefore there

		was no need to change the status quo [7].
Sunk cost fallacy	Bhopal	The Bhopal facility had a history of financial losses and equipment failures resulting in leaks. The facility had been shut down for a period of time and a decision was made to restart it to consume the feedstock, rather than dispose of it. During the final operation the toxic cloud was released. The decision to do the final produce run is an example of the sunk cost fallacy, continuing to run the facility, when it should have been shut down [8].
Framing	Challenger Space Shuttle	Temperature criteria had not been established for the solid rocket booster o-rings, but there was concern that the o-rings were brittle at low temperature. The initial framing suggested that the temperature effect was inconclusive, leading to concerns about the potential blow by of the o-rings being discarded [9].
Anchoring	Texas City Refinery	During the start up of the ISOM unit, the Raffinate Splitter Tower was filled with heated raffinate. The level indicator in the tower only covered the bottom section of the tower, and was also providing faulty readings. The operators became anchored to the level indicator results, not realizing the tower was in fact over filling [10].
Confirmation bias	Macondo	During the abandonment activities a cement barrier was installed. At the time it was believed that the installation went well and therefore the barrier would work. This led to test results being misinterpreted when assessing the barrier integrity. The belief that the barrier installation had been successful led to confirmation bias discounting the failed test results as incorrect [11].

Halo effect	Dreamworld	The Halo effect can exist when a junior or inexperienced worker is working at the direction of a more experienced person or being trained. As the junior person does not have the knowledge themselves, they will often do as they are told, deferring to the experience – the halo. The junior operator was told during her training she did not need to operate the emergency stop as that was the senior operator's role. When the raft became stuck on the rails no one hit the emergency stop, allowing the following raft to collide with it. Four people were killed when the second raft flipped as a result of the collision [12].

3.2 Hindsight bias

Hindsight bias often inhibits our ability to learn from an incident, because "the very outcome knowledge which gives us the feeling that we understand what the past was all about may prevent us from learning anything from it." [13, p. 311] In summary, we do not believe we would make the same mistakes. This phenomenon has led the IChemE Safety Centre to develop interactive case studies, where an attempt is made to remove the hindsight related to an incident, by telling the story without disclosing the outcome. Anecdotally these case studies appear to create an incident experience for the participant, without the consequences, which may assist with the learning being retained [14].

4 Ideas to avoid bias

The term often used for avoiding biases is called "debiasing". When considering the idea of system 1 and system 2 in our minds, to effectively debias we need to learn to engage system 2 to verify, correct and override where necessary our system 1 response. System 2 can be engaged by being asked questions that require thought, as this engages it to start to process. It has been found in studies that frowning while undertaking a task can increase the vigilance of system 2. This is because frowning usually occurs with cognitive strain, and this strain can be induced by frowning [2]. There are several ideas that can be tried to counter biases shown in Table 3.

Table 3. Ideas to counter bias

Bias	**Ideas to counter it**
Status quo bias	Appoint a person in the discussion to have the role of challenging the ideas, ensuring they have psychological safety to speak up without fear. This is akin to the "black hat" idea [15]. In this way there is someone free to ask questions about why the status quo exists as it is. The exploration of the status quo may highlight aspects that need changing.
Sunk cost fallacy	Engage someone from outside the issue, a coworker who is not tied to the outcome, to challenge what the next steps should be. They can help establish the pros and cons for each action, leading to a risk based decision on progression or not.
Framing	Reframe the data and discuss both aspects, identifying the counterfactual and assess both sides. From a priming aspect, it is possible to place subconscious thought in the mind prior to making decision. [3] An example might be that prior to undertaking a high risk task, the hazards and examples when things have previously gone wrong are specifically discussed. This act may prime the mind to look for and identify similar patterns.
Anchoring	When a decision needs to be reached, instead of a free discussion, have each person write down their initial ideas first. This way they have their thoughts that are not influenced by others during the discussion.
Confirmation bias	Engage someone from outside the issue, a coworker who is not tied to the outcome, to challenge what the data is telling you – do they reach the same conclusion without the background confirmation?
Halo effect	Attempting to slow down the thought process can assist with the time to look for more data to draw a conclusion from. If you are aware of the halo effect, you can try to think of different impressions of the person. This may be done by asking questions to try and establish a better understanding of them.
Hindsight bias	Attempt to understand the context in which decisions are made, not just the decision. This is useful during incident investigation, understand why an action was taken rather then what the action was. When we understand the context of

	decisions, we have the potential to see how it occurred instead of dismissing and thinking we would not have done it.

5 Conclusion

All actions and decisions we make are subject to biases, and we are often not aware of them. While most of the time this does not have an impact on the safety at a facility, sometimes the biases can lead to decisions which cause incidents. For this reason, it is important to understand the biases that may be present, as well as how our minds work with system 1 and system 2 and develop strategies to check if our system 1 decisions are likely to result in an incident.

6 References

[1] M. Haselton, N. D and A. PW, "The evolution of cognitive bias," in *The Handbook of Evolutionary Psychology*, Hoboken NJ, Wiley, 2005, pp. 724-746.

[2] D. Kahneman, Thinking, Fast and Slow, New York: Farrar, Straus and Giroux, LLC, 2011.

[3] M. Gladwell, Blink, Boston: Little, Brown and Company, 2005.

[4] B. Benson, "Cognitive Bias Codex," September 2017. [Online]. Available: https://www.visualcapitalist.com/wp-content/uploads/2017/09/cognitive-bias-infographic.html . [Accessed 19 July 2020].

[5] W. Hersing, "Managing cognitive bias in safety decision making: Application of emotional intelligence competencies," *Journal of Space Safety Engineering,* vol. 4, pp. 124-128, 2017.

[6] N. Roese and K. Vosh, "Hindsight bias," *Perspectives on Psychological Science,* pp. 411-426, 2012.

[7] Columbia Accident Investigation Board, "Columbia Accident Investigation Board," Columbia Accident Investigation Board, Washington DC, 2003.

[8] K. Bloch, Rethinking Bhopal, Amsterdam: Elsevier, 2016.

[9] Roger's Commission Report, "Report of the Presidential Commission on the Space Shuttle Challenger Accident," Presidential Commission on the Space Shuttle Challenger Accident, Washington DC, 1986.

[10] CSB, "Investigation Report Refinery Explosion and Fire," CSB, WashingtonDC, 2007.

[11] CSB, "Investigation Report Executive Summary Drilling Rig Explosion and Fire at the Macondo Well," CSB, Washington DC, 2016.

[12] Coroners Court of Queensland, "Inquest into the deaths of Kate Goodchild, Luke Dorsett, Cindy Low and Roozbeh Araghi at Dreamworld, October 2016," Queensland Courts, Southport, 2020.

[13] B. Fischhoff, "Hindsight ≠ foresight: the effect of outcome knowledge on judgement under uncertainity," *Quality and Safety in Health Care,* vol. 12, pp. 304-312, 2003.

[14] T. Kerin, "Accounting for hindsight bias: improving learning through interactive case studies," *Loss Prevention Bulletin,* no. 264, pp. 17-20, 2018.

[15] E. de Bono, Six thinking hats, Boston: Little, Brown and Company, 1985.

Unusual Hydrogen Explosions Due to Unanticipated Metal-Water Reactions

Robert Zalosh
Firexplo LLC
Wellesley, MA 02481
bzalosh@firexplo.com

Prepared for Presentation at
American Institute of Chemical Engineers
2020 Spring Meeting and 17th Global Congress on Process Safety
Virtual
April 18 - 22, 2021

Unusual Hydrogen Explosions Due to Unanticipated Metal-Water Reactions

Robert Zalosh
Firexplo LLC
Wellesley, MA 02481
bzalosh@firexplo.com

Keywords: hydrogen explosions, electrochemical reactions, metal-water reactions.

Abstract

Inadvertent generation of hydrogen from ambient temperature metal-water chemical and electrochemical reactions have been responsible for a surprising number of hydrogen explosions. This paper describes three sets of incidents involving different metals in applications that are normally considered free of explosion hazards. The first set of incidents involves wet alkaline incinerator ash with an aluminum content of at least two percent. The second set of incidents involves copper conductor electrochemical reactions in subsea electrical cable. The third set involves electrochemical corrosion reactions in galvanized dry sprinkler pipe. In all three applications, test data are discussed to elucidate the specific reactions and potential preventive and mitigation measures.

1 Introduction

Inadvertent generation of hydrogen from reactive powdered metal-water reactions and molten metal-water reactions is a well-known hazard in metal production and processing facilities. Prevention and protection measures for this hazard are described, among other places, in the NFPA 484 Standard on Combustible Metals [1]. However, inadvertent hydrogen generation from ambient temperature metal-water chemical and electrochemical reactions can also occur in many unanticipated settings and scenarios, and such scenarios have been responsible for a surprising number of hydrogen explosions.

Each of the following three sections of this paper provides details of one incident and then summarizes several other incidents for a given metal and application. It then describes the specific reactions likely responsible for the inadvertent hydrogen generation, including pertinent test data. The first set of incidents involve hydrogen generated from reactions of water with Municipal Solid Waste incinerator ash containing aluminum. The second type of metal-water electrochemical reaction involves copper cable in a seawater atmosphere. The third set of incidents entails hydrogen generation from galvanized steel corrosion reactions in dry pipe sprinkler systems.

2 Water-Incinerator Ash Reaction Incidents

2.1 Nortrader Cargo Vessel Explosion

A prime example of this type of incident is the January 2017 explosion on the dry cargo ship *Nortrader* that was the subject of a comprehensive investigation and report by the British Marine Accident Investigation Branch (MAIB) [2]. The *Nortrader* cargo of unprocessed incinerator bottom ash was generated in a nearby Plymouth, England power plant burning Municipal Solid Waste (MSW). Prior to the January 2017 incident, there had been 34 uneventful shipments of MSW bottom ash on 26 vessels similar in size to the 88 m length, nominal 1900 metric ton capacity *Nortrader*.

On January 12, 2017, over a period of about eight hours, 2,333 metric tons of wet MSW bottom ash were loaded into the *Nortrader's* 3409 m^3 cargo hold during steady heavy rain. After loading, the vessel left the dock and anchored overnight in Plymouth Sound while it waited for the weather to improve. During this time, the cargo hold ventilation fans and inlets were closed because of the heavy rain.

A large propagating explosion occurred the next day while the *Nortrader* was still at anchor. The ignition source for the explosion was the actuation of a stop button on an electrical panel situated in a ship compartment immediately forward of the large cargo hold. The ship's chief engineer, who was standing in front of the electrical panel suffered first degree burns to his face and second degree burns to his body, both hands, and lower extremities. The explosion then propagated into the cargo hold and lifted and bent all nine of the cargo hold steel latch covers, which had been held in place by latch cleats. The displaced hatch covers over the cargo hold the are visible in Figure 1.

Figure 1. *Nortrader* damaged cargo hold latch covers after explosions.

After the incident, the MSW from the Nortrader cargo hold was unloaded for inspection and testing. Employees of a consulting company retained by one of the ship's insurers drilled holes into the unloaded pile of MSW bottom ash and tested for the presence of flammable gas. They detected 1.4% hydrogen and 0.2 % carbon monoxide in the pile void spaces.

The MAIB hired another consulting company to acquire samples of the unloaded bottom ash for two types of water reactivity tests. One test series was UN Test N.5, the test method for substances which in contact with water emit flammable gases. It is described in section 33.4.1.4 (Annex E) of the UN Manual of Tests and Criteria [3]. The second test series involved an ad-hoc test with a much higher water-to-ash ratio (6:1) than in UN Test N.5 (1.5:1). Sample preparation for both tests entailed grinding the sample particulate to less than 250 micrometers for UN Test N.5, and to less than 1 mm for the ad-hoc reactivity test.

Several of the ash subsample water reactivity tests resulted in measured hydrogen production rates between 0.41 l/kg-hr and an average of 0.77 l/kg-hr. These rates are slightly less than the UN Test N.5 threshold of greater than 1 l/kg-hr of flammable gas to classify a material as a Division 4.3, substance which emits flammable gas when in contact with water [3]. Nevertheless, these hydrogen production rates indicate a potential for hydrogen accumulation in cargo holds and other enclosures with negligible ventilation and large quantities of bottom ash. Perhaps the UN Manual test criteria should be re-evaluated to provide a separate, less restrictive, category for this type of material that generates lower rates of flammable gas when in contact with large quantities of water.

The unventilated *Nortrader* cargo hold containing water-soaked bottom ash allowed the slowly generated hydrogen to accumulate over a period of about 19 hours while it was anchored in Plymouth Sound. An unsealed cargo hold light fixture access cover provided a path from the cargo hold to the adjacent ship compartment, which was also unventilated during this period. According to the MAIB report, the electrical panel in this compartment had an unsecured door, which allowed hydrogen to enter the panel interior. When an electrical circuit within the panel was energized and an arc or spark developed across a pair of contacts, the hydrogen-air mixture within the panel was ignited. The resulting explosion propagated from the panel into the ship compartment and then, via open light fixture cover, into the large cargo hold. An analysis by a MAIB contractor concluded that the hydrogen concentration in the cargo hold at the time of the explosion was in the range 6% to 9%, corresponding to a hydrogen volume between 114 m^3 and 171 m^3. The two explosions felt and heard by the *Nortrader* crew were the explosions in the electrical panel compartment and in the cargo hold itself.

2.2 MSW incinerator facility explosions

There have been several reported hydrogen explosions at MSW incinerator facilities. One such incident in Japan in 1995 is well documented [4,5,6]. A large amount of aluminum in the waste stream (9.6% Al) produced a clinker, i.e., a hard chunk of solidified molten metal and bottom ash, that was blocking flow in an ash discharge chute leading from the bottom of the incinerator to an ash hopper. Employees attempted to remove the clinker by striking it with a chisel and intermittently pouring water on it. This caused an explosion with ejected hot gas and ash burning three of the workers, one of whom subsequently died from the burn injuries [4,6]. The explosion was attributed to hydrogen formed from the aluminum-water reaction, with the subsequent hydrogen-air mixture in the chute being ignited by the chisel impacting the still hot clinker and ash.

A similar explosion occurred in a Japanese MSW incinerator in 1983 [6]. This explosion occurred in a dust bunker when an iron bar was used to crush some blocks of ash. The worker suffered severe injuries when he was thrown from the explosion site. The ash moisture content was about 10%, and there were significant amounts of aluminum in the ash.

Larson [7] briefly described a 2003 explosion in the basement ash storage bunker at a Swedish MSW incinerator due to hydrogen generation. Fortunately, there were no injuries, but the explosion damaged the bunker walls. Other incinerator facility explosions in Sweden have occurred during conveyor transport of warm wet bottom ash [8].

2.3 Other ash-water reaction explosions

Arm and Lindeburg [8] mention the occurrence of hydrogen explosions due to ash-water reactions in some Swedish ash storage facilities such as rock caverns and former oil wells. They also mention some explosion incidents associated with wetted fly ash in landfills.

There have been some flash fire and explosion incidents involving hydrogen generated from the ash aggregate used to make lightweight/foamed concrete. One such incident in 2009 in the UK occurred in a former dry well that had recently been filled with foamed concrete [9]. A grinder being used to cut reinforcing rods above the concrete ignited the hydrogen-air mixture that had accumulated there. This and other UK incidents involving foamed concrete and bricks caused the UK Highways Agency to ban the use of foamed concrete made with incinerator bottom ash.

2.4 Ash-water reactions producing hydrogen.

Although a variety of metals can react with water to produce hydrogen, most of the incinerator ash generated hydrogen has been found to be due to aluminum-water reactions. The primary reaction responsible for hydrogen generation is believed to be [10]:

$$2Al + 6H_2O \rightarrow 2Al(OH)_3 + 3H_2 \quad (1)$$

This reaction is promoted in an alkaline environment, with the major alkali source in MSW incinerator ash being calcium hydroxide. According to Nithiya et [10], the calcium hydroxide is consumed and regenerated in the following reactions, which combine to produce reaction (1).

$$2Al + 6H_2O + 2Ca(OH)_2 \rightarrow 2CaAl(OH)_5 + 3H_2 \quad (2)$$

$$2CaAl(OH)_5 \rightarrow 2Ca(OH)_2 + 2Al(OH)_3 \quad (3)$$

The presence of these compounds in ash-water reaction products has been confirmed by Nithiya et al [10], and the promoting effect of pH values > 10 in ash has been demonstrated in test data reviewed by Larsen [7] and others. Several researchers have reported significantly greater hydrogen generation rates when the ash is mixed with alkaline solutions compared to mixing with distilled water. Nithiya et al [10] cite research showing that the alkaline ash conditions cause the oxide layer on the aluminum surface to be perforated and dissolved, allowing direct contact and reaction of aluminum and water.

As implied by reactions (1) and (2), hydrogen generation rates from ash-water reactions increase with increasing aluminum content and increasing pH of the ash. The reactions have also been shown to be faster for freshly generated warm ash, and to decay after several days of ash exposure to ambient air (allowing greater oxide layer formation). These effects produce a wide variation in reported hydrogen generation yields obtained in tests with MSW ash samples.

Besides hydrogen generation rate data, reported test data include total hydrogen generated per unit mass of ash. The hydrogen yield data show the yield increasing with increasing ash aluminum content as well as other factors mentioned above. Yield data have been compiled from a variety of sources including data for bottom ash from Japanese incinerators [10], fly ash from Japanese incinerators [6], and both bottom ash and fly ash from Swedish incinerators [8]. The hydrogen generation data cumulative distribution for ambient temperature reactions obtained by mixing with distilled water are shown in Figure 2. The median hydrogen yield is 2.4 liter-H_2 per kg-ash for bottom ash and 10 l/kg-ash for fly ash. The highest measured hydrogen yields are 24 l/kg-ash for bottom ash and 37 l/kg-ash for fly ash.

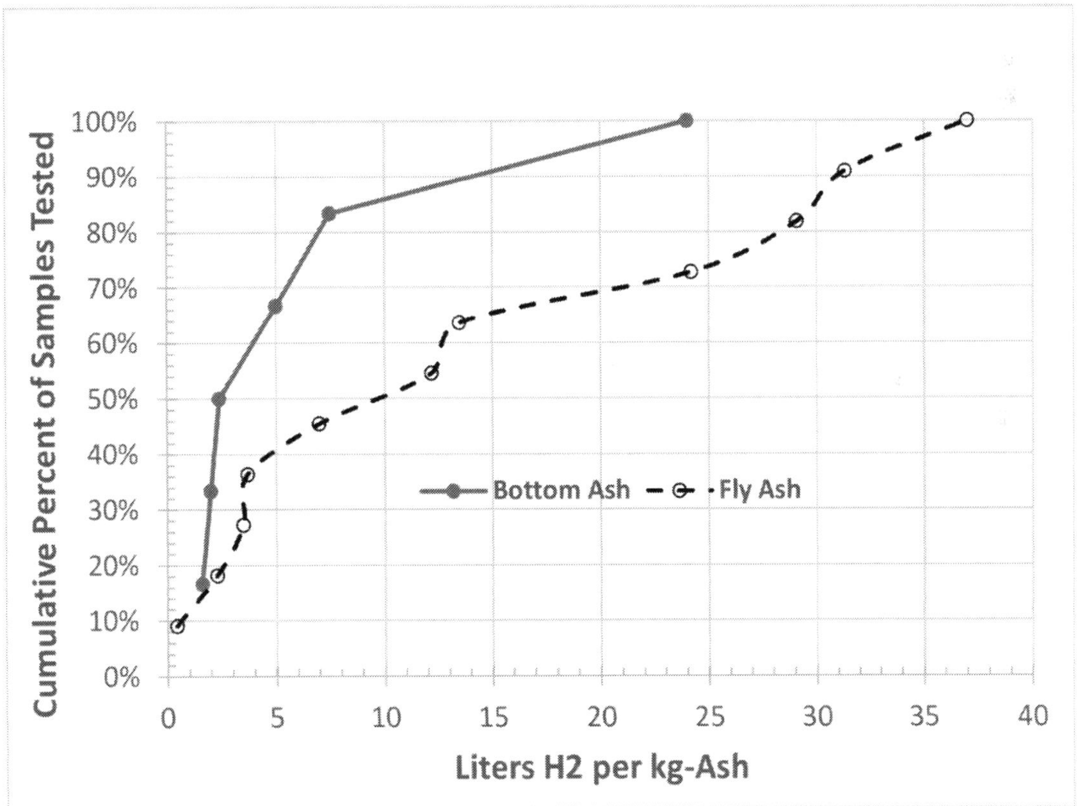

Figure 2. Hydrogen generation data cumulative distribution for bottom ash and fly ash reactions with water.

The data shown in Figure 2 can potentially be used to provide a risk-based estimate of the hydrogen accumulation hazard potential in MSW incinerator ash bunkers and storage and transportation facilities. Daily generation data reported in the cited references show that most of the hydrogen is generated in the first few days after ash formation and contact with copious quantities of water. (Based on the test methods cited above, copious water implies having a water/solid ratio of about two or greater at some point during handling and storage.) Therefore, this data can also be used as a basis for estimating MSW ash storage facility ventilation requirements to prevent flammable mixture formation. These requirements are not needed for ash obtained from other fuels providing those fuels do not have aluminum contents of about 2 percent or higher.

3 Electric Cable – Seawater Reaction Hydrogen Incidents

3.1 Offshore Junction Box Explosions

Offshore oil and gas exploration and production platforms have many electrical junction boxes exposed to sea air and aerosols that produce a saline environment within the box. Several of these boxes have experienced explosions that damage the box. Incident investigations have determined that hydrogen generation within the box produced a flammable atmosphere that was probably ignited by an arc, spark, or electrostatic discharge.

One junction box explosion incident [11] occurred in a box installed above the boat landing level of an unmanned platform in a tropical environment. The box had an IEC 60079 [12] designation "Ex e", meaning that it is intended for normally non-sparking electrical equipment with a rated voltage no greater than 11 kV, to be installed in a potentially flammable gas atmosphere. This particular box contained a 6.6 kV submarine power cable, and had breather ventilation.

During a maintenance crew visit about a year after the cable was first energized, the junction box hinged door was observed to be bulged outward and blown open, as can be seen in Figure 3. There was no interior damage or char deposits and the electrical connections remained intact. The fact that the box did not withstand the explosion is not surprising since it was not constructed or rated as an "Ex d" box.

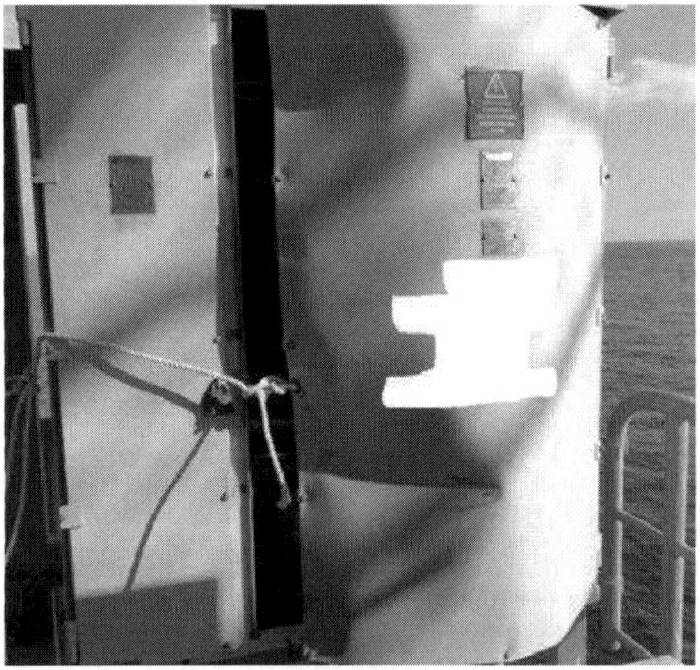

Figure 3. Damaged junction box door on offshore platform [11].

There have been at least two similar reported junction box explosion incidents at other platforms and locations. After one of those incidents, flammable gas was detected in the box [11].

Hydrogen generation from the copper conductor within seawater cables is a recognized phenomenon in subsea electrical cables with the generation and associated copper loss being more pronounced when there is cable insulation damage. This is illustrated in Figure 4, which shows the hydrogen bubbles originating from damaged cables in simulated seawater [13]. Hydrogen is also formed from seawater reactions with the copper and steel components of optical fiber cable [14]. Some seawater cable manufacturers provide hydrogen absorbing materials in their cables to prevent the hydrogen from being released from the cables.

Figure 4. Hydrogen bubbles from a damaged cable in simulated seawater [13]

3.2 Buried Cable Seawater Reaction Hydrogen Incident

An unusual electric cable-seawater reaction hydrogen explosion occurred at a Rhode Island beach in July 2015 [15]. The hydrogen was generated by the seawater corrosion of an abandoned electrical cable that had been used until 2007 to power Coast Guard navigational lights on rock jetties near the beach. People at the beach were not aware of the cable which was buried in the sand. When a hydrogen-air mixture was somehow ignited, the explosion blew a woman out of her beach chair. She suffered broken ribs and bruises upon landing on a jetty about 10 feet away.

Figure 5 shows a portion of the cable after the sand that had obscured it was removed. University of Rhode Island geochemists tested sand samples from the area and measured hydrogen concentrations as high as 5%, i.e., slightly above the Lower Flammable Limit. They also observed that the corroded cable did not have a green patina, indicating that the corrosion was due to oxide, rather than chloride formation.

Figure 5. Corroded electric cable under beach sand [15]

3.3 Seawater Induced Copper Hydrogen Evolution Reactions

Copper exposure to a seawater alkaline environment promotes hydrogen evolution via two different electrochemical reactions. First, copper serves as a catalyst for the cathodic reaction

$$2H_2O + 2e^- \rightarrow H_2 + 2[OH]^- \quad (4)$$

Sherif and Park [17] maintain that this reaction is more likely to occur in de-aerated solutions, whereas aerated NaCl solutions are more likely to produce the water and oxygen reduction reaction

$$O_2 + 2H_2O + 4e^- \rightarrow 4[OH]^- \quad (5)$$

However, Martins et al, [18] have shown that reaction (4) can produce copious amounts of hydrogen when the polycrystalline copper surface allows the formation of a copper-hydroxide activated complex. They demonstrate that copper surface treatment via polishing and annealing allows this activated complex to form and render the copper surface more hydrophilic.

The second seawater induced hydrogen evolution pathway is the following direct copper reaction with water [17]

$$2Cu + H_2O \rightarrow Cu_2O + 2H^+ + 2e^- \rightarrow Cu_2O + H_2 \quad (6)$$

Reaction (6) is suppressed in the absence of seawater because of the compact film of the cuprous oxide reaction product. Upon seawater exposure, chloride ions cause the formation of a cuprous chloride ion that partially dissolves the cuprous oxide film and forms the more porous Cu_2O and layer of atacamite ($Cu_2(OH)_3$- Cl), which readily admits more chlorine ions to reach the copper underneath it [17, 19]. The dissolution/porosity of the oxide film allows the oxidation reaction to continue and also results in further copper dissolution. Thus, copper corrosion can be often seen in the form of the green patina due to the cuprous chloride, and sometimes seen merely as a loss of copper without any discoloration. The latter was reported to be the case in the buried cable beach incident summarized above [16].

Reported test data on copper seawater corrosion rates can be used in conjunction with Reaction (6) to quantify the corresponding expected hydrogen generation rates. Corrosion rate data used for this purpose was obtained from copper exposure laboratory tests with natural and artificial and aerated seawater [19, 20], and from field tests with copper samples in seawater and sea aerosols in North Atlantic and tropical environments [21, 22]. The calculated hydrogen generation rates for exposures of several months varied from about 2 cc-H_2-per-hr-m^2-Cu to almost 14 cc-H_2 per hr-m^2. The cumulative distribution of the calculated results is shown in Figure 6. These results can be used to determine ventilation rates needed to prevent flammable hydrogen-air mixtures in electrical junction boxes or other small electrical enclosures in a seawater or sea atmosphere environment.

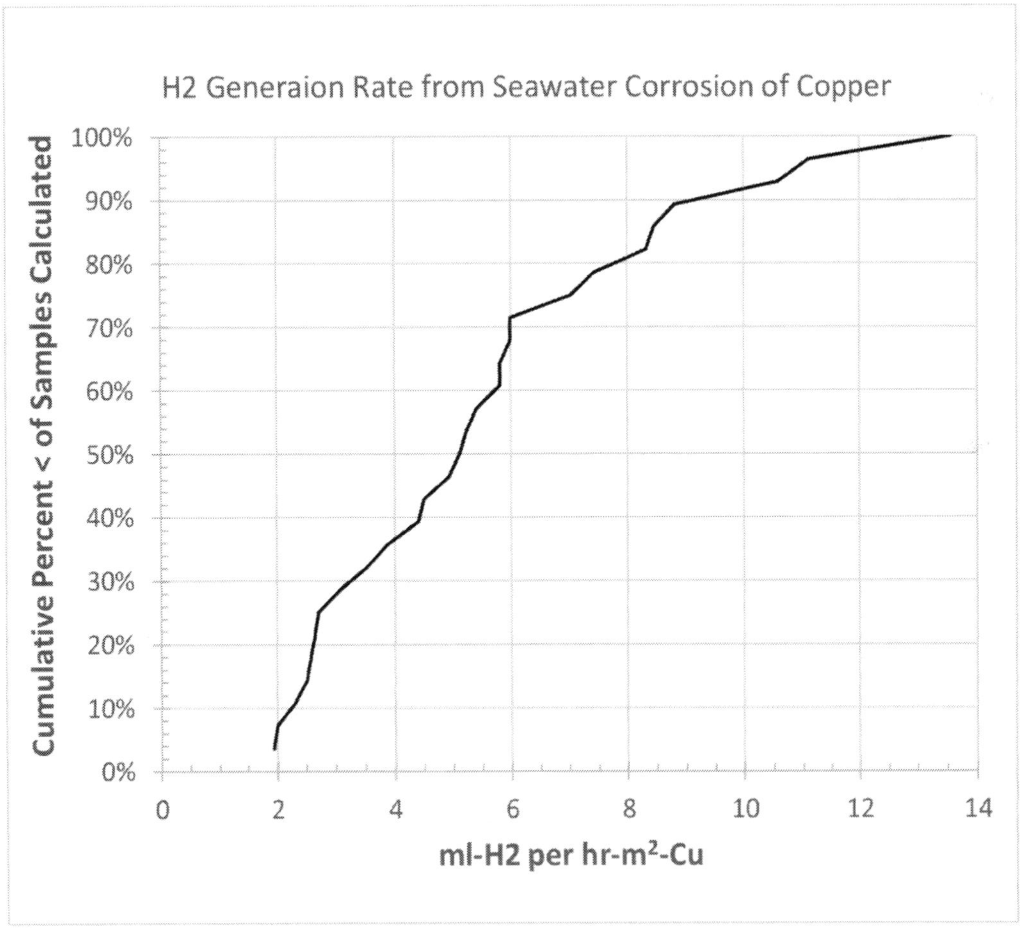

Figure 6. Calculated hydrogen generation rates from seawater exposure of copper.

4 Galvanized Sprinkler Piping Hydrogen Incidents

4.1 Helsinki Theater Sprinkler Room Explosion

On June 29, 2020, a sprinkler contractor was conducting routine maintenance on a dry pipe sprinkler system in the basement of a Helsinki theater. Shortly after the contractor opened the pipe, an explosion occurred that caused burn injuries to the contractor, who required hospitalization. The explosion caused minor damage to the stairwell adjacent to the sprinkler room [23].

An investigation by the Finnish Safety and Chemical Agency (Tukes) determined that the hydrogen was generated from a corrosive reaction involving zinc, water, and air within the galvanized steel pipe [24]. It was the first incident of its kind in Finland, but Tukes was sufficiently concerned about possible future incidents to issue a ban on the use of zinc-coated pipes in new sprinkler systems. Tukes also issued new requirements for maintenance involving opening of sprinkler piping; ventilation and a hydrogen detector are now required.

4.2 Other Sprinkler Pipe Hydrogen Fire and Explosion Incidents

Although the Helsinki theater explosion was the first incident in Finland involving hydrogen generated in galvanized sprinkler piping, there have been several such incidents in other Scandinavian countries, where galvanized dry pipe sprinkler systems are common. The following is a list of some European incidents.

- A 2014 incident occurred at a Danish industrial company when a sprinkler contractor was burned by a hydrogen flame vented from sprinkler piping while it was being drained [22].
- A second 2014 incident in Denmark; this one involved a hydrogen release from a pressurized water tank that had recently been drained. A sprinkler contractor suffered burn injuries from the resulting room explosion which may have been ignited by sparks from the contractors' tools [25].
- There was an explosion in a sprinkler fire pumphouse in France in October 2016 after a sprinkler contractor had drained the sprinkler piping in preparation for doing work on the system. The explosion occurred about 15 minutes after the drain was opened and the contractor left the pumphouse. Fortunately, there were no injuries, but the pumphouse was destroyed [26].

There have also been similar reported incidents in Japan. Yamate et al. [24] mention ignitions occurring when wet galvanized pipe have been cut or subjected to grinding during sprinkler system renovations.

4.3 Hydrogen generation reactions in sprinkler piping

The electrochemical reactions responsible for hydrogen generation in galvanized sprinkler pipe are analogous to the reactions described above for copper conductor exposure to seawater. The reactions described by Yamate et al. [27] are:

$$2H_2O \rightarrow 2H^+ + 2[OH]^- \quad (7)$$

$$2H^+ + 2e^- \rightarrow 2H \rightarrow H_2 \quad (8)$$

$$Zn \rightarrow Zn^{2+} + 2e^- \quad (9)$$

$$Zn^{2+} + 2[OH]^- \rightarrow Zn(OH)_2 \quad (10)$$

Reactions (9) and (10) produce the zinc corrosion that results in formation of the white zinc hydroxide layer seen below the indicated water line in Figure 7. Su and Fuller [28] did not observe this corrosion when the air above the water was replaced by compressed nitrogen. This suggests that oxygen is required for accelerated corrosion in galvanized pipe, and that oxygen-water electrochemical reaction (5) is a key aspect of both corrosion and hydrogen production in galvanized dry pipe sprinkler systems in which there is residual water together with air in the piping. This often occurs when the piping is not completely drained following a dry pipe valve trip test or inadvertent actuation.

Figure 7 Zinc corrosion produced in 6 weeks in half-filled pipe [25]

Yamate et al. [24] have shown that rapid hydrogen generation can also occur from galvanic corrosion associated with water filled galvanized sprinkler pipe in contact with another metal component such as a stainless steel flexible pipe section. Gas samples from the galvanic coupling pipe revealed a hydrogen concentration of 20% and significant concentrations of zinc in the water. They also obtained visual evidence of hydrogen bubbles leading to a flammable atmosphere when the piping was opened, as shown in Figure 8.

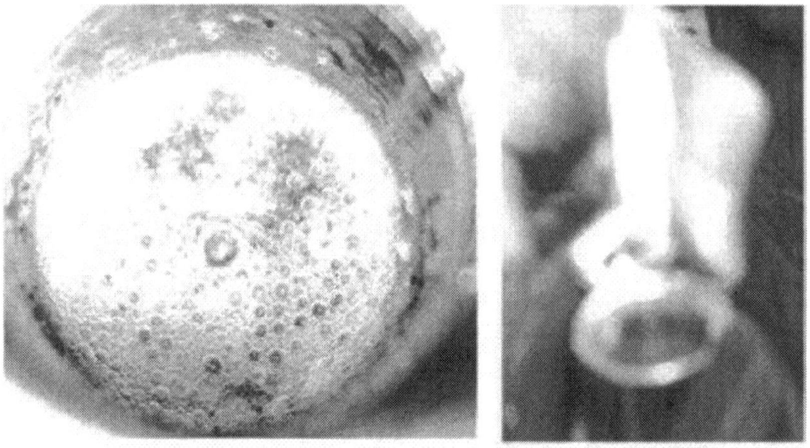

Figure 8 Hydrogen bubbles and flames in opened galvanic coupled piping [24]

Although the hydrogen generation incidents and accelerated corrosion observations have led to some bans on galvanized sprinkler piping, FM Global [28] notes that corrosion products in ordinary black carbon steel dry piping have been responsible for the vast majority of their fire losses due to obstructed sprinkler pipes. Therefore, FM Global continues to recommend use of galvanized steel pipe for dry and pre-action sprinkler systems. In view of the exposure test results noted above, FM Global recommends use of compressed nitrogen, rather than just air, in these dry pipe systems. This is different from the recommendations and requirements of many other organizations, as summarized in a recent Sprinkler Age article [29].

5 Conclusions

Metal-water chemical and electrochemical reactions have been responsible for hydrogen generation and subsequent ignition of hydrogen-air mixtures in many explosion incidents. Many incinerator ash-water incidents have occurred when the ash had a substantial aluminum and alkaline content. A compilation of test data has led to a statistical distribution of the expected hydrogen generation per unit mass of ash, with the median rate being greater for fly ash than for bottom ash.

Copper-seawater electrochemical reactions in offshore facility electrical junction boxes have led to hydrogen accumulations and subsequent ignitions associated with electrical connector arcs and sparks. Copper corrosion test data together with reported electrochemical reactions have allowed calculations of hydrogen generation rates per unit of copper cable surface area. Special cable features, such as hydrogen absorption additives, and flameproof junction box construction can help prevent and mitigate the consequences of these incidents.

Electrochemical corrosion reactions in galvanized dry pipe sprinkler systems containing residual water have led to many reported hydrogen fire and explosion incidents, particularly in Scandinavia, where there are many of these sprinkler systems. Testing has shown that the accelerated corrosion in galvanized dry pipe can be greatly reduced by replacing air in the pipe with compressed nitrogen, and by avoiding the galvanic corrosion resulting from contact between dissimilar metal components in sprinkler piping.

6 References.

[1] NFPA 484 Standard on Combustible Metals, National Fire Protection Association, 2019.

[2] Marine Accident Investigation Branch, Serious Marine Casualty Report Number 26/2017, December 2017.

[3] UN, Recommendations on the transport of dangerous goods, Manual of tests and criteria, United Nations, 5th Edition, 2009.

[4] https://h2tools.org/lessons/hydrogen-gas-explosion-municipal-refuse-incineration-facility. Accessed on 2/8/2021.

[5] http://www.shippai.org/fkd/en/cfen/CC1200025.html. Accessed 2/8/2021.

[6] S. Mizutani,, S. Sakai,, and H. Takatsuki, Investigation of hydrogen generation from municipal solid waste incineration fly ash, J Mater Cycles Waste Manag (2000) 2:16–23.

[7] R. Larsen, Energy recovery of metallic aluminium in MSWI bottom ash, Umea University M.S. Thesis, 2014.

[8] M. Arm, and J. Lindeberg, Gas generation in incinerator ash, 2015, Available at https://www.researchgate.net/publication/266467653. Accessed 2/8/2021.

[9] UK Highways Agency Interim Advice Notice 127/09, Appendix B: The Use of Foamed Concrete, October 2009.

[10] A. Nithiya, A.Saffarzadeh, and T. Shimaoka, Hydrogen gas generation from metal aluminum-water interaction in municipal solid waste incineration (MSWI) bottom ash, Waste Management 73 (2018) 342–350.

[11] Rupturing of 6.6kV Power Junction Box installed at an offshore oil production Well Head Tower, https://www.eng-tips.com/viewthread.dm?qid=434654. Accessed 2/12/2021.

[12] IEC 60079-7, Explosive atmospheres - Part 7: Equipment protection by increased safety "e", IEC, 2015.

[13] N. Douglas, Copper Loss from Electrical Conductors in a Subsea Environment as a Result of Low Insulation Resistance, Oilfield Technology, January 2019. Available at https://www.researchgate.net/publication/331951501.

[14] ITU-T Recommendation L.27, Method for estimating the concentration of hydrogen in optical fibre cables, International Telecommunication Union, 1997.

[15] State: Hydrogen from corrosion of copper cable caused Salty Brine beach blast, Providence Journal, July 24, 2015.

[16] Case Closed: Hydrogen Gas Combustion Caused Salty Brine Beach Blast, National Public Radio, July 24, 2015. Accessed 2/12/2021.

[17] E-S. Sherif, and S-M. Park, Inhibition of Copper Corrosion in 3.0% NaCl Solution by N-Phenyl-1,4-phenylenediamine, Journal of The Electrochemical Society, 152 10 B428-B433 (2005).

[18] P. Martins, et al, Hydrogen evolution reaction on copper: Promoting water dissociation by tuning the surface oxophilicity, Electrochemistry Communications 100 (2019) 30–33.

[19] E-S Sherif, R. Erasmus, and J. Comins, Corrosion of copper in aerated synthetic sea water solutions and its inhibition by 3-amino-1,2,4-triazole, Journal of Colloid and Interphase Science, 309 (2007) 470–477.

[20] F. Mansfield, et al., The Corrosion Behavior of Copper Alloys, Stainless Steels, and Titanium in Seawater, Corrosion Science, Vol. 36, No. 12, pp. 2063-2095, 1994.

[21] A. Drach, et al., Field studies of corrosion behaviour of copper alloys in natural seawater, Corrosion Science 76 (2013) 453–464.

[22] L. Nunez, et al, Corrosion of copper in seawater and its aerosols in a tropical island, Corrosion Science 47 (2005) 461–484.

[23] Explosion in Helsinki City Theater, https://www.foreigner.fi/content/print/explosion-in-helsinki-theater-at-least-one-person-in-hospital/202006291506576691. Accessed 11/17/2020.

[24] Hydrogen explosion caused the Helsinki City Theatre incident, https://tukes.fi/en/-/hydrogen-explosion-caused-the-helsinki-city-theatre-incident. Accessed 2/14/2021.

[25] Explosions in sprinkler system probably caused by hydrogen gas, https://dpa-e.eu/explosions-in-sprinkler-system-probably-caused-by-hydrogen-gas/ Accessed 2/14/2021.

[26] Internally Galvanized Steel Pipe Blamed for Explosions, https://www.complianceonegroup.com/ lifesafety/ internally-galvanized-steel-pipe-blamed-for-explosions. Accessed 2/14/2021.

[27] T. Yamate, et al., Experimental Study on Hydrogen Generation Phenomenon in Fire Piping System in Buildings, Corrosion Engineering, 59, 436-439, 2010.

[28] P. Su, and D. Fuller, Corrosion and Corrosion Mitigation in Fire Protection Systems, FM Global Report, July 2014.

[29] Galvanized Steel Sprinkler Piping, Sprinkler Age, November 2018, Available at https:// sprinkle rage.com/ galvanized-steel-sprinkler-piping/ Accessed 2/15/2021.

619874 Port of Beirut Lebanon Explosion – Analysis and Impacts

Abstract:

On 4 August 2020, 2,750 tons of ammonium nitrate stored at the Port of Beirut (POB), Lebanon, exploded, reportedly resulting in over 190 deaths, several thousand injuries, $10-15 billion USD in property damage, and leaving an estimated 300,000 people homeless. The cargo of ammonium nitrate had been stored in Warehouse 12 at the POB, after having been confiscated by the Lebanese authorities from an abandoned ship, the MV *Rhosus,* in September, 2013. It is assumed that the ammonium nitrate was stored in the warehouse without proper security and safety measures, and without consideration for the need for segregation from other incompatible materials. From recorded videos of the blast, the explosion was preceded by a fire in the warehouse, but the exact cause of the detonation is still under investigation. This was one of the largest non-nuclear explosions in modern history and the lessons learned have widespread implications for facility siting, the handling of explosives and other hazardous materials, the management of chemicals in transportation, emergency management and recovery, and building resilience for large-scale emergencies.

In response to this event, AcuTech participated as part of an assessment team, and was tasked with examining current and previous practices for the handling of chemicals and hazardous materials at the POB, taking an all-hazards approach to examine safety, security, and emergency management concerns in a unified manner.

This paper examines the multifaceted impacts of the blast, the most critical factors that may have contributed to the explosion, and best practices and lessons learned that are applicable not only to ports and facilities handling ammonium nitrate, but all facilities handling hazardous materials operating in challenging urban environments, and which maintain complex interdependencies with critical infrastructure.

AUTHOR INDEX

AIChE
120 Wall Street, FL 23
New York, NY 10005-4020

ISBN 978-1-7138-5728-0

9 781713 857280